Polymer Surface Modification to Enhance Adhesion

Scrivener Publishing
100 Cummings Center, Suite 541J
Beverly, MA 01915-6106

Adhesion and Adhesives: Fundamental and Applied Aspects

The topics to be covered include, but not limited to, basic and theoretical aspects of adhesion; modeling of adhesion phenomena; mechanisms of adhesion; surface and interfacial analysis and characterization; unraveling of events at interfaces; characterization of interphases; adhesion of thin films and coatings; adhesion aspects in reinforced composites; formation, characterization and durability of adhesive joints; surface preparation methods; polymer surface modification; biological adhesion; particle adhesion; adhesion of metallized plastics; adhesion of diamond-like films; adhesion promoters; contact angle, wettability and adhesion; superhydrophobicity and superhydrophilicity. With regards to adhesives, the Series will include, but not limited to, green adhesives; novel and high-performance adhesives; and medical adhesive applications.

Series Editor: Dr. K.L. Mittal
Email: raareviews@gmail.com

Publishers at Scrivener
Martin Scrivener (martin@scrivenerpublishing.com)
Phillip Carmical (pcarmical@scrivenerpublishing.com)

Polymer Surface Modification to Enhance Adhesion

Techniques and Applications

Edited by

K.L. Mittal

and

A.N. Netravali

Scrivener Publishing

WILEY

This edition first published 2024 by John Wiley & Sons, Inc., 111 River Street, Hoboken, NJ 07030, USA and Scrivener Publishing LLC, 100 Cummings Center, Suite 541J, Beverly, MA 01915, USA

For more information about Scrivener publications please visit www.scrivenerpublishing.com.

Wiley Global Headquarters
111 River Street, Hoboken, NJ 07030, USA

For details of our global editorial offices, customer services, and more information about Wiley products visit us at www.wiley.com.

Library of Congress Cataloging-in-Publication Data

ISBN 978-1-394-23100-3

Cover image: Pixabay.Com
Cover design by Russell Richardson

Set in size of 11pt and Minion Pro by Manila Typesetting Company, Makati, Philippines

Printed in the USA

10 9 8 7 6 5 4 3 2 1

Contents

Preface

Polymers are used vastly in many industries for a variety of applications. However, their surfaces are intrinsically non-reactive due to their low surface free energy and absence of reactive chemical groups. Concomitantly, their surface modification (also known as surface treatment or surface activation) is a *sine qua non* to render them adhesionable or bondable.

Even a cursory look at the literature will evince that there is tremendous research activity in devising new and improved ways of ameliorating the existing techniques or coming up with novel ways to modify polymer surfaces. A legion of techniques ranging from mundane to sophisticated, vacuum to non-vacuum, dry to wet, and inexpensive to sumptuous have been documented for surface modification of a variety of polymer surfaces. The choice for an optimum technique depends on a host of factors, *inter alia*, the chemical structure of the polymer substrate, nature of mating phase, level of adhesion desired and cost. Also, the longevity of treatment is a very significant consideration. There have been different attempts to prolong the life of treatment as well.

The information on the topic of this book is scattered in many different publication media and there is no single, easily accessible, and comprehensive source available. This lacuna in the literature provides the vindication and motivation for bringing out this book.

This book containing 13 (a lucky number) chapters is divided into two parts: Part I: Energetic Treatments; and Part II: Chemical Treatments. Topics covered include: atmospheric pressure plasma treatment of polymers to enhance adhesion; corona treatment of polymer surfaces to enhance adhesion; flame surface treatment of polymers to enhance adhesion; vacuum UV photo-oxidation of polymer surfaces to enhance adhesion; optimization of adhesion of polymers using photochemical surface modification; UV/Ozone surface treatment of polymers to enhance adhesion; adhesion enhancement of polymer surfaces by ion beam treatment; polymer surface modification by charged particles; laser surface modification of polymeric materials; competition in adhesion between polysort and monosort functionalized polyolefinic surfaces; amine-terminated dendritic materials for polymer surface modification;arginine-glycine-aspartic acid (RGD) modification of polymer surfaces; and adhesion promoters for polymer surfaces.

The chapters are written by renowned researchers actively engaged in surface modification of polymers. This book is profusely referenced and copiously illustrated.

This book should of great appeal and interest to polymer scientists, surface scientists, adhesionists, materials scientists, plastics engineers, and to those involved/interested in adhesive bonding, packaging, printing, painting, metallization, biological adhesion, biomedical devices, and polymer composites.

Now it gives us great pleasure to acknowledge all those who played essential roles in giving this book a body form. Naturally, first and foremost, our profound thanks go to the authors for their keen interest, sustained enthusiasm, unwavering cooperation, and sharing their valuable research experience in the form of written accounts (which essentially provided the grist for this book), without which this book could not be materialized. Also, the steadfast interest and whole-hearted support of Martin Scrivener (publisher) in this book endeavor is highly appreciated.

Kash Mittal
Hopewell Jct., NY, USA
E-mail: RAAreviews@gmail.com

Anil Netravali
Cornell University
December 2023

Part I

ENERGETIC TREATMENTS

1

Atmospheric Pressure Plasma Treatment of Polymers to Enhance Adhesion

K. Lachmann*, M. Omelan, T. Neubert, K. Hain and M. Thomas

Fraunhofer Institute for Surface Engineering and Thin Films IST, Braunschweig, Germany

Abstract

In this chapter, we present an overview on recent research and development work in the area of atmospheric pressure plasma treatments (APPTs) to generate adhesion-promoting surfaces of polymers used in various applications in automotive, aerospace, packaging, and medical fields. In comparison to the classical "corona treatment" the APPTs provide access to a broader range of industrially interesting surface modifications that are normally better controlled with respect to their physicochemical nature. Thus, application of APPTs may become a superior option for preparing polymer surfaces for adhesive bonding, adhesive-free low-temperature bonding involving homogeneous and heterogeneous substrates, lacquering, or coupling of specific biomolecules, proteins, or cells. APPT technology is, however, not only flexible for tuning surface chemistry but also is flexible with respect to plasma source and equipment design. Versatility of the APPT technology facilitates its integration into a variety of process chains. An example presented here is a hybrid technology combining both APPT and additive manufacturing based on 3D printing processes using fused filament deposition. Inline treatment by quasi-simultaneous execution of printing and APPT can, for instance, increase the adhesion of 3D printed products in print direction (z-Axis) and thus increase the mechanical stability of the printed part. In the medical field such a technology may be attractive for cell growth by promoting treatment of internal surfaces of printed porous scaffolds. In future, products made from bio-based or recycled polymers will become increasingly important. APPT technology could become an important enabler for meeting the technical requirements for the adhesion of such products.

**Corresponding author*: kristina.lachmann@ist.fraunhofer.de

K.L. Mittal and A.N. Netravali (eds.) Polymer Surface Modification to Enhance Adhesion: Techniques and Applications, (3–44) © 2024 Scrivener Publishing LLC

Keywords: Atmospheric pressure plasma treatment, chemical groups, functional coatings, bonding, adhesive-free bonding, additive manufacturing, binding of biomolecules

1.1 Introduction

Polymers play an important role in packaging, automotive, medical, biomedical, and aerospace applications. Polymer market still experiences growth, driven by the increased use of polyolefins and also of new bio-based and functional polymers. Most of these polymers exhibit inert surfaces that need to be treated to provide sufficient adhesion for lacquers, paints, and adhesive bonding. Treatment can be carried out using a wide range of surface-functionalizing processes including wet-chemical etching [1, 2], plasma [3–5], and laser modification [4, 6, 7], as well as ozone and UV treatment [4], many of which are nowadays well established in industry. In recent years, application of cold Atmospheric Pressure Plasma Treatments (APPTs) in industry has experienced very strong growth attributed to the high versatility in applications, low investment cost and the very good integration of the technology in existing process chains [8].

Changes in technical, social, and legal requirements will inevitably lead to changes in processes also in the adhesive bonding industry which are therefore the focus of this review. Adhesive bonding technology is used in a wide range of products as it offers various material combinations, long-term stability and safety while maintaining material properties and creating additional functions in the bonded product. Continuous inventions and innovations in the development of raw materials, adhesives and bonded products have enabled a dynamic development in the past decades.

Innovative bonding technology enables new, more environmentally-friendly and sustainable products and processes in many areas. For instance, improved resource efficiency can be achieved in the production of renewable energy power plant components where sealing of solar cells or joining of wind turbine rotor blades has become possible. Another example is the electromobility area with the sealing of battery cells, heat management of batteries with heat-conducting adhesives and hermetic sealing of fuel cells. In addition, adhesives with improved product properties can lead to savings, even in food packaging to increase shelf-life with resource-efficient materials.

In Europe the "European Green Deal" with its agenda responding to current socio-political changes will bring additional requirements and thus new challenges for bonding technology [9, 10]. The main point is the transition from a linear to a circular economy, minimizing the use of resources, generation of waste and emissions, and the inefficient use of energy. This will be achieved by considering resource needs, life extension, repair, refurbishment, recycling and closing energy and material loops.

The eco-friendly APPT technology with its innovations in combining suitably functionalized surfaces with bonding technology certainly has the potential to meet the new set of requirements that may result from implementation of the agenda of the European Green Deal.

These new challenges must be realized in terms of closed-loop systems along the value chain. Therefore, life cycle assessments (LCAs) enable a holistic view of adhesively bonded products with regard to ecological and economical improvements along the complete value chains [11, 12]. In this context atmospheric pressure plasma treatments (APPTs) of polymers and other materials like paper, rubber, fabrics, steel, glass and different composite materials have gained considerable importance in recent years due to their technological and economical capabilities and thus offer the potential to become the future leading technology for surface functionalization to improve adhesive bonding technology. The main advantages of APPTs are that they need no expensive equipment, are easy to handle, have a very good scalability and can simply be integrated in existing process lines.

Considering that in most applications materials are joined with other structural parts, the adhesion behavior is of great significance. The adhesion property is strongly affected by the chemistry and morphology of the surfaces. APPTs alter surface morphologies and chemical composition of different materials without affecting their bulk properties. The plasma treatment can also change the surface from hydrophobic to hydrophilic state and vice versa, depending on the type of monomer or process gas used for surface activation. Thus, plasma treatment can alter the inert polymer surface to have greater chemical/physical affinity by incorporating chemical functional groups. This leads to an increase in surface free energy and enlargement of the contact area and thus improved adhesion property. Different research groups have carried out a variety of APPT processes to improve the surface properties of materials [13–19].

These cold plasma treatments have gained increasing interest as they represent environmentally-friendly solution and can even be performed under atmospheric pressure conditions at relatively low cost [20–22]. The plasma treatment can also affect the morphology by exerting the etching effect on the surface and changing the surface roughness and wettability. This results in improved adhesion of metal/plastic compounds [23]. The cold plasma treatment of various plastics such as polypropylene, polyethylene, poly(ethylene terephthalate), poly(etheretherketone), etc. has been investigated [24–26].

In this chapter, we will focus on the use of APPTs to improve the adhesion by chemical functionalization for different applications. Firstly, we will give a short overview about the historical development of APPTs. The second section is about the establishment of functional surfaces with a high number of different chemical groups. In the third section, the adhesion improvement by APPT promoted joining of materials is discussed. In the fourth section we will show the relevance of APPTs for biomedical application for improving e.g. cell adhesion. Finally, we will give an overview on the use of APPTs to promote adhesion in additive manufacturing processes.

1.2 Historical Development of APPTs

The industrialization of APPTs started in 1951, when A. W. Eisby, a Danish engineer, invented a high-frequency treatment with a dielectric barrier discharge in air for polymer foils to improve the adhesion of inks [27]. This so-called "corona treatment" is nowadays commercially well established in the packaging sector for surface activation and cleaning. In the following decades different kinds of plasma sources for APPT working with air as process gas were developed, enabling the treatment of flat and complex surfaces [5]. In the meanwhile, APPTs have been established in the biological, pharmaceutical and medical fields, in automotive, aerospace, and electronics industries and have become important for the alternative energy sector, additive manufacturing as well as for the pretreatment of renewable and sustainable materials. Figure 1.1 shows the most common sources like dielectric barrier discharges, stabilized corona systems, and plasma jets or torches which can be operated using different types of excitations, e.g., alternating current (AC), pulsed direct current (DC), low-frequency (kHz), radio frequency (RF, 13.56 MHz) and microwaves (MW) with a frequency of 2.45 GHz.

Today APPTs can be used to functionalize a surface with various chemical groups including e.g. -OH, -COOH, -NH_2, -NO_x to improve the adhesion of, e.g., adhesives, lacquers and paints to polymers. Using nitrogen and nitrogen-hydrogen mixtures as process gases, for example, imines, amines or amides can be introduced on the polymer surface, frequently leading to higher surface free energy, better long-term stability and in many applications the desired high level of adhesion. DBD-based APPT using nitrogen or its mixtures with other gases like

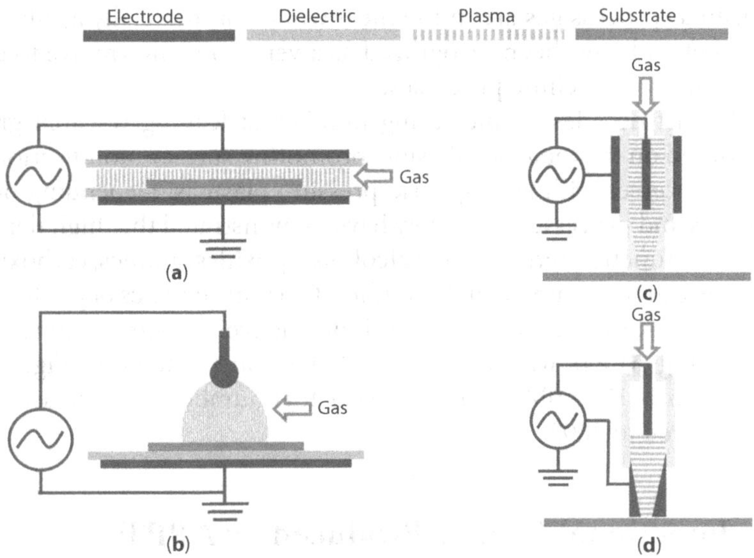

Figure 1.1 Examples of different types of plasma sources for APPTs; (a) Dielectric Barrier Discharge (DBD), (b) Stabilized Corona, (c) DBD jet, and (d) Arc jet.

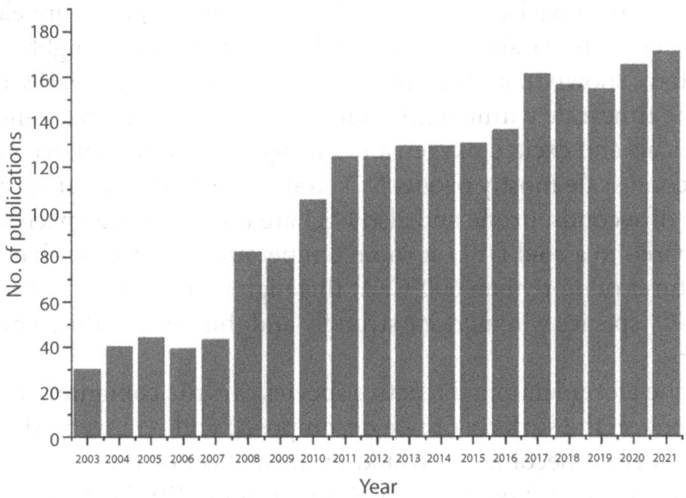

Figure 1.2 Number of publications on surface functionalization with APPT with relevance to adhesion (from Web of Science – date Nov. 15, 2022).

hydrogen as process gas are in the meanwhile commercially available on the market and have been recognized as a very good alternative to existing wet-chemical etching processes.

In the last decades an increasing number of leading research groups have worked on the topic of adhesion-promoting surface modifications for coating processes with atmospheric-pressure plasmas. Depending on the precursors and processes used, they have demonstrated that high densities of chemical functionalities such as alcohols, epoxides, amines, carboxylates or silanols can be obtained on the surface. Current statistics of publications as available via the website from "Web of Science" for surface functionalization using APPTs with relevance to adhesion is shown in Figure 1.2. With presently 170 publications per year this number has increased five-fold compared to 2003.

1.3 Functional Groups Produced by APPTs

Depending on the atmospheric pressure plasma system used, process gases, precursors and process parameters, the surfaces of several materials can be cleaned, activated or equipped with a wide range of functional layers.

Figure 1.3 shows the generated chemical species by a microdischarge in air of a dielectric barrier discharge (DBD), which has been modelled by Eliasson and Kogelschatz [28]. The predominant species are electrons, ionized species, free radicals, ozone and NO_x as well as UV light.

They have shown that the composition of species present varies depending on the timescale within a microdischarge of a DBD. On a nanosecond scale radicals and excited oxygen and nitrogen are dominant, whereas on a microsecond scale mostly ozone, NO_x and excited nitrogen are present. In the scale of seconds, ozone and also N_2O are the dominant species.

Compared to a cold DBD at room temperature, arc torches have higher temperatures of more than 600°C. At this high temperature the concentration of NO_x species will increase strongly and thus lead to different surface chemistry.

With the aforementioned reactive species organic contaminants, generally present on the surface, can be decomposed and removed. At the same time the surface becomes activated by formation of functional moieties such as nitrogen- and oxygen-containing groups. This leads to significant changes in the chemical composition of the topmost layer and thus in the surface free energy of the treated polymer. Direct electrical charging of the polymer surface by trapping ions by CH_2 groups of the polymer can also contribute to the increase of surface free energy [29].

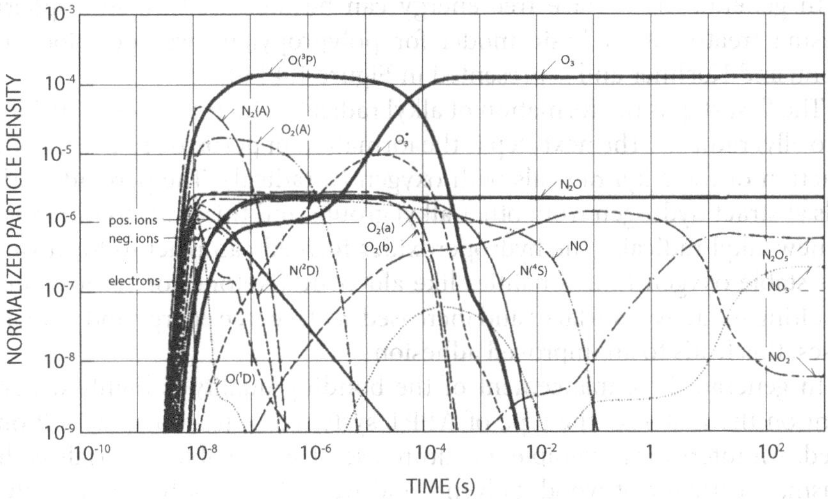

Figure 1.3 Chemical species generated by a microdischarge in air (from [28]).

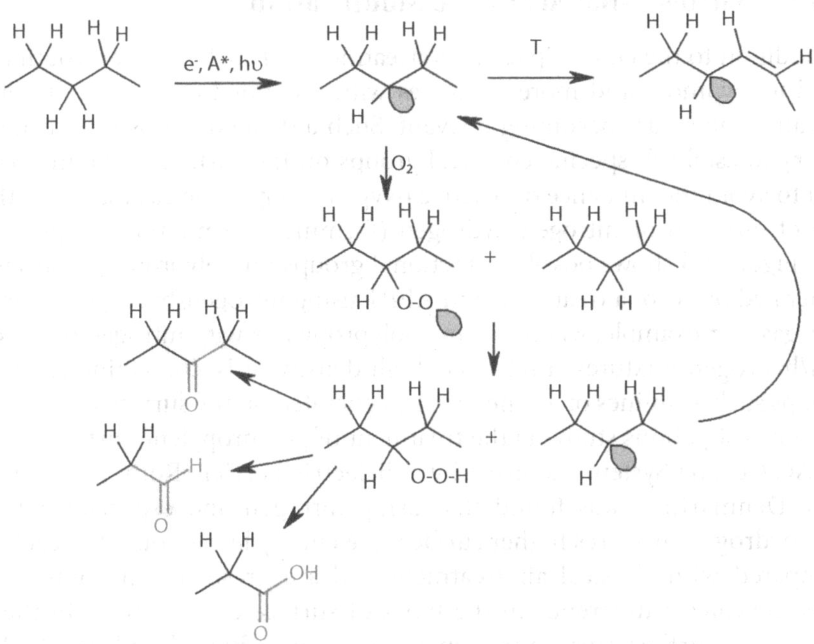

Figure 1.4 Important reaction steps for the activation of polypropylene surface (adapted from [30]).

In general, the surface free energy can be increased by atmospheric plasma treatment. A basic model for polypropylene was described by Dorai and Kushner and is presented in Figure 1.4 [30].

The first step is the formation of alkyl radicals which can also transform into allyl radicals. The next step is the formation of peroxide radicals by the reaction of the alkyl radicals with oxygen (diradical). The peroxide radicals abstract hydrogen from other alkyl groups and this leads to formation of new alkyl radicals. The hydroperoxides formed can react to form various stable oxygen functionalities like aldehydes, ketones or carboxylates, resulting in "active" surfaces and increased surface free energy and in many cases, this leads to an improved adhesion.

In general, the enhancement of the bonding quality is highly dependent on the material, the type of APPT system and treatment conditions used. An interesting example for future use of sustainable materials is the plasma treatment of wood. Solvent- and water-based polyurethane (PU) coatings showed deeper penetration in pre-treated wood in comparison to untreated wood substrates [31].

1.3.1 Nitrogen-Based Surface Modification

In addition to the classic "plasma pretreatment" of surfaces in air routinely used today, more and more processes with a defined oxygen-free (inert) gas atmosphere are becoming relevant. Such a defined atmosphere is necessary to establish specific chemical groups on the surface of the material and to avoid the influence of reactive oxygen groups. The easiest way is the use of nitrogen or nitrogen-hydrogen (forming gas) mixtures as process gas. Figure 1.5 shows possible functional groups on polypropylene surface generated by corona treatment and APPT using nitrogen/hydrogen as process gas. For example, when treating polypropylene with nitrogen or nitrogen/hydrogen mixtures, a relatively high density of N-containing reactive groups such as imines or amines can be generated on the surface [5, 32, 33].

In investigations studying the treatment of polypropylene surfaces with an AC Corona System stabilized by a dielectric barrier (RotoTEC, Tantec A/S, Denmark) it was found that using nitrogen and especially nitrogen-hydrogen mixtures higher surface free energy levels could be reached compared with classical air treatment and are presented in Figure 1.6. This indicates a difference in the types of surface compositions in these two cases. Further, surface free energy changes achieved with the N_2/H_2 mixtures were observed to be stable over a period of several weeks [34]. Moreover, in the case of adding hydrogen a long-term stability of the surface free energy for several weeks was shown.

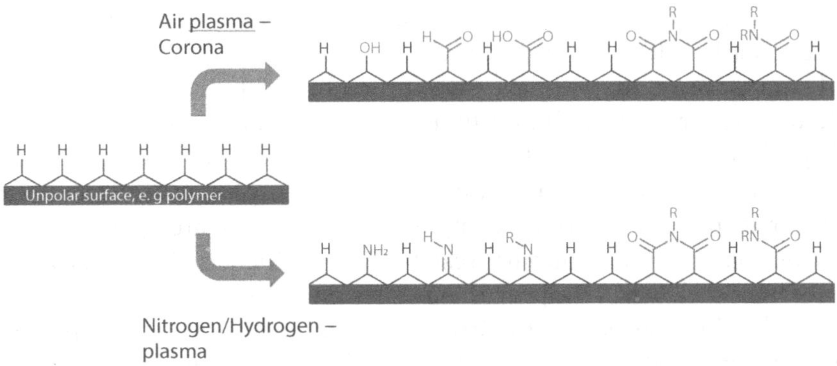

Figure 1.5 Functional groups by treatment of polypropylene surface in air (Corona) and in nitrogen-hydrogen atmosphere (APPT) (from [32]).

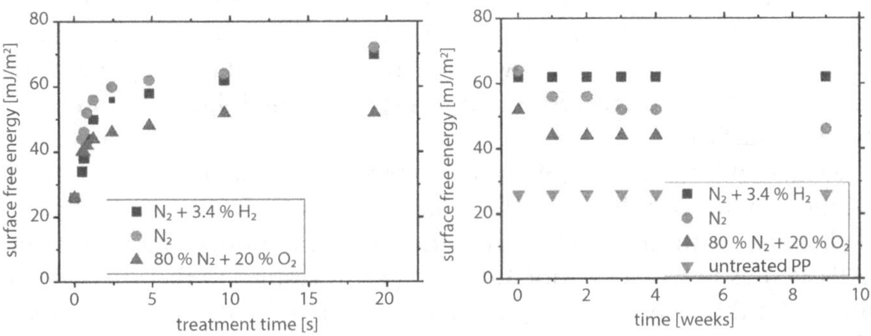

Figure 1.6 Left: Surface free energy of biaxially oriented polypropylene (BOPP) in dependence of treatment times for different process gases. Right: Stability of surface free energy of BOPP in dependence of the process gas at a treatment time of 9.6 s (from [34]).

Klages and coworkers [35] found that that polyolefin surfaces plasma treated in a nitrogen atmosphere showed an amphiphilic character, i.e., the surface may react as electrophile or as a nucleophile. The polymers treated in post-discharges of nitrogen-hydrogen DBDs not only show reactivity towards the electrophilic 4-(Trifluoromethyl)benzaldehyde (TFBA) but also react as electrophiles themselves, i.e., the afterglow-plasma-treated surfaces react also with nucleophilic chemicals. In further work Klages and coworkers [36] examined polyolefin surfaces treated using nitrogen-containing plasmas. They used chemical derivatization analysis to determine the generated functional groups. Using nitrogen or nitrogen-hydrogen containing plasmas, imines are generated on the surface, which show a very varied chemical reactivity towards aromatic aldehydes, carboxylic acid anhydrides, pyrylium dyes, isothiocyanates and with fluorescent markers like fluorescamine.

Dimic-Misic *et al.* [37] have investigated nitrogen plasma treatment of micro/nanofibrillated cellulose films, which results in increase of wettability of the surface by both polar water and nonpolar hexadecane. The total surface free energy increases with nitrogen plasma treatment with a major increase in the polar component. The surface area per unit mass was increased by the N-plasma treatment due to increased roughness on a nanometer scale which improves the adhesion property.

Besides the treatment with nitrogen-containing gases also coating processes were investigated. Chen *et al.* [38] have described a partial modification of BOPP film via a hydrophilic modification using allylamine (ALA) monomer as precursor. ALA was polymerized on the surface by Ar/O_2 or He/O_2 plasma. The results showed a permanent hydrophilic modification of the BOPP by grafting amine groups on the surface via the ALA process. The plasma modification created many micro/nano-sized holes in the BOPP film, which increased the surface roughness dramatically and the increased roughness strengthened the adhesion to the polymer surface.

Klages and coworkers [39] have shown the potential of surface modification by barrier discharge for a pure amination of the surface using allylamine, diaminocyclohexane, and 3-Aminopropyltrimethoxysilane (APTMS) as precursors. Morand *et al.* [40] investigated the direct deposition of a thin film using a plasma jet system and APTMS as precursor. The quantification of NH_2 showed that polymerization of APTMS and degradation of NH_2 groups on the surface directly correlated with Yasuda's parameter. NH_2 concentration up to (3.7 ± 1.3) groups/nm were found on the coated surfaces.

1.3.2 Oxygen-Based Surface Modification

Using corona treatment, a mixture of oxygen-based groups such as hydroxides, aldehydes, ketones and carboxylates is established on the surface of the polymers. Wagenaars *et al.* [41] used an atmospheric pressure plasma jet (APPJ) to increase the wettability of polypropylene (PP) films. Reduced water contact angle of PP was observed as a function of distance between APPJ and PP surface and oxygen gas admixture (helium gas with up to 1% oxygen), suggesting that the surface reaction mechanism is related to the atomic oxygen density produced in the plasma. Hence, multiple oxygen-species can be generated on the polymeric surface offering widespread applications especially with regard to adhesion. He-based process gas flow can also be used for hydrophilic surface modification by locally generating reactive oxygen-species, which lead to increased wettability [42].

Of high interest is the use of precursors with a single oxygen containing group to equip the surface with a monofunctional reactive group. Chen *et al.* [43] describe an atmospheric pressure plasma process with the capability of rapidly depositing carboxylic acid groups by using acetic acid/isopropanol solution. It was demonstrated that with this technique to functionalize polymer surface 7.8% carboxylic groups were found on poly(ethylene terephthalate) (PET) and 6.9% on poly(lactic acid) (PLA), which can be used for further applications. Thomas *et al.* [44] have investigated pulsed plasma processes using maleic acid anhydride (MAA) as precursor and achieved coatings with more than 10 % carboxylic groups.

Coatings with a large number of epoxy groups were also described, mostly with glycidyl methacrylate (GMA) as precursor. Using pulsed plasma processes the degree of monomer retention can be controlled. The pulsed plasma process can be divided into two phases - the plasma pulse (plasma on), where radicals will be produced by fragmenting the precursor into radicals and the pause time (plasma off), where reactions of the radicals with GMA take place with a high retention of the epoxy group of the monomer [45, 46].

1.4 Adhesion Improvement for Bonding

As described before, the variety of chemical functional groups created by plasma surface functionalization is very broad. For improving the adhesion to polymers two different ways will be presented. In the first part the classical adhesive bonding using chemically reactive surfaces combined with a subsequent adhesion process is described. By adjusting the chemical interaction between the surface and the adhesive very strong improvements of adhesively bonded joints of various materials can be achieved. In the second part an adhesive-free joining process will be presented. In this process the reactive surfaces interact with each other without any adhesive. The high surface reactivity leads to very good adhesion results between different types of materials.

1.4.1 Adhesive Bonding by Functional Groups

In recent studies [47, 48] an atmospheric pressure plasma jet was used for surface activation by plasma treatments and coatings by plasma-polymerization on flat and mechanically roughened elastomeric styrene-butadiene-rubber to improve the bond strength of SBR/leather joints based on PU-adhesive. For the coating treatment, the amine-based precursors such as 3-aminopropyltriethoxysilane (APTES) or 3-aminopropyltrimethoxysilane (APTMS) were selected to obtain functional nitrogen-groups on the surfaces.

Inoue *et al.* [48] demonstrated improved plasma-enhanced adhesion of conducting polymers such as poly(3,4-ethylenedioxythiophene): poly(styrene sulfonate) (PEDOT:PSS), polypyrrole (PPy), and polyaniline (PAni) which have been widely explored as coatings for electrodes. They report a simple method to achieve strong adhesion of various conducting polymers including PEDOT:PSS, PPy, and PAni on diverse commonly used insulating and conductive substrates including glass, polyimide, poly(dimethylsiloxane) (PDMS), indium tin oxide (ITO) and gold. Based on a hydrophilic PU-adhesive layer and an APTMS-coating, the adhesive layer strongly adheres to the substrates and moreover forms an interpenetrating polymer network with the conducting polymer.

To improve the load resistance of joints such as carbon fiber reinforced polymers (CFRPs), atmospheric-pressure plasma surface treatments can be performed to increase the adhesion using O_2/He as process gas. It was found that the load at break for CFRP joints increased with increasing plasma power and the total treatment time for plasma conditions (approx. 10 min, 50 W) at which a selective ablation of epoxy resin and uncovering of carbon fibers resulted [49].

A combination of DBD-plasma treatment and chitosan coating can be performed on ultrahigh molecular weight polyethylene (UHMWPE) fibers in order to improve adhesion and dyeing properties, as described by Ren *et al.* [50]. Also low molecular weight oxidized epoxy adhesive mediated PP/Al-joints can be plasma-activated by the use of Ar, Ar/O_2 and Ar/N_2, where especially the nitrogen functional groups promote adhesion. This increase in adhesion is based on N-rich groups, leading to an improvement in wettability and chemical reactivity towards the epoxy adhesive to create PP-Al joints and is presented in Figure 1.7. Hence, the tensile strength of the adhesive bond can be increased by 2.5 to 4.5 times compared to the untreated material [51].

A further example is based on the treatment of poly(vinylidene difluoride) (PVDF) with N_2-H_2 gas mixture. The adhesion with a hot-melt adhesive for a PVDF-steel composite could be increased to more than 1.6 N/mm by N_2-H_2 plasma treatment. This value is four times as high as the value using air-plasma. The FTIR spectra after the peel test show from the absorption peaks of the hot-melt adhesive on the PVDF surface that a cohesive fracture had occurred in the hot-melt adhesive [32].

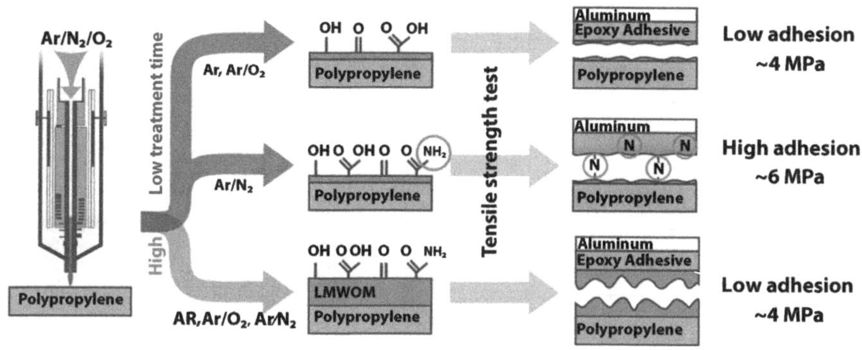

Figure 1.7 Overview of various treatment parameters to functionalize PP-substrates for epoxy adhesive joining (from [51]).

Fischer and coworkers [52] have investigated the promotion of adhesive polymer bonding by plasma modification using defined ambient conditions and process gases. The results show that, except for PTFE, for all polymer-adhesive combinations the absence of oxygen in the functionalization of polymer surfaces has a significant influence on the surface properties. In particular, the N_2-H_2 treatment in a N_2 atmosphere leads to a strong improvement in the bond strength and mostly to adhesion failure (AF), in some cases to a cohesive failure (CF) and for PA6 to a substrate failure (SF), as shown in Figure 1.8.

In Figure 1.9, at optimized process conditions a polar surface with excellent adhesive bonding property is demonstrated, which can be generated

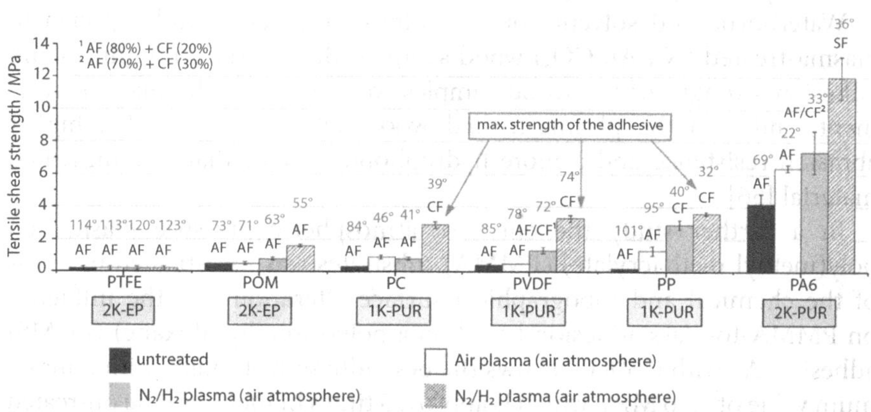

Figure 1.8 Influence of plasma gas and the surrounding atmosphere on adhesion between different polymers and adhesives (adapted from [52]).

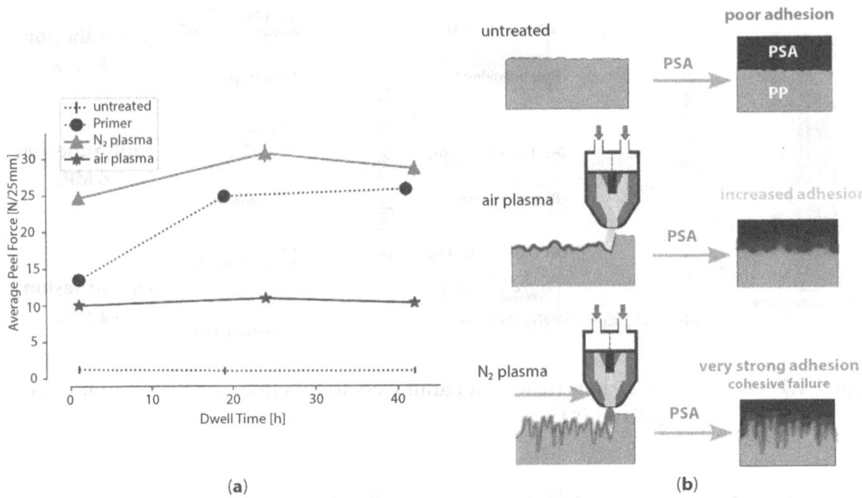

Figure 1.9 (a) Comparison of the average peel strength on untreated, primer treated and plasma treated (air, N_2) PP as a function of the dwell time of the PP–PSA system after bonding. (b) A schematic of the PP surface modification using APPJ treatment and the resulting adhesion improvement for the PP–PSA bonding (from [53]).

and can also be regarded as a promising ecological alternative to the application of liquid primer systems [53]. With a cold plasma treatment, remarkable adhesion enhancement using pure N_2 as process gas on PP with an acrylic pressure-sensitive adhesive (PSA) could be achieved.

Further improvements in the strength of adhesively bonded carbon fiber reinforced polymers and aluminum alloys were compared for three different process gases (Ar, N_2, and air) [54, 55].

Waterborne and solvent-borne coatings were successfully applied to plasma-treated (N_2, Ar, CO_2) wood samples. Both surface roughness and adhesion strength of the wood samples were increased by plasma treatment, and also the plasma-treated wood substrates showed a higher abrasion resistance and a more hydrophobic surface than the untreated material [56].

In a further study the effect of atmospheric pressure plasma on poly(methyl methacrylate) (PMMA) substrates was investigated in terms of the chemical and topographical surface alteration and the influence on PMMA-to-glass adhesion based on a poly(dimethylsiloxane) (PDMS) adhesive. Ar with 0.5% O_2 shows the best adhesion strength with a maximum value of 1.56 MPa, which is approx. 5 times higher than the untreated sample [57].

1.4.2 Adhesive-Free Joining by Functional Coatings

Innovative materials form the basis for modern industrial products in all areas of life. The market for products such as specialty films for food packaging, flexible printed circuit boards or decorative and protective films is constantly expanding. These materials must meet a variety of requirements such as optical transparency, temperature stability, good processability, and permeability for oxygen and water vapor. If such complex functions are necessary, composite films are particularly suitable, because they consist of different layers and thus combine different material properties. Despite extensive work on materials joining, there is still a great need for further development of advanced adhesive bonding processes to overcome undesirable side effects of the application of conventional bonding processes such as the consumption of large quantities of adhesive, a lack of long-term stability, a tendency to creep, and migration behavior. For these reasons, alternative methods for joining different materials, e.g., plastics and metal films are investigated.

Unpublished first results from our group by Lachmann *et al.* considered this adhesive-free, low-temperature (< 100 °C) joining process at moderate pressure (1.85 N/mm^2). Chemically reactive ultra-thin films are deposited on polyethylene (PE) and aluminum surfaces by means of atmospheric PECVD (plasma-enhanced chemical vapor deposition) as a nanometer-thick "adhesive layer" which subsequently could be bonded. They have shown adhesion strength of more than 1.3 N/mm for joined polyethylene-aluminum sheets. N_2 activation processes prior to the coatings based on dimethylacrylamide (DMAA) or APTMS led to increased bonding strength up to failure of the PE substrate (see Figure 1.10).

Recently studies were also published regarding stainless steel/plastics joints such as fluorocarbon plastic or polycarbonate [58]. The effects of plasma treatment on PTFE sheets on the bond strength using acrylic acid as precursor [59] or Ar/H_2O as process gas [60] were demonstrated. Particularly for Ar/H_2O surface treatment a high bond strength, which is based on generation of oxygen-containing functional groups, was achieved without the use of any adhesive.

The adhesive-free bonding of different web-like plastic-metal combinations at low temperatures was investigated on copper foil by Eichler *et al.* using APTMS as precursor and industrially comparable line speeds of plasma process up to 10 m/min [61].

Also, regarding medical and microfluidic applications, improved adhesion processes particularly to PDMS without any adhesives are of great interest [62–64]. Tyler *et al.* [62] investigated a co-polyacrylate coating

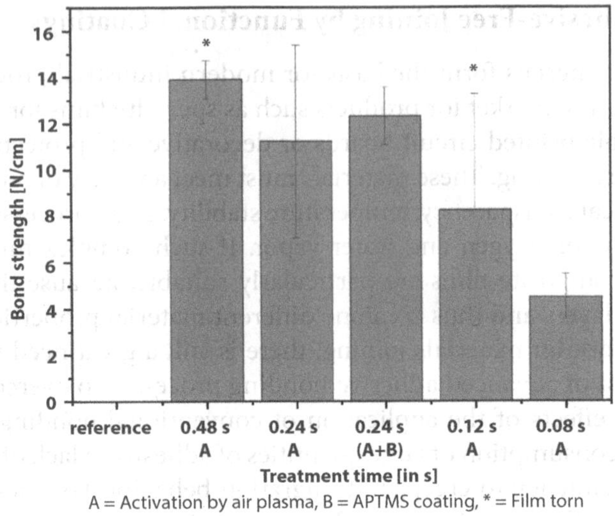

Figure 1.10 Bond strength between aluminum and PE depending on the surface treatment.

from ethylene glycol dicyclopentenyl ether acrylate (EGDPEA) and di(ethyleneglycol) methyl ether methacrylate (DEGMA) for a PDMS catheter. This coating showed much higher bacterial resistance compared to commercial catheter materials. Using oxygen plasma treatment to improve the wettability and reactivity of the PDMS catheter material, the adhesion with an EGDPEA-co-DEGMA coating could be demonstrated. A further study by Sivakumar and Lee [64] showed an oxygen-based surface modification for microfluidic devices using additional anhydride- and aminosilanes to realize bonding between PDMS and non-silicon thermoplastics such as polystyrene (PS), poly(methyl methacrylate) (PMMA), polycarbonate (PC) and poly(ethylene terephthalate) (PET). Here, the bond strength values of PDMS–thermoplastic assemblies, measured by tensile tests for PDMS–polystyrene, and PDMS–polycarbonate composites were found to be (519.5 ± 6) kPa and (476.6 ± 8) kPa, respectively.

1.5 Targeted Adhesion for Biomedical Applications

For many years, atmospheric pressure plasma processes have been a versatile tool to enhance the adhesion of biomolecules. Plasma pretreatment or corona treatment is the state-of-the-art to improve wettability of disposables made from plastics, such as microtiter plates or petri dishes [65–68]. Even argon or ammonia were used for plasma pretreatment to influence adhesion of cells [69]. Alavi *et al.* [70] describe the reagent-free covalent attachment of biomolecules for bioprinting. A double electrode plasma jet system was used to treat LDPE and polycaprolactone (PCL) with helium as process gas followed by incubation of the samples in a solution of bovine serum albumin (BSA) and fibrinogen (FG). Even rigorous washing with sodium dodecyl sulfate (SDS) was not able to remove the proteins from the surface. Also, the attachment of hydrogels, here acrylamide, was possible. Bilek *et al.* [71] used a similar approach, where PTFE was used as substrate. Using the PlasmaPlus® technology (Plasmatreat GmbH, Germany) with nitrogen as process gas followed by a spray process, Tan and Al-Rubea [72] described the attachment of tissue-specific extracellular matrix (ECM) proteins such as collagen or laminin on polystyrene and titanium (Ti) alloys.

Šrámková *et al.* [73] used plasma pretreatment - here a diffuse coplanar surface dielectric barrier discharge (DCSDBD), to activate PTFE for further attachment of poly(2-oxazoline) (POx) from the liquid phase. This polymer is discussed as an alternative to the frequently used poly(ethylene glycol) (PEG) derivatives. POx can be produced by cationic ring-opening polymerization which leads to well-defined properties as the process and the molecular structure are well controllable. The activation of PTFE was performed at ambient air and an input power of 400W at 15 kHz. After the attachment of poly(2-methyl-2-oxazoline)-*stat*-(2-(3-butenyl)-2-oxazoline) PMEOx from the liquid phase these coatings were also treated by a DCSDBD, now using ambient air or argon as process gas. In contrast to the pristine POx coatings grafted from the liquid, which exhibited antifouling property, the plasma post-treated PMEOx coatings showed enhanced adhesion of fibroblasts.

Suzuki *et al.* [74] combined UV/ozone with atmospheric pressure plasma treatment to improve the cultivation of pluripotent embryonic stem cells to achieve a feeder free and matrix coating free culture system. This is quite favorable as the risks of animal derived factors caused by conventionally used feeder cell lines or cell-adhesive matrices are circumvented. A combination of UV irradiation followed by plasma treatment of polystyrene in nitrogen gave quite promising results.

Next to surface treatment, deposition of thin films is a versatile way to enhance the adhesion of biomaterials. In most cases, the deposition of chemically functionalized films containing amino groups [75–77], carboxylic groups [78] or thiols is investigated. As described before, the plasma sources vary from atmospheric pressure plasma jet systems to surface DBD or linear electrode geometries. The precursor is introduced as a gas or as an aerosol, depending on the vapor pressure.

An example for amine-rich coatings is the use of cyclopropylamine deposited by a dielectric barrier discharge [75]. The amount of primary amino groups was determined after chemical derivatization using XPS, depending on the Yasuda parameter (W/FM, W: discharge power, F: monomer flow rate, M: molecular weight of the monomer). Cell attachment was investigated by the adhesion of human foreskin fibroblasts (HFFs). As water stability of such hydrophilic plasma-polymerized films is one of the major issues for biological application, this was one of the major focuses in the already mentioned publication [75]. It was demonstrated that there is an optimum W/FM ratio where partial desorption (low W/FM ratio) and full delamination (high W/FM ratio) occur, so that after immersion in water only a very thin layer of the plasma polymerized film remains on the substrate. Nevertheless, this is enough to achieve strong adhesion of human fibroblasts.

Coatings deposited by atmospheric pressure plasmas using a DBD could also be used to prepare surface-based cell cultivation systems for 3D cell culture in hanging droplets. This was demonstrated by Gepp *et al.* [79] by using a patterned coating prepared at Fraunhofer IST. This pattern is based on hydrophilic and superhydrophobic areas on a thin polypropylene (PP) foil. Superhydrophobicity was achieved by a double layer of a particulate SiO_x layer covered by a hydrophobic coating based on plasma-polymerized tetramethylsilane (TMS). With this coating a water contact angle > 150° could be reached and also cell culture medium was repelled from this layer. A hydrophilic dot pattern is generated by an area-selective plasma-polymerized APTMS layer, which is deposited under oxidizing conditions. In this way, a dot pattern is created on the hydrophilic dots to obtain defined droplets on the polymer surface. The resulting droplets are shown in Figure 1.11. Thus, novel closed cell culture bags to cultivate

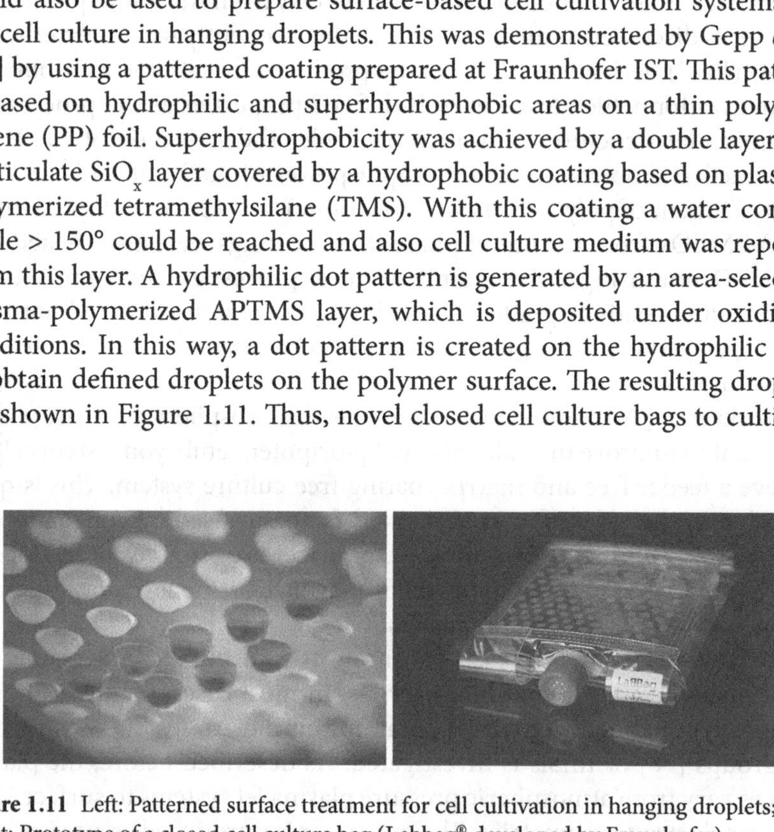

Figure 1.11 Left: Patterned surface treatment for cell cultivation in hanging droplets; Right: Prototype of a closed cell culture bag (Labbag® developed by Fraunhofer).

e.g. induced pluripotent stem cells in hanging droplets to form three-dimensional aggregates for further differentiation (Labbag®, Fraunhofer Gesellschaft, Germany) were demonstrated.

More complex plasma polymers were investigated by Czuba *et al.* [80] using a DBD set-up in combination with monomer nebulization. Copolymerization of L-3,4-dihydroxyphenylalanine (L-DOPA), methyl DOPA methacrylate (DOMAm) and ethylene glycol dimethacrylate (EGDMA) led to catechol/quinone-rich thin films (see Figure 1.12 and Figure 1.13). These films are described as an alternative to conventional wet-chemical deposited polydopamine films. With further protein immobilization the authors describe antibiofilm activity against *Staphylococcus epidermidis* (Gram-positive bacterium), and the adhesion of the human preosteblastic cell line MG-63 was possible.

Even nanoparticles (NPs) can be deposited using an atmospheric plasma. Patelli *et al.* [81] investigated an aerosol-assisted process to tune nanoroughness, surface chemistry and drug delivery on implantable devices using a plasma jet system (see Figure 1.14). Polycaprolacrone (PCL) was used as a model polymer due to its biodegradability. In a two-step process, nanoparticles (silica nanoparticles and poly(lactic-co-glycolide) PLGA nanoparticles) and a plasma polymerized layer to fix the particles on the surface were deposited. Cell attachment to PCL was determined using

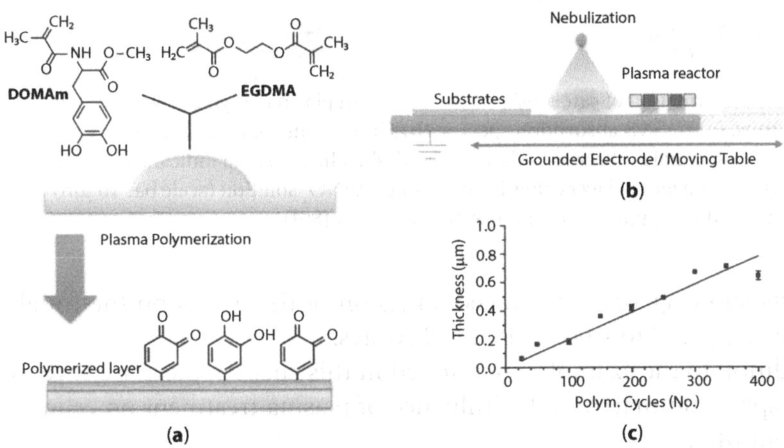

Figure 1.12 Schematic representation of the deposition of polymethacrylate-based bioinspired layers containing catechol/quinone moieties (a) and (b) DBD set-up with nebulization used for the deposition process. Thickness versus number of polymerization cycles (c) on the plasma deposition of EDGMA/DOMAm mixtures. Mean ± SD (n = 6) for 100, 200, and 300 polymerization cycles (from [80]).

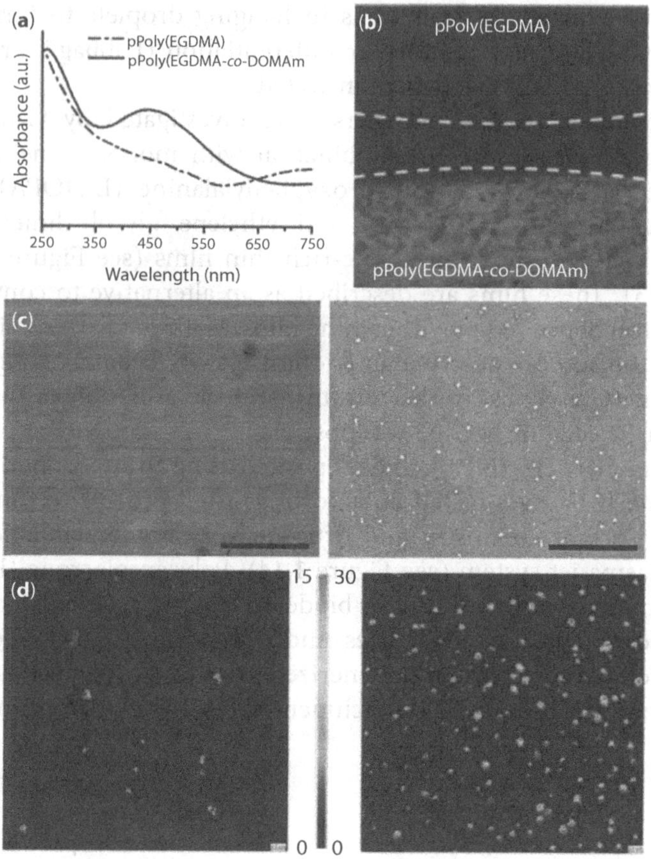

Figure 1.13 Confirmation of catechol/quinone groups in plasma polymer layers containing DOMAm units. (a) UV–vis absorption spectra. (b) Fluorescent confocal microscopy image of two-disc samples observed side-by-side (dashed lines to help indicate disc sample borders). (c) SEM images of layers after immersion in $AgNO_3$ solution (scale bar 10 μm) and (d) Nano-SIMS mapping of silver (10 × 10 μm) (from [80]).

fibroblasts showing 100% enhanced adhesion of these cells on the rough structure compared to smooth PCL substrates.

In addition to the examples mentioned in this chapter, there are several review papers summarizing the influence of plasma treatment on cellular attachment [82].

Further examples to improve cell adhesion to polymeric substrates are given in the next sections, as promising examples to tailor 3D substrates for additive manufacturing.

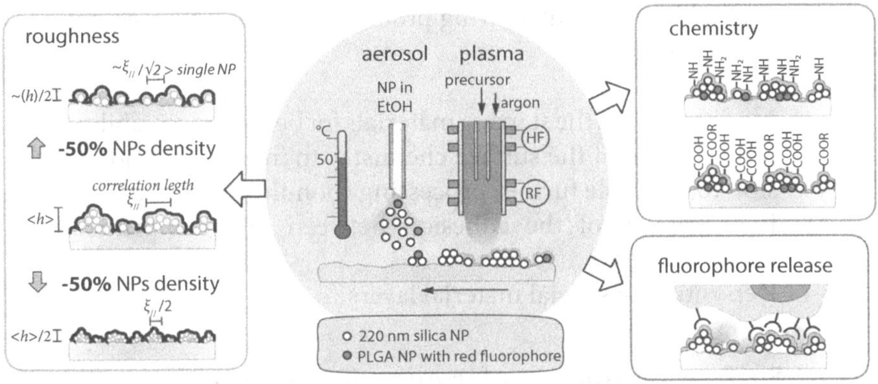

Figure 1.14 Scheme representation of the proposed technology and its control on roughness, chemistry, and molecules release. In the center, the deposition process is summarized with the aerosol deposition of the NPs and the following deposition of a fastening layer by APPJ, scanning the samples. The film properties are controlled independently: the roughness is controlled laterally and vertically by NPs size and density in the submicrometer range; the surface chemistry of the deposited APPJ plasma polymer can offer amine or carboxylic functionalities by the precursor choice; during cell growth, when PLGA NPs with a red fluorophore embedded are used, the fluorophore is released in the culture medium (from [81]).

1.6 Relevance of Adhesion in Additive Manufacturing

In additive manufacturing (AM) or 3D printing, components are created by applying layers of materials layer by layer under computer control. In this way, complex 3-dimensional components can be fabricated from a wide range of materials (polymers, metals, ceramics). There are a variety of different AM techniques such as fused deposition modeling (FDM), laser sintering or stereolithography which are suitable for different materials. Since the 3D printed components are built from individual layers, changes in the surface chemistry can be used to influence and thus enhance and improve the adhesion of layers and stability of produced components. The most important problems in additive manufacturing for polymer materials are a weak interlaminar adhesion, the printability of different polymers onto each other, adhesion failures in post-processing like adhesive bonding or painting and the missing biocompatibility with printed polymer structures. Each of these problems can be tackled by functionalization via atmospheric pressure plasma treatment. The surface modification achieved to create different chemical groups on the polymer surface can contribute to additive manufacturing processes in different ways, depending on the gases used:

1. Modification of the starting materials for better processability
2. Modification of the surface chemistry to improve the adhesion for possible further processing (bonding, painting).
3. Improvement of the adhesion between different printed material layers
4. Deposition of special material layers

1.6.1 Surface Modification for Adhesion Improvement

Interlaminar adhesion is essential for the mechanical stability and thus the usability of additively manufactured components. For many polymers processed with extrusion-based 3D printing techniques such as fused deposition modeling (FDM), the interlaminar adhesion is often insufficient. Plasma treatments of each printed polymer layer offer an interesting way to improve adhesion. This was investigated by Narahara [83] by placing an atmospheric pressure plasma jet parallel to the extruder in an FDM printer (Figure 1.15).

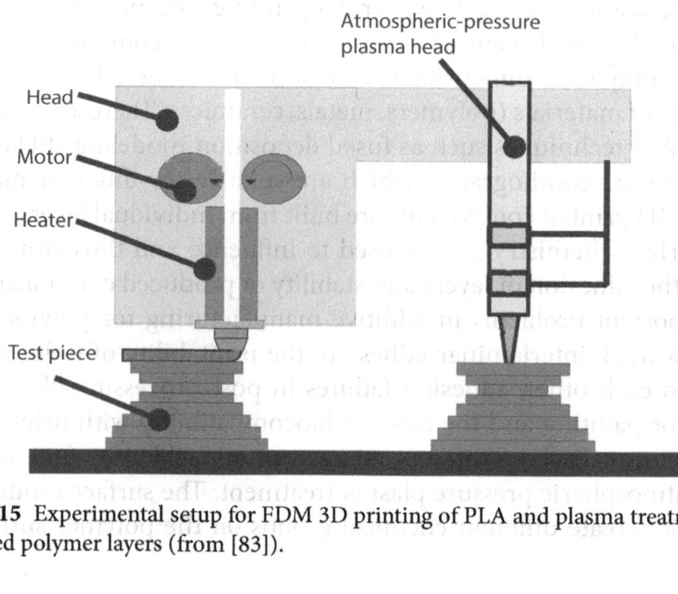

Figure 1.15 Experimental setup for FDM 3D printing of PLA and plasma treatment of the printed polymer layers (from [83]).

The atmospheric plasma jet was operated with Helium gas. It was shown in the work that the surface free energy of poly (lactic acid) (PLA) polymer can be increased by plasma treatment with a plasma jet (Figure 1.16).

Test specimens for rupture tests were printed and the rupture stress was measured. Figure 1.17 shows the positive effect of the plasma treatment on the mechanical stability of the printed part for two different building speeds.

Laser beam melted (LBM) components do not have the same mechanical properties as injection molded components. This depends in particular on the wettability of the particles to be melted, which can be varied by plasma treatment. This question was investigated by Sachs *et al.* [84]. Polyamide (PA12) and high-density polyethylene (HDPE) particles with a size of approx. 25 μm to 100 μm were functionalized by atmospheric pressure plasma treatment. The particles were treated with a plasma jet in a fluidized bed reactor (Figure 1.18). The influences of the oxygen concentration and the treatment time on the water contact angle were investigated.

For PA12 with air plasma, there is an optimal treatment time of 0.075s, for which the water contact angle reduces from approx. 90° to 60° (Figure 1.19). For longer treatments, an increase in the water contact angle was observed, which is attributed to etching effects. For the treatment of HDPE, a reduction of the water contact angle was only observed from 0.3 s onwards. For PA12, an optimal oxygen concentration was found at 5% for the treatment (Figure 1.20). For HDPE, on the other hand, higher oxygen

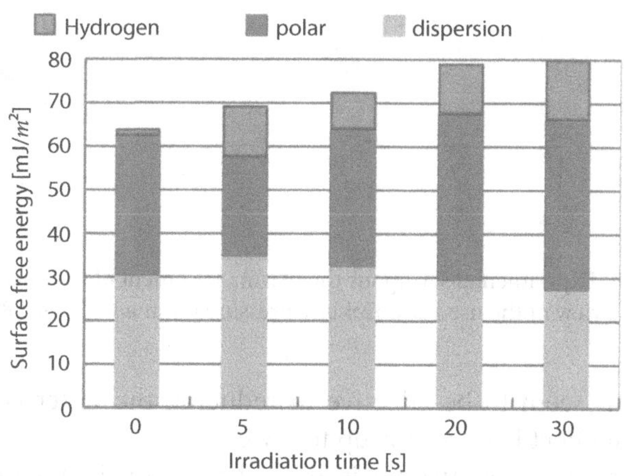

Figure 1.16 Surface free energy of PLA as a function of treatment time with a plasma jet (from [83]).

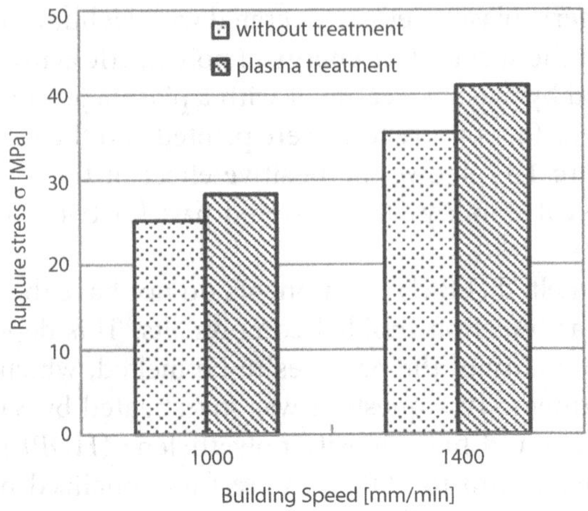

Figure 1.17 Rupture stress of FDM printed PLA test samples without and with plasma treatment (from [83]). The building speed describes the movement speed of the polymer extruder in the printing process.

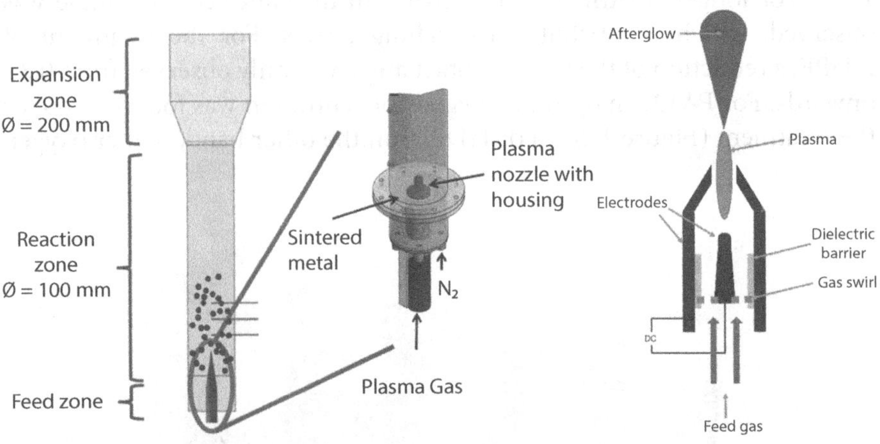

Figure 1.18 Left: Experimental set-up for the plasma treatment of polymer particles; Right: Schematic view of the used atmospheric pressure plasma jet (from [84]).

concentrations seem to be effective in reducing the water contact angle. The treatment effect is stable for up to 4 weeks.

Adhesion to polymer surfaces is also relevant for the repair of polymer components. McLouth *et al.* [85] investigated the atmospheric pressure plasma treatment of the high-performance plastic polyetherimide (PEI)

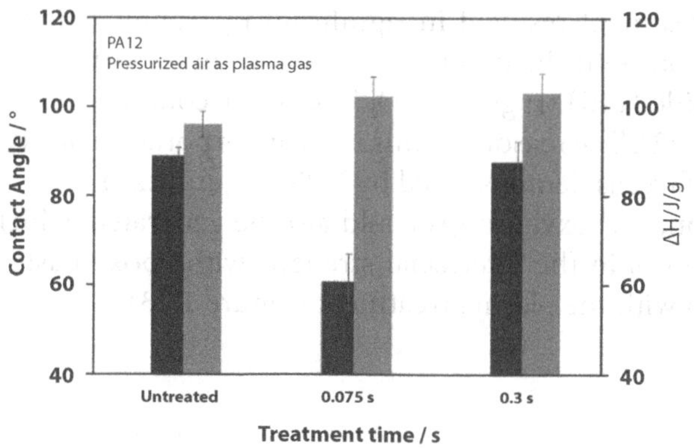

Figure 1.19 Water contact angle and melting enthalpy of PA samples for different treatment times with air plasma (from [84]).

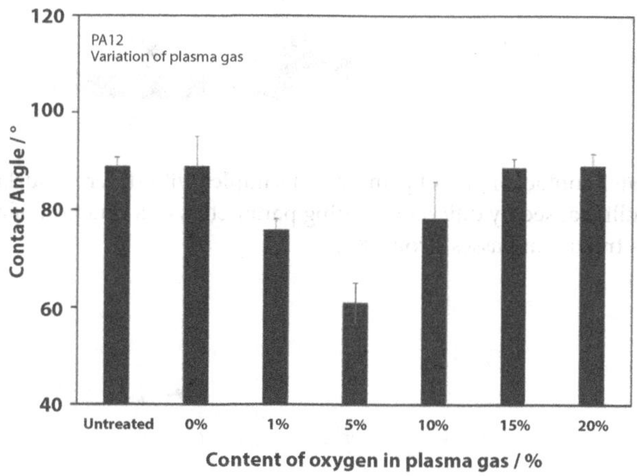

Figure 1.20 Water contact angles of PA samples treated in Ar/O_2 plasma as a function of O_2 content (from [84]).

(Ultem 9085), which has a high temperature and chemical resistance as well as high tensile and flexural strengths, with the aim of enabling the repair of FDM-printed components. In the work, FDM printed PEI was treated with helium-oxygen plasma and the influences of the plasma treatment on the water contact angle, the surface chemistry determined by XPS and tensile tests on Ultem samples bonded with a two-component epoxy paste adhesive were investigated.

Plasma treatment resulted in significant hydrophilization of the polymer surface on both the outer contours and inner surfaces (infill) of the polyetherimide (PEI) (Figure 1.21). The water contact angle was reduced from 115° to 22°. The reason for this is the incorporation of oxygen into the polymer surface, as demonstrated by XPS (Figure 1.22).

In addition, carboxyl groups could also be generated, which caused an increase of 35 % in the interfacial strength with epoxy paste adhesive in combination with the plasma treatment (Figure 1.23).

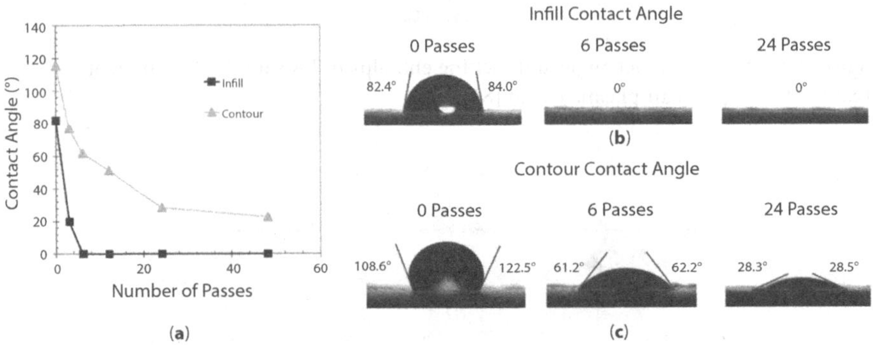

Figure 1.21 Water contact angles of printed PEI samples with different surface textures (contour and infill) caused by different printing parameters as a function of atmospheric pressure plasma treatment passes (from [85]).

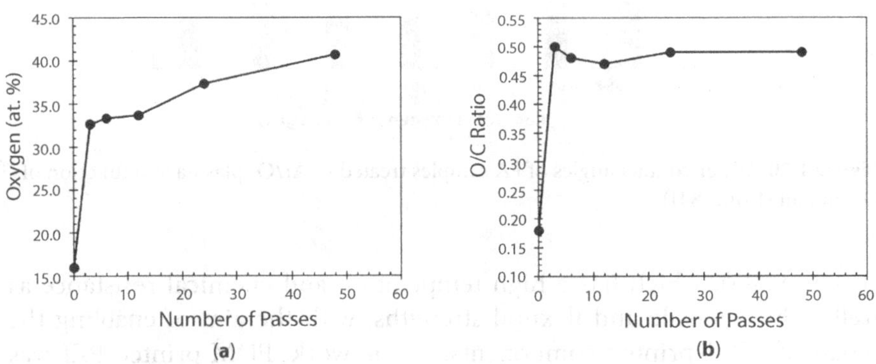

Figure 1.22 Oxygen content and O/C ratio of plasma treated PEI samples as a function of treatment passes (from [85]).

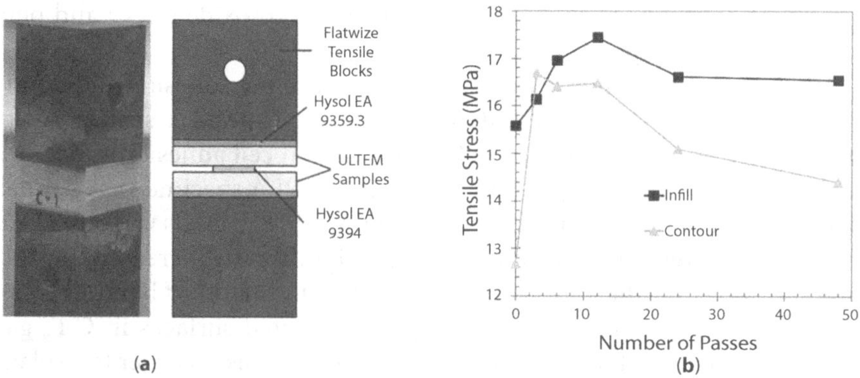

Figure 1.23 Tensile stress results of bonded plasma treated PEI samples as a function of treatment passes (from [85]).

1.6.2 Enhanced Cell Adhesion and Growth on Additive Manufactured Parts

In recent years, the additive manufacturing (AM) of PLA-based bone scaffolds has gained increased attention, because of its good processability, mechanical properties and biocompatibility. The possibilities to tune the structures of PLA-scaffolds and the potential of this biomaterial for bone regeneration makes it even more relevant. Though the use of PLA in regenerative medicine is limited due to its hydrophobicity, which hinders cell adhesion and proliferation. Another drawback of the hydrophobic property of PLA is the delayed release of acidic by-products during the degradation process.

To optimize these characteristics and increase the cell adhesion and osteoconductivity of PLA bone scaffolds, certain plasma processes can be used. The aim of plasma treating PLA scaffolds is to modify their surface chemistry and topography. Main goal of these modifications is to make the scaffolds more hydrophilic and porous, thus inducing a positive effect on the attachment of cells and biological compounds [86, 87]. In the following a variety of plasma treatment techniques and how they are applied to achieve the mentioned effects are described.

Nakagawa *et al.* [88] treated PLA specimens, which were injected from pellets. They were cleaned via ultrasonic cleaning in distilled water and stored in a desiccator until used. Surface treatment was carried out with a Plasma-Jet at 20 mm distance from the specimen and a treatment speed of 5 mm/s with a beam voltage of 10 kV AC. The gases used for plasma ignition at atmospheric pressure were air, carbon dioxide (CO_2) and perfluoropropane (C_3F_8).

After plasma treatment a significant increase in the adhesion and bonding capability of MC3T3-E1 cells was achieved. Especially surface treatments in air or in CO_2 gas showed an increase in cell adhesion capability of around 40 % (measured as optical density on PLA specimen). A similar tendency for the proliferation potency of the MC3T3-E1 cells was observed, where plasma treatment in air and CO_2 gas significantly increased the cultured cell mass on the surface of the sample. On the other hand, the cell adhesion and cell mass culturing on plasma treated surfaces in C_3F_8 gas were lower compared with treatments in air or CO_2 gas. In order to explain these effects surface roughness, contact angle and XPS measurements were used. Based on the measurement results, it was elucidated that the adhesion capability and cell mass culturing of MC3T3-E1 cells could be improved by plasma treatment in air and CO_2 gas, due to the following:

1. Increased wettability; contact angle with water decreased from 77.4° to 39.8° with air and 42.2° with CO_2.
2. Reduction of C1s content; clear inverse correlation with cell adhesion capability was observed.
3. Increased O1s content and O1s/C1s ratio showed significant positive correlation with cell adhesion capability.

Even though the increased hydrophilicity has a great impact on cell adhesion it is difficult to evaluate cell adhesion by only measuring the contact angle with water. Adhesion capability is significantly affected by the physicochemical properties of the surface, such as the molecular structure, functional groups, surface charge, surface roughness, topography, etc. One of the well-known benefits of 3D printing is the fabrication of complex structures that are optimized for their use. However, the resolution of printed elements is limited if the printed features such as pores, are relatively large compared to the size of the cells. This leads to cell adhesion problem and ultimately to low rates of tissue formation and cell differentiation [89, 90].

Wang *et al.* [91] investigated how 3D printed PLA scaffolds could be treated with atmospheric plasma for use as bone tissue replacement. The goal was to modify the surface roughness and topography to mimic the nanoscale and extracellular matrix properties of the bone tissue they intended to replace. The plasma jet used for the treatment is described by Shashurin *et al.* [92] and is schematically shown in Figure 1.24a. It consists of a Pyrex syringe equipped with a pair of high voltage electrodes, a central ceramic isolated electrode of 1 mm diameter, and an outer grounded ring electrode. A typical photograph of the plasma jet is shown in Figure 1.24b for a voltage of 5 kV and helium flow rate of 17 L/min. The length of the visible plasma jet was approximately 50 mm.

The scaffold samples were treated for 1, 3 and 5 minutes at a distance of 12 mm below the plasma source. Before cell seeding, all scaffolds were characterized and sterilized under UV light for half an hour.

Results revealed that all plasma treated scaffolds had significantly enhanced fibroblast cell adhesion compared to scaffolds without surface modification. Cell adhesion of the one-minute treated scaffolds was increased by about 40 % and five minutes treated scaffolds had an increase of about 80 % compared to non-treated scaffolds. Cell proliferation tests showed a significantly increased number of cells after 5 days of culture when compared to those on the PLA control.

These improvements may be attributed to changes in surface chemistry and surface roughness. As the plasma flame is exposed to air, various reactive nitrogen and oxygen species interact with the surface and result in oxidation processes, thus increasing oxygen concentration while reducing carbon concentration at the surface. Although the overall atomic percentage of O1s did not change with 5 min plasma treatment, it created new O1s bonding states related to new C=O-containing groups. These oxygen-containing groups on the surface of the PLA scaffolds led to higher polar

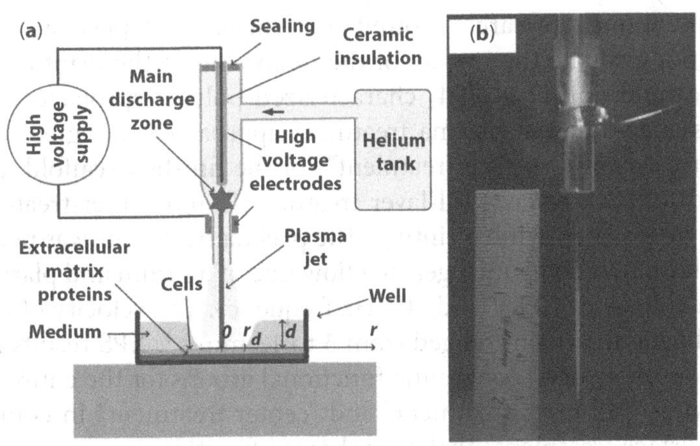

Figure 1.24 (a) Schematic depiction of the plasma jet set-up used to treat 3D printed PLA scaffolds for better cell adhesion, (b) typical photograph of plasma jet at U=5 kV and He flow rate of 17 L/min (from [92]).

surface free energy and thus to better surface wettability, which is reflected in the contact angle measurements. Contact angle dropped from 70° to 24° right after 5 min plasma treatment, while plasma treatments for 1 and 3 min decreased the contact angle to 49° and 41°, respectively. The contact angle gradually recovered from 24° to around 40° after three days of storage in a dry container at room temperature, which is still significantly lower than the contact angle of untreated PLA scaffold. The decrease of surface wettability after storage can be explained, as O=C bonds can be easily oxidized.

While the plasma treatment changes the surface chemistry it also changes the surface roughness due to etching processes and the local heating by the plasma jet. The surface will partially re-form, allowing for the transformation from micro- to nano-scale topography. These changes in surface topography can be seen in Figure 1.25 with scanning electron microscopy (SEM). Measurements with atomic force microscopy confirm the optical increase in roughness seen in SEM images. Roughness without treatment is 1.2 nm and after treatment for 1, 3, and 5 min roughness increased to 10.5, 22.9, and 27.6 nm, respectively. The created nano-scale topography and increased roughness will result in higher surface free energy and wettability once the scaffold interacts with tissue. Furthermore, the well-structured nano-scale surface topography showed beneficial cell behavior and promoted improved cell clustering and spreading [93]. This was further confirmed by cell proliferation results, which showed higher cell viability on plasma treated samples.

The mentioned research showed the efficiency of plasma treatment, but did not evaluate the impact of the treatment on the printing process of the scaffold. Liu *et al.* [94] characterized poly(ϵ-caprolactone) (PCL) scaffolds after different plasma treatment approaches, a regular uniform surface treatment ("surface treatment") of the finished scaffold, uniform treatment of each layer ("full-layer treatment") and center-treated layers ("center treatment") while printing. The plasma treatment was performed with a plasma-jet using nitrogen at a flow rate of 5 L/min and plasma ignition at a voltage of 10 kV and 50 kHz frequency. The velocity of the plasma-jet during treatment ranged from 3 to 12 mm s^{-1}. XPS results indicate generation of nitrogen containing functional groups for the entire scaffold on applying "full-layer treatment" and "center treatment" in comparison to the "surface treatment" that only showed functional groups on the top surface of the scaffold. A more homogeneous treatment of the entire scaffold is also reflected in the decreased water contact angle of the full-layer- and center-treated scaffolds. *In vitro* biological assessment showed the best results on surface-treated and full-layer-treated scaffolds. The Alamar

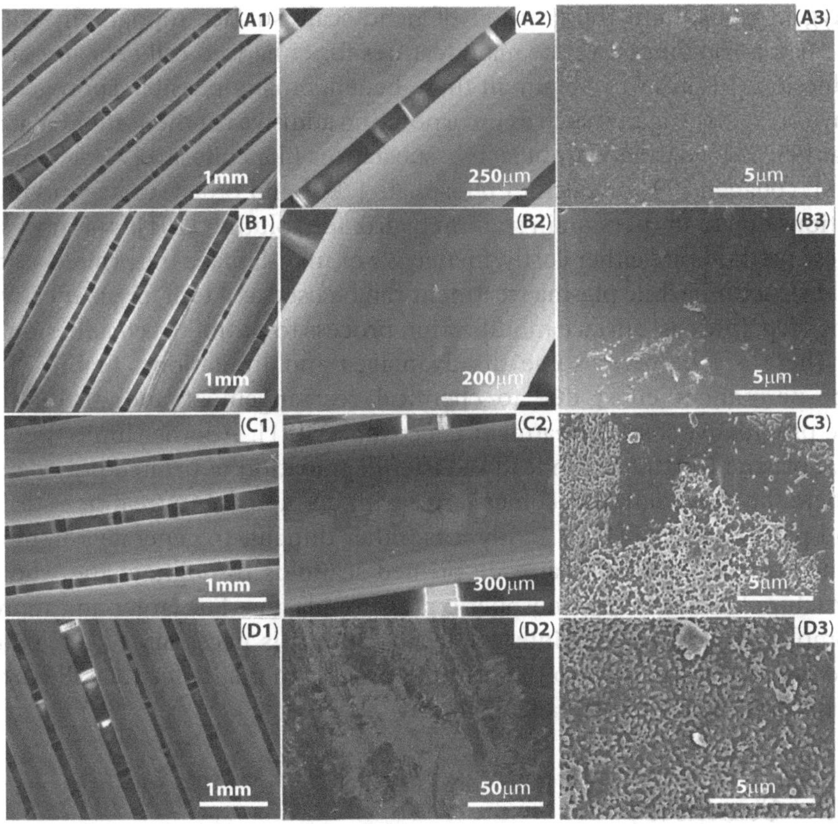

Figure 1.25 Scanning electron microscopy images of untreated (A1-3) and plasma treated PLA scaffolds at different treatment times: 1 min (B1-3), 3 min (C1-3) and 5 min (D1-3). Optical roughness of the PLA scaffold and newly emerged nano-structures can be detected with increased treatment time (Images from [91]).

Blue assay also confirmed this observation with no statistical difference between them. While full-layer and surface-treated scaffolds show the same cell adhesion, the application of selective and a more homogeneous plasma treatment may benefit further developments and possibilities for scaffolds with gradient functionalities.

The previously mentioned papers [89–94] concluded that atmospheric pressure plasma treatments are an effective way to modify the surface of complex 3D-printed scaffold structures to enhance their biocompatibility and cell adhesion. Especially suited are plasma jets for the ever-changing scaffold geometries originating from additive manufacturing processes. Increased wettability, suitable surface chemistry with functional oxygen-groups and the creation of nano-scale surface topography with increased roughness are the main features for enhanced cell behavior and adhesion. In order to obtain all these benefits many possible approaches have been investigated such as incorporating additives in the scaffold material [95, 96], coating with bioactive substances like chitosan [97], alginate [98], collagen [99] or calcium phosphate [100] and other surface modifications due to immersion in sodium hydroxide solution [101]. But most of these methods are either costly, ineffective or may damage the printed scaffold structure, while plasma treatment can be used as a tool for an effective one-step, low cost surface modification process to promote cell adhesion.

There are however some disadvantages and limitations, mainly the non-permanent effect due to chemical rearrangement and, therefore, the progressive loss of treatment effectiveness [91]. Another limitation is that plasma treatment can lead to faster degradation of printed polymers, where the degradation becomes faster with an increase in treatment time and plasma energy [102]. Lastly it is rather difficult to generate a homogeneous modification of already printed scaffolds throughout the whole structure by means of atmospheric pressure plasma. For larger and more complex structures with internal architecture it is necessary to integrate the plasma treatment into the printing process to guarantee that all parts of the scaffold are treated adequately.

1.7 Summary

The industrial interest in the use of atmospheric pressure plasma treatments (APPTs) is constantly increasing due to the cost advantage this technology may potentially offer for various industrial applications. It has the potential to be used as a leading alternative to classical corona treatment and wet-chemical processes in the future. Like the well-established corona treatment, APPT is characterized by its very good scalability, integrability and flexibility in terms of materials to be treated. Compared with the corona treatment, however, APPTs which are not limited to air as process gas are more flexible with respect to the nature or type of surface modification. Thus, the surface of materials including e.g. plastics, paper, rubber, fabrics, steel, glass and various composites can be specifically activated, chemically functionalized or coated to optimize surface free energy and adhesion property. For instance, APPTs can be used to prepare a surface for an adhesive bonding process by the formation of specific chemical functional groups such as amines, imines, aldehydes, epoxides which can form covalent bonds with adhesives resulting in a strong adhesion improvement.

This mechanism enables strong bonding between homogeneous and heterogeneous material compounds for different adhesives. APPT application is further interesting in order to meet sustainability requirements since it may contribute to avoid environmentally unfriendly applications of adhesives by tailored surface properties thanks to plasma treatment. Thus, a homogeneous and efficient surface modification or coating process based on reactive functional monomers can be used to join materials without adhesives at temperatures below 80 °C. This so-called low-temperature adhesive-free joining process has the potential for an eco-friendly joining of a spectrum of different materials.

The ability to endow surfaces with specific chemical functionalities also plays an important role in biomedical applications that require strong coupling of proteins and cells. Protein adsorption is normally the first step in cell adhesion. It is driven by an improved wetting behavior and its strength depends on the nature of surface modification (type of functional groups, electrostatic and chemical interactions) and the type of protein and cell. Especially amine and carboxylic groups, as they are accessible by APPT, according to numerous reports in the literature can be used to fulfill this target. In recent years, it has become apparent that research in biomedical applications is also looking at plasma-assisted deposition of more complex molecules to achieve more specific cell-surface interaction. In the future, such research will be increasingly important for the reduction of animal testing in the development of new drugs.

The use of APPTs in additive manufacturing (AM) systems can significantly enhance the adhesion in AM parts and thus potentially increase the range of applications for these parts. By using specially designed small microplasma nozzles, APPTs can be well integrated into extrusion-based 3D printing processes such as fused deposition modeling (FDM). The functionalization of the printed polymer surfaces improves the adhesion of e.g. adhesives or lacquers on the printed components. The method is also suitable to produce treated internal surfaces, e.g. in the printing of porous structures that are no longer or only insufficiently accessible for surface treatment after completion of the printing process. Further, there is also evidence of an improvement in the mechanical stability of printed components in the z-direction through a plasma-induced improvement in wettability. With the features described above it is expected that the hybrid technology combining printing and APPT may become interesting also in the biomedical printing field e.g. for the production of bone replacements.

In the wake of the implementation of the United Nations sustainability goals, bio-based or biodegradable polymer materials beyond the medical field will also be increasingly used in a wide range of applications in the future. Here also APPT may play an important role in achieving adhesion enhancement. Finally, APPT is also expected to become an option in the debonding technology useful e.g. in re-manufacturing processes.

The APPT applications presented in this chapter are intended to provide an idea of the potential of this technology as a key enabler in the efficient control of adhesion, especially in the treatment of polymers.

1.8 Acknowledgements

The authors would like to thank the publishers for the opportunity to include this chapter in their book. Thanks to all team members in the Department of Interfacial Chemistry and Adaptive Adhesion at the Fraunhofer Institute for Surface Engineering and Thin Films IST and all our partners and researchers from industry and university. Without their experience, contribution and support, the development of APPTs, and thus this chapter would not have been possible.

References

1. K. W. Lee and A. Viehbeck, Wet chemical modification of polyimide surfaces: Chemistry and application, in: *Polyimides; Fundamentals and Applications*, M. Ghosh and K.L Mittal (Eds.), pp. 505-532, CRC Press, Boca Raton, FL (1996).
2. L. M. Siperko and R. R. Thomas, Chemical and physical modification of fluoropolymer surfaces for adhesion enhancement: a review, *J. Adhesion Sci Technol.* 3, 157-173 (1989).
3. M. Strobel, C. S. Lyons and K. L. Mittal (Eds.), *Plasma Surface Modification of Polymers: Relevance to Adhesion*, CRC Press, Boca Raton, FL (1994).
4. K. L. Mittal (Ed.), *Polymer Surface Modification: Relevance to Adhesion, Vol. 5*, CRC Press, Boca Raton, FL (2009).
5. M. Thomas and K. L. Mittal (Eds.), *Atmospheric Pressure Plasma Treatment of Polymers: Relevance to Adhesion*, Wiley-Scrivener, Beverly, MA (2013).
6. K. L. Mittal and T. Bahners (Eds.), *Laser Surface Modification and Adhesion*, Wiley-Scrivener, Beverly, MA (2015).
7. K. L. Mittal and W.-S. Lei (Eds.), *Laser Technology: Application in Adhesion and Related Areas*, Wiley-Scrivener, Beverly, MA (2018).
8. U. Kogelschatz, Atmospheric-pressure plasma technology, *Plasma Phys. Control. Fusion* 46, B63-B75 (2004).
9. C. Fetting, The European Green Deal, *ESDN Report*, ESDN Office, Vienna, Austria (2020).

10. S. Wolf, J. Teitge, J. Mielke, F. Schütze and C. Jäger, The European Green Deal — More than climate neutrality, *Intereconomics* 56, 99–107 (2021).
11. S. Spierling, V. Venkatachalam, M. Mudersbach, N. Becker, C. Herrmann and H. J. Endres, Review: End-of-life options for bio-based plastics in a circular economy — Status quo and potential from a life cycle assessment perspective, *Resources* 9, 90 (2020).
12. J. Rickert, F. Cerdas and C. Herrmann, Exploring the environmental performance of emerging (chemical) recycling technologies for post-consumer plastic waste, *Procedia CIRP* 90, 426–431 (2020).
13. P. Bourke, D. Ziuzina, D. Boehm, P. J. Cullen and K. Keener, The potential of cold plasma for safe and sustainable food production, *Trends Biotechnol.* 36, 615–626 (2018).
14. C. Chaiwong, P. Rachtanapun, P. Wongchaiya, R. Auras and D. Boonyawan, Effect of plasma treatment on hydrophobicity and barrier property of polylactic acid, *Surf. Coat. Technol.* 204, 2933–2939 (2010).
15. K. G. Kostov, T. M. C. Nishime, A. H. R. Castro, A. Toth and L. R. O. Hein, Surface modification of polymeric materials by cold atmospheric plasma jet, *Appl. Surf. Sci.* 314, 367–375 (2014).
16. M. Helena, M. E. R. Heidemann, J. B. L. Dotto, A. M. C. Bruno and C. Cristiane, Cold plasma treatment to improve the adhesion of cassava starch films onto PCL and PLA surface, *Colloids Surfaces* A 580, 123739 (2019).
17. M. J. P Macedo, G. S. Silva, M. C. Feitor, T. H. C. Costa, E. N. Ito and J. D. D. Melo, Composites from recycled polyethylene and plasma treated kapok fibers, *Cellulose* 27, 2115–2134 (2020).
18. M. Mozetič, Surface modification to improve properties of materials, *Materials* 12, 441 (2019).
19. A. Jordá-Vilaplana, V. Fombuena, D. García-García, M. D. Samper and L. Sánchez-Nácher, Surface modification of polylactic acid (PLA) by air atmospheric plasma treatment, *Eur. Polym. J.* 58, 23–33 (2014).
20. F. Arán-Ais, C. Ruzafa-Silvestre, M. P. Carbonell-Blasco, M. A. Pérez-Limiñana and E. Orgilés-Calpena, Sustainable adhesives and adhesion processes for the footwear industry, *Proc. Institution Mechanical Engineers.* Part C, *J. Mech. Eng. Sci.* 235, 585-596 (2021).
21. Y. Chang, T. Sun, C. Fan and X. Zhou, The effect of surface modification on the properties of sisal fiber and improvement of interfacial adhesion in sisal/starch composites induced by starch nanocrystals, *Composite. Interfaces* 25, 981-994 (2018).
22. V. P. Romani, B. Olsen, M. P. Collares, J. R. M. Oliveira, C. Prentice and V. G. Martins, Cold plasma and carnauba wax as strategies to produce improved bi-layer films for sustainable food packaging, *Food Hydrocolloids* 108, 106087 (2020).
23. C. Lambare, P.-Y. Tessier, F. Poncin-Epaillard and D. Debarnot, Plasma functionalization and etching for enhancing metal adhesion onto polymeric substrates, *RSC Advances* 5, 62348–62357 (2015).

24. A. Saranko, Z. Szakal, G. Kalacska, P. Samyn, J. Sukumaran, S. Klebert and Z. Karoly, Adhesion and sliding tribological properties of polyolefins treated by diffuse coplanar surface barrier discharges, *Express Polym. Lett.* 12, 972–985 (2018).
25. C. Mandolfino, E. Lertora and C. Gambaro, Influence of cold plasma treatment parameters on the mechanical properties of polyamide homogeneous bonded joints, *Surf. Coat. Technol.* 313, 222–229 (2017).
26. J. López-García, F. Cupessala, P. Humpolíček and M. Lehocký, Physical and morphological changes of poly(tetrafluoroethylene) after using non-thermal plasma-treatments, *Materials* 11, 2013 (2018).
27. A. W. Eisby, Method of surface treatment of plastic films for increasing the adhesion of printing inks to the film surface, as well as electrical systems for carrying out the method, Patent DK107312C (1962).
28. B. Eliasson and U. Kogelschatz, Modeling of silent discharge plasmas, *IEEE Trans. Plasma Sci.* 19, 309-323 (1991).
29. E. Bormashenko, G. Whyman, V. Multanen, E. Shulzinger and G. Chaniel, Physical mechanisms of interaction of cold plasma with polymer surfaces, *J. Colloid Interface Sci.* 448, 175–179 (2015).
30. R. Dorai and M. J. Kushner, A model for plasma modification of polypropylene using atmospheric pressure discharges, *J. Phys. D: Appl. Phys. 36*, 666–685 (2003).
31. J. Žigon, S. Dahle, M. Petric and M. Pavlic, Enhanced abrasion resistance of coated particleboard pre-treated with atmospheric plasma, *Drvna Industrija* 71, 129-137 (2020).
32. U. Vohrer, R. Bosse, T. Hoder, M. Cernak and M. Thomas, Cleaning, functionalization and coating – product optimization with atmospheric pressure plasma, *Vakuum in Forschung und Praxis* 28, 26-32 (2016).
33. M. Thomas, M. Eichler, K. Lachmann, J. Borris, A. Hinze and C.-P. Klages, Adhesion improvement by nitrogen functionalization of polymers using DBD-based plasma sources at ambient pressure, in: *Atmospheric Pressure Plasma Treatment of Polymers*, M. Thomas and K. L. Mittal (Eds.), pp. 251-271, Wiley-Scrivener, Beverly MA (2013).
34. M. Thomas, K. Lachmann, A. Dohse, M. Thrane, J. Morlock and C.-P. Klages, Surface amination of three-dimensional substrates by RotoTEC plasma treatment under defined gas atmospheres at ambient pressure, in Proceedings of the 19th International Symposium on Plasma Chemistry, http://www.ispc-conference.org/ispcproc/papers/351.pdf (2009).
35. S. Kotula, M. Lüdemann, J. Philipp, M. Thomas and C.-P. Klages, Plasma nitrogenation of polymer surfaces with a new type of combinatorial plasma-printing reactor, *Plasma Process Polym.* 14, 1600137 (2017).
36. C.-P. Klages, Z. Khosravi and A. Hinze, Some remarks on chemical derivatization of polymer surfaces after exposure to nitrogen-containing plasmas, *Plasma Process Polym.* 10, 307–312 (2013).

37. K. Dimic-Misic, M. Kostić, B. Obradović, A. Kramar, S. Jovanović, D. Stepanenko, M. Mitrović-Dankulov, S. Lazović, L.-S. Johansson, T. Maloney and P. Gane, Nitrogen plasma surface treatment for improving polar ink adhesion on micro/nanofibrillated cellulose films, *Cellulose* 26, 3845–3857 (2019).
38. W. X. Chen, J. S. Yu, W. Hu and G. L. Chen, Partial hydrophilic modification of biaxially oriented polypropylene film by an atmospheric pressure plasma jet with the allylamine monomer, *Appl. Surf. Sci.* 387, 957-964 (2016).
39. J. Borris, M. Thomas, C.-P. Klages, F. Faupel and V. Zaporojtchenko, Investigations into composition and structure of DBD-deposited amino group containing polymer layers, *Plasma Process Polym.* 4, S482-S486 (2007).
40. G. Morand, C. Guyon, P. Chevallier, M. Saget, V. Semetey, D. Mantovani and M. Tatoulian, Investigation of 3-aminopropyltrimethoxysilane for direct deposition of thin films containing primary amine groups by open-air plasma jets, *Plasma Process Polym.* 19, 2200047 (2022).
41 . E. Wagennars, D. R. Shaw, P. M. Gyuk, A. T. West and M. Momoh, Surface modification of polymer films using an atmospheric-pressure plasma jet, in Proc 22nd International Plasma Chemistry Society Meeting (2015).
42. P.-Y. Cheng, J.-H. Tsai and J.-Z. Chen, Hydrophilic patterning of octadecyltrichlorosilane (OTS)-coated paper via atmospheric-pressure dielectric-barrier discharge jet (DBDjet), *Cellulose* 27, 10293-10301 (2020).
43. W.-Y. Chen, A. Matthews, F. R. Jones and K.-S. Chen, Immobilization of carboxylic acid groups on polymeric substrates by plasma-enhanced chemical vapor or atmospheric pressure plasma deposition of acetic acid, *Thin Solid Films* 666, 54-60 (2018).
44. M. Thomas, M. von Hausen, C.-P. Klages and P. Baumhof, Generation of stable coatings with carboxylic groups by copolymerization of MAA and VTMS using DBD at atmospheric pressure, *Plasma Process Polym.* 4, S475-S481 (2007).
45. C.-P. Klages, A. Hinze, K. Lachmann, C. Berger, J. Borris, M. Eichler, M. von Hausen, A. Zänker and M. Thomas, Surface technology with cold microplasmas, *Plasma Process Polym.* 4, 208-218 (2007).
46. A. Herrmann, K. Lachmann, K. Fischer, J. Kovac and M. Thomas, Area-selective epoxy coatings by DBD-PECVD in 3D cavities for protein coupling, *Surface Innovations* 3, 206-214 (2015).
47. R. Múgica-Vidal, J. Mercadal-Guillén, F. Alba-Elías and E. Sainz-García, Improvement of the adhesive capacity of SBR for footwear outsoles by surface activation and coating deposition with atmospheric pressure plasma, *Plasma Process Polym.* 18, 2100046 (2021).
48. A. Inoue, H. Yuk, B. Lu and X. Zhao, Strong adhesion of wet conducting polymers on diverse substrates, *Science Advances* 6, eaay5394 (2020).
49. J. Kim, R. Mauchauffé, D. Kim, J. Kim and S. Y. Moon, Mechanism study of atmospheric-pressure plasma treatment of carbon fiber reinforced polymers for adhesion improvement, *Surf. Coat.Technol.* 393, 125841 (2020).

50. Y. Ren, Z. Ding, C. Wang, C. Zang, Y. Zhang and L. Xu, Influence of DBD plasma pretreatment on the deposition of chitosan onto UHMWPE fiber surfaces for improvement of adhesion and dyeing properties, *Appl. Surf. Sci.* 396, 1571-1579 (2017).
51. K. Polášková, M. Klíma, Z. Jeníková, L. Blahová and L. Zajíčková, Effect of low molecular weight oxidized materials and nitrogen groups on adhesive joints of polypropylene treated by a cold atmospheric plasma jet, *Polymers* 13, 4396 (2021).
52. V. Fischer, E. Stammen, K. Dilger, M. Eichler, S. Paulmann and C.-P. Klages, Promotion of adhesive polymer bonding by plasma modification using defined ambient conditions and process gases, *J. Energy Eng.* 11, 135-139 (2017).
53. M. Kehrer, A. Rottensteiner, W. Hartl, J. Duchoslav, S. Thomas and D. Stifter, Cold atmospheric pressure plasma treatment for adhesion improvement on polypropylene surfaces, *Surf. Coat. Technol.* 403, 126389 (2020).
54. X. Wang, J. P. Lin, J. Y. Min, P. C. Wang and C. C. Sun, Effect of atmospheric pressure plasma treatment on strength of adhesive-bonded aluminum AA5052, *J. Adhesion* 94, 701-722 (2018).
55. J. Lin, C. Sun, J. Min, H. Wan and S. Wang, Effect of atmospheric pressure plasma treatment on surface physicochemical properties of carbon fiber reinforced polymer and its interfacial bonding strength with adhesive, *Composites B.* 199, 108237 (2020).
56. H. Gholamiyan, B. Gholampoor and R. Hosseinpourpia, Application of waterborne acrylic and solvent-borne polyester coatings on plasma-treated fir (Abies alba M.) wood, *Materials* 15, 370 (2022).
57. V. Bagiatis, G. W. Critchlow, D. Price and S. Wang, The effect of atmospheric pressure plasma treatment (APPT) on the adhesive bonding of poly(methyl methacrylate) (PMMA)-to-glass using a polydimethylsiloxane (PDMS)-based adhesive, *Int. J. Adhesion Adhesives* 95, 102405 (2019).
58. K. Takenaka, R. Machida, T. Bono, A. Jinda, S. Toko, G. Uchida and Y. Setsuhara, Development of a non-thermal atmospheric pressure plasma-assisted technology for the direct joining of metals with dissimilar materials, *J. Manuf. Process.* 75, 664-669 (2022).
59. K. Hori, S. Fujimoto, Y. Togashi, T. Kuroki and M. Okubo, Improvement in molecular-level adhesive strength of PTFE film treated by atmospheric plasma combined processing, *IEEE Trans. Ind. Applications* 55, 825-832 (2019).
60. M. Nishino, Y. Okazaki, Y. Seto, T. Uehara, K. Endo, K. Yamamura and Y. Ohkubo, Adhesive-free adhesion between plasma-treated glass-cloth-containing polytetrafluoroethylene (GC–PTFE) and stainless steel: Comparison between GC–PTFE and Pure PTFE, *Polymers* 14, 394 (2022).
61. M. Eichler, A. Lehm, D. Romstedt, V. Schoenberner and H. Meyer, Adhesive-free bonding of web-like plastic metal combinations at low temperatures, *Key Eng. Materials* 809, 392-397 (2019).

62. B. J. Tyler, A. Hook, A. Pelster, P. Williams, M. Alexander and H. F. Arlinghaus, Development and characterization of a stable adhesive bond between a poly(dimethylsiloxane) catheter material and a bacterial biofilm resistant acrylate polymer coating, *Biointerphases* 12, 02C412, 1-12 (2017).
63. B. Pecar, M. Mozek and D. Vrtacnik, Thermoplastic - PDMS polymer covalent bonding for microfluidic applications, *Informacije Midem* 47, 147-154 (2017).
64. R. Sivakumar and N. Y. Lee, Chemically robust succinimide-group-assisted irreversible bonding of poly(dimethylsiloxane)–thermoplastic microfluidic devices at room temperature, *Analyst* 145, 6887-6894 (2020).
65. R. Bitar, M. Asadin, S. Van Vrekhem, P. Cools, H. Declerq, R. Morent and N. De Geyter, Local plasma activation of PS films with a defined design for biomedical use, *Surf. Coat. Technol.* 350, 985-996 (2018).
66. K. Cheng, C. Chang, Y. Yang, G. Liao, C. Liu and J. Wu, Enhancement of cell growth on honeycomb-structured polylactide surface using atmospheric-pressure plasma jet modification, *Appl. Surf. Sci.* 394, 534-542 (2017).
67. K.G. Doherty, J.S. Oh, P. Unsworth, C.M. Sheridan, P. Weightman, J.W. Bradley and R.L. Williams, Plasma polymerization using helium atmospheric-pressure plasma jet with heptylamine monomer, *Plasma Process Polym. 16*, e1800185 (2019).
68. F.C. Kung, Y.L. Kuo, O. Gunduz and C.C. Lin, Dual RGD-immobilized poly(L-lactic acid) by atmospheric pressure plasma jet for bone tissue engineering, *Colloids Surfaces B 178*, 358-364 (2019).
69. R.A. Surmenev, R.V. Chernozem, D.S. Syromotina, C. Oehr, T. Baumbach, B. Krause, A.N. Boyandin, L.M. Dvoinina, T.G. Volova and M.A. Sumeneva, Low-temperature argon and ammonia plasma treatment of poly-3-hydrohybutyrate films: Surface topography and chemistry changes affect fibroblast cell *in vitro*, *European Polym. J.* 112, 137-145 (2019).
70. S. K. Alavi, O. Lotz, B. Akhavan, G.C. Yeo, R. Walia, D.R. McKenzie and M.M.M. Bilek, Atmospheric pressure plasma jet treatment of polymers enables reagent-free covalent attachment of biomolecules, *ACS Appl. Mater. Interfaces 12*, 38730-38743 (2021).
71. M.M.M. Bilek, M. Vabdrovcová, A. Shelemin, A. Kuzminova, O. Kylián, H. Biederman, L. Bačáková and A.S. Weiss, Plasma treatment in air at atmospheric pressure that enables reagent-free covalent immobilization of biomolecules on polytetrafluorethylene (PTFE), *Appl. Surf. Sci.* 518, 146128 (2020).
72. F. Tan and M. Al-Rubeai, Customizable implant-specific and tissue-specific extracellular matrix protein coatings fabricated using atmospheric plasma, *Frontiers. Bioeng. Biotechnol.* 7:247 (2019).
73. P. Šrámková, A. Zahoranová, J. Kelar, Z.K. Tučeková, M. Stupavská, R. Krumpolec, J. Jurmanová, D. Kováčik and M. Černák, Cold atmospheric pressure plasma: simple and efficient strategy for preparation of

poly(2-oxazoline)-based coatings designed for biomedical applications. *Sci Reports* 10, 9478 (2020).
74. H. Suzuki, K. Kasai, Y. Kimuara and S. Miyata, UV/ozone surface modification combined with atmospheric pressure plasma irradiation for cell culture plastics to improve pluripotent stem cell culture, *Mater. Sci. Eng. C* 123, 112012 (2021).
75. K.V. Chan, Y. Onyshchenko, M. Asadian, A.Y. Nikiforov, H. Declerq, R. Morent and N. De Geyter, Investigating the stability of cyclopropylamine-based plasma polymers in water, *Appl. Surface Sci.* 517, 146167 (2020).
76. J. Ibrahim, S.A. Al-Babtaineh, S. Cousens, A. Michelmore, C. Corr and J. Whittle, A surface dielectric barrier discharge for deposition of allylamine polymer costings, *Appl. Surface Sci.* 544, 148826 (2021).
77. K. Lachmann, A. Dohse, M. Thomas, S. Pohl, W. Meyring, K.E.J. Dittmar, W. Lindenmeier and C.P. Klages, Surface modification of closed plastic bags for adherent cell cultivation, *European Physical Journal – Applied Physics* 55, 13812 (2011).
78. P. Cools, H. Declercq, N. DeGeyter and R. Morent, A stability study of plasma polymerized acrylic acid film, *Appl. Surface Sci.*, 432, 214-223 (2018).
79. M.M. Gepp, R. Duckstein, F. Kayatz, N. Rodler, Z. Scheuerer, J.C. Neubauer, K. Lachmann, C. Stramm, A. Liebmann, M. Thomas and H. Zimmermann, Labbag® - a versatile bag-based cultivation system for expansion, differentiation and cryopreservation of human stem cells, *Current Directions in Biomedical Engineering* 3, 371–374 (2017).
80. U. Czuba, R. Quintana, M.C. De Pauw-Gillet, M. Bourguignon, M. Moreno-Couranjou, M. Alexandre, C. Detrembleur and P. Choquet, Atmospheric plasma deposition of methacrylate layers containing catechol/quinone groups: An alternative to polydopamine bioconjugation for biomedical applications, *Advanced Healthcare Mater.* 1701059, 1-11 (2018).
81. A. Patelli, F. Mussano, P. Brun, T. Genova, E. Ambrosi, N. Michieli, G. Mattei, P. Scopece and L. Moroni, Nanoroughness, surface chemistry, and drug delivery control by atmospheric plasma jet on implantable devices, *ACS Appl. Mater. Interfaces*, 10, 39512-39523 (2018).
82. H.T. Sasmazel, M. Alazzawi and N.K.A. Alsahib, Atmospheric pressure plasma surface treatment of polymers and influence on cell cultivation, *Molecules* 26, 1665 (2021).
83. H. Narahara, Improvement and evaluation of the interlaminar bonding strength of FDM parts by atmospheric-pressure plasma, *Procedia CIRP* 42, 754-759 (2016).
84. M. Sachs, J. Schmidt, W. Peukert and K.-E. Wirth, Treatment of polymer powders by combining an atmospheric plasma jet and a fluidized bed reactor, *Powder Technol.* 325, 490-497 (2018).
85. T.D. McLouth, S.M. Gustafson, H.I. Kim and R.J. Zaldivar, Enhancement of FDM ULTEM® 9085 bond strength via atmospheric plasma treatment, *J. Manufacturing Processes* 66, 179–188 (2021).

86. B.M.P. Ferreira, L.M.P. Pinheiro, P.A.P. Nascente, M.J. Ferreira and E.A.R. Duek, Plasma surface treatments of poly(L-lactic acid) (PLLA) and poly(hydroxybutyrate-co-hydroxyvalerate) (PHBV), *Mater. Sci. Eng. C* 29, 806-813 (2009).
87. J. Yang, G. Shi, J. Bei, S. Wang, Y. Cao, Q. Shang, G. Yang and W. Wang, Fabrication and surface modification of macroporous poly(L-lactic acid) and poly(L-lactic-co-glycolic acid) (70/30) cell scaffolds for human skin fibroblast cell culture, *J. Biomed. Mater. Res.* 62, 438-446 (2002).
88. M. Nakagawa, F. Teraoka, S. Fujimoto, Y. Hamada, H. Kibayashi and J. Takahashi, Improvement of cell adhesion on poly(L-lactide) by atmospheric plasma treatment, *J. Biomed. Mater. Res.* 77A, 112-118 (2006).
89. J.M. Sorbal, S.G. Caridade, R.A. Sousa, J.F. Mano and R.L. Reis, Three-dimensional plotted scaffolds with controlled pore size gradients: Effect of scaffold geometry on mechanical performance and cell seeding efficiency, *Acta Biomaterialia* 7, 1009-1018 (2011).
90. S.H. Oh and J.H. Lee, Hydrophilization of synthetic biodegradable polymer scaffolds for improved cell/tissue compatibility, *Biomed. Mater.* 8, 014101 (2013).
91. M. Wang, P, Favi, X. Cheng, N.H. Golshan, K.S. Ziemer, M. Keidar and T.J. Webster, Cold atmospheric plasma (CAP) surface nanomodified 3D printed polylactic acid (PLA) scaffolds for bone regeneration, *Acta Biomaterialia* 46, 256-265 (2016).
92. A. Shashurin, M. Keidar, S. Bronnikov, R.A. Jurjus and M.A. Stepp, Living tissue under treatment of cold plasma atmospheric jet, *Appl. Phys. Lett.* 93, 181501 (2008).
93. A. San-Miguel and S.H. Behrens, Influence of nanoscale particle roughness on the stability of Pickering emulsions, *Langmuir* 28, 12038-12043 (2012).
94. F. Liu, W. Wang, S. Hinduja and P.J. Bártolo, Hybrid additive manufacturing system for zonal plasma-treated scaffolds, *3D Print. Additive Manuf.* 5, 205-213 (2018).
95. C.E. Corcione, F. Scalera, F. Gervaso, F. Montagna, A. Sannino and A. Maffezzoli, One-step solvent-free process for the fabrication of high loaded PLA/HA composite filament for 3D printing, *J. Therm. Anal. Calorim.* 134, 575-582 (2018).
96. C.E. Corcione, F. Gervaso, F. Scalera, S.K. Padmanabhan, M. Magaghiele, F. Montagna, A. Sannino, A. Licciulli and A. Maffezzoli, Highly loaded hydroxyapatite microsphere/PLA porous scaffolds obtained by fused deposition modelling, *Ceram. Int.* 45, 2803-2810 (2019).
97. S. Zeng, Z. Cui, Z. Yang, J. Si, Q. Wang, X. Wang, K. Peng and W. Chen, Characterization of highly interconnected porous poly(lactic acid) and chitosan-coated poly(lactic acid) scaffold fabricated by vacuum-assisted resin transfer molding and particle leaching, *J. Mater. Sci.* 51, 9958-9970 (2016).
98. I. Fernández-Cervantes, M.A. Morales, R. Agustín-Serrano, M. Cardenas-García, P.V. Pérez-Luna, B.L. Arroyo-Reyes and A. Maldonado-García,

Polylactic acid/sodium alginate/hydroxyapatite composite scaffolds with trabecular tissue morphology designed by a bone remodeling model using 3D printing, *J. Mater. Sci.* 54, 9478-9496 (2019).

99. Y. Chen, A.F.T. Mak, M. Wang, J. Li and M.S. Wong, PLLA scaffolds with biomimetic apatite coating and biomimetic apatite/collagen composite coating to enhance osteoblast-like cells attachment and activity, *Surf. Coatings Technol.* 201, 575-580 (2006).
100. S.H. Kim, S.A. Oh, W.K. Lee, U.S. Shin and H.W. Kim, Poly(lactic acid) porous scaffold with calcium phosphate mineralized surface and bone marrow mesenchymal stem cell growth and differentiation, *Mater. Sci. Eng. C.* 31, 612-619 (2011).
101. V. Martin, I.A. Ribeiro, M.M. Alves, L. Gonçalves, R.A. Claudio, L. Grenho, M.H. Fernandes, P. Gomes, C.F. Santos and A.F. Bettencourt, Engineering a multifunctional 3D-printed PLA-collagen-minocycline-nano hydroxyapatite scaffold with combined antimicrobial and osteogenic effects for bone regeneration, *Mater. Sci. Eng. C.* 101, 15-26 (2019).
102. Y. Wan, C. Tu, J. Yang, J. Bei and S. Wang, Influences of ammonia plasma treatment on modifying depth and degradation of poly(L-lactide) scaffolds, *Biomaterials* 27, 2699-2704 (2006).

2

Corona Treatment of Polymer Surfaces to Enhance Adhesion

N. Dole[1]*, K. Ahmadi[2], D. Solanki[2], V. Swaminathan[3], V. Keswani[4] and M. Keswani[5]

[1]*Electrical and Computer Engineering Department, University of Houston, Houston, TX, USA*

[2]*Materials Science and Engineering Program, University of Houston, Houston, TX, USA*

[3]*Bioengineering Department, University of Illinois at Urbana-Champaign, Urbana, IL, USA*

[4]*Department of Information Technology, G.H. Raisoni College of Engineering, Nagpur, India*

[5]*Materials Science and Engineering Department, University of Arizona, Tucson, AZ, USA*

Abstract

Polymers offer a wide range of bulk physical, chemical and mechanical properties, are inexpensive and are relatively easy to synthesize. However, very often these polymer surfaces do not possess properties needed for application in certain external environments which could be one of the limiting factors in their usage. Polymer surface characterization and modification is of fundamental importance for the good functioning of a composite material and plays a key role in determining successful implementation of the polymer. Tailoring of polymer surface properties such as adhesion, wettability, surface roughness and chemical reliability (including heat-soak) is critical for several applications like decorative coatings, protective films, thin film technologies and biomaterials, to name a few. For these reasons, surface modification techniques which can transform these inexpensive materials into highly valuable finished products have become an important part of the polymer and nanocomposite industries. This chapter reviews the importance and applications of polymer surface modification using corona treatment to enhance adhesion by reporting most recent advancements in the field. Novel

**Corresponding author*: nikhil.dole@gmail.com

K.L. Mittal and A.N. Netravali (eds.) Polymer Surface Modification to Enhance Adhesion: Techniques and Applications, (45–76) © 2024 Scrivener Publishing LLC

corona treatment (type of plasma, pressure, and power) and its potential in several applications is briefly discussed in latter part of this chapter.

Keywords: Adhesion, surface modification, plasma, corona treatment, polymers

2.1 Introduction

Polymers, due to their low cost and wide range of mechanical, chemical, electrical, and optical properties, have replaced metals, wood, and glasses in numerous applications and nowadays play vital roles in many industries. However, they typically have low surface free energy, Table 2.1, and lack polar functional groups on the surface that may adversely impact their wettability, printability, bondability, and compatibility with other materials which, in turn, compromises their usability in many decorative and engineering applications. Surface modification can address these issues while maintaining desired bulk properties and adding more functionality to polymers by altering the surface chemistry, morphology, or both. Such modification occurs either by altering the polymer surface characteristics or by applying a coating on top of it. Various techniques for modifying polymer surfaces are documented in literature [1–5], but traditionally they are placed into two main categories: chemical and physical methods.

Table 2.1 Surface free energies of common polymers.

Polymer	Surface free energy γ_s (*mJ/m²*)	Reference(s)
PTFE (Polytetrafluoroethylene)	~19-24	[9]
PE (Polyethylene)	~27-33	[10], [11]
PP (Polypropylene)	~26-30	[12], [13]
PS (Polystyrene)	~37-42	[10], [13]
PET (Poly(ethylene terephthalate))	~40-47	[12]
PC (Polycarbonate)	~40-48	[14], [15]

2.1.1 Chemical *versus* Physical Methods in Polymer Surface Modifications

Chemical modification such as solution-based, sol-gel, and chemical vapor deposition (CVD), as apparent from the name, usually involves exposing the surface to chemical agents in the form of liquid, gel, or vapor that can render the surface with desired properties. Due to the variety of chemical agents, a wide range of surface properties, such as wear-resistant, antistatic, antibacterial, and electrically conductive characteristics, can be achieved using chemical treatments. Although chemical methods have proved effective, they may impose environmental costs by generating chemical wastes that require further handling. Besides, some of these methods, such as CVD, can also be costly and may have constraints on the sample size which ultimately limits their scalability. In addition, surface modification, if achieved by applying a coating, may not address intrinsic wettability and adhesivity issues of common polymers, which can impact the durability of the coating. Therefore, the base polymers in such modifications may need prior treatments to improve their adhesivity. Chemical methods to improve surface adhesion are generally based on etching, which has operational hazards, generates waste, and typically requires rigid process control and several rinsing steps to achieve optimum results. An example of such surface treatment is sodium etch of PTFE, involving multiple steps, including immersion of the part in a sodium-containing solution to de-fluorinate the surface and introduce functional groups that can form hydrogen bonds and improve the wettability of the surface [6–8]. Because of the surface chemistry modification challenges associated with these chemical methods, polymer scientists are inclined to utilize physical methods instead of chemical techniques.

Physical methods such as laser, plasma, and corona treatments, compared to chemical methods, are typically more environmentally benign, scalable, and have facile process control [16]. Therefore, on occasions when alternative physical methods are available, particularly for improvement of surface wettability, they are generally considered more favorable [16]. Among these methods, corona treatment, which utilizes corona discharge plasma to alter surface properties, is an inexpensive, clean, fast, and mature technology that is widely used in the polymer industry. This treatment is used to enhance the surface free energy of polymer to allow improved wettability and adhesion of inks, coatings, and adhesives [17].

2.1.2 Corona Treatment and Impact on Polymers

Corona (air plasma) treatment method was invented in 1951 by a Danish Engineer, Verner Eisby, to address the challenge of printing on polymers [18]. The method is based on passing films in high-frequency corona electric discharge between an electrode with a sharp tip and a grounded roller counter electrode, placed 1-3 mm apart, Figure 2.1.

The air in high electric discharge between the electrodes is ionized and creates plasma, a reactive state of matter containing ions, electrons, and radicals. When the polymer surface is exposed to the discharge, the imposed energy can break the surface molecules' bonds and, depending on voltage, frequency, and duration of treatment, as well as the nature and geometry of polymer, can impact the polymer surface in several ways:

- The broken bonds result in a high-energy active surface, which, in the presence of oxygen and free radicals, can generate various polar functional groups, such as carbonyl, carboxyl, and hydroxyl [19–22]. These polar groups increase polymer surface free energy [23]. Therefore, corona treatment of low surface free energy polymers can improve their wettability and render them printable. However, this increase in surface free energy tends to diminish over time [23–25]. Therefore, it is suggested to use the treated film for further processing immediately after the treatment.
- The treatment can introduce roughness to the surface [26–28]. Introducing roughness increases the surface area that

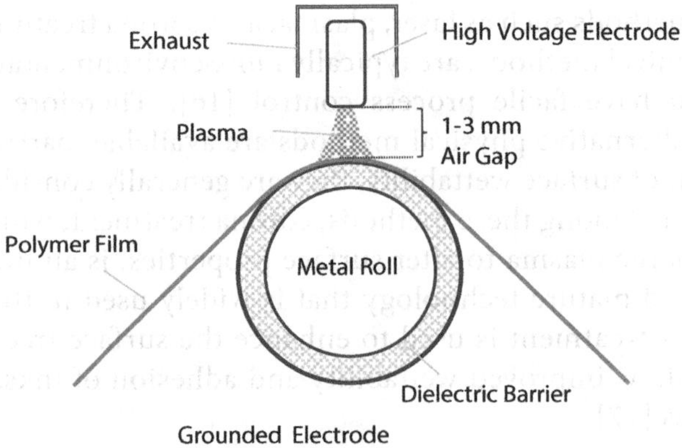

Figure 2.1 Schematic of a basic corona-treater configuration.

interacts with the adhesive or coating and has an anchoring effect. However, this effect depends on the type of polymer and can be negligible.

- Polymer chain scission and generation of low-molecular-weight oxidized materials (LMWOMs) can occur as a side effect of corona treatment [29]. LMWOMs that are loosely bonded to the surface generate weak boundary layers and compromise the adhesion property of the surface [25].
- Corona treatment can also crosslink surface regions and, therefore, increase the molecules' molecular weight and film's cohesive strength by cross-linking low-molecular-weight compounds [30, 31].
- The treatment can decontaminate and disinfect the surface by oxidizing contaminants and biomolecules [32–34].

The depth of these changes is only a few nanometers and typically does not influence the bulk of materials. Based on these effects, each surface may have optimum treatment conditions. Overtreatment of the surface may lead to its degradation by increasing the roughness and densities of LMWOMs, and undertreatment can result in the generation of insufficient functional groups that can adversely affect surface properties.

2.1.3 Corona Treatment Applications and Limitations

Corona treatment is typically used on polymers to increase their surface free energy and, in turn, their wettability. A wide range of polymers, such as PP, PET, and PVC, can be rendered printable and adhesion able using this method (refer to section 6 for more details). Benefits, cost, ease of use, and efficiency of corona treaters in large-scale and in-line production encouraged various industries to also treat textiles [35], papers [36], wood [37], metals [38], and glass [39] in addition to polymers, to impart desired changes on the surface, including decontamination, introduction of functional groups, and growth of oxide layer. Remeika *et al.* [40] studied the feasibility of using high throughput corona treatment as a substitute for UV-Ozone treatment to remove surface contamination in industrial perovskite-based solar cell fabrication. Burkitt *et al.* [41] also reported using corona treatment in a roll-to-roll fabrication of perovskite solar cells. These examples indicate the potential of corona treatment methods in applications beyond traditional industries.

Despite the advantages of the corona treatment, it has a few shortcomings: typical corona treaters are designed for the treatment of two-dimensional

(2D) webs and cannot treat three-dimensional (3D) objects. However, the advent of 3D corona treaters enabled to perform corona treatment on 3D objects [40]. Corona treatment has a small penetration depth, so it is a surface treatment and does not affect bulk of the material [43]. Besides, corona treatment is limited to the air, and gas type, pressure, and flow in a typical corona treater cannot be controlled. Therefore, in applications where a higher degree of control on the plasma conditions is required for tuning the surface chemistry and properties, other plasma surface modification methods, such as low and atmospheric pressure, chemical, and flame plasma methods, can be utilized [31, 44].

2.2 Mechanism of Corona Treatment

From the mid-19th century, plasma has been a topic of interest for theoretical and experimental physicists, finding research motivation with regards to applications in fusion energy. Since past few decades, plasma processing has become an essential surface modification and fabrication method for many modern industries [45–47]. Plasma can be defined as ionized gas, where temperature could be very high (fusion plasma) or moderate to room temperature (gas discharge plasma) depending on application [48]. Variables used to generate and sustain/control plasma are given by fundamental parameters such as density of particles. Thermal energy from flame through exothermic reaction could ionize gas to generate plasma, or compressing gas by applying pressure could also generate plasma, or application of external electric field could generate plasma from neutral gas at low temperature [49, 50]. Particularly, various methods used for electric field can subcategorize plasma generation mechanisms such as dc discharge, pulsed dc discharge, RF discharge by capacitive or inductive coupling, microwave discharge, and dielectric discharge [50]. Design of plasma reactor is generally based on plasma generation method and applications of plasma [51, 52].

Various types of polymer substrates have shown an essential role in numerous applications ranging from medical industry, food industry, and modem IC technological development [28, 53, 54]. However, such polymer surfaces often require treatments to improve adhesion/surface wetting [48–50]. Polymer surface analysis has been carried out since the late 20th century [55]. Adhesion phenomenon studied by several authors [56–58] reports that surface free energy of polyethylene increases from 32 mJ/m^2 to 42 mJ/m^2 post corona treatment enabling stronger adhesion possible. The phenomenon of corona can be described as state of gas which contains excited molecules, positive ions and electrons, which is generated when current/voltage is passed through two electrodes with a polymer film placed in between. Here, when excited electrons travel in a chamber with air molecules or moisture/humidity, they knock out orbital electrons and hence create ions as well as high energy electrons and this cycle continues. When such ions are created, generally polymer surface is endowed with polar groups and depending on energy of corona and moisture conditions, surface crosslinking could be effected at various levels resulting in surface modifications. Typical examples of oxidized groups on a polymer surface are hydroxyl, carboxyl, carbonyl, and ester. Overall, corona treatment is used as an oxidation mechanism to introduce polar groups into the surface which can improve various mechanical and surface properties such as wettability, adhesion, and surface free energy. Many alternative theories are also proposed in literature [59–61] to explain principle of corona treatment, while increase in surface free energy post corona treatment is a widely accepted mechanism. The effectiveness of corona treatment is determined by activation energy [63, 64] E_j, which can be calculated as:

$$E_j = \frac{P}{Lv} \tag{6.1}$$

Where, E_j: Activation energy [J/m^2], P: Power of corona discharge [W]; L: Length of discharge electrode [m]; v: Speed of material shift [m/s]. The calculation of actual activation energy also requires additional information regarding type of polymer substrate, thickness, width, required depth of corona treatment on surface, moisture/humidity, etc. Billmeyer [57] proposed that free radicals are generated by oxygen atoms breaking the molecules of polymer as the main reason for oxidation, but Sun *et al.* [58] suggest the effects of various media for corona treatment such as air, O_2, N_2 and further suggest that oxygen could be from multiple sources including impurity in gas, from polymer film itself or from the apparatus design components. Corona treatment has evidently illustrated changes in surface morphology of the polymer substrate. Kim and Goring [28] studied the effect of time and temperature on surface roughness and demonstrated control over bump diameter by corona treatment. Dong *et al.* [59] demonstrated that yield strength and ductility of polymer were significantly improved by corona treatment in air.

2.2.1 Equipment and Operation Details for Corona Treatment

A simplified construction of equipment for corona treatment system consists of high frequency discharge between two electrodes. The plasma in air is generated due to discharge between electrodes. The polymer substrate is placed between the two electrodes where surface is oxidized due to energized electrons [56]. Relative simplicity of the process and easy upscaling for mass production make corona treatment an attractive surface modification option for industrial scale. Initial investigation by Steinhauser and Ellinghorst [60] used a lab scale stationary corona discharge system which involved two copper disk electrodes in a bell jar reactor, which required higher voltage to maintain a homogeneous field for ignition. Modern equipments used in industries have sophisticated controls to regulate plasma uniformity across large substrate area which enables improved yield.

Several modifications in the corona treatment equipment have been made to improve efficiency, achieve higher penetration depth and for application to various types of polymers. Plasma dielectric barrier discharge system uses an electrode covered with glass, ceramic, or similar type of dielectric barrier layer [61, 62]. The major advantage of such system is that plasma discharge can be controlled for a very small amount of time and hence a very small dose of corona can be delivered efficiently. 230 V, 50 Hz step-up transformer with 10 mm gap at 48% humidity is reported for atmospheric air plasma dielectric barrier discharge system [61]. The schematic diagram for such electrode arrangement is represented in Figure 2.2(a).

Another variation of the corona generation system is atmospheric pressure glow microdischarge as illustrated in Figure 2.2(b). Here, an air-locked quartz glass chamber is used to visualize the glow as well as to accommodate characterization equipment such as spectrometer [63, 64]. These authors investigated plasma generated by a small diameter anode and relatively larger cathode with helium, nitrogen or air flow through the chamber. They used two different step-up transformers as power supply

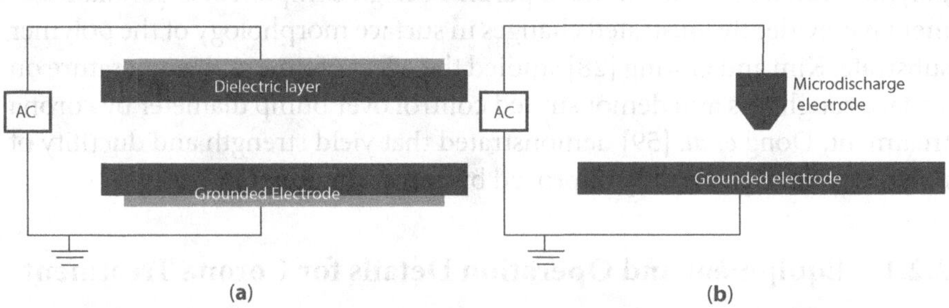

Figure 2.2 (a) Simplified schematic of electrode arrangment for dielectric barrier discharge system; (b) Schematic of atmospheric glow microdischarge electrode setup.

to generate low or high discharge current and found that stable plasma was generated at low emission currents and considering smaller diameter of electrode used in this configuration, this becomes major limiting step in operation at the cost of longer duration of corona treatment requirement. Modern industrial scale corona treatment machines are equipped with a few more features which enable single or double side treatment for polymer substrates and also feature precise control of voltage/current to achieve desired penetration depth. They also contain ozone extraction system through electrode housing, thus enabling safe operations for mass production of corona treated polymers. Such equipments can also treat large size polymer substrates and ensure uniformity of plasma distribution throught the chamber. Conventional low pressure plasma treatments are studied in literature which enable functionalization of polymer surface by various groups. The objective of such low pressure plasma treatment is similar to atmospheric pressure plasma. By application of low pressure plasma, surface can be cleaned from contaminants, or functionalized by polar groups to increase surface free energy or even can be etched to increase roughness. Park *et al.* [65] and Sapieha *et al.* [66] studied the effect of low pressure plasma vs corona treatment for polyethelyne. Low pressure plasma treatment was performed in a steel vessel with 2.45 GHz microwave power at 500 mTorr pressure. They concluded that the degree of oxidation was comparable from these two methods, while reduction in peel strength was observed only for corona treatment. Jha *et al.* [67] studied effects of atmospheric and low pressure plasmas on poly(etheretherketone) (PEEK) film surface modification where air with a flow rate of 51 L/minute at a pressure of 4.5 bar was used for atmosphere pressure treatment and capacitively coupled RF power electrode with 10^{-2} Torr base pressure of air was used for low pressure plasma treatment. Autors concluded that atmospheric pressure plasma improved surface free energy of PEEK comparable to low pressure plasma for similar duration of plasma treatment. Overall, atmospheric plasma requires addditional ignition mechanism compared to the low pressure plasma, but it also has small mean free path and hence magnetic field is not required for atmospheric plasma at the cost of less coverage on substrate for plasma processing [68, 69]. Schutze *et al.* [70] investigated breakdown voltage required for various plasma methods and concluded that low pressure discharge requires breakdown voltage of less than 1 kV compared to 10-50 kV for corona and 2-25 kV for dielectric barrier discharge. However, due to complexity involved in low pressure plasma generation, so for industrial applications, atmospheric plasma treatment is favored.

2.2.2 Effect of Plasma Source on Efficiency of Corona Treatment

The effects of inductively coupled plasma (ICP) and capacitively coupled plasma (CCP) on etching of poly (methyl methacrylate) (PMMA) and polycarbonate (PC) are studied in literature [65]. The authors used custom-built diffusion pump-based system for both ICP and CCP along with 13.56 MHz frequency power supply as shown in Figure 2.3.

They evaluated etch rates by both ICP and CCP methods at 40-60 mTorr pressure and 5 sccm O_2 flow rate. The authors reported that for a range of 0-350 W for ICP and 25-150 W power range for CCP, etch rate for ICP was almost double compared to CCP for both PMMA and PC substrates. Authors found that PMMA surface roughness increased from 1.5 nm to 2.3 nm by ICP and to 2.9 nm by CCP etching. They demonstrated control over surface roughness for both polymers by tuning process parameters like chamber pressure and plasma power enabling ICP and CCP methods to be useful for surface modification of thick polymer substrates. Effect of plasma distance on surface modification of poly(ethylene terephthalate) (PET) and thermoplastic crystalline polymer like poly(etheretherketone) (PEEK) was studied using Ar and O_2 plasmas [71, 72]. The authors studied direct Ar plasma at 0 cm distance followed by remote plasma at 80 cm distance with identical process parameters and concluded that remote plasma avoids surface degradation reaction which would take place if treatment

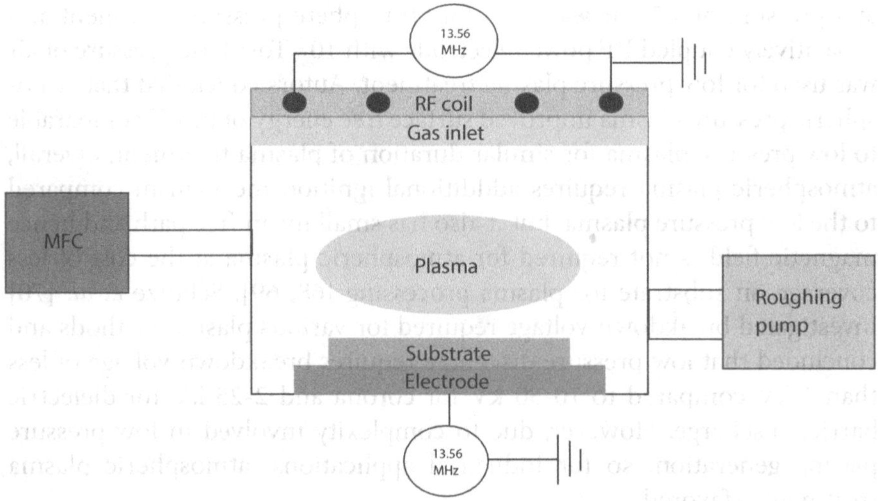

Figure 2.3 Schematic representation of ICP system.

were performed by plasma generation at polymer surface. The authors reported continuous decrease in water contact angle for plasma treated PEEK films for up to 10 s of treatment.

Other authors [72, 73] also report similar findings regarding surface degradation with direct plasma compared to remote plasma for oxygen and hydrogen plasma sources, concluding that remote plasma treatment provides a hydrophilic surface. Plasma downstream reactor (PDR) was investigated for continuous treatment of powders [74], schematic of PDR assembly is illustrated in Figure 2.4. The authors studied surface modification of powder substrate by introducing gas from top of chamber at 2 mbar pressure and using filter for solid-gas separation post treatment. During thermal characterization, the authors found that highly undesirable hot spots generated in conventional setup can be avoided by PDR and hence it provides precise control for plasma treatment of powder substrates for large volume manufacturing facilities in industry.

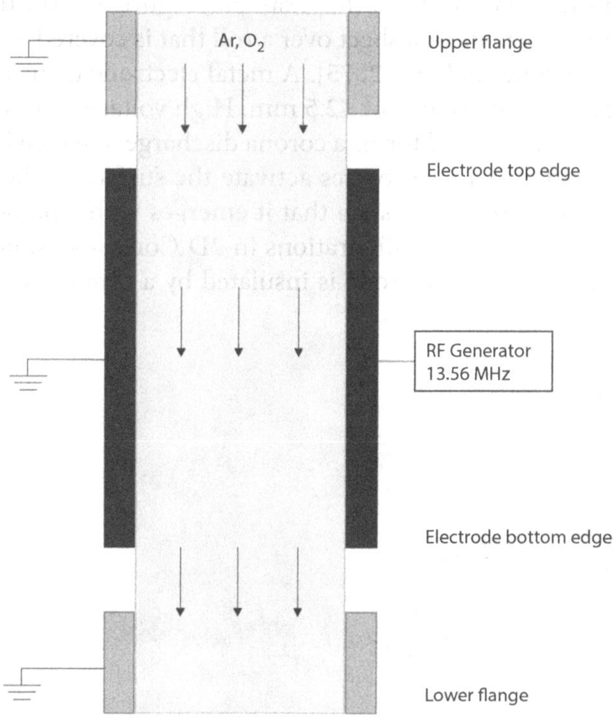

Figure 2.4 Schematic representation of PDR.

2.3 Factors Affecting Performance of Corona Treatment

Corona discharge treatment has emerged as a very versatile, cost-effective, and scalable technique for surface activation and modification of various polyolefin plastics. In this section, we will discuss the parameters affecting surface preparation for this technique and the mechanistic understanding that helps implement this technique to multiple industrial applications.

2.3.1 Effect of Material Surface Preparation: 2-D vs. 3-D

A corona treatment system essentially is comprised of three parts: an electrode, an insulating dielectric, and a ground return path. The material to be treated needs to be encapsulated within the corona discharge that occurs in this setup under controlled conditions. The discharge that occurs in this apparatus is typically 1-dimensional, so that surface coverage is achieved by shifting the material through the corona at a constant rate and dosage. In most 2-dimensional systems (for example, Figure 2.5(a)), the “web” is fed in the form of a plate or a sheet over a roll that is covered with an insulator such as silicone rubber [42, 75]. A metal electrode is suspended over the roll to create a narrow gap of ≤2.5 mm. High voltage across the air gap ionizes the atmosphere and forms a corona discharge in which high energy electrons and other reactive species activate the surface of the sheet. The sheet is fed at a constant rate such that it emerges with a homogeneously treated surface. Alternate configurations in 2D Corona activators include “bare roll” in which the electrode is insulated by a ceramic while the roll

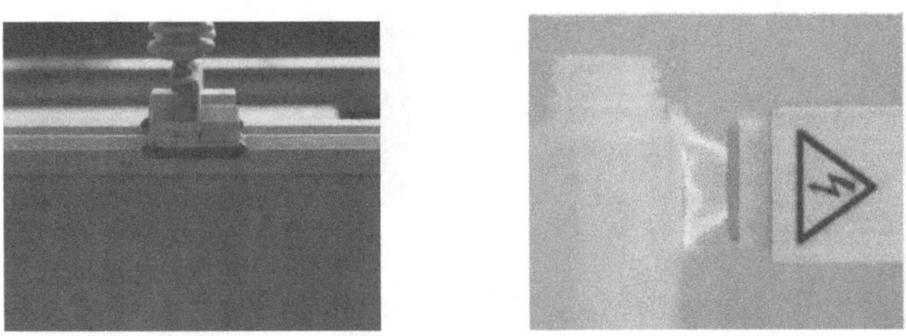

(a) (b)

Figure 2.5 Images of (a) 2D corona treatment and (b) 3D corona treatment of plastic part. Reprinted from [42, 75].

is comprised of anodized aluminum, and the "double dielectric" system wherein both the roll and electrodes are insulated by dielectrics [42].

As most polymer materials find application in versatile shapes and form factors, surface treatments can be greatly beneficial for 3-dimensional coverage. This has led to the emergence of 3D corona discharge systems. While the system can be used to treat various shapes, the construction and operational schematic is similar to 2D coronas as the discharge is localized between the source and grounded electrodes that are constructed from moderately thick aluminum or stainless-steel wires. The key difference here is that instead of a single electrode and roll, 3D corona activator has two or more heads. An arc is created by a high voltage across the gap that exceeds the air breakdown value (~30 kV/cm). Although the discharge is effectively a 1D wire-like shape, treatment of 3D objects is accomplished by blowing-out the arc using the airstream to a length of several centimeters to strike the 3D surface, and travel over and around it to the return path. Instead of the treatment surface moving, the heads are moved to achieve the desired coverage on the specimen (Figure 2.5(b)). While it may be rather straightforward to treat convex 3D surfaces that are curved outwards, there are limitations to access concave and re-entrant surfaces with internal features and cavities. An added challenge with 3D coronas is that it can be particularly difficult to activate "plasmaphobic" or heat-sensitive materials without causing damage. While this can be somewhat mitigated with multiple discharge heads, the effective power dissipated in the blown-out arc can be significantly lower than with 2D systems. 3D coronas are also known to generate large amounts of ozone, which may require special filters for certain sensitive applications and environments [42].

2.3.2 Mechanistic Discussions of Corona Parameters

Plasma in typical corona activators is generated in gaseous atmospheric conditions by radio-frequency (RF) electromagnetic fields in the 10-40 kHz range by preferentially stripping electrons and causing ionization, excitation, and dissociation effects. The reactive species created in these processes include predominantly ozone, atomic oxygen, and oxygen free radicals. Under these conditions, electron temperature can locally exceed 10^5 K in the inter-electrode space; however, the kinetic energies of larger and heavier ions, atoms and other gaseous particles in the reactant mix are 2-4 orders of magnitude lower than that of electrons. At the same time, the heat capacity of electrons is much smaller as well. By virtue of this differentiation, the ionized air is not generally heated, and the average surface temperature of the treated plastic film is not expected to significantly differ from the ambient conditions during the process. Consequently, albeit sensitive to temperature, the layers of plastic or polymer films being treated do not necessarily undergo adverse changes. It is, however, worth noting that the kinetic energy of electrons in the plasma generally exceeds the energies required to break the covalent bonds in the polymer chains. When these bonds break, the presence of radicals and other atomic groups in the gas stream can initiate various reactions, including most commonly oxidation, cross-linking or degradation and surface modification through free radical reactions to create oxidizing functional groups (hydroxyl, carboxyl, carbonyl, ester, etc.). The net effect of all these mechanisms are the changes in the surface free energy, surface roughness, wettability, and adhesion characteristics [56, 76].

2.3.3 Influence of Physical Factors and Equipment Design

The basic parameters determining the effectiveness of a corona discharge treatment are the output power from the generator, P (W), the length of the discharge electrode, L (m), and the interaction time based on the speed of material shift through the activator, v (m/s), which determine the energy dose carried and transmitted by the ionized particles. This corona treatment dose is expressed in terms of activation energy, E_j [J/m^2], as shown in equation (2.1). E_j scales with power coupled into the discharge, and inversely with the size and speed of material shift. It is also dependent on the properties of the activated material and its level of crystallinity: the surface free energy is also known to increase more slowly with more crystalline polymers (polyester < polyethylene (PE) < polypropylene (PP) < biaxially oriented polypropylene (BOPP). Thicker materials can also affect the treatment, particularly in the case of films containing lubricants as they can migrate from a larger area. More stretched films tend to be less responsive to treatment and this has been linked to low content of polar groups. The trade-off between threshold E_j and material shift speed can be significant for production environments based on required level of activation. While E_j in the range of 1-3 kJ/m^2 is known to be optimal, it has also been shown that low E_j (0.5kJ/m^2) favors better discharge uniformity [29].

With regards to activator design, choice of materials for both the discharge electrode, and (to a lesser extent) the grounded electrode, can influence the composition of the discharge space. The geometry and size can impact the distribution of the electromagnetic field as greater electrode wire curvature is known to cause higher electron density and more positive ions, while the nature of the insulation and its thickness are limiting factors for the electrical capacity. Further, corrosion and sputtering of the electrodes can negatively impact the process and their useful life. The gap between electrodes is an important design parameter as it controls the electric field intensity. It is usually configured at 1-3 mm, and nominally is thicker than the treated film by 0.3-2 mm, to achieve desirable corona discharge. In general, low E_j and low inter-electrode gap are beneficial for controlling more uniform corona discharges. However, the current-voltage characteristics are not constant and can be affected by the electrode temperature. Upon initiation of discharge, the gas and electrode temperatures increase rapidly to cause increase in average current. Subsequently, the ignition voltage drops and the current voltage of corona discharge increases. On the generator side, higher voltage can increase the activation level and cause more surface oxidation. Increasing the supply frequency is also known to be beneficial for uniformity, whereas lowering the frequency is known to provide more targeted surface treatment and improved wettability. The corona discharge polarity is also known to impact the effectiveness of surface treatment. On a counter-intuitive note, positive coronas have been shown to be more effective even though they contain less free electron density than negative coronas due to their preferred energy level distribution.

2.3.4 Influence of Plasma Chemistry and Gas Composition

The tendency to create extremely oxidizing species during a corona discharge can have both positive and detrimental impacts to the treated sample. In the minimally ionized atmosphere of a corona discharge, the role of ions is less significant than that of free radicals. The latter have longer lifetimes, and greater surface mobility, which are both correlated to the homogeneous and largely uniform outcomes of surface modification experiments [29]. These reactive species can enable cleaning and removal of amorphous and low molecular weight surface contaminants by chemical etching and ablation processes, and subsequently increase the surface free energy by reaction with polar functional groups. However, the ozone that is evolved during the discharge can induce surface cracks and cuts such that excessively oxidizing conditions can also cause severe surface roughness and polymer degradation [77]. It is also to be noted that, due to lack of a consensus, the depth of impact of the corona treatments on the surface modified layers from various reports spans a rather broad range from 0.5nm – 10nm or deeper.

There are numerous other factors that can exacerbate or mitigate the harmful effects of treatment. For example, with the polymer film serving as a source, additives such as lubricants, stabilizers, and anti-blocking and anti-oxidation agents can either react with the plasma constituents or impede the treatment by migrating to the surface [56, 58]. Ambient humidity also strongly influences the corona treatment, as it is known to enhance the surface oxidation reactions and cause increased roughness. For example, during the deposition of low molecular weight oxidized material (LMWOM) on PP surface, low relative humidity does not alter the roughness irrespective of the voltage. However, at high humidity and even under low treatment voltage, the LMWOM aggregates cause increased roughness. The potential formation of nitric acid is also indicated as an undesirable product under humid conditions. While the atmospheric oxygen acts as a source for the highly reactive ozone and atomic oxygen radicals, the incorporation of inert gas or nitrogen environment can significantly suppress the reactivity and reduce the degradation of the polymer.

2.3.5 Effects of Process Control Methods

As the process of corona treatment is rather sensitive to various factors, precise control against too low or too high treatment is essential for maintaining a high-quality surface. On one hand, insufficient activation can lead to poor adhesion and on the other hand, excessive treatment may cause surface defects, damage to the mechanical properties and in some instances lead to other undesirable chemical pathway. Figure 2.6 is a

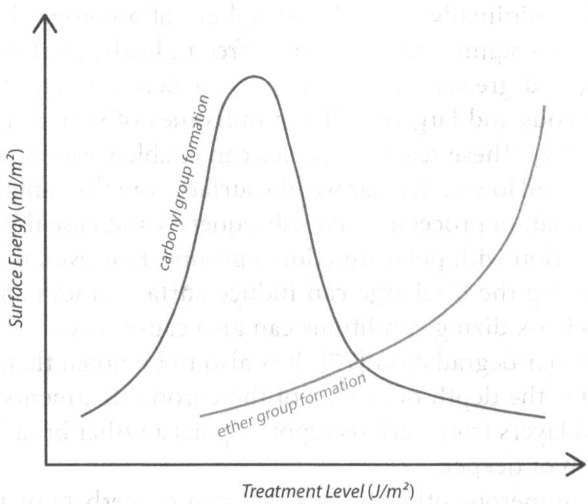

Figure 2.6 Shows trend curves depicting the effect of treatment level on surface energy and functional group formation. Adapted from ref. [58].

schematic showing the variation of surface energy and conversion of preferred carbonyl (polar) group to ether (non-polar) group under excessive treatment levels.

To calibrate and maintain a robust and scalable process, it is possible to monitor the treatment quality through offline sampling and determination of surface free energy from measured contact angles of appropriate probe liquids (suited for laboratory and small-scale settings), or by continuous measurement of discharge accumulated in the surface layer of the film (more suited for industrial applications). In an alternate approach that has been proposed, the energy transmitted during treatment can be calculated and controlled by measuring in-line and regulating the real current and voltage flowing through the sample during the process.

2.3.6 Hydrophobic Recovery and Mitigation by Additives

Immediately after corona activation, films may demonstrate strong effects of chemical activation, high surface free energy and enhanced roughness and adhesion. However, over time, the effects wane and the films gradually show signs of reversal towards complete inversion of the surface property from hydrophilic to hydrophobic (like untreated surface). This adverse

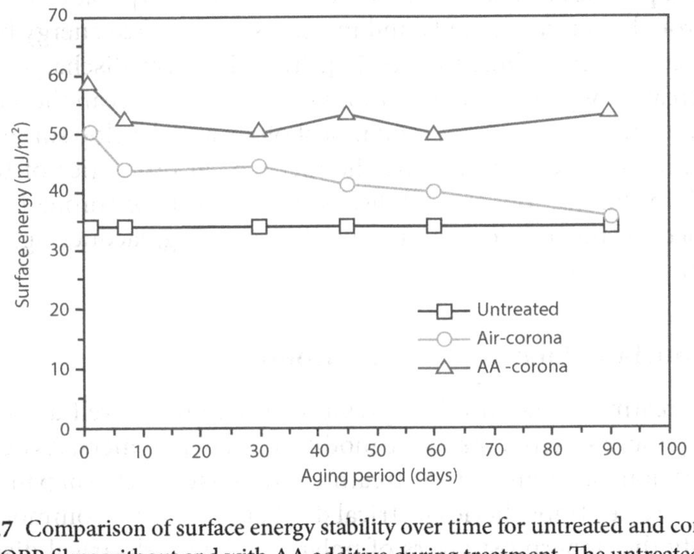

Figure 2.7 Comparison of surface energy stability over time for untreated and corona treated BOPP films without and with AA additive during treatment. The untreated film has surface energy of 32.8 mJ/m^2. Adopted from [79].

phenomenon, which renders the film surface practically useless beyond ~30 days in many cases, is called "hydrophobic recovery" [42].

The mechanism for this reversal of behavior is believed to be a combination of the small size of the polar functional groups, high mobility of the polymer chains near the surface, and thermodynamic activation that can cause the polar groups to invert inward and migrate below the film surface. Although steric workarounds such as grafting of bulky functional groups have been developed, the additional process steps involved can lengthen the process and increase the film costs. However, the introduction of an *in-situ* additive such as vapor phase acrylic acid (AA) monomer into the corona region during treatment has proved successful in stabilizing the film surface characteristic against transient behavior. Figure 2.7 shows an example of AA-corona treated BOPP film that could maintain high surface energy against aging in comparison to Air-corona treated film that practically reverts to its original untreated level after 90days [78].

2.4 Surface Effects of Corona Treatment

Corona treatment is gaining popularity for surface modification of polymers by creating surface ligands with high binding capacity to other functional groups. The corona process modifies surface morphology by altering surface roughness, crystallinity and increases surface free energy by introducing surface cross-linking [76, 77]. The electrical discharges during corona treatment generate polar groups on the surface, clean the surface of material to enhance roughness and morphology and develop surface properties without directly modifying the mechanical properties of the bulk material. In this section, we will discuss these effects of corona treatment on surfaces of polymeric materials used in printing, lacquering, lamination, adhesive bonding and so on.

2.4.1 Surface Polar Functional Groups

Corona treatment may use low pressure (vacuum) as well as air/atmosphere pressure system for surface modification of polymers. As explained in the previous sections, corona treatment uses asymmetrical pair of electrodes to create atmospheric electrical discharges that are commonly used to modify the uppermost layers of polymers. These electrical discharges are created near the discharge electrode and spread in the direction of the grounded one, with usual appearance of discharges near sharp edges and thin wires of electrodes [56]. This differentiates corona discharge from other sources of plasma under atmospheric pressure which include cold plasma torch, gliding arc, or dielectric barrier discharge. Depending on the polarity of the active electrode, polarity of the corona discharge may be positive, negative, or bipolar. Overall, conduction in corona discharge may be unipolar or bipolar depending on the contribution of both ionic dipoles.

Even though there are less numbers of free electrons as compared to a negative discharge, positive corona discharge is more effective due to higher electron energy. The effectiveness of surface modification is also influenced by the gas chemistry in which discharges occur. For example, oxygen in air between the electrodes transforms into highly reactive ozone and atomic oxygen. On the contrary, nitrogen ions are more inert and have less significant impact on the polymer surface modification. Changes in the uppermost layer occur quickly for polyesters than for PE whereas PP is the slowest to undergo surface modification [76]. Higher crystallinity, surface thickness and flexibility inhibit surface modification due to low polar constituent to begin with and hence hard to change with treatment [77].

The physical and chemical etching causes ablation and cleaning of the surface to remove organic contaminants. For low pressure, low ionization plasmas, chemical etching is more dominant than the physical one. It is significant to improve the adhesion property because amorphous and low molecular weight components cause poor adhesion. The electrons in the plasma are used to reduce the O_2 and H_2O to O^- and OH^- ions during discharge [56, 77]. These negative ions interact with positive ions in the plasma creating neutral species. With some water vapor content in the air, active ingredients such as O, OH, O_2, H_2O_2 are present [29]. For example, production of ozone depends on polarity of the discharge, relative humidity, and gas chemistry. Presence of ozone in the discharge leads to surface cracks and cuts caused mainly due to propagation of oxygen through growth stress in the surface layer [77].

The range of impact of corona treatment on the uppermost surface functionality has been proposed to be around 10nm by Gooch [77] and up to a several tens of nanometers by Zenkiewicz [76] depending on E_j value. Along with surface effect of ozone, the electron collision impact causes chemical bonds C-C, C-H to break in the polymer chain as shown in Figure 2.8. Free radicals are formed by breaking these chemical bonds which later take part in recombination, unsaturation, branching or cross-linking. The cross-linking leads to an improvement of bond strength because of improvement of cohesive strength on the surface of

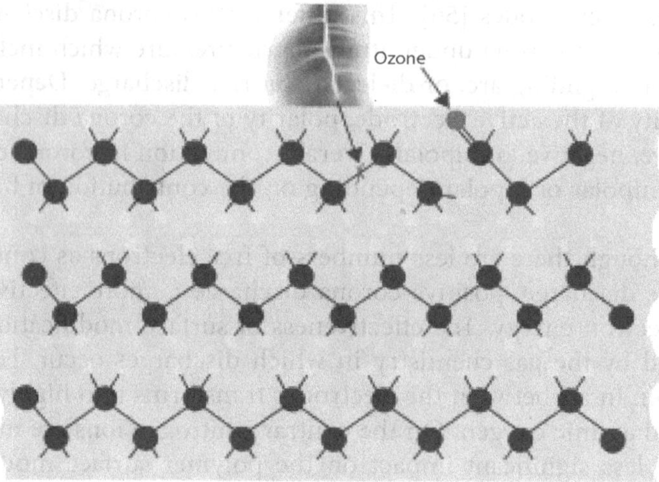

Figure 2.8 Penetration of ozone reactive species through the uppermost layer of the polymer surface after corona treatment.

material [7]. The free radicals react with other atmospheric reactive particles such as oxygen, ozone, water vapor, OH groups leading to formation of polar compounds in the surface of the material. Main groups emerging on the uppermost surface were found to be C-OH, C=O and COOOH by Chan [79], Desai and Singh [80], -OH or -CH—O by Brzezinski *et al.* [81], and peroxide, hydroxyl, carbonyl and carboxyl esters and acids by Pascual *et al.* [82].

Nowadays, the use of corona treatment to generate surface polar groups is expanding to thicker flat materials as well as three-dimensional objects, for example, pipes, fiber objects, tubing. One of the main advantages of this process are no requirements regarding specific conditions during modification (like low temperature, plasma treatment, surface priming etc.) [56]. In addition, low investment cost in terms of active electrode, process, and energy consumption makes corona treatment a possible solution for creating surface polar groups on plastics and polymers to improve their adhesion and surface morphology.

2.4.2 Modifying Surface Wettability

Wettability of polymeric surfaces is highly influenced by surface morphology, roughness, and functional groups (C-O, C-N, C-F and C-H) on the surface. Polymer surfaces are usually hydrophobic and their contact angle with water amounts from 60° to 90°. Corona treatment paves the way to modify surface properties that have an impact on the contact angle and surface free energy [76].

There are various approaches to determine surface free energy of solids from measured contact angles of appropriate probe liquids [83, 84], here the Owens-Wendt approach was used. Contact angles on a polymeric surface with water and diiodomethane allow for determining the surface free energy of the material.

Drops of constant volume are placed on the substrate and the measurement of contact angle is done using a goniometer. Surface free energy is a key parameter driving the spreading of ink on material and determines its wettability [56]. The polar and dispersion components of the surface free energy are determined by the Owens-Wendt method [56]:

$$\left(\gamma_s^d\right)^{0.5} = \frac{\gamma_d \quad (cos\theta_d + 1) - \gamma_w(cos\theta_w + 1)\sqrt{\frac{\gamma_d^p}{\gamma_w^p}}}{2\left[\sqrt{\gamma_d^d} - \sqrt{\frac{\gamma_w^d}{\gamma_w^p}\gamma_d^p}\right]} \tag{2.2}$$

$$\left(\gamma_s^p\right)^{0.5} = \frac{\gamma_w \quad (cos\theta_w + 1) - 2\sqrt{\gamma_s^d\gamma_w^d}}{2\sqrt{\gamma_w^p}} \tag{2.3}$$

$$\gamma_s = \gamma_s^d + \gamma_s^p \tag{2.4}$$

Where,

γ_s^d: dispersion component of surface free energy of the film,
γ_s^p: polar component of surface free energy of the film,
γ_w: surface tension of water,
γ_s^w: dispersion component of surface tension of water,
γ_w^p: polar component of surface tension of water,
θ_w: contact angle of water,
γ_d^d: dispersion component of surface tension of diiodomethane,
θ_d: contact angle of diiodomethane,
γ_p^d: polar component of surface tension of diiodomethane,
γ_s: surface free energy of the film.

Figure 2.9 shows example of significant reduction in contact angle with corona treatment of poly(lactic acid) (PLA) film, from 76.2° before the treatment to 58.0° after surface modification.

Impact of corona treatment on the value of surface free energy and its polar and dispersion components are evaluated by Izdebska [56]. Figure 2.10 shows comparison of the surface free energy components without and with Design of Experiments (DOI) for corona treatment. The observed trend below is in favor of improving the polymer substrate wettability.

Contact angle of the substrate with an appropriate probe liquids is used to determine the work of adhesion, (W_{adh}) which is a thermodynamic parameter related to the adsorption property of polymers. W_{adh} is expressed as:

$$W_{adh} = \gamma_L (1 + cos\theta) \tag{2.5}$$

Where, θ: contact angle with liquid, γ_L: surface tension of liquid. The work of adhesion after corona treatment is enhanced by formation of polar surface groups and surface morphology modification on the uppermost

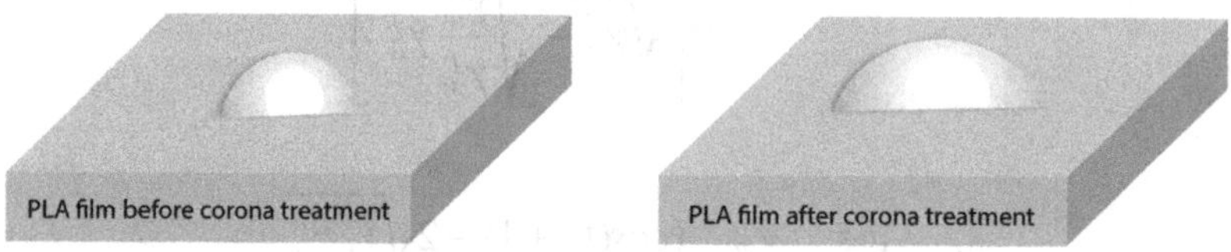

Figure 2.9 Comparison of contact angle before and after corona treatment of PLA film. Reprinted from Ref. [56].

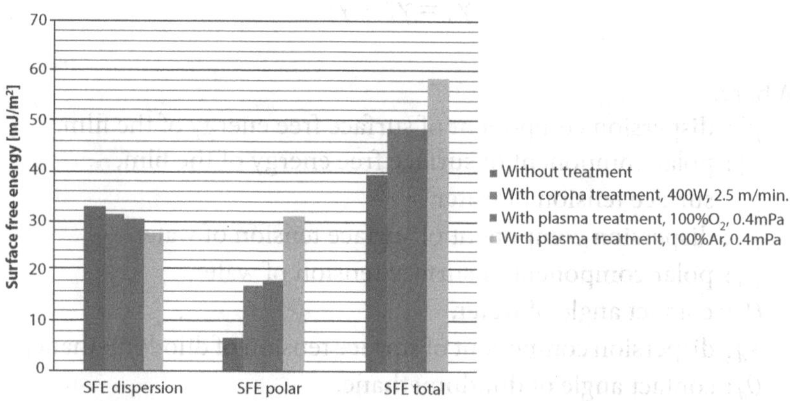

Figure 2.10 Impact of corona treatment on surface free energy and its polar and dispersion components [56].

layers [66]. In a study by Sapieha *et al.* [66], corona treatment improved the W_{adh} of polymer substrate from 89mJ/m^2 to 110mJ/m^2.

2.5 Adhesion Improvement by Corona Treatment

The bond strength and adhesion are the most relevant mechanical properties of polymeric materials which can be altered and improved by surface treatment using corona electrical discharge. The several factors influencing the effects of corona treatment are corona discharge, nature of the polymer substrate, storage time before and after corona treatment, and kind of additives in the matrix polymer [23]. In this section, impact of corona discharge on polymers such as polypropylene (PP), poly(ethylene terephthalate) (PET), poly(vinyl chloride) (PVC), ethylene vinyl acetate (EVA) and polystyrene (PS) is discussed in detail. The above-mentioned polymers are very inert and have low surface free energies and hence it is necessary to treat the surfaces of these polymers prior to processes such as printing, laminating or bonding.

2.5.1 Polypropylene (PP)

The uppermost layer of the PP film needs modification to improve its adhesion and wettability. Before the corona treatment, PP surface is typically characterized by surface free energy in the range 25-30mJ/m^2 whereas processes like printing and lamination require at least 37mJ/m^2 and 42mJ/m^2, respectively [85]. Oxidation with corona discharges helps improve wettability and adhesion of the liquid toward modified upper layer of PP films. Oxygen atoms from ozone and/or oxygen in the plasma are used to modify the uppermost PP layer by forming functional groups such as ketone, hydroxyl, peroxide, ester, and carbonates. The presence of such functional groups helps to increase the surface free energy due to their polar nature, [56]. The impact of the corona treatment is dependent largely on the activation energy E_j as shown in equation 2.1. In the range of $E_j < 1$kJ/m^2, the surface free energy increases slowly, whereas in the range of E_j 1-4kJ/m^2 there is a rapid increase in the surface free energy of PP [86]. If the activation energy exceeds 5kJ/m^2, the geometric structure of the uppermost surface layer is modified that causes increase in the surface roughness and results in amorphous phase formation of PP film. With a higher activation energy >8kJ/m^2, chemical bonds in the PP chaina are broken causing formation of LMWOM on the surface of activated films. Studies with activation energy >10kJ/m^2 by Caceres [85] and Zenkiewicz [86] have reported sever damages of surface layers rendering the PP films useless.

2.5.2 Polyethylene (PE)

Corona treatment not only increases the surface free energy but also increases the average molecular weight and causes cross-linking of macromolecules in the layer [56]. Like PP films, PE films have very low surface free energy in the order of 25-30mJ/m^2. Hence, they need to be surface modified by corona treatment to increase their surface free energy to a level of 45-52mJ/m^2 to enable printing, lamination, or bonding process. With post corona treatment, the uppermost layer of PE undergoes surface modification with emergence of polar groups such as hydroxyl, carboxyl, carbonyl, to name a few. The activation influences the value of surface free energy to increase it to about 48-55mJ/m^2 with the polar component amounting to roughly 11.6 – 17.8 mJ/m^2. The contact angle with water showed dramatic decrease post corona treatment, with 93.5° before treatment and reduced to 51.4° after treatment. For other liquids such as diiodomethane, formamide and glycerol, a nominal decrease in contact angle i.e. 50%, 33% and 26% respectively was observed.

2.5.3 Poly(ethylene terephthalate) (PET)

PET films are typically characterized by higher surface free energy than polyolefin films, although they also often require surface modification process by corona treatment. Surface free energy of PET films reported by Dobrzanski and Dobrzanska-Dankiewicz is in range of 40-42mJ/m^2 [87].

Corona treatment of PET causes emergence of high numbers of various functional groups on the uppermost layer as reported by Zenkiewicz

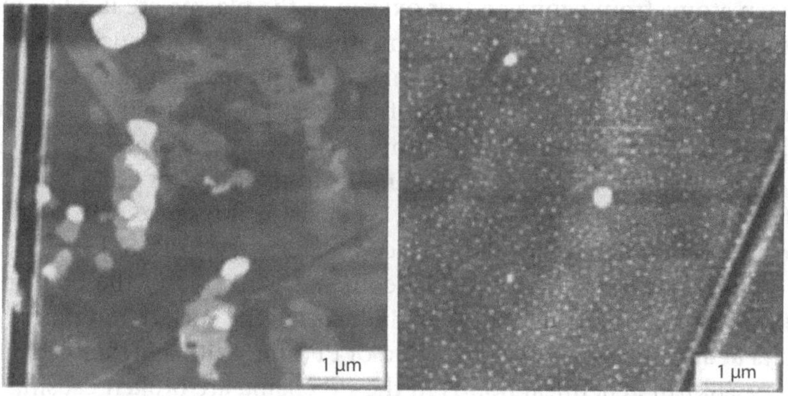

(a) (b)

Figure 2.11 AFM images of PET surface (a) untreated (b) modified with corona discharge of activation energy ~ 10kJ/m^2. Reprinted from [76].

[76]. LMWOMs formed during the surface modification and their impact on the geometric surface modification play a significant role. As shown in AFM images in Figure 2.11, there is a visible change in the surface morphology caused by ablation of the macromolecules caused by corona treatment. In case of higher concentration of C-O groups (hydroxyl, ether, epoxy) and lowest concentration of O-C-O (carboxyl) groups, the bond breaking forces for the surface group were the highest and effective surface modification of the uppermost layer was observed by Sapieha *et al.* [66].

2.5.4 Poly(vinyl chloride) (PVC)

Surface free energy of non-activated PVC largely depends on the composition of the film. For example, in hard PVC films, the surface free energy is roughly ~ 39mJ/m^2 and for softened PVC films, it is in the range of 32-38mJ/m^2. Corona treatment of PVC yields similar effect as PE and does not result in any major problems. However, if the PVC film is softened, modification of the uppermost layer is quite difficult due to migration of the plasticizers to the uppermost layer leading to non-uniform bending/twisting of polymer surfaces [56].

2.5.5 Polystyrene (PS)

PS films are typically amorphous, transparent, and fragile with low surface free energy approximately 43mJ/m^2. There are two types of PS films, high impact PS and general-purpose PS which are differentiated by their response to activation energy. Corona treatment can be used to modify both PS surfaces as well as the surface morphology. Typically, activation energy in the range of 0.5-3kJ/m^2 along with oxidation is sufficient to see decrease in contact angles with water and diiodomethane. The durability of this treatment was typically observed to last 2 weeks by Zienkiewicz [76].

2.6 Summary

As explained throughout this chapter, corona treatment is a very effective process for modifying the uppermost layer in polymer surfaces, and hence it finds wide application in the polymer and textile industries.

A major advantage of the treatment in addition to surface enhancement is the low cost, in-line treatment. We have discussed multiple studies to

Table 2.2 Modification of surface free energies of polymers after corona treatment [56, 66].

Material	Before corona treatment	After corona treatment
Polyethylene	25 – 30 mJ/m^2	48 – 55 mJ/m^2
Poly(ethylene terephthalate)	32 – 38 mJ/m^2	42 – 58 mJ/m^2

determine impact of corona discharges on topography and morphology of the material, surface chemistry and contact angles. The dosage applied, treatment time and pulsed discharge on the film surface are optimized for maximum enhancement of the surface free energy. Table 2.2 shows a summary of change in surface free energy after corona treatment. Functional groups formed as a result of discharges improve the polymer surface wettability and adhesion due to modification of uppermost layers in the polymer films. This change in the uppermost layer of the surface is attributed to the formation of polar groups and dispersive bonds indirectly affecting the surface bond strengths. However, the surface modification by corona treatment may also increase the risk of dust contamination and may have a negative impact on the printing and laminating process.

References

1. M. Strobel, C. S. Lyons, and K. L. Mittal (Eds), *Plasma Surface Modification of Polymers: Relevance to Adhesion*. CRC Press, Boca Raton, FL (1994).
2. K. L. Mittal (Ed), *Polymer Surface Modification: Relevance to Adhesion*, Vol. 5, CRC Press, Boca Raton, FL (2004).
3. S. K. Nemani, R. K. Annavarapu, B Mohammadian, A. Raiyan, J. Heil, Md. A. Haque, A. Abdelaal and H. Sojoudi, Surface modification of polymers: Methods and applications, *Advanced Materials Interfaces*, 5, 1801247 (2018).
4. K. L. Mittal and W.-S. Lei (Eds.), *Laser Technology: Applications in Adhesion and Related Areas*. Wiley-Scrivener, Beverly, MA (2018).
5. Y. Kusano, Atmospheric pressure plasma processing for polymer adhesion: A review, *J. Adhesion*, 90, 755–777 (2014).
6. L. W. McKeen, Powder coating and films, in *Fluorinated Coatings and Finishes Handbook (Second Edition)*, L. W. McKeen (Ed.), pp 185–209, William Andrew Publishing, Norwich, NY (2016).
7. J. P. Badey, E. Espuche, Y. Jugnet, B. Chabert, and T. M. Duc, Influence of chemical and plasma treatments on the adhesive properties of PTFE with an epoxy resin, *Int J Adhesion Adhesives*, 16, 173–178 (1996).

8. S. R. Kim, Surface modification of poly(tetrafluoroethylene) film by chemical etching, plasma, and ion beam treatments, *J Appl. Polym. Sci*, 77, 1913–1920 (2000).
9. M. Lewin, A. Mey-Marom, and R. Frank, Surface free energies of polymeric materials, additives and minerals, *Polymers for Advanced Technologies*, 16, 429–441 (2005).
10. D. K. Owens and R. C. Wendt, Estimation of the surface free energy of polymers, *J. Appl. Polym. Sci.*, 13, 1741–1747 (1969).
11. C.-S. Ren, K. Wang, Q.-Y. Nie, D.-Z. Wang, and S.-H. Guo, Surface modification of PE film by DBD plasma in air, *Appl. Surf. Sci.*, 255, 3421–3425 (2008).
12. J. Izdebska, Printing on polymers: Theory and practice, in: *Printing on Polymers: Fundamentals and Applications*, J. Izdebska and S. Thomas (Eds.), pp 1-20, William Andrew Publishing, Norwich, NY (2016).
13. R. N. Shimizu and N. R. Demarquette, Evaluation of surface energy of solid polymers using different models, *J. Appl. Polym. Sci.*, 76, 1831–1845 (2000).
14. K. G. Kostov, Y. A. A. Hamia, R. P. Mota, A. L. R. dos Santos, and P. A. P. Nascente, Treatment of polycarbonate by dielectric barrier discharge (DBD) at atmospheric pressure, *J. Phys.: Conf. Ser.*, 511, 012075 (2014).
15. R. O. F. Verkuijlen, M. H. A. van Dongen, A. A. E. Stevens, J. van Geldrop, and J. P. C. Bernards, Surface modification of polycarbonate and polyethylene naphtalate foils by UV-ozone treatment and μPlasma printing, *Appl. Surf. Sci*, 290, 381–387 (2014).
16. M. Ozdemir, C. U. Yurteri, and H. Sadikoglu, Physical polymer surface modification methods and applications in food packaging polymers, *Critical Reviews in Food Science and Nutrition*, 39, 457–477 (1999).
17. K. L. Mittal, The role of the interface in adhesion phenomena, *Polym. Eng. Sci.*, 17, 467–473 (1977).
18. C. Oehr and M. R. Wertheimer, Industrial plasma applications, *Plasma Process Polym.*, 19, 2277025 (2022).
19. D. Briggs, D. G. Rance, C. R. Kendall, and A. R. Blythe, Surface modification of poly(ethylene terephthalate) by electrical discharge treatment, *Polymer*, 21, 895–900 (1980).
20. D. Briggs and C. R. Kendall, Chemical basis of adhesion to electrical discharge treated polyethylene, *Polymer*, 20, 1053–1054 (1979).
21. D. Briggs and C. R. Kendall, Derivatization of discharge-treated LDPE: an extension of XPS analysis and a probe of specific interactions in adhesion, *Int J Adhesion Adhesives*, 2, 13–17 (1982).
22. E. Földes, A. Tóth, E. Kálmán, E. Fekete, and Á. Tomasovszky–Bobák, Surface changes of corona-discharge-treated polyethylene films, *J. Appl. Polym. Sci.*, 76, 1529–1541 (2000).
23. M. Lindner, N. Rodler, M. Jesdinszki, M. Schmid, and S. Sängerlaub, Surface energy of corona treated PP, PE and PET films, its alteration as function of storage time and the effect of various corona dosages on their bond strength after lamination, *J. Appl. Polym. Sci.*, 135, 45842 (2018).

24. J. M. Strobel, M. Strobel, C. S. Lyons, C. Dunatov, and S. J. Perron, Aging of air-corona-treated polypropylene film, *J Adhesion Sci Technol*, 5, 119–130 (1991).
25. S. Guimond and M. R. Wertheimer, Surface degradation and hydrophobic recovery of polyolefins treated by air corona and nitrogen atmospheric pressure glow discharge, *J. Appl. Polym. Sci.*, 94, 1291–1303 (2004).
26. D. Kowalczyk and I. Kamińska, Effect of corona discharge on the stability of the adhesion of thin silicone-organic coating to polyamide fiber surface made by the sol–gel method, *J Coat Technol Res*, 14, 1115–1125 (2017).
27. A. Popelka, I. Novák, M. A. S. A. Al-Maadeed, M. Ouederni, and I. Krupa, Effect of corona treatment on adhesion enhancement of LLDPE, *Surf. Coat. Technol.*, 335, 118–125 (2018).
28. C. Y. Kim and D. A. I. Goring, Surface morphology of polyethylene after treatment in a corona discharge, *J. Appl. Polym. Sci.*, 15, 1357–1364 (1971).
29. M. Strobel, V. Jones, C. S. Lyons, M. Ulsh, M. J. Kushner, R. Dorai and M. C. Branch, A comparison of corona-treated and flame-treated polypropylene films, *Plasmas Polym.*, 8, 61-95 (2003).
30. D. Zhang, Q. Sun, and L. C. Wadsworth, Mechanism of corona treatment on polyolefin films, *Polym. Eng. Sci.*, 38, 965–970 (1998).
31. K. S. Johansson, Surface modification of plastics in: *Applied Plastics Engineering Handbook (Second Edition)*, M. Kutz (Ed.), pp. 443-487, William Andrew Publishing, Norwich, NY (2017).
32. I. V. Timoshkin, M. Maclean, M. P. Wilson, M. J. Given, S. J. MacGregor, T. Wang, and J. G. Anderson, Bactericidal effect of corona discharges in atmospheric air, *IEEE Trans. Plasma Sci.*, 40, 2322–2333 (2012).
33. S. Li, I. V. Timoshkin, M. Maclean, S. J. MacGregor, M. P. Wilson, M. J. Given, J. G. Anderson, and T. Wang, Steady-state corona discharges in atmospheric air for cleaning and decontamination, *IEEE Trans. Plasma Sci.*, 41, 2871–2878 (2013).
34. V. Scholtz, J. Pazlarova, H. Souskova, J. Khun, and J. Julak, Nonthermal plasma — A tool for decontamination and disinfection, *Biotechnology Advances*, 33, 1108–1119 (2015).
35. N. Carneiro, A. P. Souto, E. Silva, A. Marimba, B. Tena, H. Ferreira and V. Magalhães, Dyeability of corona-treated fabrics, *Color. Technol.*, 117, 298–302 (2001).
36. M. Pykönen, H. Sundqvist, J. Järnström, Otto-Ville Kaukoniemi, M. Tuominen, J. Lahti, J. Peltonen, P. Fardim and M. Toivakka, Effects of atmospheric plasma activation on surface properties of pigment-coated and surface-sized papers, *Appl. Surf. Sci.*, 255, 3217–3229 (2008).
37. I. Sakata, M. Morita, N. Tsuruta, and K. Morita, Activation of wood surface by corona treatment to improve adhesive bonding, *J. Appl. Polym. Sci.*, 49, 1251–1258 (1993).
38. D. Minzari, P. Møller, P. Kingshott, L. H. Christensen, and R. Ambat, Surface oxide formation during corona discharge treatment of AA 1050 aluminium surfaces, *Corrosion Sci.*, 50, 1321–1330 (2008).

39. Corona Treatment of Glass Sheets Prior to Coating Application, *Tantec.* https://tantec.com/corona-treatment-of-glass-sheets-prior-to-coating-application/ (accessed Sep. 17, 2022).
40. M. Remeika, L. K. Ono, M. Maeda, Z. Hu, and Y. Qi, High-throughput surface preparation for flexible slot die coated perovskite solar cells, *Organic Electronics*, 54, 72–79 (2018).
41. D. Burkitt, R. Patidar, P. Greenwood, K. Hooper, J. McGettrick, S. Dimitrov, Matteo Colombo, V. Stoichkov, D. Richards, D. Beynon, M. Davies and T. Watson, Roll-to-roll slot-die coated P–I–N perovskite solar cells using acetonitrile based single step perovskite solvent system, *Sustainable Energy Fuels*, 4, 3340–3351 (2020).
42. S. Ebnesajjad and A. H. Landrock, Material surface preparation techniques, in: *Adhesives Technology Handbook (Third Edition)*, S. Ebnesajjad and A. H. Landrock (Eds.), pp. 35-66, William Andrew Publishing, Norwich, NY (2015).
43. R. Shishoo, Introduction – The potential of plasma technology in the textile industry, in : *Plasma Technologies for Textiles*, R. Shishoo (Ed.), pp. 15-30, Woodhead Publishing (2007).
44. R. A. Wolf, Plastic surface modification in: *Plastic Surface Modification: Surface Treatment and Adhesion*, R. A. Wolf (Ed.), pp. 1-9, Hanser (2010).
45. I. Novák, A. Popelka, Z. Špitalský, I. Krupa, and S. Tavman, polyolefin in packaging and food industry, in: *Polyolefin Compounds and Materials: Fundamentals and Industrial Applications*, I. Krupa and M. Al-AliMa'adeed (Eds.), pp. 181–199, Springer (2016).
46. E. Kontou and M. Niaounakis, Thermo-mechanical properties of LLDPE/SiO2 nanocomposites, *Polymer (Guildf)*, 47, 1267–1280 (2006).
47. Y. G. Hsuan, Geotextile resins and additives: *Geotextiles: From Design to Applications*, R.M. Koerner and R. Bygness (Eds.), pp. 17–23, Woodhead Publishing (2016).
48. R. Stewart, V. Goodship, F. Guild, M. Green, and J. Farrow, Investigation and demonstration of the durability of air plasma pre-treatment on polypropylene automotive bumpers, *Int J Adhesion Adhesives*, 25, 93–99 (2005).
49. V. Scholtz, J. Julák, and V. Kříha, The microbicidal effect of low-temperature plasma generated by corona discharge: Comparison of various microorganisms on an agar surface or in aqueous suspension, *Plasma Processes and Polymers*, 7, 237–243 (2010).
50. M. Stepczyńska and M. Zenkiewicz, Effects of corona treatment on the properties of the surface layer of polylactide, *Polimery*, 59, 220–226 (2014).
51. D. Briggs, New developments in polymer surface analysis, *Polymer (Guildf)*, 25, 1379–1391 (1984).
52. S. Abbott, *Adhesion Science: Principles and Practice*. DEStech Publications, Lancaster, PA (2015).

53. M. Stradal and D. A. I. Goring, Corona-induced autohesion of polyethylene: Dependence of bonding on frequency and power consumption in various gases, *Can J Chem Eng*, 53, 427–430 (1975).
54. L. J. Gerenser, J. F. Elman, M. G. Mason, and J. M. Pochan, ESCA studies of corona-discharge-treated polyethylene surfaces by use of gas-phase derivatization, *Polymer (Guildf)*, 26, 1162–1166 (1985).
55. H. Schonhorn and F. W. Ryan, Surface crosslinking of polyethylene and adhesive joint strength, *J Appl Polym Sci*, 18, 235–243 (1974).
56. J. Izdebska, Corona Treatment in: *Printing on Polymers: Fundamentals and Applications*, J. Izdebska and S. Thomas (Eds.), pp. 123-142, Elsevier, Waltham, MA (2015).
57. F. W. Billmeyer, *Textbook of Polymer Science*, John Wiley & Sons (1984).
58. C. Sun, D. Zhang, and L. C. Wadsworth, Corona treatment of polyolefin films—A review, *J. Polymer Processing Institute*, 18, 171–180 (1999).
59. S. Dong, S. Sapieha, and H. P. Schreiber, Mechanical properties of corona-modified cellulose/polyethylene composites, *Polym Eng Sci*, 33, 343–346 (1993).
60. H. Steinhauser and G. Ellinghorst, Corona treatment of isotactic polypropylene in nitrogen and carbondioxide, *Angew. Makromolekulare Chemie*, 120, 177–191 (1984).
61. S. K. Pankaj, C. Bueno-Ferrer, N. N. Misra, L. O'Neil, B. K. Tiwari, P. Bourke and P. J. Cullen, Dielectric barrier discharge atmospheric air plasma treatment of high amylose corn starch films, *LWT - Food Sci. Technol.* 63, 1076–1082 (2015).
62. H. E. Wagner, R. Brandenburg, K. v. Kozlov, A. Sonnenfeld, P. Michel, and J. F. Behnke, The barrier discharge: basic properties and applications to surface treatment, *Vacuum*, 71, 417–436 (2003).
63. V. I. Arkhipenko, A. A. Kirillov, Y. A. Safronau, and L. v. Simonchik, DC atmospheric pressure glow microdischarges in the current range from microamps up to amperes, *European Physical J D* 60, 455–463 (2010).
64. V. I. Arkhipenko, A. A. Kirillov, Y. A. Safronau, L. v. Simonchik, and S. M. Zgirouski, Plasma non-equilibrium of the DC normal glow discharges in atmospheric pressure atomic and molecular gases, *European Physical J D* 66, 1–11 (2012).
65. J. H. Park, S. H. Lee, K. H. Choi, H. S. Noh, J. W. Lee, and S. J. Pearton, Comparison of dry etching of PMMA and polycarbonate in diffusion pump-based O2 capacitively coupled plasma and inductively coupled plasma, *Thin Solid Films*, 518, 6465–6468 (2010).
66. S. Sapieha, J. Cerny, J. E. Klemberg-Sapieha and L. Martinu, Corona versus low pressure plasma treatment: Effect on surface properties and adhesion of polymers, *J Adhesion*, 42, 91–102 (1993).
67. S. Jha, S. Bhowmik, N. Bhatnagar, N. K. Bhattacharya, U. Deka, H. M. S. Iqbal and R. Benedictus, Experimental investigation into the effect of adhesion

properties of PEEK modified by atmospheric pressure plasma and low-pressure plasma, *J. Appl. Polym. Sci*, 118, 173 - 179 (2010).
68. L. Bardos and H. Barankova, Plasma processes at atmospheric and low pressures, *Vacuum*, 83, 522-527 (2008).
69. V. Majidi and D.M.Coleman, Magnetic confinement and control of trajectories of discharge species in atmospheric pressure analytical spark discharges, *Appl. Spectroscopy.*, 41, 200-207 (1987).
70. A. Schutze, J.Y.Jeong, S.E.Babayan, J. Park, G.S.Selwyn and R.F.Hicks, The atmospheric-pressure plasma jet: a review and comparison to other plasma sources, *IEEE Trans. Plasma Sci.*, 26, 1685-1694 (1998).
71. N. Inagaki, S. Tasaka, and S. Shimada, Comparative studies on surface modification of poly(ethylene terephthalate) by remote and direct argon plasmas, *J Appl Polym Sci*, 79, 808–815 (2001).
72. N. Inagaki, S. Tasaka, T. Horiuchi, and R. Suyama, Surface modification of poly (acryl ether ketone) film by remote oxygen plasma, *J Appl Polym Sci*, 68, 271–279 (1998).
73. Y. Yamada, T. Yamada, S. Tasaka, and N. Inagaki, Surface modification of poly(tetrafluoroethylene) by remote hydrogen plasma, *Macromolecules*, 29, 4331–4339 (1996).
74. C. Roth, A. Spillmann, A. Sonnenfeld, and P. R. von Rohr, Thermal characterization, and optimization of a plasma downstream reactor for particle surface modification, *Plasma Processes and Polymers*, 6, S566–S570 (2009).
75. D.P. Ferreira, J. Cruz, and R. Fangueiro, Surface modification of natural fibers in polymer composites, in: *Green Composites for Automotive Applications*, G. Koronis and A. Silva (Eds.), pp. 3-41, Elsevier, Cambridge, MA (2019).
76. M. Zenkiewicz, Corona discharge in an air as a method of modification of polymeric materials' surface layers, *Polimery*, 53, 1-13 (2008).
77. J. W. Gooch, *Encyclopedic Dictionary of Polymers*, Springer, New York (2011)
78. N. Kalapat, T. Amornsakchai and T. Srikhirin, Surface modification of biaxially oriented polypropylene (BOPP) film using acrylic acid treatment: Part II. Long term aging surface properties, *Surface Coatings Technol.*, 234, 67-75 (2013).
79. C-M. Chan, Surface treatment of polypropylene by corona discharge and flame, in: *Polypropylene Poly Sci and Techno*, J. Karger-Kocsis (Ed.), 2, pp 800 - 805, Springer (1999).
80. S.M. Desai and R. P. Singh, Surface modification of polyethylene, *Advances Polym. Sci*, 169, 231-293 (2004).
81. S. Brzezinski, M. Zenkiewicz, S. Polowinski, D. Kowalczyk, I. Kawbownik, S. Lutomirski and G. Malinowska, Application of corona discharge to modify the surface layer of polyester textile, *Polimery*, 54, 552-558 (2009).
82. M. Pascual, R. Sanchis, L. Sanchez, D. Garcia and R. Balart, Surface modification of low-density polyethylene (LDPE) film using corona discharge plasma for technological applications, *J Adhesion Sci Technol.*, 22, 1425-1442 (2012).

83. F.M. Etzler, Determination of the surface free energy of solids: A critical review, *Rev. Adhesion Adhesives*, 1, 3-45 (2013).
84. F.M. Etzler, Characterization of surface free energies and surface chemistry of solids, in: *Contact Angle, Wettability and Adhesio*n, volume 3, K. L. Mittal (Ed.) pp. 219-264, CRC Press, Boca Raton, FL (2003).
85. C. A. Caceres, Controlling in-line the energy level applied during the corona treatment, *Polymer Testing*, 31, 505-511 (2012).
86. M. Zenkiewicz, Some effects of corona discharge treatment of biaxially-oriented polypropylene film, *J Adhesion*, 77, 25-41 (2001)
87. A. L. Dobrzanski and A. D. Dobrzanska-Danikiewicz, (2011). Struktura i własnosci materiałoẃ polimerowych obrabianych powierzchniowo oraz pokryc polimerowych (Online) Available from: http://www.openaccesslibrary.com/vol05/7.pdf. Accessed 05.03.15.

3

Flame Surface Treatment of Polymers to Enhance Their Adhesion

Joseph DiGiacomo* **and LaWayne Johnson**

Flynn Burner Corporation, Mooresville Blvd., Mooresville, NC, USA

Abstract

Adhesion promotion techniques have wide applications in numerous industries for a wide range of plastic parts, and polyolefin films such as those made of PE, PP, and PET.

One method used to modify the surface of these and other polymer products to promote adhesion of inks, coatings, adhesives, metallized paper and films, microorganisms or cells is flame plasma treatment. In addition, flame plasma is used to clean the surface of metals and to roughen the surface of paperboard to enhance adhesion of the polyolefin coating.

This chapter describes the theory behind hydrocarbon fired flame plasma surface treatment to promote adhesion of water-based inks, coatings, adhesives, labels and other laminates to polyolefin-based substrates.

Critical parameters in flame plasma treatment are flame chemistry, flame geometry, plasma output, and distance of the burner from the part. The interrelationship between these variables, and how to control them for optimum surface treatment, are discussed.

A completely new patented process design has been developed and successfully implemented providing significantly improved control of the flame chemistry, while at the same time simplifying the process control and mechanical hardware required.

In addition, the new design improves the overall efficiency of the flame plasma treatment process by more accurately controlling air/gas chemistry, simplifying the control valve piping, incorporating a variable frequency drive (VFD) combustion air blower to more accurately control and vary the burner firing rate, i.e., amount of plasma generated, and refining the control algorithms.

**Corresponding author*: joedigiacomo28@gmail.com

K.L. Mittal and A.N. Netravali (eds.) Polymer Surface Modification to Enhance Adhesion: Techniques and Applications, (77–118) © 2024 Scrivener Publishing LLC

Keywords: Flame treatment, polymers, adhesion strength, polyolefins, equivalence ratio, stoichiometric ratio, surface free energy, surface tension

3.1 Introduction

The converting industry uses a significant number of polyolefins ranging from polyethylene coated paperboard, metalized paper & film, polyolefin films, to exotic combinations of polymers to obtain different package characteristics, such as barrier property, packaging for retort pouches and containers, strength and flexibility.

Polyolefins are those polymers produced from an olefin or an alkene monomer with the general formula C_nH_{2n}. They are composed of non-polar, long chain molecules, with low surface free energy and lack chemically reactive groups. They are inert molecules. Most polyolefins are inherently hydrophobic Unlike materials with high surface free energies like metals and ceramics, virgin plastics do not contain the available bond sites provided by charged ions distributed over the surface. These provide for van der Waals bonding forces. Without this molecular attraction (van der Waals forces), liquids fail to "wet out", or adhere to the surface of an untreated polyolefin, resulting in poor adhesion of an ink, coating, metal, adhesive, or lamination of one substrate to another, i.e., polyethylene to paperboard to aluminum.

As long as solvent-based coatings, inks, and adhesives were used in converting processes, surface modification or surface treatment was not usually required, or if it was, the level of treatment was relatively low. This is because the surface tension of a solvent based liquid is low compared to the surface free energy of the polymer

As a general rule, acceptable adhesion is achieved if the surface free energy of the substrate i.e, the ability to attract a liquid, is 10 dynes/cm$^+$ higher than the surface tension, i.e., the resistance to spreading of the liquid coating, ink, etc. In this situation, the liquid is said to "wet out" or adhere to the surface.

The reason why some polyolefins do not require surface modification is that their inherent surface free energy is such that the difference between the surface free energy of the polyolefin and the surface tension of the ink, coating, etc., is > 10 dynes/cm.

Surface tension is a measure of surface free energy, a property of liquids, due to molecular forces by which all liquids through contraction of the surface tend to bring the contained volume into a shape having the minimal area. A liquid placed on a flat horizontal surface will either lay flat on the surface, called "wetting out" or form a spherical drop, know as "dewetting" An in-between state can also exist. The nomenclature is *surface tension* for liquids, and *surface free energy* for solids. The units are force/unit length. For liquids, the unit of measurement is mN/m; for solids it is mJ/m^2.

$$^{+}1 \text{ dyne/cm} = 1 \text{ mN/m} = 1 \text{ mJ/m}^2$$

A more complete discussion of surface tension and surface free energy can be found in a paper by Hansen [1].

The angle that the droplet or edge of the liquid makes with the solid is called the contact angle. Contact angles are both advancing (forward) and receding (backward). The lower the contact angle, the better the wettability. This is illustrated in Figure 3.1.

Table 3.1 shows the surface free energies of common substrates. Table 3.2 lists the surface tension of liquids.

As long as solvent-based coatings were used in converting operations or that the surface free energy of the substrate was high enough, as for SBR as compared to polypropylene, for example, surface treatment while important was not critical because of the 10 dyne/cm rule, the difference between the surface free energy of the substrate and the surface tension of the liquid is ≥ 10 dynes/cm. This began to change with the passage of the Clean Air Act of 1979 and subsequent amendments.

In order to reduce VOC emissions, converters had to either install air pollution abatement equipment or switch to water-based coatings. Switching to water-based coatings was the best alternative based on economic

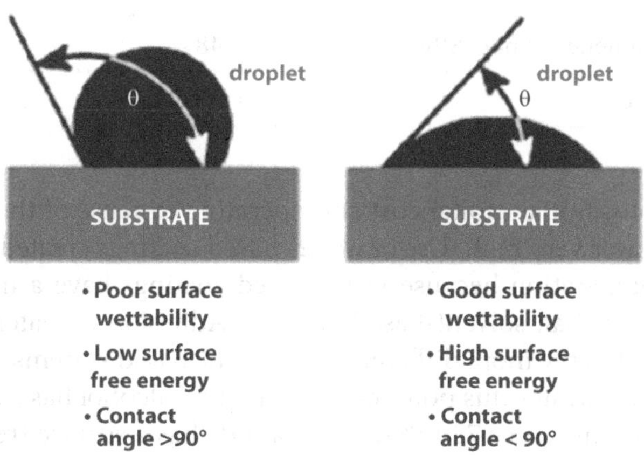

Figure 3.1 Contact angle vs. surface parameters.

Table 3.1 Surface free energy of polymer substrates.

Polymer substrate	Surface free energy (mJ/m^2)
Polypropylene	29 – 31
Polyethylene	30 – 31
Polystyrene	38
Low ionomer polystyrene	33
Polyester	41 – 44
Rigid Poly(vinyl chloride)	39
Plasticized Poly(vinyl chloride)	33 – 38
Epoxy	< 36
Poly(vinyl acetate)/PE copolymer	33 – 44
Acrylonitrile butadiene styrene (ABS)	35 – 42
Polyamide	41 – 47
Poly(vinyl fluoride)	25
Fluorinated polyethylene and polypropylene	18 – 22
Polytetrafluoroethylene (Teflon)	19 -20
Poly(ethylene terephthalate) (PET)	41 – 44
Polycarbonate	46
Silicone	24
Styrene butadiene rubber (SBR)	48
Polyurethane	N/A

considerations. Initial capital cost and operating expense of the early fume incinerators was very high. Use of water- based coatings created a criticality in surface preparation because water-based coatings have a much higher surface tension than solvent-based systems. Adhesion of water-based systems is considerably more difficult than solvent-based systems.

Table 3.2 illustrates this point quite well. Ethyl alcohol has a surface tension of 22 mN/m, water 72 mN/m; polyamide has a surface free energy of 41-47 mJ/m^2.

Table 3.2 Surface tension of liquids.

Liquid	Surface tension (mN/m)
Ethyl alcohol	22
n-Butanol	25
Toluene	29
Formamide	58
Glycerol	63
Water	72

Using an ethyl alcohol-based coating with polyamide, 41-47 dynes/cm – 22 dynes/cm = 19-25 dynes/cm > 10 dynes/cm. Good adhesion.

Using a water-based coating with polyamide, 41-47 dynes/cm - 72 dynes/cm = -31 - -25 dynes/cm < 10 dynes/cm. Poor adhesion

There are several techniques [2–7] used to modify polymer surfaces. These are:

- Flame Plasma
- Corona Discharge
- Cold Plasma
- Chemical Priming
- UV radiation
- Ozonation
- Combination of UV radiation & ozone
- Laser

A brief description of each technique is given below, with a detailed discussion of flame plasma treatment.

- Flame plasma, often referred to as the "fourth state of matter", contains positively and negatively charged ions that are highly reactive, as well as C-H-O functional organic groups. The ions are formed by the combustion of hydrocarbon gases, such as natural gas, propane, butane, acetylene, at high temperature
- Corona discharge uses electrical energy to dissociate O_2 in air or other process gas molecules. These species react with the polymer surface forming reactive chemical sites

- Cold plasma is electrical discharge, under vacuum, dissociating air or other inert gases, such as nitrogen, helium, argon, etc.
- Chemical priming with silane coupling agents promotes cross-linking between the substrate, the organofunctional portion of the primer and the ink, coating, laminate, etc., to form a bond.
- UV radiation, ozone and the combination of the two polarize the polymer surface.
- Laser is used for surface cleaning and preparation, coating removal, surface roughness modification as well as for polymer surface modification

A comparison of the advantages and disadvantages of each technique is shown in Table 3.3.

Table 3.3 Comparison of adhesion promotion techniques.

Technique	Advantages	Disadvantages
Chemical Priming (Coupling Agents)	• Good Results	• High cost • Potentially toxic, hazardous
Corona	• Widely used • Low initial cost	• Produces ozone • High maintenance • Potential for pinholes
Atmospheric Plasma (possesses higher ionized gas plasma density compared to corona)	• Good Results • Less likely to create pinholes	• High maintenance
Direct Flame	• Higher treatment levels	• Higher capital cost
Glow Discharge	• Good results	• High capital costs
Ozone, UV, Ozone + UV	• Excellent for 3D shapes	• Long treatment time
Laser	• Good for removal of surface coatings • Good for polymer surface modification	• Higher cost • Requires specialized training

3.2 Chemistry of Flame Treatment

With the exception of chemical priming, all of these techniques impart polar groups to the polymer surface. These "surface" treatments affect only a few molecular layers and do not alter the bulk properties of the polymer. Because of this surface phenomenon, the treatment level can be adversely affected by excessive handling or rubbing of the surface.

The combustion of a hydrocarbon-based fuel gas, natural gas, propane, LPG, butane or mixtures of these results in complex chemical reactions in addition to "final"" end products of CO_2, H_2O, N_2 and oxides of nitrogen such as, NO and NO_2,(NO_x) The nitrogen is a carryover from the nitrogen in the oxidant (air.) The NO_x is formed at the high temperature of the combustion reaction.

Flame treaters use premixed laminar flames, i.e., the ratio of oxidant (air) to fuel is proportioned and mixed prior to ignition and combustion. Figure 3.2 shows the main zones in a laminar flame profile.

In direct flame plasma treatment, the adiabatic flame temperature is approximately 1816° C, and this high temperature gas, as a result of the combustion process, contains oxygen molecules, oxygen atoms and ions,

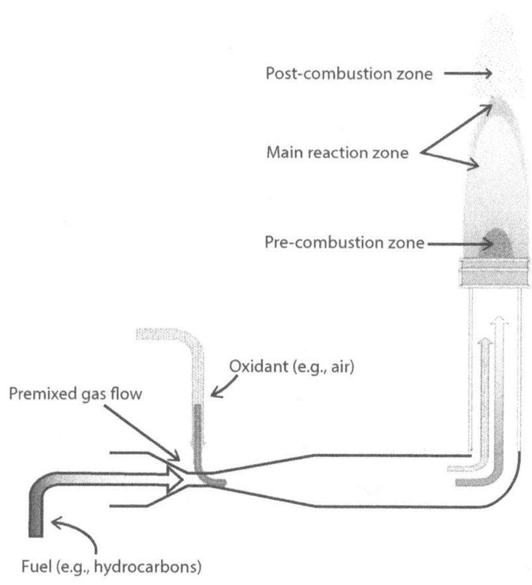

Figure 3.2 Different zones in a laminar flame profile.

free electrons, nitrous oxides as well as carbon-oxygen and oxygen-hydrogen groups. This gas is highly reactive and is called a plasma.

A schematic representation of the reaction, without the formation of nitrous oxides, is shown in Figure 3.3.

Strobel and coauthors [2] postulated that the mechanism of surface oxidation involved polymer radical formation by O and OH, followed by rapid reaction of the polymer radicals with O, OH and O_2.

Studies [8–13] have shown that the flame plasma treatment of polyolefins results in high levels of oxidation with the formation of hydroxyl (OH), carbonyl (COOH) and carboxyl (O=CO) functionalities. Levels of ether (C-O-C) and ester (O=C-O) functional groups are also formed.

The depth of oxidation in flame plasma treatment is 5-10 nm. With flame treated polyolefins, this extensive oxidation is concentrated in a shallow region resulting in a highly wettable surface that is relatively stable upon aging. This is the main reason that flame plasma treatment, when compared to corona treatment, lasts longer.

The species present in the flame plasma are shown in Table 3.4.

These species impart polar groups on the polymer surface. Since H_2O is a polar molecule, by polarizing the polymer surface, van der Waals forces

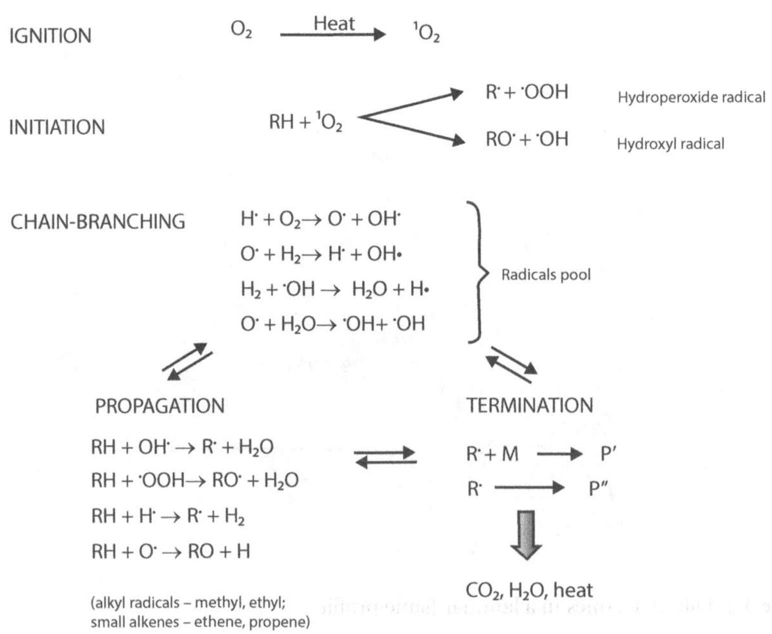

Figure 3.3 Overview of the combustion process.

Table 3.4 Species present in flame plasma.

Electrons, e^-
Oxygen atoms/ions, O, O^-, O^+
Carbonyl, carboxyl, hydroxyl functional groups 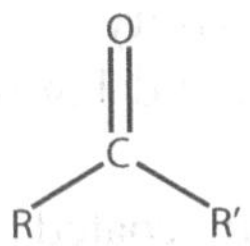 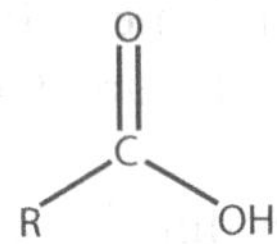 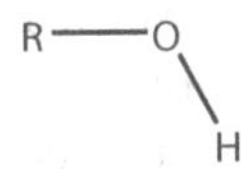
Ether ester functional groups 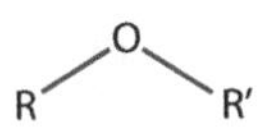 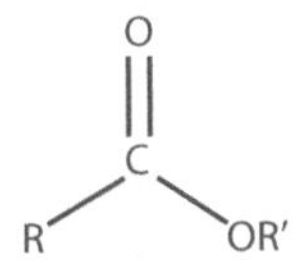
Free radicals containing an unpaired electron

are created resulting in an attractive molecular force between the coating and the polymer surface.

A more complete discussion of the combustion mechanism and formation of the reactive species is given by Glassman and Yetter [9] and Garbassi, *et al.* [10].

3.3 Flame Treatment Equipment

The equipment used for 3D polyolefin parts (complex geometries) includes a conveyor which transports the objects to be treated through the flame, an air/gas mixing station or plasma generator, oxidant, air and gas flow control valves, gas train with safety components and a venturi mixer. The plasma generator provides the flame plasma to the part. Electronic ignition and flame monitoring devices are part of the plasma generator. If precise control of the air/fuel ratio is required because of the nature of the polymer, or for quality control or data acquisition, a flame plasma analyzer may be incorporated into the plasma generator.

A critical component of the system is the burner. Different types of burners are used in flame treating 3D parts: the drilled port burner and the ribbon burner. Depending on the geometry of the part, small nozzle burners may be used, either separately or together with the other types. The conveyor travels at a speed from 3-30 m/min.

The flame geometry is critical in insuring that the surface of part is treated uniformly. Ribbon burners provide an exceptionally uniform flame geometry compared to drilled port burners. For low conveyor speeds, less that 8m/min, a drilled port burner should be satisfactory; for higher speeds, ribbon burners should be used for best flame uniformity.

The different types of burners are shown in Figure 3.4, and Figure 3.5 shows a complete flame treating station.

The system for flame treating products such as thin films, PE coated paperboard, metalized paper and film, and foils is different from treating three

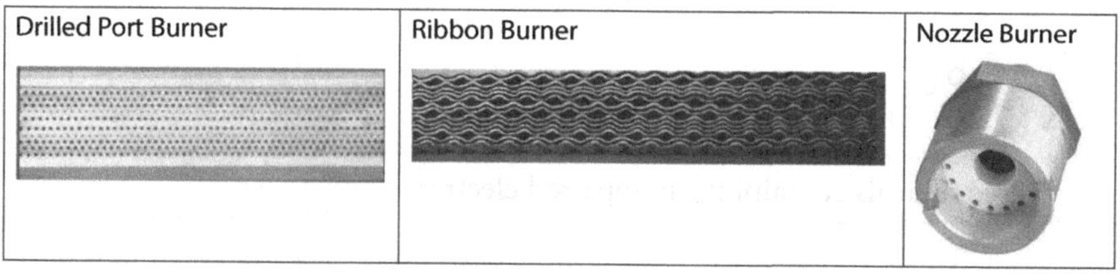

Figure 3.4 Burners used in flame treating 3D geometries.

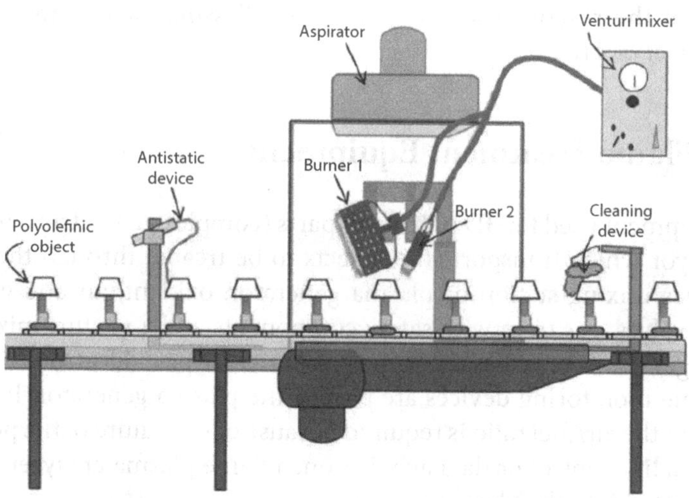

Figure 3.5 Schematic representation of a flame treating station for treating 3D geometries with antistatic and cleaning device to remove dust.

dimensional parts because treatment uniformity is critical. These webs travel at speeds as high as 1500 m/min and as low as 60 m/min during ramp up. The system must be capable of treating the substrate web within this range of web line speeds uniformly, and without thermal degradation at low line speeds.

The system consists of the plasma generator similar to that used for treating 3D objects but uses a separate combustion air blower since these systems must generate orders of magnitude more flame plasma. A flame plasma analyzer is required. An important component is the water cooled back-up roll, together with a nip roll (or rolls). The back-up roll and nip roll (rolls) serve two functions: one the water in the roll, usually at an inlet temperature of 27-30^{0}C, absorbs the heat energy from the combustion of the fuel gas, and two the nip roll (rolls) ensures that the substrate is held tightly to the back-up roll [8]. Note that the sheet has a 180^0 wrap on the back-up roll. The burner should fire vertically down onto the sheet. This makes exhausting the hot products of combustion easier than if the burner were to fire at any other position relative to the sheet and the reactive species are deposited in the most direct pathway to the sheet, see Figure 3.6.

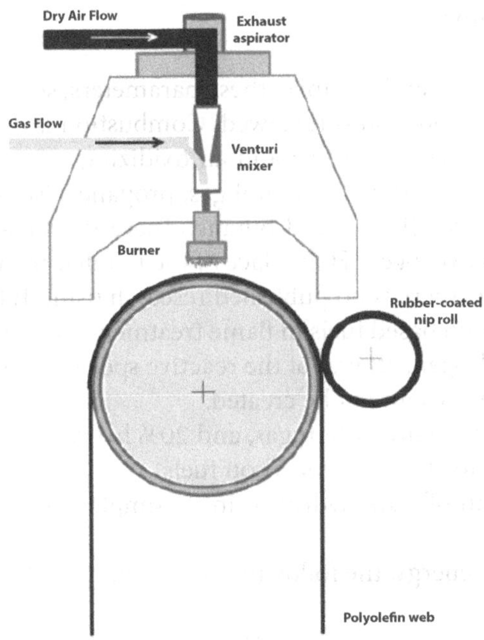

Figure 3.6 Schematic representation of a flame treating station for treating web fed paperboard, PE coated board, flexible films, foils, and metalized films & paper.

It is critical in the flame treatment of films that the substrate does not "flutter", doing so changes the distance of the flame to the polyolefin surface. This will be discussed later in this chapter

3.4 Factors Controlling Flame Plasma Surface Treatment

The combustion of a hydrocarbon fuel under controlled conditions generates a flame plasma which modifies the surface of a polyolefin. The depth of modification is 5 – 10 nanometers (10^{-9} meter). Since it is a surface modification, the bulk properties of the polymer are not affected.

There are five key variables that optimize flame plasma surface treatment of polymer substrates.

- Flame Chemistry (Fuel/Air Ratio)
- Amount of Plasma Generated
- Flame Geometry
- Distance of the Substrate from the Flame
- Dwell Time

In order to control and optimize these parameters, some basic information relating to combustion is reviewed. Combustion is defined as a rapid chemical reaction between a fuel and an oxidizing agent. The fuel is any clean hydrocarbon gas such as natural gas, propane, also described as liquefied petroleum gas (LPG), and butane. There is increasing discussion for using gaseous hydrogen, H_2 in place of the familiar hydrocarbon-based fuel gases. To date, there is no published research using 100% hydrogen to replace hydrocarbon based fuels in flame treatment. Since there are no carbon atoms in hydrogen, several of the reactive species formed with hydrocarbon-based fuels would not be created.

Mixtures of 80% hydrocarbon gas, and 20% hydrogen would behave in a similar manner to 100% hydrocarbon fuels.

The combustion of pure hydrogen in its simplest form is: $2H_2 + O_2 \rightarrow 2H_2O$.

With sufficient energy, the following can occur:

$$H_2 \xrightleftharpoons{\Delta T} 2H$$

Hydroxyl radicals can be formed as follows: H· + O_2 → HO· + O and react with H_2 to form O + H_2 → HO· + H·

How this will affect flame plasma treatment of polyolefins is not known at this time, but the present authors offer the theory that the hydroxyl radical may behave in a manner similar to the hydroxyl radical formed during the combustion reaction of a hydrocarbon gas.

The oxidizing agent is the O_2 present in air. The air for the combustion of the hydrocarbon gas must be supplied to the reaction zone, i.e., burner nozzle, at the required rate. There are two basic types of flames involved in combustion: the diffusion flame (laminar flow) and the premixed (turbulent flow) flame. The diffusion flame is characterized by a gas-rich area into which O_2 must diffuse. The reaction kinetics of diffusion flames is slow compared to premixed flames, which are characterized by proportioning and mixing the air and gas before ignition and combustion take place. Premixed flames consist of an inner or primary cone (pre-combustion zone), a secondary (main reaction zone) and an outer cone, (post-combustion zone). The optimum flame plasma resides in this secondary cone (main reaction zone).

See Figure 3.2.

For natural gas, the following describes the combustion reaction:

$$CH_4 + 2O_2 + 8N_2 \rightarrow CO_2 + 2H_2O + 8N_2 + \textit{flame plasma (reactive species)}$$

The reaction is exothermic, i.e., heat is produced. The ideal air/fuel ratio, i.e., the exact amount of oxygen required to burn the fuel completely is defined as the stoichiometric air/fuel ratio. There is no excess oxygen or fuel remaining after combustion has taken place. A lower air/gas ratio is called sub-stoichiometric, or "rich", containing more fuel than there is oxygen; a higher ratio is "lean", containing excess air, and thus more oxygen needed for the complete combustion of the fuel.

Another way to express air/fuel ratio is the "equivalence ratio".

The equivalence ratio (Φ) of an air/gas mixture is defined as the ratio of the mass of hydrocarbon (fuel) to the mass of oxidizer ratio to the stoichiometric mass of hydrocarbon to the stoichiometric mass of oxidizer ratio.

$$\Phi = \frac{\left(A/F\right)_{stoic}}{A/F} = \frac{F/A}{\left(F/A\right)_{stole}}$$

Fuel rich: $\Phi > 1$ Fuel lean: $\Phi < 1$

A is the mass of air; F is the mass of hydrocarbon fuel
λ is the mass ratio of air/fuel divided by the stoichiometric ratio. $\lambda = 1/\Phi$.

3.4.1 Flame Chemistry

Flame chemistry is determined by the air/fuel ratio. In general, the stoichiometric ratio (air/gas) is *approximately* 10:1 (air to fuel) for natural gas. It is 24:1 for propane (LPG) and 32:1 for butane, on a volumetric basis. Note: Butane is rarely used in most developed countries. Mixtures of propane and butane are used in several Asian counties.

DiGiacomo [16] has shown that the optimum flame chemistry is that which provides an O_2 concentration *in the flame plasma, i.e.*, after the combustion reaction, 0.1% – 0.5% Φ = 0.99 - 0.976). A "lean" flame is one in which there is excess O_2 (Φ <1) A "rich' flame is one with excess fuel (Φ >1). A rich flame is not effective in enhancing or promoting adhesion, because the concentration of reactive species produced is low; similarly a lean flame is also not effective, as the concentration of reactive species is low; due to dilution of the plasma from the excess air in the air/fuel mixture.

Simply setting the optimum air/fuel ratio will not assure optimal surface treatment for the following reasons. In the combustion industry, air and fuel are measured by volume, however, the combustion reaction is based on mass. Controlling air and gas on a volume basis does not take into account the change in O_2 concentration in the air due to the temperature of the combustion air, i.e., density changes, and humidity (water molecules displace a portion of the oxygen molecules in a fixed volume of air). A combustion air blower provides a fixed volume of air regardless of the air density, the O_2 concentration will be affected by temperature and the humidity of the air. Higher temperature air contains less mass of O_2, at the same volume due to the decrease in the density of the air at the higher temperature H_2O molecules will displace some of the oxygen molecules in a given volume of air. If the flame chemistry is optimized on a cool dry day, or on a hot, humid day it will not be at that same optimum ratio.

Similarly, gas composition changes based on the gas field from which the gas is drawn.

Again, optimizing at one gas composition does not mean optimal flame chemistry at a different gas composition. The change in the O_2 concentration with air temperature and humidity is illustrated in Table 3.5. Table 3.6 illustrates the differences in gas composition based on the location of the gas field.

Table 3.5 Effect of temperature & humidity on O_2, and other gases concentration (%).

Temperature (K)	Humidity (%)	O_2	N_2	Ar_2, CO, H_2	H_2O	Density (g/cm³ x 10⁻³)
289	0	20.99	78.03	0.98	0	0.37263
289	80	20.70	79.64	0.97	1.4	0.37067
289	100	20.62	76.67	0.96	1.75	0.37014
305	20	20.79	75.97	0.97	0.95	0.35100
305	80	20.19	75.06	0.90	3.81	0.34719
305	100	19.99	74.32	0.90	4.76	0.35969

Table 3.6 Gas composition (%) from different gas fields.

Component	Alabama	Ohio	Pennsylvania	Alaska
Methane (CH_4)	90.0	94.1	83.4	99.6
Ethane (C_2H_6)	5.0	3.0	16	-----
Propane (C_3H_8)	-----	0.4	-----	-----
Volume of air/ Volume of gas (Stoichiometric)	10.47	10.72	11.70	10.52

Although the differences in oxygen concentration based on temperature and humidity, and gas composition based on the amount of the main hydrocarbon seem relatively small, if one is trying to control the oxygen concentration in the flame plasma to a range of 0.1-0.5% O_2. These seemingly small differences actually have a significant effect on the concentration of oxygen in the flame plasma.

A review by Farris and co workers [8] suggests that temperature and pressure of the ambient air are significant in the flame chemistry. If this air is being used as the oxidant, then we agree; if they are postulating that external air has a significant effect on air/fuel chemistry, then we do not agree. It is the composition of the oxidant that has orders of magnitude more influence over the plasma flame chemistry than ambient conditions.

Figure 3.7 shows the effect of temperature and humidity on λ (1/ϕ), which supports Table 3.5.

The first generation of flame treaters tried to compensate for the changes in flame chemistry due to differences in air temperature, humidity and gas composition using flowmeters. However, flowmeters cannot compensate for these differences since they measure volume, not mass flow. Mass flowmeters, measuring O_2 concentration in the pre-burnt air/fuel mixture, were among other methods used to measure and control the flame chemistry. Unfortunately, none of them were practical from a cost and/or accuracy basis.

It was not until the Flame Plasma Analyzer was developed by a leading combustion equipment manufacturer, in the US, that a cost effective, easy to use, accurate instrument was developed to precisely measure and control the flame chemistry [16].

Figures 3.9 and 3.10 illustrate the functioning of the flame plasma analyzer. Further explanation is given later in this chapter.

Why not use an off-the-shelf, low cost oxygen analyzer instead of the flame plasma analyzer one might ask? The reason is the accuracy of the flame plasma analyzer compared to a commercially available oxygen analyzer.

Figure 3.8 compares the scale for a typical O_2 analyzer and that for the flame plasma analyzer and illustrates the accuracy and precise control provided when using the flame plasma analyzer. The scale of a commercially available oxygen analyzer is shown on the top scale, labeled % oxygen; the scale, labeled plasma value, of the flame plasma analyzer is shown at the bottom. The flame plasma analyzer measures an oxygen concentration of

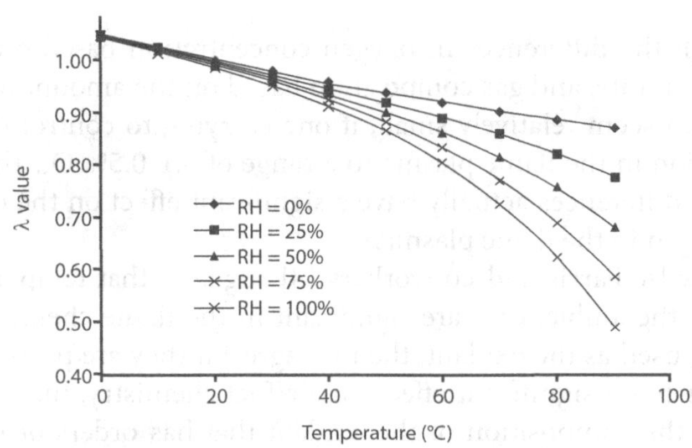

Figure 3.7 Influence of temperature and humidity on λ (1/ϕ), taken from [8].

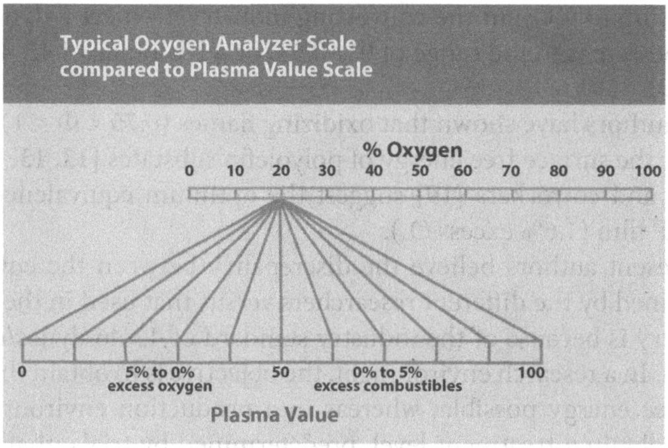

Figure 3.8 Comparison of typical O_2 analyzer scale to plasma value scale.

0-5% and expands this range to a scale of 0-100 plasma value, where a plasma value equal to 50 indicates 0% oxygen and 0% fuel in the post-combustion gases, or products of combustion (POCs), i.e., a stoichiometric air/fuel ratio. A plasma value of 0 indicates 5% excess oxygen in the post combustion gases. An oxygen concentration of 0.1-0.5% is equivalent to a plasma value of 45-49. Plasma values greater than 50 indicate a gas-rich air/fuel ratio, $\Phi > 1$.

Surface treatment involves a flame plasma at the surface of the substrate to be treated. The approach is to simulate the reaction occurring at the surface by taking a continuous sample of the air/fuel mixture and catalytically initiating the combustion reaction of the sample in a sealed chamber.

A zirconium oxide O_2 sensor measures the precise amount of excess O_2 or the existence of an excess fuel air/gas ratio. Since effective flame treating occurs when the air/fuel ratio is "lean", measuring the amount of excess fuel is not meaningful. One needs to know the amount of excess O_2 present in the post-combustion gases, since these gases contain the reactive species that react with the surface of the polymer.

This can be displayed as % O_2 or a numerical value called Plasma Value or PV.

A PV = 50 means complete combustion, i.e., no excess fuel or O_2 in the flame plasma. Each unit below 50 represents 0.1% excess oxygen. A PV = 45 means there is 0.5% excess O_2 in the flame plasma. A reading above 50 represents excess fuel and is not effective in increasing the surface free energy of the polymer.

The optimum $\%O_2$, in the converting industry, to meet a dyne level of 42-46 dynes/cm is in the range of 01.-0.5% or a PV range of 45-49. ($0.976 < \Phi < 0.99$).

Many authors have shown that oxidizing flames ($0.75 < \Phi < 1$) result in increasing the surface free energy of polyolefin substates [12, 13, 17, 18].

Strobel and coworkers [19] suggest the optimum equivalence ratio is 0.93 for PP film (1.6% excess O_2).

The present authors believe the discrepancy between the equivalence ratio obtained by the different researchers versus that used in the converting industry is because of the industry standard of 42-46 dynes/cm treatment level. In a research environment, the objective is to obtain the highest surface free energy possible; whereas in a production environment, the goal is to obtain a treatment level, predetermined by trials, that provides an acceptable bond.

Using a flame plasma analyzer is an accurate, easy to operate, and reproducible measure of the flame chemistry. The analog output can be used in a control loop for automatic control of the flame chemistry. Or it can be displayed for information and/or to assist in manual control of the flame chemistry. When compared to measuring the actual fuel/air ratio using air and gas flowmeters, this is no compensation for combustion air temperature, and humidity and well as gas composition and temperature. In an industrial environment, this is simply not practical.

The flame plasma analyzer functions as follows and is illustrated in Figure 3.9.

Figure 3.10 shows the internal components of the Analyzer.

A sample of the air/gas mixture is piped to the analyzer cell at a rate of $0.6m^3/hr$ ($2\ ft^3/hr$). The catalytic reactor is maintained at approximately 593°C) The sample is catalytically ignited, combusted and the hot products of combustion i.e., POCs flow past the ceramic/platinum sensor and produce an electrical voltage proportional to the amount of excess O_2 or excess hydrocarbon gas. The signal is amplified, conditioned, i.e., changed in such a way that it will be converted at the next stage in processing, and linearized, i.e., the output signal produced by the sensor and the corresponding nonlinear value of the oxygen concentration produced is mapped. The signal is an analog input to the PLC which through proprietary algorithms utilize an analog PID control loop providing a 4-20 milliamp signal to modulate a bypass combustion air flow control valve to maintain the specified % O_2, or Plasma Valve, regardless of the gas composition, temperature, combustion air temperature or humidity.

A discrete step function occurs when excess hydrocarbon (unburnt fuel), $\Phi > 1$ is present in the POCs. Since it is well known [12, 13, 17–19]

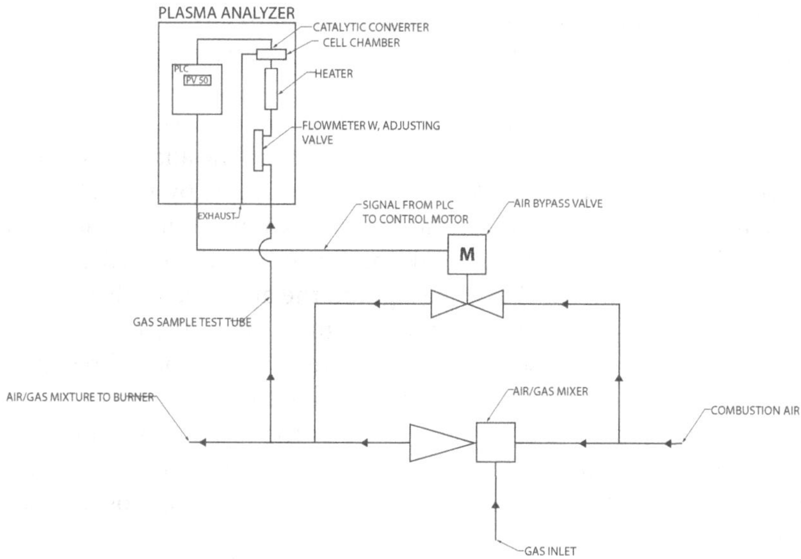

Figure 3.9 Schematic of the flame plasma analyzer & air bypass flow control valve.

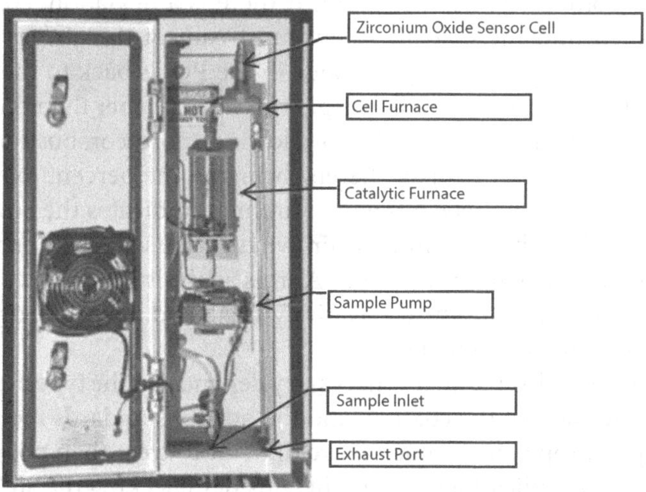

Figure 3.10 Flame plasma analyzer.

that oxidizing flames enhance adhesion, the flame plasma analyzer does not measure excess fuel in the flame plasma. An indication that excess fuel is present is displayed.

Previous versions of the flame plasma analyzer controlled the gas flow to the air/fuel burner mixer. The current version of the analyzer maintains the preset plasma value, i.e., % oxygen in the POCs, by modulating the air bypass valve which introduces air from the combustion air blower into the air/gas mixture downstream of the air/gas mixer. If the analyzer senses that the mixture is richer than the setpoint, the air bypass valve will open to bring the ratio to the setpoint. If the mixture is lean, the valve will close to return the ratio to the setpoint. See Figure 3.9. This control results in a much faster response time and closer control of the air/fuel ratio.

The use of a plasma analyzer ensures that the preset optimum air/fuel ratio, usually determined by trials, is consistently maintained regardless of changes in combustion air temperature, humidity, gas composition or gas temperature.

Once the optimum value for % excess oxygen is determined, the operator inputs the value (setpoint) into the HMI, i.e., human-machine interface. Once the flame is established, a sample of the air/gas mixture is continuously aspirated, i.e., drawn into the analyzer cell chamber, where it is catalytically ignited and burned. The amount of oxygen or fuel is determined and compared to the setpoint. If there is a deviation, combustion air is injected into the oxidant stream upstream of the air/gas mixer in order to bring the percent oxygen value in the POCs back to the setpoint. If the measurement indicates the air/gas mixture is richer than the setpoint value, the air bypass valve will open allowing more combustion air, i.e., oxidant, to flow into the mixer thereby bringing the percent oxygen back to the setpoint. Conversely, if the measurement indicates the air/gas mixture is leaner than the setpoint, the air bypass valve will close bringing the percent oxygen valve to the setpoint. As this analysis is done continuously, deviations from the setpoint are relatively small and have little or no effect on the adhesion enhancement

Whether such close control is necessary depends on the type of substrate, the surface tension of the coating, and the quality standards required. For example, polypropylene is much more difficult to treat than polyethylene. For less critical applications one might consider checking the air/fuel ratio upon loss of surface treatment, from one day to the next. For low speed lines (<30m/min) used for 3D polymers this is cost effective; for web fed polymers, precise control of the air/fuel ratio is required

The effect of air/gas stoichiometric ratio on treatment level (in terms of adhesion strength) is shown in Figure 3.11 for natural gas. This illustrates

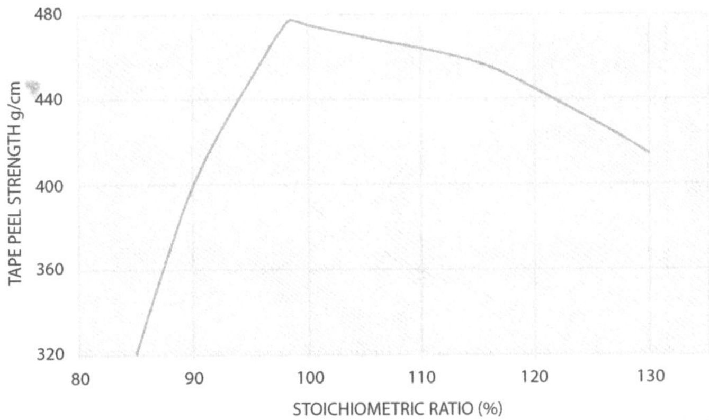

Figure 3.11 Tape peel strength vs. Stoichiometric ratio (%).

that a lean air/gas mixture ($\Phi < 1$) provides the highest treatment level. At a sub-stoichiometric air/fuel ratio, the treatment is not as high as the "ideal" ratio which while lean, depends on the substrate, and line speed. If the air/gas mixture is too lean, the treatment drops off. This shows how critical the air/gas ratio is in regard to optimizing flame surface treatment. The same general curve holds for other fuel gases [20].

Overtreatment using flame plasma is usually accompanied by thermal degradation of the polymer if too much energy is applied. High substrate temperatures can initiate the migration of chemical additives in the polyolefin substrate such as plasticizers, heat & light stabilizers, release agents, anti-statics and UV stabilizers. These compounds can significantly reduce the surface free energy of the substrate.

If the water temperature inside the back-up roll is too high so that not enough heat energy from the combustion process can be absorbed by the water, the internal temperature of the substrate will increase causing migration of the additives in the polyolefin. It can also occur if there is a high concentration of ether groups compared to carbonyl groups in the flame plasma as illustrated in Figure 3.12.

This occurs because there are two competing reactions in the formation of these functional groups during combustion of the hydrocarbon fuel gas. These reactions occur at different rates, the faster reaction creates carbonyl, carboxyl and hydroxyl groups which are polar and attract the polar ink, coating, etc., enhancing adhesion, A second competing reaction which occurs simultaneously, but at a slower reaction rate, converts the carbonyl groups into ether groups which are non-polar and do not attract the polar water molecules from the water-based ink, coating, etc.

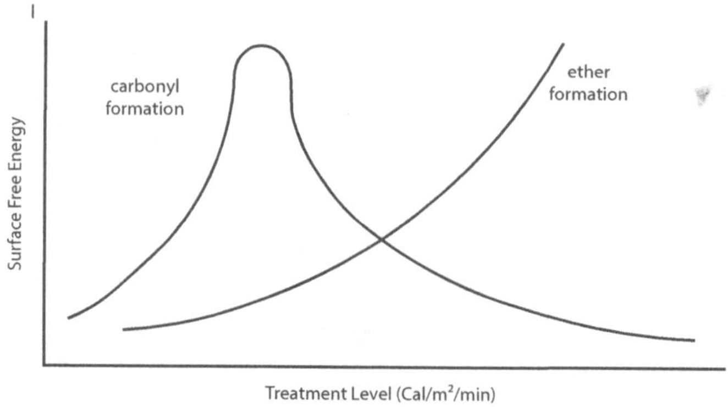

Figure 3.12 Relationship between carbonyl and ether formation vs. treatment level.

At low treatment levels, there are not enough carbonyl groups created to provide polar sites on the surface of the polymer; with increasing energy, more carbonyl groups are created increasing the surface free energy of the polymer. At very high treatment levels the reaction kinetics is reversed and the conversion of the polar carbonyl groups to ether groups occur sand the result is a decrease in surface free energy. Figure 3.12 illustrates the reaction kinetics [21].

In flame plasma surface treatment, thermal degradation usually occurs before the conversion of the carbonyl to ether group happens and is rarely seen.

3.4.2 Amount of Plasma Generated

The amount of plasma generated is determined by the air/gas mixture, combustion air pressure, burner discharge area, and the flame retention feature of the burner. Since line speeds vary, the system must be designed to provide optimum flame treatment over the entire range of process line speeds.

Modern flame treaters control the amount of plasma generated as a function of line speed. The drilled port burner consists of a series of staggered drilled ports, see Figure 3.13. The other type of line burner, which is mostly used for high speed lines >250 ft/min) (>61 m/min) is the ribbon burner, shown in Figure 3.14. Ribbon burners are designed for about an 8:1 turn-down ratio, i.e., the ratio of maximum plasma generated to minimum plasma generated.

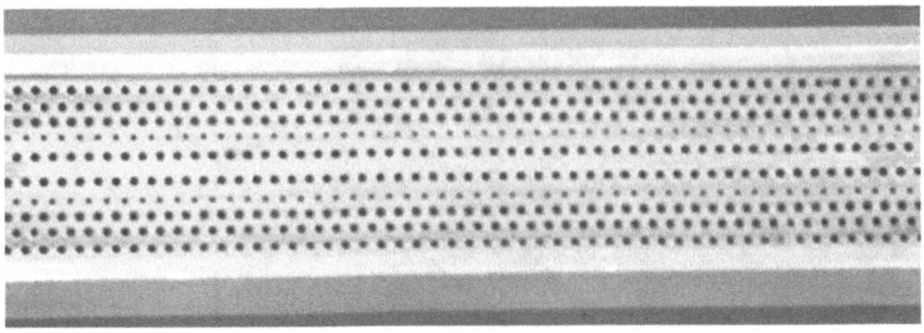

Figure 3.13 Drilled port burner.

Figure 3.14 Ribbon burner.

The effect of the amount of plasma generated (air/gas mixture flow rate) for different line speeds on surface free energy is shown in Figure 3.15. Higher line speeds require more plasma generation to achieve the same level of surface free energy of the substrate [8]. This will be limited by the size of the air/gas mixer, combustion air blower, and the air and gas flow control valves used. It is important that the equipment be sized correctly to ensure the system has sufficient capacity to generate the required amount of flame plasma. The minimum amount of plasma that the system is designed to generate is important. It must be low enough so that at low operating line speeds, or during ramping up to operating line speed, it is low enough so as not to cause thermal degradation of the substrate. Burner turndown, TD, defined as maximum burner output to minimum burner output is determined by the maximum and minimum air and gas flow rates of the air/gas mixer and burner design to result in a stable flame front at all burner firing rates.

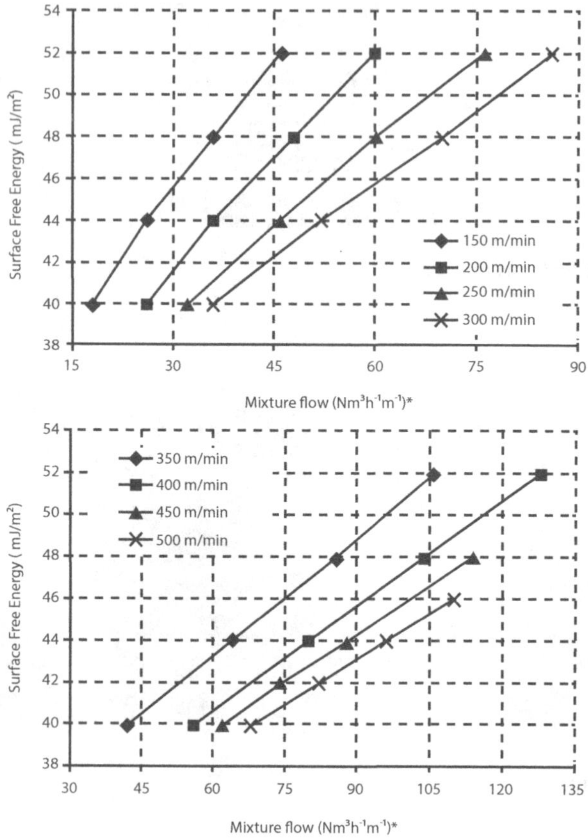

Figure 3.15 Effect of Air and gas mixture flow at low line speeds (top) and at high line speeds (bottom) on surface free energy of biaxially oriented polypropylene (BOPP) film per unit of burner length. Adapted from [8]. **Normal cubic meter/hr equals to one cubic meter at 0^0 C & 1 atm (101.3kPa). Mixture flow Nm^3/ hr/burner flame length.*

3.4.3 Flame Geometry

In order to obtain uniformity of surface treatment, the plasma must react with the surface of the substrate uniformly, without discontinuity. This uniformity of treatment is achieved using a ribbon burner, which provides an extremely uniform flame front geometry. A ribbon burner consists of a series of stainless steel "ribbons" the crimp and angle of which are specifically designed for the type of polyolefin substrate, to ensure uniformity of treatment and totality of treatment.

The drilled port burner if used on high speed web lines can result in "laning" or striations across the width of the sheet. Laning indicates differences in surface free energy across the web width.

A paper from the University of Colorado, Department of Mechanical Engineering, investigating different types of ribbon burners concludes that the 8-port ribbon burner is the most effective for flame treatment [22]. Figure 3.16 shows the ribbon configuration of an 8-port ribbon burner.

Figures 3.17 a & b show the uniform geometry of the flame of an 8-port ribbon burner. Flame uniformity ensures consistent surface activation across the width of the substrate.

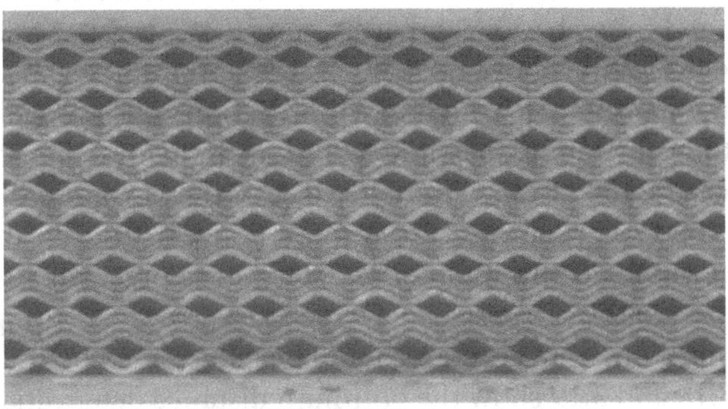

Figure 3.16 8-Port ribbon burner.

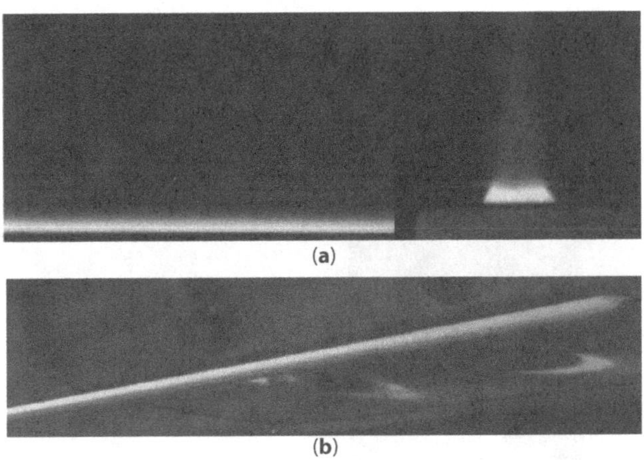

Figure 3.17 (a) Flame uniformity of an 8-port ribbon burner--front & side views. (b) Flame uniformity of an 8-port ribbon burner--front view.

3.4.4 Distance of the Substrate from the Flame

It is generally recognized that the gap between the flame and the substrate is important in determining the extent of the activation realized from the flame plasma treatment.

Many researchers have carried out empirical tests to determine the optimum distance between the flame and the web.

Ayers and Shofner suggested the optimum distance between the flame and the film substrate was 0-6 mm above the main reaction zone [23] Seng and co-workers suggest the optimum distance is 5-10 mm above the main reaction zone for PP [24]. See Figure 3.2 for different reaction zones in a flame. Other authors concluded that the optimum distance between the tip of the flame and the web was less than 10mm [12]. Strobel's work concludes that a distance of 2mm above the flame is the optimum distance [19].

Measuring the distance from the main reaction zone of the flame to the surface of the substrate is not practical in an industrial application. It is easier to measure the distance from the face of the burner to the substrate surface. As the line speed increases or decreases, the height of the flame will change proportionally. In order to determine a practical distance from the burner face to the substrate, an average distance considering the firing rate from minimum to maximum should be determined by trials.

Figures 3.18 and 3.19 show the optimum distance determined by Flynn Burner Corporation for PE coated paperboard and polymer films [16, 20]

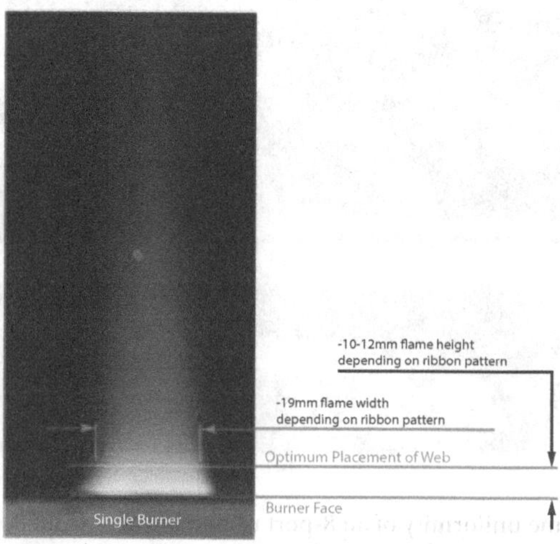

Figure 3.18 Optimum distance of the substrate from the burner face.

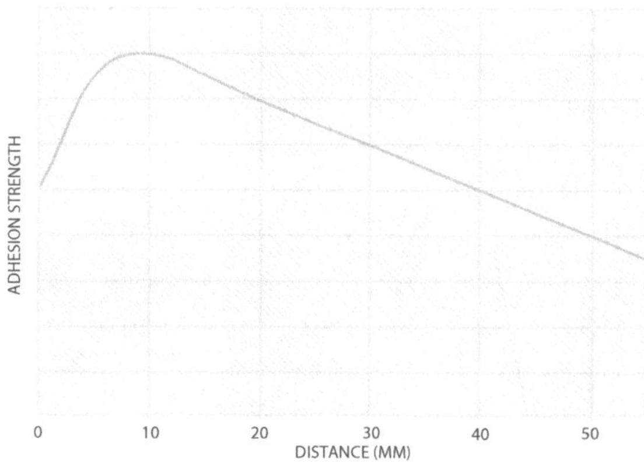

Figure 3.19 Adhesion strength vs. distance of the substrate from the burner face, adapted from [20].

There are highly sophisticated flame treatment systems available that mechanically move the burner closer to, or further away from, the web as a function of line speed, using an algorithm similar to that used for burner output (amount of flame plasma generated) to line speed.

Farris and coworkers [8] have presented the graph in Figure 3.20, which reinforces the fact that there is an optimum distance of the flame from the substrate. The distance is from the burner tip to the substrate. Figure 3.19 is from the burner face to the substrate.

If the position of the substrate is below the optimum distance, the substrate is within that part of the flame which has a negligible concentration of oxidizing species, OH radicals and O atoms. The optimum distance is within the luminous zone of the flame where the concentration of the oxidizing species is highest.

3.4.5 Dwell Time

Dwell time is an important factor in optimizing the surface treatment. The web (or 3D object) must be in contact with the flame plasma for sufficient time for the reaction kinetics to be maximized. If the exposure time to the flame plasma is too short, the adhesion enhancement will be low. If it is too long, the carbonyl-ether functionalities play a role in the decrease in surface free energy as well as resulting in thermal degradation.

Figure 3.21 shows the effect of dwell time on treatment level.

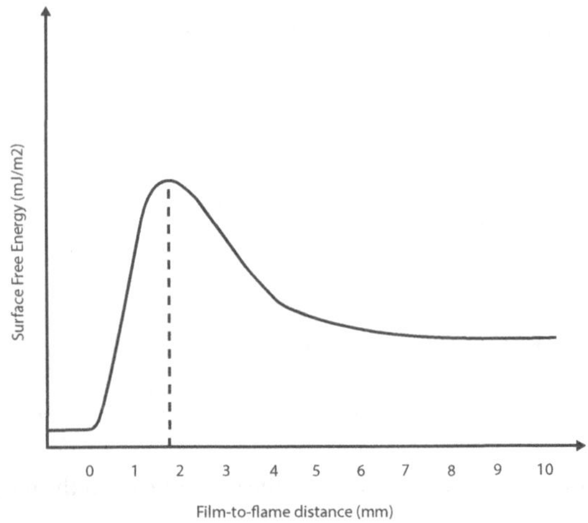

Figure 3.20 Surface free energy vs. film-to-flame distance (adapted from [8]).

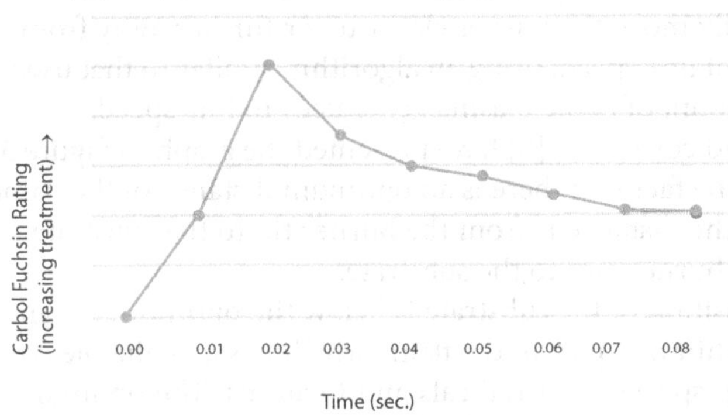

Figure 3.21 Effect of dwell time on treatment level.

A typical flame treater includes:

- Treater Station
- Air/Gas Mixing Station (Plasma Generator)
- PLC Control/Electrical Cabinet

The Treater Station includes the enclosure, exhaust plenum, burner, water cooled back-up roll, and nip roll. Examples of Treater Stations are shown in Figure 3.22. This shows the treater station with the water cooled back-up roll, burner, nip roll, and exhaust plenum.

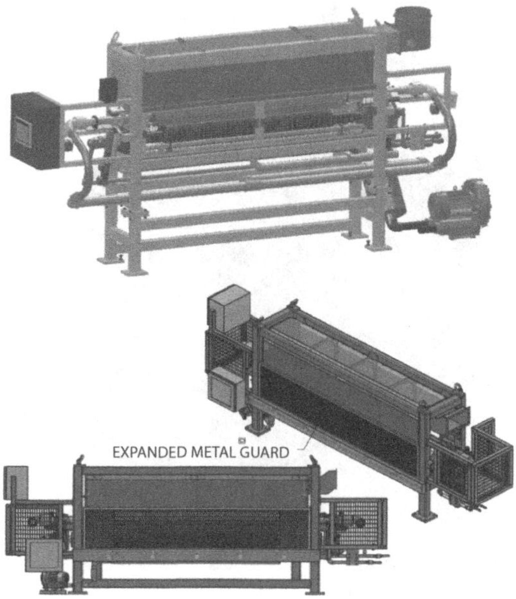

Figure 3.22 Treater stations.

A word about the water cooled back-up roll. The web path is designed so that the web has a 180^0 degree wrap around the roll. See Figure 3.6. A nip roll is provided to ensure that the web is held tight to the roll. The burner is positioned directly above the roll and fires straight down. Approximately 90% of the heat transferred from the combustion process is absorbed by the water flowing through the roll. The rest of that energy is absorbed by the sheet. The water is maintained at a temperature of 30^0C by means of a water temperature control unit. The heat energy from the burner increases the temperature of the substrate by approximately 5.5 ^{0}C. The balance of the heat generated is absorbed by the water flowing through the back-up roll or in the exhaust fan flow from the treater station.

The Air/Gas Mixing Station with the fuel gas train, combustion air components as well as the bypass air control valve is shown in Figure 3.23.

HMI /Electrical Cabinet which contains the PLC, flame control relays, motor starters is shown in Figure 3.24.

Strobel *et al.* [25] have compared the two most commonly used surface enhancement methods: flame plasma surface treatment and corona treatment on PP films. In their studies, AFM (atomic force microscopy, X-ray photoelectron spectroscopy (XPS or ESCA (Electron Spectroscopy for

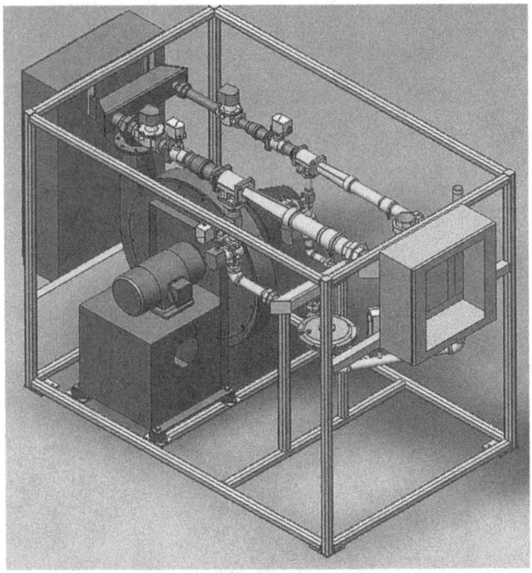

Figure 3.23 Air/gas mixing station.

Figure 3.24 Electrical cabinet.

Chemical Analysis) were used. Effects of flame plasma and corona treatments on modified polypropylene films are shown in Table 3.7. Compared to corona treated polymers, the flame treated films were not affected by water washing and their wettability and O/C ratio did not change, as shown in Table 3.7. The most interesting difference between the two adhesion enhancement methods is the effect of water washing. In the case of flame treated PP, washing had no effect on the wettability or O/C ratio [26]. The authors conclude that long term aging of flame treated PP does not result in a significant loss of wettability or in the extent of oxidation. Comparing flame treated PP to corona treated PP, the corona treated PP showed more

Table 3.7 Surface properties of corona and flame plasma treated polypropylene film before and after water washing.

	XPS O/C atomic ratio		Θ_a (°) Advancing water contact angle		Θ_r (°) Receding water contact angle	
Treatment	**Unwashed**	**Washed**	**Unwashed**	**Washed**	**Unwashed**	**Washed**
None	0.0	0.0	110	109	83	83
Flame (0.03s)	0.07	0.07	97	96	41	42
Flame (0.06s)	0.12	0.12	94	94	29	29
Flame (0.09s)	0.18	0.18	84	85	21	21
For RH = 2-5%						
Corona (0.17 J/cm²)	0.07	0.04	93	97	56	58
Corona (1.7 J/cm²)	0.12	0.06	76	96	51	53
Corona (17 J/cm²)	0.23	0.08	59	88	35	36
For RH = 95-100%						
Corona (0.17 J/cm²)	0.08	0.06	91	97	54	54
Corona (1.7 J/cm²)	0.12	0.07	75	96	50	51
Corona (17 J/cm²)	0.23	0.09	54	90	90	30

significant changes with time. While both treated substrates are quite stable, flame treated PP is definitely more stable than corona treated PP.

Potts and coworkers [27] applied different levels of flame and corona treatments to aluminum foil and then extrusion coated with LDPE (low density polyethylene). In general, the bond between the LDPE and the aluminum foil improved with both flame and corona treatments.

The role of acid-base interactions in LDPE/aluminum foil adhesion has been previously explored [28]. Higher adhesion strengths can be obtained between these two materials by maximizing the basic sites on one surface and the acidic sites on the other. Both flame and corona treatments increased the basic characteristics of the foils. This increase in surface basicity correlated to increased adhesion. In addition, corona, and to a larger extent, flame treatment volatizes the rolling oils on the aluminum foil. Table 3.8 presents contact angles of pH1 and pH 10 solutions on treated foils and shows that the highest level of flame plasma treatment produced the best overall adhesion to LDPE.

Previous work has described a simple technique using the contact angles of basic and acidic solutions on the surface of the substrate being studied. Changes in contact angles provide an indication of surface acidity or basicity [28].

The samples were aged 15 days after the flame treatment in sealed plastic containers which were stored at ambient temperature of 22^0 C and 45% relative humidity.

Table 3.8 Contact angles (degrees) of pH1 and pH 10 solutions on treated aluminum foils.

	Initial, pH 1	**Aged, pH 1**	**Initial, pH 10**	**Aged, pH 10**
No Treatment	80	81	77	78
Corona Low	66	66	66	67
Corona Medium	65	66	72	70
Corona High	61	61	69	71
Flame Low	68	70	71	72
Flame Medium	57	55	59	58
Flame High	42	42	48	46

3.5 Measurement of Treatment Level

Treatment level is measured by several methods. The simplest, quickest and lowest cost is to determine the surface free energy of the plastic substrate using distilled water. If after treatment, the water "wets out", i.e., lays flat on the surface for 3 seconds, the surface has been treated to a level of approximately 72 dynes/cm. A dyne level of 72 is the maximum treatment possible, in most instances, a treatment level this high is usually not required, as well as being difficult to achieve for most polyolefins, with current techniques

A more practical but also quick, simple and inexpensive approach is to use dyne level test solutions. These are mixtures of two chemicals, formamide and cellosolve. Different concentrations of each result in solutions with different surface tensions.

Available surface tensions range from 30 dynes/cm to 72 dynes/cm, in 2 dyne/cm increments

ASTM standard D2578 -17 [14], equivalent to ISO 8296-2003 [15], describes the test method. It is generally as follows:

- Wet the tip of a cotton applicator with one of the formamide/cellosolve mixtures. They will have a designated dyne level surface tension, i.e., 38, 40, 58, etc.
- Spread the liquid over the surface
- Note the time required for a continuous film of liquid to break into droplets. A continuous film should hold for 2 seconds or longer for the surface free energy of the polymer substrate to be equivalent to the surface tension of the test solution
- If the continuous film holds for more than 2 seconds, proceed to the next highest level solution, if not use the next lowest level.
- Repeat this procedure until the film is continuous for 2 seconds. The surface free energy of the substrate is equivalent to the surface tension of the test solution.

Note: A new applicator must be used for each test and the test solutions have a specific shelf-life and should not be used after the expiration date.

Another test incorporates the use of a Mayer Rod, a stainless steel rod tightly wrapped with stainless steel wires of different gauges. See Figure 3.25 for a typical Mayer rod. This test is used for flat polyolefin sheets. The test procedure is as follows:

- Deposit 2-3 drops of dyne level test solution on a flat, level surface.
- Place the Mayer rod on the polyolefin surface near and lengthwise to the series of drops. Carefully pull down the Mayer rod into the solution.
- With a smooth continuous movement, move the Mayer rod so as to lay down a thin continuous layer of liquid. Several solutions can be applied at the same time for a simultaneous drawdown as long as there is no mixing of the different solutions.
- Observe the film(s) until the continuous film breaks into droplets. Record which solution comes nearest to wetting (laying flat) on the substrate surface for a period of 2 seconds. Repeat this test until a solution is found that meets the 2 second standard. Carefully clean and dry the Mayer rod between each test.

The Tape Peel test, described by ASTM standard D2141, measures the amount of force required to peel a pressure-sensitive tape from a polyolefin surface. The ability to obtain an acceptable bond between the polymer surface and the coating or adhesive depends on the level of treatment, i.e., the

Figure 3.25 Typical Mayer rod.

dyne level. The force required to peel a pressure- sensitive tape is directly related to the degree of adhesion.

The Carbol Fuchsin Dye test determines the degree to which the dye will stain the surface of a treated substrate. Once a standard color intensity (based on the dye penetration) is determined to indicate that the substrate is treated sufficiently high enough to promote acceptable adhesion, subsequent samples are compared to the color intensity of this standard.

More sophisticated tests use a Goniometer, which measures the contact angle of water on the surface of the treated polymer. The lower the angle, the higher the treatment level.

The use of ESCA, (electron spectroscopy for chemical analysis), XPS, (X-ray photoelectron spectroscopy), and SSIMS, (static secondary ion mass spectroscopy), are highly sophisticated analytical techniques used to determine the degree of surface oxidation and are yet another means to measure the level of the surface modification.

The analytical equipment required to perform contact angle measurements and/or to measure the amount of surface oxidation are expensive and the testing time is drawn out. Most manufacturing operations use a dyne liquid or dyne pens to measure the degree of surface modification after surface treatment.

3.6 Safety and Other Considerations

People unfamiliar with gas combustion systems are concerned about safety, as they should be. Industrial combustion systems have been utilized since the advent of significant gas usage for industrial processes beginning in the 1900s.

These early systems were crude and were not supplied with the array of redundant safety controls available with current day systems. An example of a redundant control is the requirement that two fuel gas safety shut off valves piped in series be installed in the gas supply to the burner.

Modern industrial combustion systems are designed to national standards specified by the governing body in the country the system will be installed in. In the United States and South America, National Fire Protection Association, NFPA 86-2023 Standard for Ovens and Furnaces provides recommended practices and guidelines, as well as engineering data related to combustion of gaseous fuels.

In the European Union (EU), the standards are harmonized for the members of the European Economic Community (EEC) and known as Europäische Norm ("European Norm" abbreviated ENs. Although similar to US standards, there are differences.

These standards include the following:

- High & Low Gas Pressure Switches - If the gas pressure falls below, or increases above, the set-point, the treater cannot be started. If during operation, either of these, or both pressures are outside the set-points, the treater instantaneously shuts down.
- Low Combustion Air Pressure Switch - If the combustion air pressure, at any time, falls below the operating set-point, the system cannot be started or will shut down if operating.
- Exhaust Fan Pressure Switch - The treater can only be started if the exhaust fan is running. If it fails during operation, the treater is shut down.
- Two normally closed gas shut-off valves with a normally open vent valve in between the two normally closed valves or two normally closed gas shut-off valves with a valve proving system- Two normally closed, zero leakage fuel gas shut-off valves with either a normally open vent valve between the two shut-off valves, or a leak test arrangement, are provided in the fuel gas train. The gas train is designed to the NFPA 86-2023 (Ovens and Furnaces) Standard in the US or EN 746-2 (Industrial Thermoprocessing Equipment) Standard in Europe. Both valves simultaneously and instantaneously shut off the flow of fuel gas to the flame treater when the starting or operating interlocks, such as high and low gas switches, exhaust fan pressure switch, combustion air pressure switch, loss of flame and any customer interlocks are de-energized through the combustion control safety system.
- Electronic Spark Ignition – Provides a 10-15 s trial for ignition of the burner. If the flame is established and verified within this time interval, the fuel gas shut-off valves remain open, if not, they close in less than 1 second.
- Electronic Flame Monitoring – The presence of flame is electronically sensed by either the principle of flame rectification, or ultraviolet (UV) radiation. and not by the heat from the combustion process. In flame rectification, the flame control imposes a low current which in the presence of flame completes the circuit. A UV sensor detects the UV radiation emitted by the flame. If no flame is present, the flame control de-energizes the fuel gas shut-off valves, which close in less than 1 second.

3.7 Adhesion Improvement

Figures 3.26, 3.27, and 3.28 show the enhancement by flame treatment of different polymer substrates as well as on aluminum foil using different analytical techniques. Figure 3.26 shows the effect of different plasma outputs (low, medium, high flame) on relative adhesion strength. The samples were initially tested within a day of treatment. The aged samples were tested after 2 weeks. The samples were stored in sealed containers under constant temperature and humidity conditions. Figure 3.27 shows ToF-SIMS spectra of both untreated and a sample treated 200 times. The presence of hydrocarbon peaks, i.e., C_nH_m due to fragmentation of the LDPE polymer, and the formation of oxygen containing functional groups can be seen in the treated sample [28]. In Figure 3.28, XPS spectra for untreated and treated injection molded polypropylene clearly show the increase in binding energy of the treated sample vs. the untreated sample. The higher the binding energy, the better the adhesion.

Tuominen and co-workers [31] investigated the correlation between surface properties and heat sealability and hot tack temperature and strength of flame treated low density polyethylene coated paperboard. It was observed that for equivalence ratios of 0.86-0.80, i.e., oxygen-rich air/fuel mixtures, flame treatment enhanced the heat sealability of LDPE coated paperboard as well as the hot tack strength, while decreasing the heat sealing temperature.

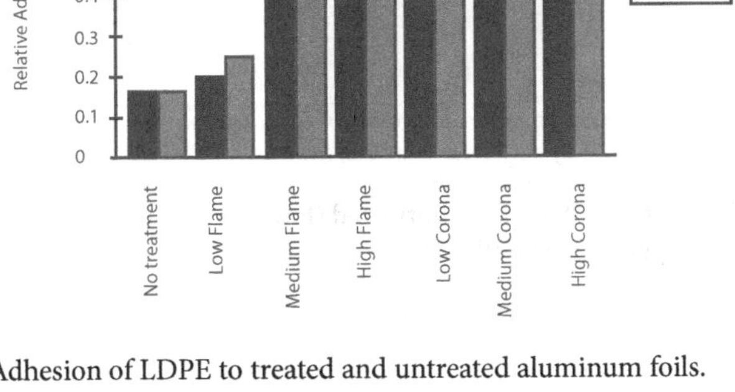

Figure 3.26 Adhesion of LDPE to treated and untreated aluminum foils.

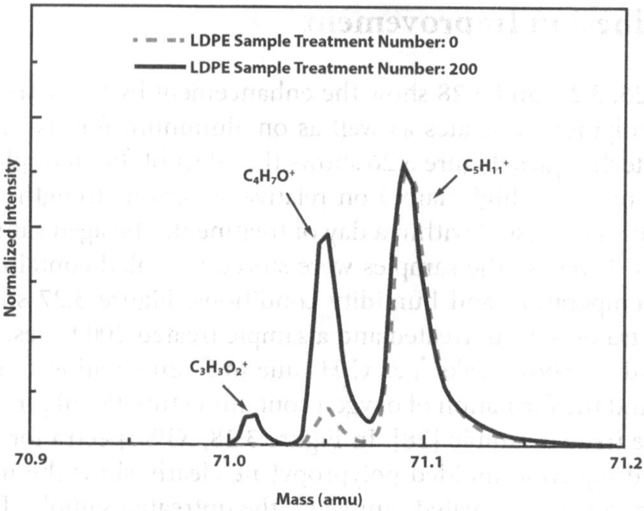

Figure 3.27 Positive ToF-SIMS spectra of untreated and flame treated LDPE [29].

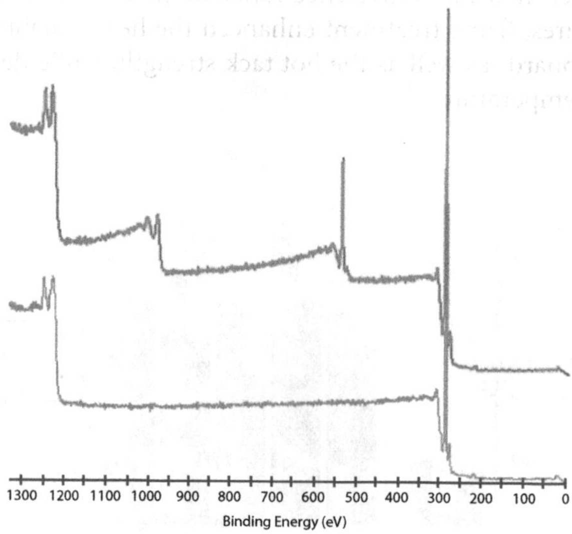

Figure 3.28 XPS spectra for Sample A untreated (lower) and flame treated (upper) Injection Molded Polypropylene [30].

3.8 Summary

Flame plasma treatment of polyolefins is a highly effective method to increase their surface free energy, thereby enhancing adhesion to water based inks, coatings, and adhesives. Use of water-based inks, in place of solvent-based inks, is environmentally beneficial, since the use of solvents releases volatile organic compounds, VOCs, into the atmosphere. Breathing VOCs can irritate the eyes, nose and throat, can cause difficulty breathing and nausea, and can damage the central nervous system and other organs. Some VOCs can cause cancer. The Clean Air Act of 1970, with revisions in 1977 and 1990, regulated the discharge of VOCs as well other substances classified as air pollutants, such as oxides of nitrogen, carbon monoxide, ground level ozone, lead, sulfur dioxide and particulate matter [32].

Thermal incineration is the most effective method to remove VOCs from industrial processes before being exhausted into the atmosphere, but this has a cost in the creation of $CO_{2,}$ which is considered a greenhouse gas that contributes to global warming. New hydrophobic polymers structures are constantly being developed and flame plasma treatment can provide an effective adhesion enhancement.

This chapter has described the reasons for flame plasma treatment and the critical process parameters that affect treatment. How these variables relate to one another and how they can be controlled has been discussed. With the successful development of the flame plasma analyzer, consistent and easy-to-use flame plasma surface treatment is now available. A formerly elusive key variable in flame plasma treatment (air/fuel chemistry) has been identified and can be measured and controlled. Modern flame plasma surface treaters provide high treatment levels reliably, consistently, efficiently and are environmentally acceptable.

References

1. F.K. Hansen, The Measurement of Surface Energy of Polymer by Means of Contact Angles of Liquids on Solid Surfaces: A Short Overview of Frequently Used Methods, Rame-Hart Instrument Co.www.ramehart.com.
2. M. Strobel, M.J., Walzak, J.M., Hill, A., Lin, E. Karbshenski and C.S. Lyons, A comparison of gas phase methods of modifying polymer surfaces, *J. Adhesion Sci. Technol.* 9, 365-383 (1994).
3. M. Strobel, C.S. Lyons and K.L. Mittal (Eds.), *Plasma Surface Modification of Polymers: Relevance to Adhesion*, CRC Press, Boca Raton, FL (1994).

4. K.L. Mittal (Ed.) *Polymer Surface Modification: Relevance to Adhesion,* Vol 5, CRC Press, Boca Raton, FL (2009).
5. K.L. Mittal and T. Bahners (Eds.) *Laser Surface Modification and Adhesion,* Wiley-Scrivener, Beverly, MA (2015).
6. M. Thomas and K.L. Mittal (Eds.) *Atmospheric Pressure Plasma Treatment of Polymers: Relevance to Adhesion*, Wiley-Scrivener, Beverly, MA (2013).
7. K.L. Mittal and W.S. Lei (Eds.) *Laser Technology: Application in Adhesion and Related Areas,* Wiley-Scrivener, Beverly, MA (2018).
8. S. Farris, S. Pozzoli, P. Biagioni, L. Duo, S. Mancinelli and L. Piergiovanni, The fundamentals of flame treatment for the surface activation of polyolefin polymers-A review, *Polymer* 51, 3591-3605 (2010).
9. L. Glassman and R. Yetter, *Combustion,* 4th Edition, Academic Press, San Diego, CA, USA (2008).
10. F. Garbassi, E. Occhiello, F. Polato and A. Brown, Surface effect of flame treatments on polypropylene, *J. Mater. Sci.*, 22,1450-1456 (1998).
11. D. Briggs, D.M. Brewis and M.B. Konieczko, X-ray photoelectron spectroscopy studies of polymer surfaces, *J. Mater. Sci.*,14,1344-1354 (1979).
12. J.G. Dillard, T.F. Conner, C.E. Burtoff, A.J. Cosentino, R.I. Cline and G.M. Maclver, Surface properties and adhesion of flame treated sheet molded compound (SMC), *J. Adhesion*, 26,181-189 (1988).
13. E. Sheng, L. Sutherland, D.M. Brewis and R.J. Heath, An X-ray photoelectron spectroscopy study of polypropylene, *Appl. Surface Sci.*, 26, 249-254 (1999).
14. Standard test method for wetting tension of polyethylene and polypropylene films. ASTM D 2578-17 (2017).
15. Plastics-film and sheeting--determination of wetting tension. ISO 8296-2003 Revised (2018).
16. J. DiGiacomo, Flame plasma treatment, in *Extrusion Coating Manual,* W. Durling (Ed.) 5th Edition, pp.111-117, TAPPPI Press, Atlanta, GA (2017).
17. I. Sutherland, D.M. Brewis, R.J. Heath and E. Sheng, Modification of polypropylene by flame treatment, *Surface Interface Anal.*17, 507-510 (1994).
18. D.M. Brewis, Pretreatments of hydrocarbon and fluorocarbon polymers, *J. Adhesion* 37, 97-107 (1992).
19. M. Stobel, M. Branch, M. Ulsh, R.S. Kapaun, S. Kirk and C.S. Lyons Flame surface modification of polypropylene film, *J. Adhesion Sci. Technol.*,10, 515-539 (1996).
20. J. DiGiacomo, Adhesion promotion using flame plasma pre & post treatment a tutorial, Proc. Annual Meeting of The Society of Plastics Engineers ANTEC (2020).
21. Private communication, T. Bezigian, Q&A PLC (1988).
22. W.T.Ramm, Optical spectroscopic diagnostics of a pre-mixed methane air flame impinging on a rotating cylinder, Master's Thesis, University of Colorado, Boulder, CO (2000).
23. R.I. Ayers and D.L. Shofner, Preparing polyolefin surfaces for ink and adhesives, *J. Soc. Plastics Engineers*, 28, 51-55 (1972).

24. E. Sheng, L. Sutherland, D.M. Brewis, R.J. Heath and R.H. Bradley. Studies of polyethylene modified by flame treatment, *J. Mater. Chem.* 4, 487-490 (1994).
25. M. Strobel, N. Sullivan, M.C. Branch, J. Park, M. Ulsh, R.S. Kapaun and B. Leys Surface modification of polypropylene film using N_2O containing flames. *J. Adhesion Sci. Technol.*,14,1243-1264 (2000).
26. A.P. Pipers and R.J. Meier, Adhesion behavior of polypropylene films after flame treatment determined by XPS (ESCA) spectral analysis. *Electron Spectroscopy Related Phenomena*, 121, 299-313 (2001).
27. M.W. Potts, M.H. Hansen, B.T. Kuetttel and J.D. Goins, Effect of corona and flame treatments on extrusion coating performance properties. Proc. Annual Conference of TAPPI Polymers, Laminations and Coating Meeting (1993).
28. M.H. Hansen, M.F. Finlayson, J.D. Goins and M. Castille, Role of corona discharge on improving polyethylene/aluminum adhesion-an acid/base perspective. Proc. Annual Conference of TAPPI Polymers, Laminations and Coating Meeting (1993).
29. L. Mazzola and A. Cusma, Flame treatment of polymeric materials--Relevance to adhesion: A critical review. *Rev. Adhesion Adhesives*, 4, 433-466 (2014).
30. Private communication, D. F. Williams, M-L. Abel, E. Grant and J. F. Watts, The flame treatment of injection molded polypropylene: New insights on an established process (1998).
31. M. Tuominen, M. Ek, P. Saloranta, M. Toivakka, and J. Kuusipalo, The effect of flame treatment on surface properties and heat sealability of low density polyethylene coating. *Packaging Sci. Technol.* 26, 201-214 (2013).
32. Title 42 U.S. Code 85, Section 7408-Air quality criteria and control techniques (1977).

4

Vacuum UV (VUV) Photo-Oxidation of Polymer Surfaces to Enhance Adhesion

Gerald A. Takacs* and Massoud J. Miri

School of Chemistry and Materials Science, Rochester Institute of Technology, Rochester, New York, USA

Abstract
A wide variety of vacuum UV (VUV) photo-oxidation techniques, with wavelengths less than 180 nm, are reviewed focusing on increasing the hydrophilicity and enhancing the adhesion to polymer surfaces, such as fluorine-containing polymers, Nafion®, polyimides, polyethylene, polystyrene, cyclo-olefin polymers, poly(ethylene terephthalate), polybenzimidazole, polypropylene, poly(ethylene 2,6-naphthalate), polyethersulfone, polyetherimide, and sustainable polymers.

Keywords: Vacuum UV, photo-oxidation, adhesion, polymers, plasmas

4.1 Introduction

Polymer surfaces are often not very reactive when they come in contact with other materials. Therefore, there is a need to modify their surfaces to enhance adhesion without altering their bulk properties. There are many wet (Table 4.1), dry (Table 4.2), and combination of wet and dry approaches to physically and chemically modify polymer surfaces. To minimize liquid waste, dry surface modification techniques (Table 4.2) using plasma and radiation sources in the presence of oxygen are often employed to increase the concentration of oxygen and wettability of the surface resulting in enhanced adhesion [1–5]. This Chapter reviews VUV surface photo-oxidation of polymers to increase adhesion.

**Corresponding author*: gatsch@rit.edu

K.L. Mittal and A.N. Netravali (eds.) Polymer Surface Modification to Enhance Adhesion: Techniques and Applications, (119–154) © 2024 Scrivener Publishing LLC

Table 4.1 Wet techniques for surface modification (Courtesy of Dr. Kash Mittal).

Wet techniques
• Chemical reactions: oxidation, sulfonation, halogenation, ozonation, amination
• Grafting
• Selective etching
• Deposition of Langmuir-Blodgett monolayers
• Deposition of coupling agents, e.g., silanes and silane deposition followed by plasma treatment
• Surfactant adsorption
• Photochemical using photoactive compounds
• Solvent surface swelling
• Dissolution (proper second phase in contact which can dissove undesirable low molecular weight materials)
• Microbial

Table 4.2 Dry techniques for surface modification (Courtesy of Dr. Kash Mittal).

Dry techniques
• Plasma (low pressure, atmospheric): radiofrequency (RF) and microwave (MW)
• Corona
• Flame
• Ultraviolet (UV)
• UV sensitization (photo-activation)
• Ozone
• UV/ozone
• X-ray
• Laser
• Electron
• Ion bombardment (also, ion-assisted reactions)
• Atomic oxygen
• Casting against other materials (e.g., polymers against Pt, Hg)
• Rubbing against other materials

4.2 Vacuum UV Photo-Oxidation Process

4.2.1 VUV Background

Early wavelength measurements were limited to longer than 200 nm because of the photoabsorption of solar radiation by air. Then Schumann constructed the first vacuum spectrometer with a fluorite prism which was later determined by Lyman, using a concave diffraction grating, to have a wavelength limit of 125 nm. Therefore, the region of the electromagnetic spectrum from 125-200 nm is now known as the Schumann region. Since air has strong photoabsorption from 0.2-200 nm, Boyce defined these wavelengths as vacuum ultraviolet (VUV) [6].

Not only is VUV photochemistry important in nature [7], but it plays a significant role in surface modification of materials causing: etching, cross-linking, formation of functional groups for further reaction, and improvement in adhesion [8–20]. Since oxygen is often a component in VUV photochemistry of polymers, this Chapter focuses on VUV photo-oxidation for improving adhesion using emissions from plasmas with wavelengths shorter than 180 nm.

4.2.2 VUV Radiation

The generation of VUV radiation requires the input of energy to form a partially ionized gas called the plasma, which is neutrally charged containing an equal number of positive and negative charges in a given volume that includes some of the following: electrons, positive ions, negative ions, radicals, and ground state and excited atoms and molecules.

4.2.2.1 Emission from Excited Atoms

One method to generate VUV photons is using a microwave (MW) discharge, with a frequency of 2.45 GHz, to excite atoms by coupling microwave power into a MW discharge cavity surrounding the quartz tubing containing the gas which is pumped to low pressure by a vacuum pump [6]. The performance of the MW discharge and resulting surface modification depends on a number of factors including: absorbed MW power; flow rate, pressure, and nature of the gas; distance of sample from cavity; treatment time; and wavelengths of photons emitted.

Table 4.3 lists wavelengths of the first resonance lines emitted from some excited gaseous atoms when at low pressure.

Table 4.3 Resonance emission wavelength lines for gaseous atoms [11].

Gas	Wavelength lines, nm
He	53.7, 58.4
Ne	73.6, 74.4
Ar	104.8, 106.7
Kr	116.5, 123.6
D	121.5
H	121.6
Xe	129.6, 147.0

For example, low-pressure helium or argon MW plasmas, (1.3 - 9.3) x 10^1 Pa, operating at a frequency of 2.45 GHz and absorbed power of 28-38 W were used to modify polymer surfaces located downstream from the plasma (Figure 4.1) [21]. The emission from the discharge is aligned to maximize the interaction of photons with the substrate. The low-pressure MW discharge produces radiation that is primarily a line source due to emission from excited rare gas (Rg*) atoms, as shown in equation (4.1).

$$Rg^* \rightarrow Rg + h\nu \tag{4.1}$$

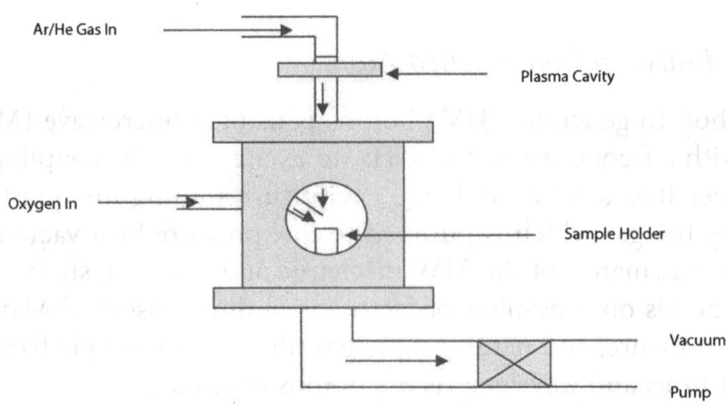

Figure 4.1 Schematic of downstream low-pressure microwave plasma VUV photo-oxidation [21].

Neutral atomic resonance lines, arising from $^3P_1 \rightarrow {}^1S_0$ and $^3P_2 \rightarrow {}^1S_0$ transitions, occur at 53.7 and 58.4 nm for He, and 104.8 and 106.7 nm for Ar [6]. At the distance the sample was placed from the discharge (23.8 cm) [21], deactivation and recombination processes involving charged particles and metastables, that occur via homogeneous and heterogeneous collisions in transit to the substrate, make only negligible contribution and thus the interaction of VUV photons with the substrate is maximized [22, 23].

Polymers treated with VUV radiation are oxidized when exposed to ambient air and reactions at the surface may change with time. Therefore, to study VUV photo-oxidation treatment, oxygen was introduced into the vacuum system about 3 cm above the sample (Figure 4.1) with the helium/argon and oxygen flow rates kept at 50 and 1-40 sccm, respectively [21].

For microwave excitation, emission continua from the rare gas excimer (Rg_2^*), that are formed by the pressure-sensitive reaction (4.2), have not been observed for He at pressures less than 5.3 x 10^3 Pa and have only been found for Ar at pressures greater than 1.7 x 10^4 Pa [24].

$$Rg^* + Rg + M \rightarrow Rg_2^* + M \tag{4.2}$$

A similar VUV photo-oxidation technique has been used with a 13.56 MHz radio- frequency (RF) discharge of Ar, Kr, or Xe where the polymer sample is contained in a separate glass chamber connected to the discharge chamber by a LiF window having a cut-off wavelength below 105 nm [25].

4.2.2.2 Emission from High Pressure Rare Gas Plasmas

At high pressures, where the excimers, Rg_2^*, are formed by reaction (4.2), a continuum of VUV radiation is emitted by reaction (4.3), having the emission maxima listed in Table 4.4, and are commercially available as excilamps [26–28].

$$Rg_2^* \rightarrow 2Rg + h\nu \tag{4.3}$$

One way to produce the excimer is to use a high pressure rotating arc plasma in a magnetic field where a 110 V potential is applied to a sharpened graphite rod cathode that is positioned coaxially inside a grounded graphite anode tube (2.5 cm long, 1.9 cm o.d, 1.4 cm i.d.) (Figure 4.2) [29, 30]. The electrode assembly is mounted in a Pyrex vacuum chamber (15 cm diameter, 30 cm high) and filled with argon or helium at a pressure of (5.3–6.7) × 10^4 Pa. At these pressures, argon and helium plasmas emit strongly

Table 4.4 Emission maximum wavelength for rare gas excimers, Rg_2^*, [26–28].

Rg_2^*	Emission maximum wavelength, nm
He_2^*	74
Ne_2^*	83
Ar_2^*	126
Kr_2^*	146
Xe_2^*	172

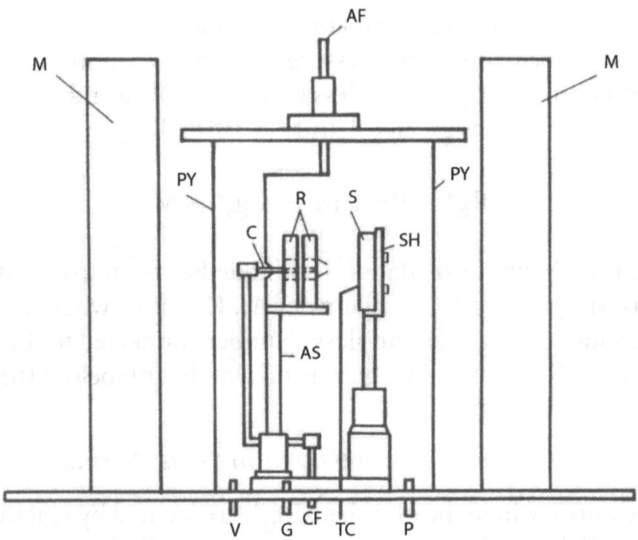

Figure 4.2 Schematic diagram of high-pressure rotating arc plasma. (A) anode, (AF) anode feed-through, (AS) anode support, (C) cathode, (CF) cathode feed-through, (G) gas inlet, (M) external magnetic coils, (P) pressure gauge, (PY) pyrex glass pipe, (R) anode ring support, (S) substrate, (SH) substrate holder, (TC) thermocouple, and (V) vacuum pump with throttle valve [29].

in the VUV, especially due to emission from the Ar_2^* and He_2^* excimers producing a continuum from 105-155 nm and 58-110 nm, respectively, that are superimposed on the resonance lines for neutral excited Ar and He atoms which occur at 104.8 and 106.7 nm, and 53.7 and 58.4 nm, respectively [6]. The singly ionized species of Ar and He are reported to emit at ca. 92.0 and 93.2 nm, and 30.4 nm, respectively [6]. The arc current was selected between 3-7 A. To form a homogeneous source of radiation at the polymer surface, the arc was made to rotate (130 Hz) at the edge of the anode by the application of a magnetic field (9.4 x 10^{-3} T). The substrate was located 5 cm from the arc and its temperature was monitored with a thermocouple [29–31].

4.2.2.3 *Emission from Rare-Gas Halides and Halogen Dimers*

Different VUV emission spectra can also be obtained from rare-gas halide exciplexes and halogen dimers having unique wavelengths of emission, e.g., F_2^* (158 nm), ArBr* (165 nm), and ArCl* (175 nm) [27, 32].

4.2.2.4 *Other VUV Radiation Sources*

Atmospheric plasma jets of Ar [33], He [34], and air have been used to generate VUV radiation to modify polymers with down to sub-110 nm wavelengths [35] while a synchrotron source produced a broadband of non-monochromatic radiation down to ca. 60 nm [36, 37]. Also, equipment imitating near-Earth space produces photons with wavelengths ranging from 90 to 115 nm [38, 39].

4.2.3 VUV Optical Filters

To vary the VUV photon energy to which polymers are exposed, a variety of filters with different cut-off wavelengths (Table 4.5) may be placed in

Table 4.5 Cut-off wavelengths for crystalline materials [6].

Material	Cut-off wavelength, nm
LiF	105
MgF_2	112
CaF_2	122
SrF_2	128
NaF	132
BaF_2	134
Sapphire	141
Fused Silica	160

front and in contact with the polymer samples. These optical filters prevent ions and metastables from interacting with the polymer substrate and only allow photons with wavelengths longer than the cut-off wavelengths to modify the polymer surface.

4.2.4 Penetration Depths of VUV Radiation with Polymers

The high VUV photoabsorption coefficients of polymers (10^4-10^5 cm^{-1}) [40] provide a penetration depth not exceeding a few hundreds of nm, hence resulting in surface modification without altering the bulk properties. Absorption of the energetic VUV photons may cause: bond breakage releasing molecular fragments (etching), cross-linking of free radicals on the surface, and formation of functional groups for further reaction, such as with oxygen-containing species for potential adhesion improvement [8–20].

4.2.5 Analytical Methods for Surface Analysis

These surface changes may be measured by using some of the analytical techniques including: weight loss; surface physical properties (Scanning Electron Microscopy (SEM) and Atomic Force Microscopy (AFM)); atomic composition, chemical states, and quantitative analysis of the top 2-10 nm of the surface (X-ray Photoelectron Spectroscopy (XPS)); and wettability (water contact angle (CA)).

4.2.6 VUV Photochemistry of Oxygen

Oxygen molecules absorb photons in the VUV region of the electromagnetic spectrum and photodissociate to ground state, ^3P, and electronically excited, ^1D and ^1S, oxygen atoms with threshold wavelengths of 242.4, 175, and 133.2 nm, respectively [7, 41, 42]. At high pressures, the O atoms may recombine with O_2 molecules involving a third molecule, M, forming ozone, equation (4.4), which may react with the polymer and/or undergo photodissociation forming excited oxygen atoms and molecules [42].

$$O + O_2 + M \rightarrow O_3 + M \tag{4.4}$$

The reaction of O and O_3 with a material can affect its adhesion.

4.2.7 Reactions of O Atoms and Ozone with a Polymer Surface

The primary reaction of O atoms with the hydrogen-containing backbone of a polymer (RH) is hydrogen abstraction, followed by oxidation with O_2 forming C-O and C=O functional groups by the following reactions [43]:

$$O + RH \rightarrow R\bullet + OH\bullet \tag{4.5}$$

$$OH\bullet + RH \rightarrow H_2O + R\bullet \tag{4.6}$$

$$R\bullet + O_2 \rightarrow RO_2\bullet \tag{4.7}$$

$$R\bullet + O \rightarrow RO\bullet \tag{4.8}$$

$$RO_2\bullet + RH \rightarrow ROOH + R\bullet \tag{4.9}$$

$$ROOH \rightarrow RO\bullet + OH\bullet \tag{4.10}$$

The alkoxyl radicals, produced in steps (4.8) and (4.10), may lead to cleavage of the polymer chain to form carbonyl groups, as shown in reaction (4.11), recombine with a free radical to form ethers (4.12), or abstract a hydrogen atom to produce an alcohol in reaction (4.13).

$$R'R''CO\bullet \rightarrow R'\bullet + R''C = O \tag{4.11}$$

$$RO\bullet + R\bullet \rightarrow ROR \tag{4.12}$$

$$RO\bullet + RH \rightarrow ROH + R\bullet \tag{4.13}$$

Oxygen atoms may also add to unsaturated C-C bonds resulting in the formation of triplet biradicals which may then follow one of three pathways: (1) rearrangement to produce a carbonyl-containing compound, (2) ring closure to form an epoxide, or (3) decomposition into free radicals [44]. The O-containing triplet biradical may react with O_2 to form a Criegee-like intermediate which may rearrange to give carbonyl and ester moieties [45, 46].

Similarly, ozone is well known to react with unsaturated carbon-carbon bonds forming a primary ozonide that decomposes into carbonyl-containing compounds and a Criegee intermediate (Figure 4.3) [47, 48].

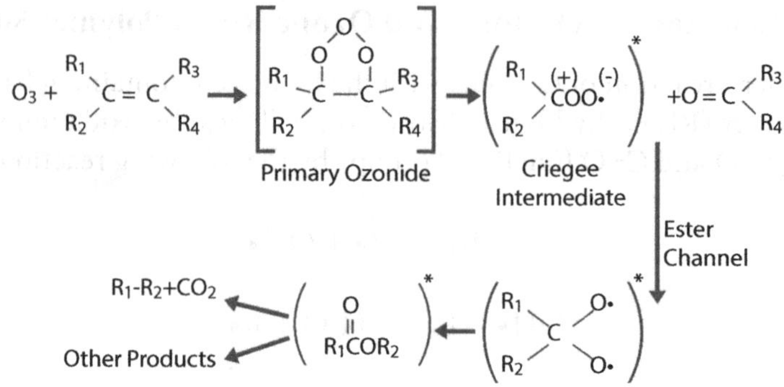

Figure 4.3 Mechanism for ester formation from the reaction of ozone with sp^2-hybridized carbon [47].

4.3 Adhesion to VUV Surface Photo-Oxidized Polymers

4.3.1 Fluorine-Containing Polymers

4.3.1.1 Fluoropolymers

Fluoropolymers, like Teflon® polytetrafluoroethylene (PTFE), -$(C_2F_4)_n$-, poly(tetrafluoroethylene-co-hexafluoropropylene) (FEP) (Figure 4.4) and polytetrafluoroethylene-co-perfluoropropyl vinyl ether (PFA) (Figure 4.5), are well known for their excellent thermal and chemical resistance, and low surface free energies which account for their nonstick behavior. They are extensively used in space applications, protective coatings, biotechnology, and microelectronic packaging [49].

Since photoabsorption by fluoropolymers occurs primarily in the VUV region of the electromagnetic spectrum, VUV photo-oxidation is an excellent means for modifying their surface properties and enhancing the adhesion of deposited metal layers [50]. Their low dielectric constants

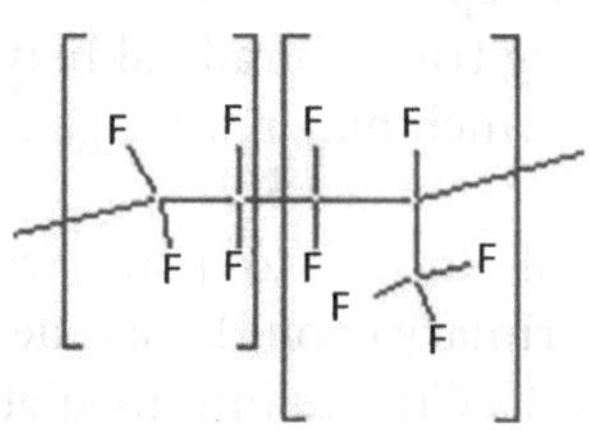

Figure 4.4 Molecular structure of FEP.

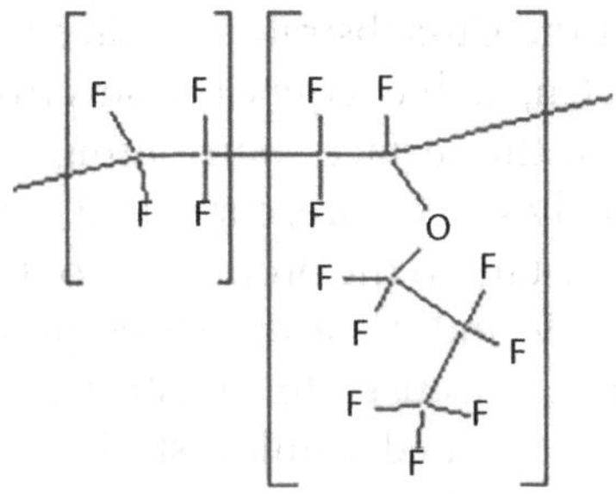

Figure 4.5 Molecular structure of PFA.

make them attractive materials for high-performance electronics applications which would allow conductors, like Cu, to be placed closer together reducing the signal interconnect delay time and producing faster microelectronic devices [51]. The VUV photo-oxidized surfaces of PTFE [13, 14, 21, 31], FEP [52–55], and PFA [55, 56] showed the following changes: (1) surface roughening; (2) defluorination; (3) incorporation of oxygen; (4) improved wettability; and (5) improved adhesion of sputtered Cu at the interface. Adhesion was measured using a tape peel test and recording the % Cu remaining on the treated section of the substrate. For the untreated samples, nearly all of the sputtered Cu remained on the surfaces of FEP and PFA [53, 56] but was removed from PTFE [21] partly due to the larger group electronegativity of CF_3 and C_3F_7O in FEP and PFA, respectively, compared to CF_2 in PTFE [53, 56]. Treatment of PTFE with VUV photo-oxidation resulted in at least 90% of the Cu remaining adhered to the modified surface [21]. For example, Table 4.6 shows the effect of the optical filter cut-off wavelength on the percentage of sputter-deposited copper removed from PTFE after treatment for 15 min with a 7 A rotating He arc plasma [31]. For FEP [53] and PFA [56], at short treatment times, there was good adhesion to the modified surfaces. However, at long treatment times, the sputtered Cu was not detected by XPS on either the Cu-side of the tape or the substrate and the oxygen concentrations were similar to the untreated samples indicating that cohesive failure had occurred within

Table 4.6 Percentage of sputter-deposited copper removed from PTFE surface as a function of optical filter (cut-off wavelength, nm) when treated for 15 min with He rotating arc plasma [31].

No Filter	LiF (105)	CaF_2 (125)	BaF_2 (135)	Crown Glass (278)
< 1	10	70	95	99

the polymer and not at the Cu-substrate interface [53, 56]. In contrast to thermal evaporation of Cu, high energy Cu atoms from sputter deposition produce fluoride ions on the surface contributing to good adhesion with FEP and PFA, causing cohesive failure, compared to the more polycrystalline PTFE which was resistant to the production of fluoride ions [55, 57].

VUV radiation from He and Ar heat-assisted plasmas (HAPs), passing through a MgF_2 substrate, was used to investigate the adhesion property of PTFE to epoxy adhesive-coated stainless steel and isobutylene-isoprene rubber for the evaluation of adhesive-bonding and adhesive-free adhesion, respectively. The surface chemical composition, especially high ratios of oxygen-containing functional groups and C-C crosslinks, significantly contributed to the adhesive-free adhesion property. The Ar-HAP treatment was recommended to improve the adhesive bonding while He-HAP should be used for adhesive-free adhesion [58].

The fluoropolymer (Noves™ 1700) coated on a glass substrate was irradiated at 172 nm through a mask pattern to obtain hydrophilic conditions. Conductive poly(3,4-ethylenedioxythiophene) was printed on to the selective-wetting surfaces resulting in the formation of patterns for use as an organic temperature sensor [59].

VUV radiation increased the hydrophobicity of the catechol-bearing fluoropolymer poly(2-(perfluorooctyl)ethyl acrylate) resulting in good thermal and healing properties which are useful in superamphiphobic fabrics and organic/inorganic hybrid coatings [60].

4.3.1.2 Nafion® (Figure 4.6)

To promote renewable energy sources and reduction in emission levels, fuel cells containing an ionic conducting polymer electrolyte membrane, like Nafion®, which is a copolymer of tetrafluoroethylene and perfluoro[2-(fluorosulfonylethoxy) vinyl]ether, have become of interest. In aqueous solutions, the activated sulfonic acid group allows protons to conduct through the Nafion® membrane by "hopping" from one acid group to the next [61]. To potentially improve conductivity, VUV photo-oxidation of Nafion® downstream from an Ar MW plasma caused defluorination and the incorporation of oxygen onto the surface [62]. The overlapped C 1s XPS spectra for Nafion® samples modified with 0, 0.5 and 1 h VUV photo-oxidation and then grafted with the proton conducting poly(acrylic acid) are shown in Figure 4.7. The peaks at high binding energies arise from carbon bonded to fluorine, while the peaks at low binding energies are due to carbon-carbon and carbon-oxygen bonding environments indicating very effective grafting of poly(acrylic acid) with this treatment [63].

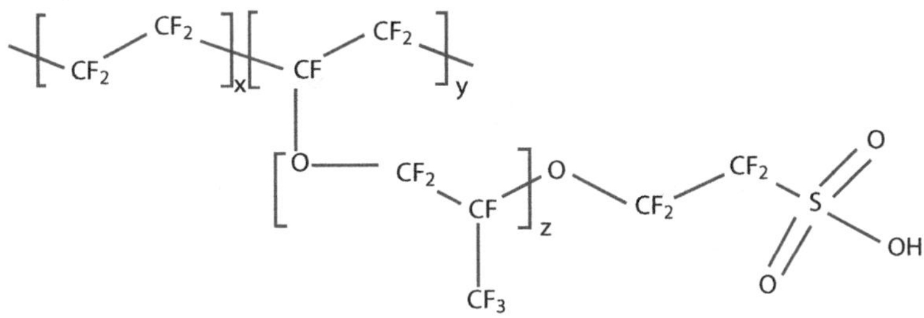

Figure 4.6 Molecular structure of protonated Nafion.

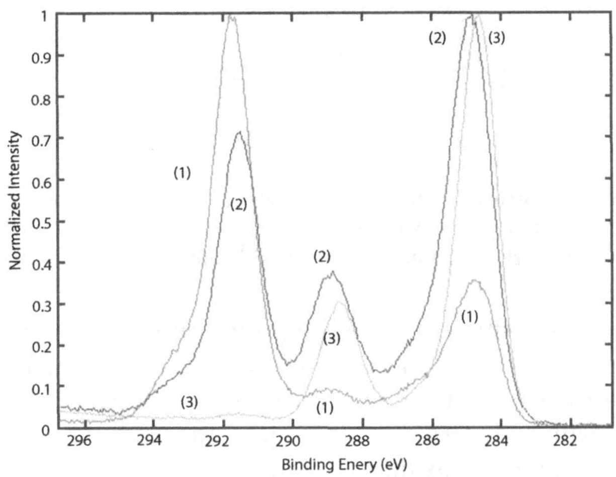

Figure 4.7 Overlapped C 1s XPS spectra for Nafion-117 after graft polymerization of acrylic acid on pre-treated Nafion-117 exposed to: (1) 0 h, (2) 0.5 h and (3) 1 h of VUV photo-oxidation [63].

4.3.2 Polyimides (PIs) (Figure 4.8)

Polyimides (PIs) have a number of attractive thermal, mechanical, and electrical properties making them valuable for the aerospace and microelectronic industries where there are needs for modifying and microstructuring the surface [16, 64]. VUV photo-oxidation of PIs causes etching and incorporation of hydrophilic oxygenated functional groups which can affect adhesion to the surface [13, 16, 29, 30].

In military aircraft and space applications, fibers are often added as reinforcements to improve the mechanical and tribological performance of PIs used in low earth orbit where they are subjected to radiation and atomic oxygen. After exposure to a low-pressure deuterium lamp (115 - 223 nm),

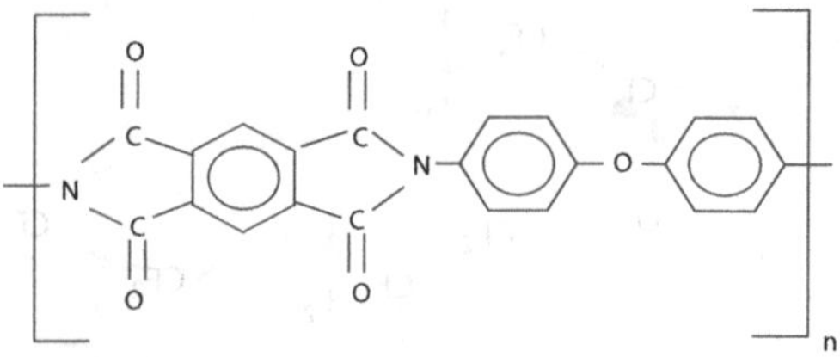

(a)

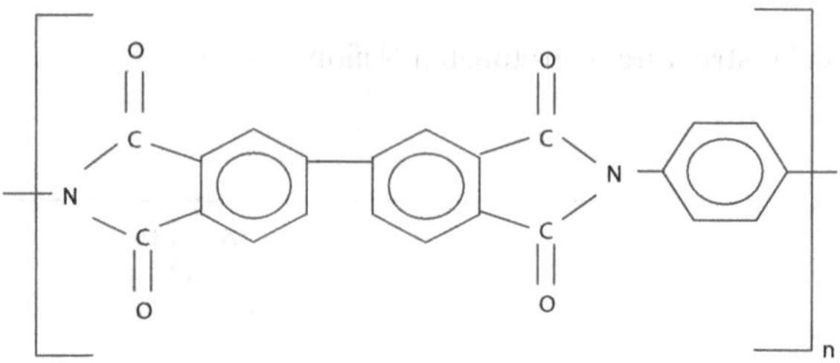

(b)

Figure 4.8 Molecular structures of some polyimides (PIs). (a) Poly(pyromellitic dianhydride-oxydianiline) (PMDA-ODA), Kapton HN polyimide. (b) Poly(biphenyl dianhydride-para-phenylene diamine) (BPDA-PDA), Upilex-S polyimide.

the chemical composition of these composites shows an increase in oxidation affecting their friction and wear behavior. The radiation breaks chains in the composite leading to an increase in the wear rate [65, 66].

A PI film with a cholesteryl-side moiety was oxidized with VUV (172 nm) treatment and the oxidized surface was then treated with photo-functional acid chloride to form ester bonding. This method led to a facile introduction of both photo-isomerizable azobenzene and photo-crosslinkable cinnamate units suggesting an alternative to self-assembled monolayer formation for controlling the alignment of nematic liquid crystals [67].

Incorporating -$C(CF_3)_2$- groups into the backbone of the aromatic PIs reduces the electronic interaction between chromophoric centers in the molecular structure, which produces slightly colored to colorless films. Their transparencies make them optimal for space applications, such as thermal control coatings, low earth orbit (LEO) radiation exposure and solar propelled missions where they are oxidized in the presence of VUV radiation and O atoms [68].

Gaseous ethanol in a nitrogen atmosphere treated with 172 nm wavelength VUV radiation oxidized the PI surface producing bonding sites between PI- poly(etheretherketone) (PEEK) (Figure 4.9) and PI-tin for use in flexible electronic packaging [69].

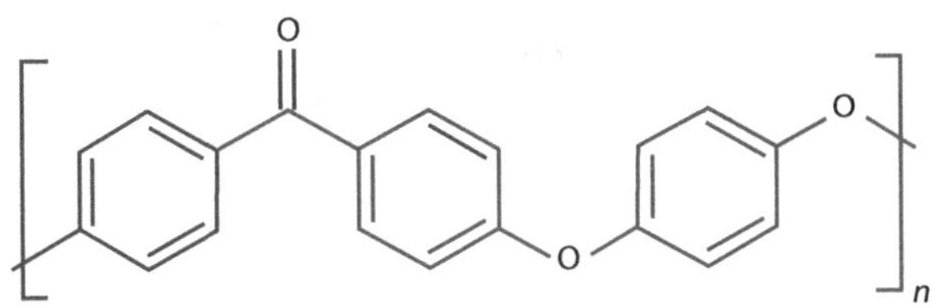

Figure 4.9 Molecular structure of poly(etheretherketone) (PEEK).

A polymer composite consisting of a polyimide track (nuclear) membrane filled with polyhedral oligomeric silsesquioxane exhibited higher resistance than pure PI when exposed to near-Earth vacuum and VUV (90-115 nm) [39].

4.3.3 Polymers and Metals

The adhesion of optical materials for lenses and windows can be achieved by placing silicone oil between the surfaces and irradiating with VUV emission from an Xe_2^* excimer lamp (172 nm) causing bonding due to the transformation of the silicone oil into inorganic glass in the presence of excited oxygen (equation (4.14)) [70, 71].

$$(\text{-O-Si(CH}_2\text{)-O-})_n + O_2 + h\nu \rightarrow SiO_2 + CO_2 + H_2O \quad (4.14)$$

The hydrophilic surfaces of PIs after VUV photo-oxidation [13, 16, 29, 30], when exposed to tetraethoxysilane or 1,3,5,7-tetramethylcyclotetrasiloxane and then irradiated again with 172 nm radiation produced extremely flat SiO_x (oxide nanoskin (ONS)) [72, 73]. Similarly, VUV photodissociation of tetraethoxyorthosilicate with Xe and Ar excimer lamps formed ONS on the polymer substrates without the need for organic solvents [18]. The adhered ONS layer allowed for both the formation of self-assembled monolayers, without any marked changes in surface morphology or bulk properties, and site-selective deposition of metal oxide and metal films for photolithography and micro-patterning [18, 72, 73]. For example, (heptadecafluoro-1,1,2,2-tetrahydrodecyl)trimethoxysilane (FAS) chemisorbed onto the ONS/PI_{ox} substrate and treated with VUV radiation showed strong adhesion of SnO_x microstructures to the SiOH-covered regions of the substrate [71].

As with PI, the hydrophilic surfaces of polystyrene (PS) [18, 20, 69, 70] (Figure 4.10) and poly(methyl methacrylate) (PMMA) [72] after VUV

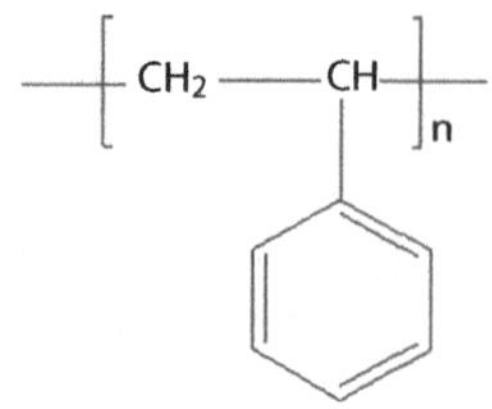

Figure 4.10 Molecular structure of polystyrene (PS).

photo-oxidation when exposed to tetraethoxysilane or 1,3,5,7-tetramethylcyclotetrasiloxane and then irradiated again produced ONS on the polymer substrates [77–79].

In contrast to PS, PMMA and PI, 172 nm irradiation of poly(1-trimethylsilyl-1-propyne) (PTMSP) [80] and poly(dimethylsiloxane) (PDMS) [81, 82], which are very permeable non-porous membranes, alters the highly hydrophilic surfaces to low hydrophilic SiO_2-rich surfaces, changing their permeability property. After VUV photo-oxidation of PDMS, a surface coating of the biomimetic co-polymer containing 2-methacryloxyethylphosphorylcholine (MPS) and 3-methacryloxytriethoxysilane (METESi) was effectively applied, which reduced adsorption of the bovine serum albumin (BSA), leading to micropatterned surfaces for selective protein adsorption and cell adhesion at desired locations on biodevices [83]. Electron paramagnetic resonance (EPR) studies showed that gaseous oxygen plays a key role in the presence of free radicals formed during VUV photolysis of PDMS [84].

VUV photo-oxidation of perhydropolysilazane at 172 nm results in silanol groups being transferred to an SiO_x network [80] forming a flexible oxygen and water-vapor barrier coating on the thermoplastic polymer poly(ethylene terephthalate) (PET) (Figure 4.11). This application may influence the technology of flexible organic light-emitting diodes and solar cells [86].

Synchrotron VUV light (62 nm) and He VUV plasma (50-60 nm) yielded improved line edge roughness of polysilicon organic underlayers during

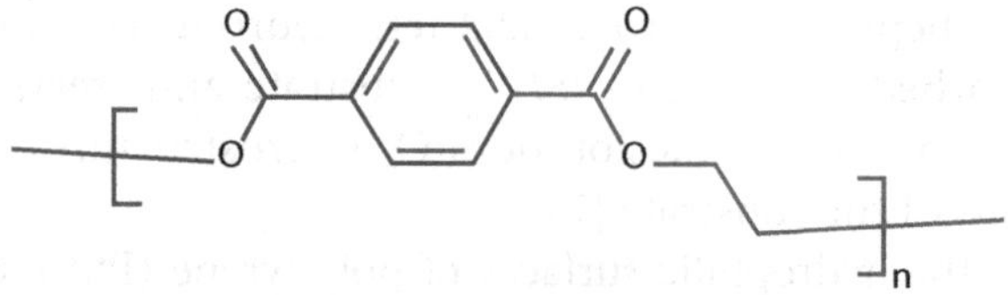

Figure 4.11 Molecular structure of poly(ethylene terephthalate) (PET).

line-space photoresist modifications [37]. Near-Earth space photons, with wavelengths 90-115 nm, were used to study the binding of amorphous and crystalline SiO_2 composites to polyalkaneimide [38].

172 nm VUV has been employed as a highly efficient photo-oxidation technique to assist in the deposition of polydopamine on poly(lactic-co-glycolic acid)-coated Zn for cardiovascular stent applications [87], as well as interaction with gaseous ethanol in a nitrogen atmosphere resulting in an oxidized PI surface producing bonding sites between PI and Sn for use in flexible electronic packaging [69].

4.3.4 Polyethylene (PE), $-(C_2H_4)_n-$

VUV photo-oxidation of low-density (LDPE) [25, 88–90] and high-density polyethylene (HDPE) was investigated and a substantial reduction in irreversibly adsorbed human serum albumin (HSA) was detected due to the surface concentration of polar C=O groups generated by the oxidation [81]. This treatment method suggests a high potential to regulate protein adsorption on polymers for biomedical applications [90].

4.3.5 Polystyrene (PS) (Figure 4.10)

PS is one of the most widely used thermoplastic polymers with applications in the aerospace, bottling, and cell culture industries but it is not a naturally reactive polymer due to lack of active sites on its surface. VUV photo-oxidation increases the hydrophilicity with oxidation [23, 25, 74, 75] resulting in the ability to thinly graft the super water absorbent polymer poly(acrylic acid) [73] and adhere SiO_x (oxide nanoskin (ONS)) to its modified surface [77–79]. Treatment of polystyrene with O atoms in the absence of radiation [91] also showed an increase in the concentration of oxygen on the surface with treatment time up to a saturation level of 23.6 at % compared to 20 at% observed by VUV photo-oxidation [74].

Treatment of PS with an atmospheric Ar plasma jet (ca. 120-130 nm) and with only VUV radiation, passing through a MgF_2 optical window (112 nm), showed that the most efficient method for oxidation was using VUV radiation only since etching of the surface occurred with the plasma species [33].

4.3.6 Cyclo-Olefin Polymers

Cyclo-olefin polymers (COPs) are used in many applications, such as camera lenses/prisms, lenses for cameras incorporated in mobile phones, pick-up lenses, as well as medical devices including vials, syringes and optical lab test cells. VUV photo-oxidation improves their wettability and adhesion property with the following advantages: (1) the equipment is simple and inexpensive; (2) the equipment can be safely and easily operated; (3) no chemical reagents are required; and (4) there are no residual polluting byproducts or time consuming after-treatments [87]. VUV treatment of COPs with 172 nm photons in air introduces oxygenated functional groups on the surface [93], which promote room temperature bonding [94] and formation of microstructured SiO_x/COP stamps for patterning TiO_2 on polymer substrates via microcontact printing [95], as well as enhancing the assembly of COP materials for microphysiological systems with applications in drug discovery [96].

4.3.7 Poly(ethylene terephthalate) (PET) (Figure 4.11)

PET is commonly used in many applications for packaging, bottling, textile, electrical insulation, photography, magnetic tape, optical, cardiovascular implants, and the aerospace industry [97, 98] and has been well studied by VUV photo-oxidation [43, 99]. Figure 4.12 displays an increase in hydrophilicity, i.e., decrease in water contact angle, with VUV photo-oxidation of PET as a function of treatment time [43]. VUV photo-oxidation results in the adhesion of SiO_x network to PET [85, 86]. PET treated with a He atmospheric pressure plasma jet at different incident angles promoted reactive oxygen species on the surface [34].

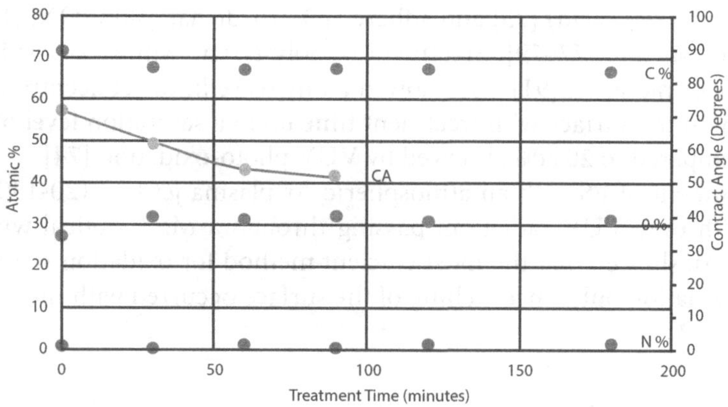

Figure 4.12 Quantitative XPS (at %) and water contact angle (CA) results for PET treated with VUV photo-oxidation as a function of treatment time [43].

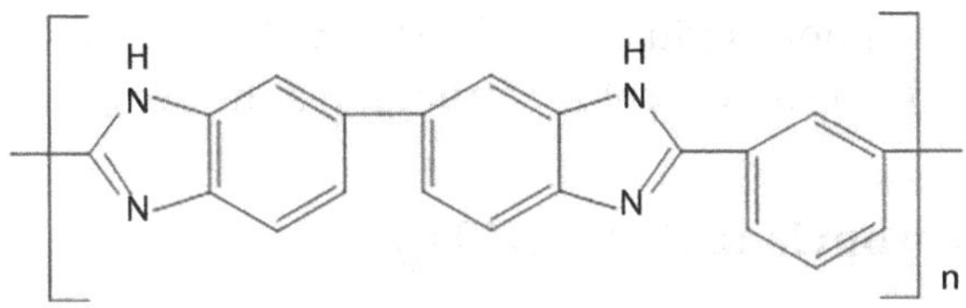

Figure 4.13 Poly 2,2′-m-(phenylene)-5,5′-bibenzimidazole or *meta*-PBI.

4.3.8 Polybenzimidazole (PBI) (Figure 4.13)

Poly 2,2'-m-(phenylene)-5,5'-bibenzimidazole, better known as polybenzimidazole (PBI) or meta-PBI, is a high-performance polymer consisting of benzimidazole units. The high thermal stability, chemical resistance and mechanical strength exhibited by PBI have allowed its suitability for different applications, such as fabrication of firefighters' suits [100, 101], aerospace purposes [102] and as a high temperature proton exchange membrane in fuel cells (HT-PEMFCs) where it is doped with phosphoric acid (H_3PO_4) [100, 103–105].

PBI treated with 104.8 and 106.7 nm photo-oxidation showed an increase in the O concentration up to a saturation level of 15 ± 1 at %; a decrease of the C concentration by about the same amount; and little change in the N concentration [106]. With increasing treatment time, there were significant decreases in the concentrations of C-C sp^2, C-C sp^3 and C = N groups, and increases in the concentrations of C = O, O-C = O, O-(C = O)-O, C-N, and N-C = O containing moieties. The water contact angle decreased from 83° for pristine PBI down to 43° making the surface more hydrophilic, primarily due to the oxidation, since AFM detected no significant changes in surface roughness. TGA analysis showed an improvement of water adhesion to the treated surface but no measurable increase of doped H_3PO_4 to the VUV surface photo-oxidized PBI samples [106] in contrast to O atom treatment which resulted in a greater decrease in water contact angle down to 34° and an increase in temperature by TGA for the loss of water from H_3PO_4-doped samples from 172°C for untreated sample to 176°C with treatment [107].

4.3.9 Poly(etheretherketone) (PEEK) (Figure 4.9)

Robust oxidation with 172 nm irradiation of a gaseous ethanol in a nitrogen atmosphere created bonding between PEEK and PI for use in micro-assembly of flexible and wearable electronics [69]. VUV treatment and low-pressure photo-oxidation of PEEK at 172 nm have been investigated as a surface pre-treatment for structural adhesive bonding [108, 109]

4.3.10 Polypropylene (PP), -$(C_3H_6)_n$-

VUV-induced oxidation of PP occurs at a lower level than that of PE because with PE there is a higher rate of chain scissions as a result of irradiation even at lower oxygen concentration [89]. While exposure of PP to a synchrotron source, generating a broadband of non-monochromatic radiation down to ca. 60 nm [36, 37], resulted in the formation of nanowrinkles due to the surface undergoing polymer degradation after exposure to oxygen in air [36].

4.3.11 Poly(ethylene 2,6-naphthalate) (PEN) (Figure 4.14)

The physical properties of PEN make it a suitable material for flexible electronic applications, solar cells, food packaging, quality fibers, magnetic recording tapes, printed circuit boards, liquid crystal displays, organic light emitting diode devices, bottle fabrication material, and a stable blending material with other polymers. Surface plasma oxidation of PEN has been conducted at atmospheric pressure [110, 111] and low pressure VUV photo-oxidation [112].

4.3.12 Polyethersulfone (PES) (Figure 4.15)

International need for water quality is placing a high demand on separation technology where PES is a membrane often used in the operation of water cleaning devices because of its superior thermal, mechanical, and chemical properties which assist in the construction of different surfaces and pore sizes. VUV treatment and low-pressure photo-oxidation of PES at 172 nm has been investigated as a surface pre-treatment for structural adhesive

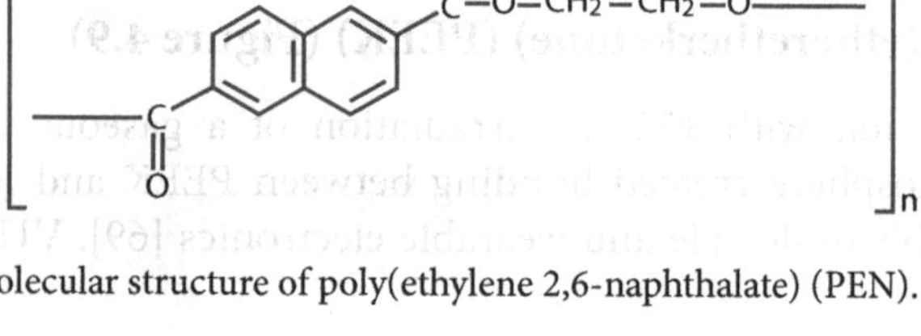

Figure 4.14 Molecular structure of poly(ethylene 2,6-naphthalate) (PEN).

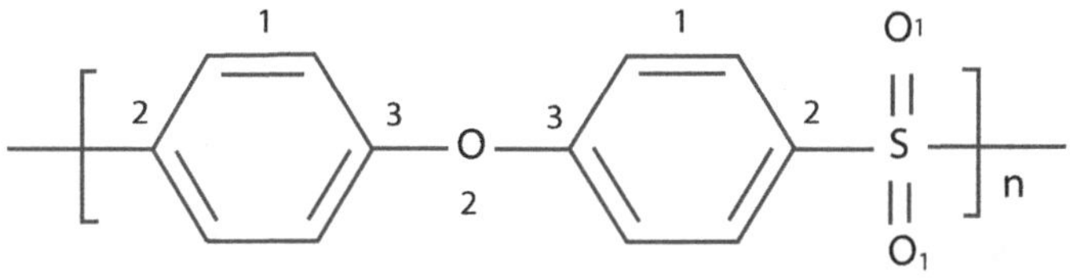

Figure 4.15 Chemical structure of polyethersulfone (PES) showing three kinds of C bonding: (1) C=C, (2) C-S, and (3) C-O and two types of O bonding: (1) O=S, and (2) O-C.

bonding [108]. VUV photo-oxidation with a low-pressure Ar plasma emitting 104.8 and 106.7 nm photons improves the hydrophilicity of PES by flowing oxygen over the surface during treatment. X-ray photoelectron spectroscopy (XPS) detected an increase in O at% (3.7 ± 1.0 at%) and water contact angle measurements showed a decrease from 71.9° for untreated PES down to a saturation level of 41.9° (Figure 4.16) with treatment.

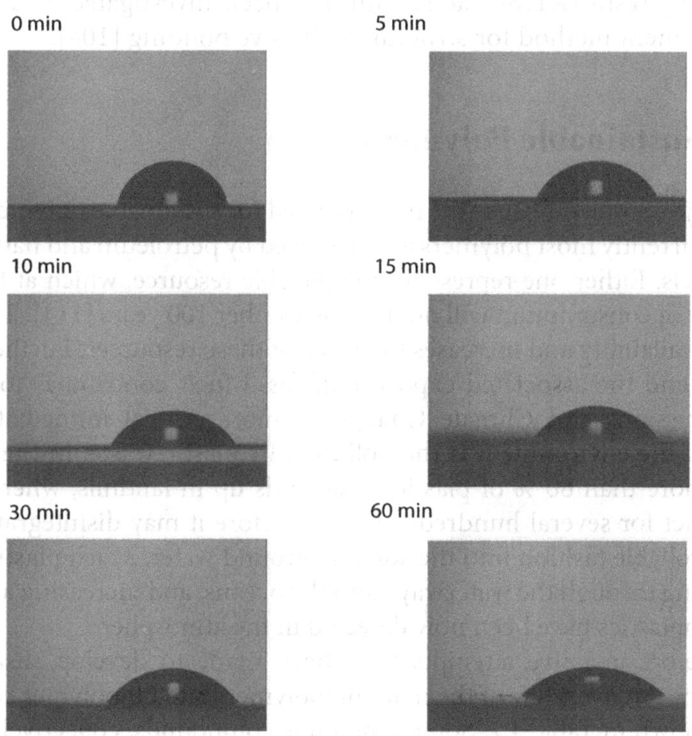

Figure 4.16 Water contact angle measurements after PES treated with VUV photo-oxidation for 0, 5, 10, 15, 30 and 60 min [113].

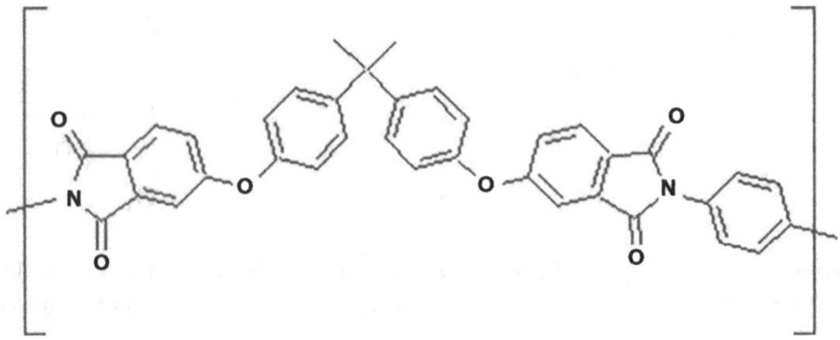

Figure 4.17 Molecular structure of polyetherimide (PEI).

Because no major changes in surface roughness were detected, the increase in hydrophilicity was mainly due to oxidation of the surface [113].

4.3.13 Polyetherimide (PEI) and Epoxy Resin (RTM6)

VUV treatment and low-pressure photo-oxidation of PEI (Figure 4.17) and epoxy resin (RTM6) at 172 nm has been investigated as a surface pre-treatment method for structural adhesive bonding [108].

4.4 Sustainable Polymers

Two major problems cause the massive need for sustainable polymers. One is that currently most polymers are produced by petroleum and natural gas fossil fuels. Either one represents a depletable resource, which at the current rate of consumption will not last for another 100 years [114]. This limits the availability and increases the price of these resources. Furthermore, the use and the associated exposure of fossil fuels contributes to higher CO_2 emissions and Climate Change. A more critical immediate issue affecting the environment is the pollution by plastic waste. In the United States more than 60 % of plastic waste ends up in landfills, where it can stay intact for several hundreds of years, before it may disintegrate in an uncontrollable fashion into the soil and ground water. Much plastic waste is leaching through the waterways into the oceans, and increasing amounts of microplastics have been now detected in the atmosphere.

Therefore, massive attempts have been made to develop sustainable polymers. An overview of the main biopolymers and biopolymer composites is shown in Table 4.7. Some renewable compounds, collectively called biomass, can be blended with synthetic polymers affording renewable alternatives. An example would be a polyethylene/starch blend obtained

Table 4.7 Overview of dominant examples of synthetic biopolymers or biocomposites.

Polymer or composite type	**Repeat unit**	**Biomass source**	**Typical composition**	**Biodegradation rate**
Polyethylene/ Starch	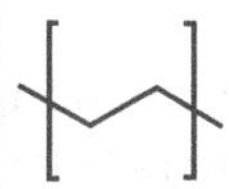	e.g. corn	typically up to 20 wt. % starch	low (only high for starch)
Carboxymethyl cellulose (CMC)	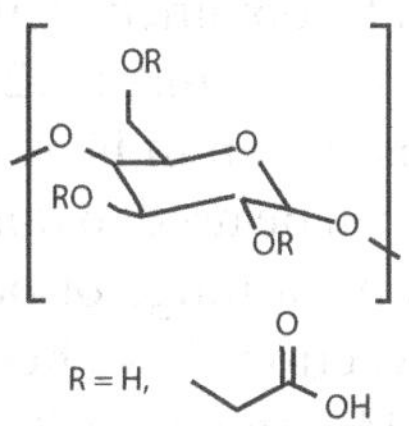	cellulose	pure	medium
Poly(lactic acid) (PLA)	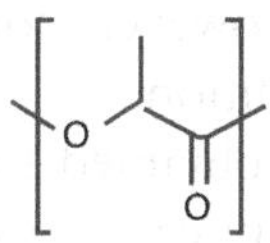	e.g. corn	up to 100 %	medium
Poly(glycolic acid) (PGA)	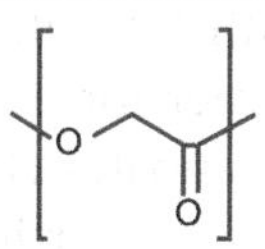	primarily petroleum based	up to 100 %	medium
Poly-(caprolactone) (PCL)	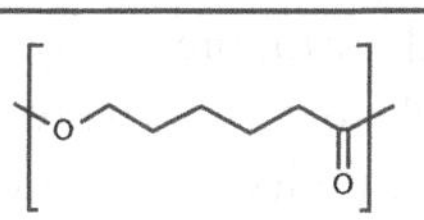	e.g. soybean oil	up to 100 %	medium
Poly(3-hydroxy-butyrate) (PHB)	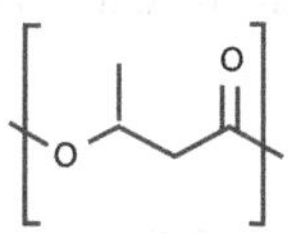	microbial biosynthesis	up to 100 %	high (2 weeks in compost)
Poly(butylene succinate) (PBS)	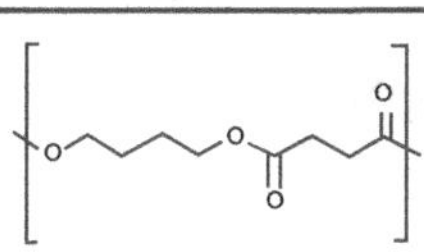	Biobased PBS: e.g. corn, sugarcane	up to 100 %	medium-high

by extrusion. Because of the simplicity of the process and the potential beneficial impact of the biomass additive, this method has been frequently applied [115–117]. Limitations to this method are that the biomass needs to be solid to avoid leaching, unwanted odor, and discoloration. The type of biomass needs to be well dispersible, and only applied to a certain extent, e.g. up to 20 weight %, in order to allow the properties of the polymer to dominate that of the blend. Carboxymethyl cellulose (CMC), which is obtained from cellulose by reaction with chloroacetic acid, has higher intermolecular forces than cellulose resulting, for example, in a higher viscosity of its solutions. CMC is being used for numerous applications, including medicine [118].

A few renewable compounds, such as isosorbide, can be applied directly as monomers for the syntheses of sustainable polymers [119, 120]. A much larger number of renewable monomers can be obtained by utilizing chemical reactions of compounds originating from biomass [121–123]. As an example, the synthesis of poly(lactic acid) is shown in Figure 4.18.

Many types of polymerization reactions can be used to obtain renewable polymers. Besides classical organic reactions, a range of other reactions, including thiol-ene click and metathesis reactions have been successfully applied to produce renewable polymers [124]. The majority of the polymerizations follow a step-growth mechanism, however, some renewable polymers also have been produced via chain-addition.

An important group of biopolymers can be obtained through monomers that can be readily produced by biochemical processes, such as succinic acid. The polymer from succinic acid with 1,4-butandiol, polybutylene succinate has excellent properties as polymer for packaging [125]. Finally, some polymers such as polyhydroxyl alkanoates (PHAs) are also predominantly produced via biochemical processes, particularly via the use of microbial enzymes, and are highly degradable [126].

Surface modification of polymers via plasma is a viable and less costly option to achieve higher adhesion, hydrophilicity and suitability for medical applications. For example, plasma surface modification can be applied

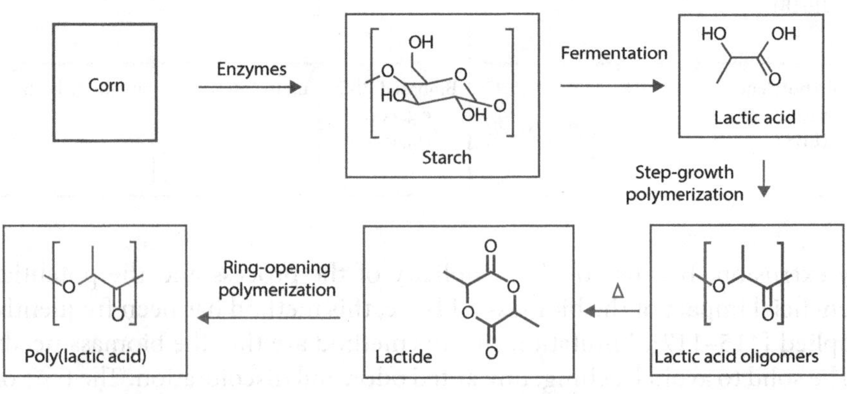

Figure 4.18 The different stages of the synthesis of poly(lactic acid) starting from corn.

to polyolefins, i.e. polyethylene and polypropylene, which represent the most produced polymers with close to 150 million metric tons globally per annum [127]. Polyolefins lack good adhesion compared to polymers containing polar groups, such as polyacrylates. Many attempts have been made to produce copolymers of olefins, such as ethylene, with comonomers bearing polar groups, at moderate and therefore less costly synthesis conditions [128–130]. However, frequently only the modification of the surface of a polar polymer will be sufficient to obtain desired adhesion or an increase in hydrophilicity [131]. The dominant form of surface modification is by treatment forming a plasma, and ion implantation and electron beam radiation representing alternative surface treatment methods.

Because of sustainable polymers becoming increasingly relevant, recent research has also been published on their surface modification using plasma at different conditions, such as low versus high temperature as mentioned in our previous review article [132]. In the following, emphasis is placed on publications that appeared in the last three years, i.e. 2019 to 2022 (patents not included).

Several review articles on the plasma surface treatment of sustainable polymers in general have been published [133–135]. The plasma treatment can alter the chemical composition, crystallinity, morphology, roughness, lubricity, hydrophilicity, and antimicrobial property of the polymers. Particularly for medical applications, cold plasma treatment appears to be the method of choice, since heat-resistant medicines may be applied. The treatment of lignocellulosic fibers with cold plasma has been performed to result in enhanced mechanical properties of related composites [136]. Few studies exist on the plasma based surface treatment of polyethylene/starch blends. In one study the polyethylene was compatibilized with the starch by adding maleic anhydride and glycerol, which resulted in an increase in the melt flow index of the blends [137]. Plasma activated surface treatment of CMC was applied to graft the CMC onto multiwalled carbon nanotubes [138], and in other work polypropylene was coated with plasma treated CMC containing essential oils [139].

The plasma induced surface treatment of poly(lactic acid) (PLA) has been published in several articles [140–142], where the formation of oxygen and nitrogen-containing groups occurred on the surface by chain scission and crosslinking. Hydrophilicity was clearly increased and water contact angles decreased upon treatment of PLA [140]. Among medical applications for plasma modified PLA are vaccine development, drug encapsulation, and gene delivery [141]. A hydrogen peroxide gas plasma (HPGP) treatment of PLA showed to be a promising sterilization method [142]. While the wettability of filaments of poly(glycolic acid) (PGA) could be enhanced by plasma surface treatment, their mechanical properties did not [143]. When oxygen plasma was applied to polycaprolactone (PCL) pellets and filaments, significant changes in chemical composition and wettability were observed [144]. Low-pressure (10 Pa) treatment of poly(butylene succinate) (PBS) was efficient in enhancing hydrophilicity by 90 %. In addition, the treatment allowed for a better adhesion of SiO_x coating onto the PBS [145]. In other work, PBS nanofibrous membranes were treated with low temperature plasma causing etching of the fibers. The etching depth depended on the treatment time. The roughness of the surface increased due to the plasma treatment, but had no effect on the water contact angle or wettability [146].

Only two articles have been published on the effect of VUV photo-oxidation of biopolymers. These investigations were by Sato and co-workers involving PLA [147, 148]. Therefore, there is apparently a need to investigate VUV photo-oxidation of many other biopolymers.

4.5 Summary

Dry techniques involving vacuum UV (VUV) photo-oxidation with wavelengths shorter than 180 nm are reviewed focusing on adhesion improvement of polymer surfaces. Future adhesion research is discussed with respect to making polymers more hydrophilic and sustainable.

References

1. M. Strobel, C.S. Lyons and K.L. Mittal (Eds.), *Plasma Surface Modification of Polymers: Relevance to Adhesion*, CRC Press, Boca Raton, FL (1994).
2. K.L. Mittal (Ed.), *Polymer Surface Modification: Relevance to Adhesion*, Vol. 5, CRC Press, Boca Raton, FL. (2009).
3. M. Thomas and K.L. Mittal (Eds.), *Atmospheric Pressure Plasma Treatment of Polymers: Relevance to Adhesion*, Wiley-Scrivener, Beverly, MA (2013).
4. K.L. Mittal and T. Bahners (Eds.) *Laser Surface Modification and Adhesion*, Wiley-Scrivener, Beverly, MA (2015).
5. K.L. Mittal and W.-S. Lei (Eds.), *Laser Technology: Applications in Adhesion and Related Areas*, Wiley-Scrivener, Beverly, MA (2018).
6. J.A.R. Samson, *Techniques of Vacuum Ultraviolet Spectroscopy*, John Wiley & Sons, New York (1967).
7. J.R. McNesby and H. Okabe, Vacuum ultraviolet photochemistry, in: *Advances in Photochemistry*, Vol. 3, W.A. Noyes, Jr., G.S. Hammond, and J.N. Pitts, Jr. (Eds.), pp. 157-240, Interscience Publishers, New York (1964).

8. S. Lazare, P.D. Hoh, J.M. Baker and R. Srinivasan, Controlled modification of organic polymer surfaces by continuous wave far-ultraviolet (185 nm) and pulsed-laser (193 nm) radiation: XPS studies. *J. Amer. Chem. Soc.* 106,4288-4290(1984).
9. K.L. Mittal (Ed.), *Physicochemical Aspects of Polymer Surfaces,* Vols. 1 and 2, Plenum Press, New York (1983).
10. S. Tazuke and H. Kimura, Surface photografting, 2. Modification of polypropylene film surface by graft polymerization of acrylamide. *Makromol. Chem.* 179,2603-2612(1978).
11. E.M. Liston, The role of vacuum ultraviolet light in low-pressure plasma processing, in: Proc. 7th Int. Symp. Plasma Chemistry, IUPAC, pp. 513-517(1985).
12. E.M. Liston, Plasma treatment for improved bonding: A review. *J. Adhesion* 30,199-218(1989).
13. F.D. Egitto, Plasma etching and modification of organic polymers. *Pure Appl. Chem.* 62,1699-1708(1990).
14. F.D. Egitto and L.J. Matienzo, Modification of poly(tetrafluoroethylene) and polyethylene surfaces downstream from helium microwave plasmas. *Polym. Degrad. Stab.* 30,293-308(1990).
15. E.M. Liston, L. Martinu and M.R. Wertheimer, Plasma surface modification of polymers for improved adhesion: a critical review. *J. Adhesion Sci. Technol.* 7,1091-1127(1993).
16. F.D. Egitto and L.J. Matienzo, Plasma surface modification and etching of polyimides. *Plastics Engineering (New York),* 36,389-452(1996).
17. C.-M. Chan. T.-M. Ko and H. Hiraoka, Polymer surface modification by plasmas and photons. *Surf. Sci. Reports.* 24,1-54(1996).
18. N. Takezoe, A. Yokotani and K. Kurosawa, Materials processing using vacuum ultraviolet rare gas excimer lamps, *Memoirs of the Faculty of Engineering, Miyazaki University* 28,1-16(1999).
19. M.R. Wertheimer, Plasma processing and polymers: A personal perspective. *Plasma Chem. Plasma Process.* 34,363-376(2014).
20. Y. Kusano, Atmospheric pressure plasma processing for polymer adhesion: A review. *J. Adhesion* 90,755-777(2014).
21. H. Desai, L. Xiaolu, A. Entenberg, B. Kahn, F.D. Egitto, L.J. Matienzo, T. Debies and G.A. Takacs, Adhesion of copper to poly(tetrafluoroethylene) surfaces modified with vacuum UV radiation downstream from He and Ar microwave plasmas, in: *Polymer Surface Modification: Relevance to Adhesion,* Vol. 3, K.L. Mittal (Ed.), pp.139-157, CRC Press, Boca Raton, FL (2004).
22. J.P. Badey, E. Urbaczewski-Espunche, Y. Jugnet, D. Sage, T.M. Duc and B. Chabert, Surface modification of poly(tetrafluoroethylene) by microwave plasma downstream treatment. *Polymer* 35,2472-2479(1994).
23. J.P. Lens, B. Spaay, J.G.A. Terlingen, G.H.M. Engbers and J. Feijen, Mechanism of the immobilization of surfactants on polymer surfaces by means of an argon plasma treatment: influence of UV radiation. *Plasmas Polymers* 4,159-182(1999).

24. R.E. Huffman, J.C. Larrabee and Y. Tanaka, Rare gas continuum light sources for photoelectric scanning in the vacuum ultraviolet. *Appl. Opt.* 4,1581-1588(1965).
25. J. Hopkins, S.H. Wheale and J.P.S. Badyal, Synergistic oxidation at plasma/polymer interface. *J. Phys. Chem.* 100,14062-14066(1996).
26. T. Oppenlander, Mercury-free sources of VUV/UV radiation: application of modern excimer lamps (excilamps) for water and air treatment. *J. Environ. Eng. Sci.* 6,253-264(2007).
27. I.W. Boyd and I.I. Liaw, Development and application of UV excimer lamps from 354 nm-126 nm, Proc. SPIE 6261,626104-1-626104-16 (2006).
28. U. Kogelschatz, Excimer lamps: history, discharge physics, and industrial applications, Proc. SPIE 5483,272-286 (2004).
29. G.A. Takacs, V. Vukanovic, F. Egitto, L.J. Matienzo, F. Emmi, D. Tracy and J.X. Chen, Modification of organic polymer surfaces with radiation from high pressure inert gas plasma rotating in a magnetic field, in: *Proc.10th IUPAC Inter. Symp. on Plasma Chemistry (ISPC-10)*, Bochum, Germany, 1, 1.4-29 pp.1-6 (1991).
30. G.A. Takacs, V. Vukanovic, D. Tracy, J.X. Chen, F.D. Egitto, L.J. Matienzo and F. Emmi, Photoetching and modification of organic polymer surfaces with vacuum UV radiation. *Polym. Degrad. Stab.* 40,73-81(1993).
31. S. Zheng, A. Entenberg, G.A. Takacs, F.D. Egitto and L.J. Matienzo, Adhesion of copper to poly(tetrafluoroethylene) surfaces modified with vacuum UV radiation from helium arc plasma. *J. Adhesion Sci. Technol.* 17,1801-1813(2003).
32. J-Y. Zhang and I.W. Boyd, Development of large area excimer VUV and UV sources from dielectric barrier discharge, *Mater. Res. Soc. Symp. Proc.*471,53-58(1997).
33. R. Zaplotnik and A. Vesel, Effect of VUV radiation on surface modification of polystyrene exposed to atmospheric pressure plasma jet. *Polymers* 12,1136-1148(2020).
34. T.M.C. Nishime, R. Wagner and K.G. Kostov, Study of modified area of polymer samples exposed to a He atmospheric pressure plasma jet using different treatment conditions. *Polymers* 12,1028-1047(2020).
35. F. Liu, L. Nie, X. Lu, J. Stephens and K. Ostrikov, Atmospheric plasma VUV photon emission. *Plasma Sources Sci. Technol.* 29,065001-065012(2020).
36. L.B. Fitaroni, T.A. Cacuro, C.A.R. Costa, E.M, Lanzoni, D. Galante, J.R. Araujo, M.G.P. Homem, W.R. Waldman and S.A. Cruz, Polymeric nanowrinkles: surface modification of polypropylene films in the VUV energy range. *J. Mater. Sci.* 56,9532-9543(2021).
37. L. Buzi, H. Miyazoe, M.P. Sagianis, N. Marchack, J.M. Papalia and S.U. Engelmann, Utilizing photosensitive polymers to evaluate UV radiation exposures in different plasma chamber configurations. *J. Vac. Sci. Technol. A* 38,033006-033013(2020).

38. V.I. Pavlenko and N.I. Cherkashina, Effect of SiO_2 crystal structure on the stability of polymer composites exposed to vacuum ultraviolet radiation. *Acta Astronautica* 155,1-9(2019).
39. V.V. Kashibadze, S.V. Zaitsev, D.S. Prochorenko and E.P. Klopot, Study of the kinetics of gas evolution of a polymer composite under the influence of vacuum and VUV. *J. Phys.: Conference Series* 1889,022008-022014(2021).
40. V.E. Skurat and Y.I. Dorofeev, The transformations of organic polymers during illumination by 147.0 and 123.6 nm light. *Angew. Makromol. Chem.* 216,205-224(1994).
41. J.G. Calvert and J.N. Pitts, *Photochemistry*, pp. 205-209, John Wiley & Sons, New York (1966).
42. H. Okabe, *Photochemistry of Small Molecules*, pp. 177-181, John Wiley & Sons, New York (1978).
43. A. Morgan, M. Cocca, K. Vega, A. Fleischer, S.G. Gupta, M. Mehan and G.A. Takacs, Vacuum UV photo-oxidation of poly(ethylene terephthalate). *J. Adhesion Sci. Technol.* 31,2542-2554(2017).
44. B.J. Finlayson-Pitts and J.N. Pitts, *Atmospheric Chemistry*. pp. 459-469, John Wiley & Sons, New York (1986).
45. D.R. Fagerburg and H. Clauberg, Photodegradation of poly(ethylene terephthalate) and poly(ethylene/1,4-cyclohexylenedimethylene terephthalate), in: *Modern Polymers*, J. Scheirs and T.E. Long (Eds.), pp. 609-641, John Wiley & Sons, Chichester, UK (2003).
46. M. Krysak, B. Parekh, T. Debies, R.A. DiLeo, B.J. Landi, R.P. Raffaelle and G.A. Takacs, Gas-phase surface functionalization of multi-walled carbon nanotubes with vacuum UV photo-oxidation. *J. Adhesion Sci. Technol.* 21,999-1007(2007).
47. B.J. Finlayson-Pitts and J.N. Pitts, *Chemistry of the Upper and Lower Atmosphere*. Academic Press, New York (2000).
48. B. Parekh, T. Debies, P. Knight, K.S.V. Santhanam and G.A. Takacs, Surface functionalization of multi-walled carbon nanotubes with UV and vacuum UV photo-oxidation. *J. Adhesion Sci. Technol.* 20,1833-1846(2006).
49. E.T Kang and Y. Zhang, Surface modification of fluoropolymers via molecular design. *Advanced Materials* 12,1481-1494(2000).
50. H. Esrom, J.Y. Zhang and U. Kogelschatz, Photochemical modification and etching of PTFE with excimer VUV/UV radiation, in: *Polymer Surfaces and Interfaces: Characterization, Modification and Application*, K.L. Mittal and K-W. Lee (Eds.), pp. 27-35, CRC Press, Boca Raton, FL(1997).
51. C.R. Davis and F.D. Egitto, Modifying fluoropolymers for electronic packaging. *Chemtech*, pp. 44-50 (March 1995).
52. J.X. Chen, D. Tracy, S. Zheng, L. Xiaolu, S. Brown, W. VanDerveer, A. Entenberg, V. Vukanovic, G.A. Takacs, F.D. Egitto, L.J. Matienzo and F. Emmi, Photoetching and modification of poly(tetrafluoroethylene-co-hexafluoropropylene) polymer surfaces with vacuum UV radiation. *Polym. Degrad. Stab.* 79,399-404(2003).

53. W. Dasilva, A. Entenberg, B. Kahn, T. Debies and G.A. Takacs, Adhesion of copper to poly(tetrafluoroethylene-co-hexafluoropropylene) (FEP) surfaces modified with vacuum UV photo-oxidation downstream from Ar microwave plasma. *J. Adhesion Sci. Technol.* 18,1465-1481(2004).
54. B. Parekh, S. Zheng, A. Entenberg, T. Debies and G.A. Takacs, Surface modification of poly(tetrafluoroethylene-co-hexafluoropropylene) with vacuum UV radiation from rotating helium dc arc plasmas. *J. Adhesion Sci. Technol.* 21,983-998(2007).
55. D.G. Abreu, W. Dasilva, A. Entenberg, T. Debies and G.A. Takacs, Chemistry at the Cu-fluoropolymer interface: Relevance to adhesion, in: *Adhesion Aspects of Thin Films*, Vol. 3, K.L. Mittal (Ed.), pp. 263-276, CRC Press, Boca Raton, FL(2007).
56. W. Dasilva, A. Entenberg, B. Kahn, T. Debies and G.A. Takacs, Surface modification of Teflon® PFA with vacuum UV photo-oxidation, *J. Adhesion Sci. Technol.* 20,437-455(2006).
57. A. Naujokas, D.G. Abreu, G.A. Takacs, T. Debies, M. Mehan and A. Entenberg, Fluoride ions at the Cu-fluoropolymer interface. *Surf. Interface Anal.* 45,1056-1062(2013).
58. Y. Ohkubo, Y. Okazaki, M. Shibahara, M. Nishino, Y. Seto, K. Endo and K. Yamamura, Effects of He and Ar heat-assisted plasma treatments on the adhesion properties of polytetrafluoroethylene (PTFE). *Polymers* 13,4266-4280(2021).
59. M. Nitani, K. Nakayama, K. Maeda, M. Omori and M. Uro, Organic temperature sensors based on conductive polymers patterned by a selective-wetting method. *Organic Electronics* 71,164-168(2019).
60. W. Ma, B. Ameduri and A. Takahara, Molecular aggregation structure and surface properties of biomimetic catechol-bearing poly[2-(perfluorooctyl) ethyl acrylate] and its application to superamphiphobic coatings. *ACS Omega* 5,8169-8180(2020).
61. K.S.V. Santhanam, R.J. Press, M.J. Miri, A.V. Bailey and G.A. Takacs, *Introduction to Hydrogen Technology*, 2nd ed., John Wiley & Sons, Hoboken, New Jersey (2018).
62. A. Khot, F. Lu, T. Debies and G.A. Takacs, Surface activation of perfluorosulfonic acid membrane used in fuel cells with vacuum UV photo-oxidation. *J. Adhesion Sci. Technol. 27,309-323*(2013).
63. A. Bailey, F. Lu, A. Khot, S. Hussain, K. Rugg, G.J. Leong, T. Debies and G.A. Takacs, Surface characterization of poly(acrylic acid) grafted to photo-oxidized perfluorosulfonic acid membrane used in fuel cells. *Mater. Res. Soc. Symp. Proc.* 1269E,1269-FF05-04(2010).
64. M.K. Ghosh and K.L. Mittal (Eds), *Polyimides: Fundamentals and Applications*, Marcel Dekker, New York (1996).
65. G. Zhao, B. Liu, Q. Wang and T. Wang, The effect of the addition of aramid fibers on tribological properties of polyimide after ultraviolet or atomic oxygen irradiation. *J. Eng. Tribol.* 226,864-872(2012).

66. G. Zhao, B. Liu, Q. Wang and T. Wang, The effect of the addition of talc on tribological properties of aramid fiber-reinforced polyimide composites under high vacuum, ultraviolet or atomic oxygen environment. *Surf. Interface Anal.* 45,605-611(2013).
67. A. Sasaki, H. Aoshima, S. Nagano and T. Seki, A versatile photochemical procedure to introduce a photoreactive molecular layer onto a polyimide film for liquid crystal alignment. *Polym. J.* 44,639-645(2012).
68. F.A. Rasoul, D.J.T. Hill, J.S. Forsythe, J.H. O'Donnell, G.A. George, P.J. Pomery, P.R. Young and J.W. Connell, Surface properties of fluorinated polyimides exposed to VUV and atomic oxygen. *J. Appl. Polym. Sci.* 58,1857-1864(1995).
69. T.H. Yang, Y-S. Chiu, C-Y. Yang, A. Shigetou and C.R. Kao, Polyimide-polyetheretherketone and tin-polyimide direct bonding via ethanol-assisted vacuum ultraviolet irradiation. *Trans. Japan Institute Electronic Packaging* 12, E19-012-1-E19-012-8(2019).
70. M. Murahara, T. Funatsu and Y. Okamoto, Photochemical adhesion of fused silica optical elements with no adhesive strain, *Proc. SPIE* 6403,640322-1-640322-8(2006).
71. M. Murahara, Y. Sato, T. Jitsuno and Y. Okamoto, Role of oxidizing agent: thin film formation by photo-oxidizing silicone oil for vacuum UV rays transmittance and high hardness, *Proc SPIE* 8530,85300S-1-85300S-6(2012).
72. N. Shirahata and A. Hozumi, Etchingless microfabrication of a thick metal oxide film on a flexible polymer substrate. *Chem. Mater.* 17,20-27(2005).
73. A. Hozumi, S. Asakura, A. Fuma, N. Shirahata and T. Kameyama, Preparation of a well-defined amino-terminated self-assembled monolayer and copper microlines on a polyimide substrate covered with an oxide nanoskin. *Langmuir* 21,8234-8242(2005).
74. X. Li, M. Toro, F. Lu, J. On, A. Bailey, T. Debies, M. Mehan, S.K. Gupta and G.A. Takacs, Vacuum UV photo-oxidation of polystyrene. *J. Adhesion Sci. Technol.* 30,2212-2223(2016).
75. A.G. Shard and J.P.S. Badyal, Plasma oxidation versus photooxidation of polystyrene. *J. Phys. Chem.* 95,9436-9438(1991).
76. A. Hozumi, T. Masuda, K. Hayashi, H. Sugimura, O. Takai and T. Kameyama, Spacially defined surface modification of poly(methylmethacrylate) using 172 nm vacuum ultraviolet light. *Langmuir* 18,9022-9027(2002).
77. A. Hozumi, H. Inagai, Y. Yokogawa and T. Kameyama, Molecular-scale growth of silicon oxide on polymer substrate through vacuum ultraviolet light-assisted photooxidation of organosilane precursor. *Thin Solid Films* 437,89-94(2003).
78. A. Hozumi, Y. Wu, K. Hayashi, H. Sugimura, O. Takai, Y. Yokogawa and T. Kameyama, Micro-wear resistance of ultrathin silicon oxide film-covered polymer substrate. *Surface Sci.* 532-535,1056-1060(2003).
79. A. Hozumi, Processing and application of "oxide nanoskin". *Advances Sci. Technol.* 45,660-667(2006).
80. S. Sato, Y. Ishiba, T. Wade, S. Kanehashi, S. Matsumoto, H. Matsumoto and K. Nagai, Hydrophilic molecular sieve surface layer formation in

hydrophobic poly(1-trimethylsilyl-1-propyne) membranes. *J. Membrane Sci.* 429,364-372(2013).

81. T. Yamamoto, Study on 172-nm vacuum ultraviolet light surface modifications of polydimethylsiloxane for micro/nanofluidic applications. *Surf. Interface Anal.* 43,1271-1276(2011).
82. V-M. Graubner, R. Jordan, O. Nuyken, B. Schnyder, T. Lippert, R. Koetz and A. Wokaun, Photochemical modification of cross-linked poly(dimethylsiloxane) by irradiation at 172 nm. *Macromolecules* 37,5936-5943(2004).
83. Y. Hamada, T. Ono, T. Akagi, K. Ishihara and T. Ichiki, Photochemical oxidation of poly(dimethylsiloxane) surface and subsequent coating with biomimetic phosphorylcholine polymer. *J. Photopolym. Sci. Technol.* 20, 245-249(2007).
84. L. Schilinsky, G. Jeschke, R. Tschaggelar, R. Wilken and D. Klose, Formation and decay of radicals during vacuum-UV irradiation of poly(dimethylsiloxane). *Polym. Degrad. Stab.* 144,497-507(2017).
85. Y. Naganuma, S. Tanaka, C. Kato and T. Shindo, Formation of silica coatings from perhydropolysilazane using vacuum ultraviolet excimer lamp. *J. Ceram. Soc. Jpn.* 112,599-603(2004).
86. L. Prager, A. Dierdorf, H. Liebe, S. Naumov, S. Stojanovic, R. Heller, L. Wennrich and M.R. Buchmeiser, Conversion of perhydropolysilazane into a SiO_x network triggered by vacuum ultraviolet irradiation: Access to flexible, transparent barrier coatings. *Chemistry – A European J.* 13,8522-8529(2007).
87. H. Fang, X. Qi, S. Zhou, S. Yang, C. Hang, Y. Tian and C. Wang, High-efficient vacuum ultraviolet-ozone assist-deposited polydopamine for poly(-lactic-co-glycolic acid)-coated pure Zn toward biodegradable cardiovascular stent applications. *ACS Appl. Mater. Interfaces* 14,3536-3550(2022).
88. A. Hollander, J.E. Klemberg-Sapieha and M.R. Wertheimer, Vacuum-ultraviolet induced oxidation of polyethylene. *Macromolecules* 27,2893-2895(1994).
89. A. Hollander, J.E. Klemerg-Sapieha and M.R. Wertheimer, Vacuum-ultraviolet induced oxidation of polymers polyethylene and polypropylene. *J. Polym. Sci. Part A: Polym. Chem.* 33,2013-2025(1995).
90. V.N. Vasilets, A.V. Kuznetsov and V.I. Sevastianov, Vacuum ultraviolet treatment of polyethylene to change surface properties and characteristics of protein adsorption. *J. Biomed. Mater. Res. A* 69,428-435(2004).
91. X. Li, F. Lu, M. Cocca, K.Vega, A. Fleischer, A. Bailey, M. Toro, S. Sachdev, T. Debies, M. Mehan, S.F. Gupta and G.A. Takacs, Surface modification of polystyrene with O atoms produced downstream from an Ar/O2 microwave plasma. *Technologies* 6,21-28(2018).
92. Y-J. Kim, Y. Taniguchi, K. Murase, Y. Taguchi and H. Sugimura, Vacuum ultraviolet-induced surface modification of cyclo-olefin polymer substrates for photochemical activation bonding. *Appl. Surf. Sci.* 255,3648-3654(2009).
93. Y-J. Kim and H. Sugimura, Effects of irradiation atmosphere on vacuum ultraviolet-induced surface modification of cyclo-olefin polymer substrates. *Appl. Phys. Express* 12,101002-1-101002-5(2019).

94. M. Gonda, T. Utsunomiya, T. Ichii and H. Sugimura, Room temperature bonding of cycloolefin polymer by vacuum ultraviolet surface photoactivation. *Int. J. Adhesion Adhesives* 100,102604-102609(2020).
95. C-T. Wu, T. Utsunomiya, T. Ichii and H. Sugimura, Microstructured SiO_x/COP stamps for patterning TiO_2 on polymer substrates via microcontact printing. *Langmuir* 36,10933-10940(2020).
96. M. Yamanaka, X. Wen, S. Imamura, R. Sakai, S. Terada and K. Kamei, Cyclo olefin polymer-based solvent mass-productive microphysiological systems. *Biomed. Mater.* 16,035009-035010(2021).
97. A.F. Whitaker and B.Z. Jang, The mass loss mechanisms of polymers in a radio frequency induced atomic oxygen environment. *J. Appl. Polym. Sci.* 48,1341-1367(1993).
98. G. Peng, D. Yang and S. He, Effect of VUV radiation on properties and chemical structure of polyethylene terephthalate film, in: *Protection of Materials and Structures from Space Environment*, J.L. Kleiman (Ed.), pp. 225-232, Springer, Netherlands (2006).
99. A. Hollander, J.E. Klemberg and M.R. Wertheimer, The influence of vacuum-ultraviolet radiation on poly(ethylene terephthalate). *J. Polym. Sci. Part A: Polym. Chem.* 34,1511-1516(1996).
100. J.S. Wainright, J-T. Wang, R.F. Savinelli and M. Litt, Acid-doped polybenzimidazoles: A new polymer electrolyte. *J. Electrochem. Soc.* 142,L121-L123(1995).
101. R. Davis, J. Chin, C-C. Lin and S. Petit, Accelerated weathering of polyaramid and polybenzimidazole firefighter protective clothing fabrics. *Polym. Degrad. Stab.* 95,1642-1654(2010).
102. N. Bhatnagar, D. Pyngrope, J. Rajiv, B. Sangeeta, P.H. Shantanu, V.T. Bui and H. Bonin, Electron beam modification of space durable polymeric nano-adhesive bonding of ultra-high temperature resistant polymer. *J. Polym. Eng.* 31,381-386(2011).
103. S.S. Araya, F. Zhou, V. Liso and S.L. Sahlin, A comprehensive review of PBI-based high temperature PEM fuel cells. *Int. J. Hydrogen Energy* 41,21310-21344(2016).
104. E. Quartarone and P. Mustarelli, Polymer fuel cells based on polybenzimidazole/H3PO4, *Energy Environ. Sci.* 5,6436-6444(2012).
105. M.S. Kondratenko, M.O. Gallyamov and A.R. Khokhlov, Performance of high temperature fuel cells with different types of PBI membranes as analyzed by impedance spectroscopy. *Int. J. Hydrogen Energy* 37,2596-2602(2012).
106. T. Kovach, S. Boyd, A. Garcia, A. Fleischer, K. Vega, R. Hilfiker, J. Shertok, M. Mehan, S.K. Gupta and G.A. Takacs, Surface modification of polybenzimidazole (PBI) with microwave generated vacuum ultraviolet (VUV) photo-oxidation. *Current Microwave Chemistry* (2021) DOI:10.2174/2213335608666210908123730.
107. K. Vega, M. Cocca, H. Le, M. Toro, A. Garcia, A. Fleischer, A. Bailey, J. Shertok, M. Mehan, S.K. Gupta and G.A. Takacs, Enhancing the wettability of polybenzimidazole (PBI) to improve fuel cell performance, in: *Advances*

in Contact Angle, Wettability, and Adhesion, Vol. 4, K.L. Mittal (Ed.), pp. 179-191, Wiley-Scrivener, Beverly, MA (2019).

108. E. Arikan, J. Holmannspotter, F. Zimmer and T. Hofmann, The role of chemical surface modification for structural adhesive bonding on polymers-Washability of chemical functionalization without reducing adhesion. *Int. J. Adhesion Adhesives* 95,102409-102421(2019).
109. E. Arikan, J. Holtmannspotter, T. Hofmann and H-J. Gudladt, Vacuum-UV of polyetheretherketone (PEEK) as a surface pre-treatment for structural adhesive bonding. *J. Adhesion* 96,917-944(2020).
110. M. Kormunda, T. Homola, J. Matousek, D. Kovacik, M. Cernak and J. Pavik, Surface analysis of poly(ethylene naphthalate) (PEN) films treated at atmospheric pressure using diffuse coplanar surface barrier discharge in air and in nitrogen. *Polym. Degrad. Stab.* 97,547-553(2012).
111. E. Gonzalez, M.D. Barankin, P.C. Guschi and R.F. Hicks, Remote atmospheric-pressure activation of the surfaces of polyethylene terephthalate and polyethylene naphthalate. *Langmuir* 24,12636-12643(2008).
112. A. Morgan, M. Mehan, T. Debies and G.A. Takacs, Vacuum UV photo-oxidation and reaction of O atoms with poly(ethylene 2,6-naphthalate), in: *Proceedings of the 9th Euro. Tech. Symp. on Polyimides & High Performance Functional Polymers (STEPI 9)*, pp.183-192(2013).
113. S. Oakes, R. Keeley, H. Heineman, T. Allston, J. Shertok, M. Mehan, G.K. Thompson and G.A. Takacs, Vacuum UV (VUV) photo-oxidation of polyethersulfone (PES). *Technologies 10*,49-60(2022).
114. J.L. Hancock, W. Wu, C.A. Hall and M. Jefferson, Forecasting the limits to the availability and diversity of global conventional oil supply: Validation. *Energy* 64,130-153(2014).
115. H. Mohit, S.M. Rangappa, S. Siengchin, S. Gorbatyuk, P. Manimaran, C. Alka Kumari, A. Khan and M. Doddamani, A comprehensive review on performance and machinability of plant fiber polymer composites. *Polym. Composites* 43,608-623(2022).
116. B. Boubekeur, N. Belhaneche-Bensemra and V. Massardier, Low-density polyethylene/poly(lactic acid) blends reinforced by waste wood flour. *J. Vinyl Additive Technol.* 26,443-45(2020).
117. C. Ge, H.N. Cheng, M.J. Miri, R.K. Hailstone, J.B. Francis, S.M. Demyttenaere and N.A. Alharbi, Preparation and evaluation of composites comprising polypropylene and cotton gin trash. *J. Appl. Polym. Sci.* 137,49151-49163(2020).
118. H. N. Cheng and A. Biswas, Chemical modification of cotton-based natural materials: Products from carboxymethylation. *Carbohydrate Polym.* 84,1004–1010 (2011).
119. A. Gandini and T.M. Lacerda, Monomers and macromolecular materials from renewable resources: State of the art and perspectives. *Molecules* 27,159-174(2021).
120. D.H. Weinland, R.-J. van Putten and G.-J. M. Gruter, Evaluating the commercial application potential of polyesters with 1,4:3,6-dianhydrohexitols

(isosorbide, isomannide and isoidide) by reviewing the synthetic challenges in step growth polymerization. *Eur. Polym. J.* 164,110964-110998(2022).
121. A. Pellis, M. Malinconico, A. Guarneri and L. Gardossi, Renewable polymers and plastics: Performance beyond the green. *New Biotechnol.* 60,146-158(2021).
122. H.R. Kricheldorf and S.M. Weidner, Syntheses of polylactides by means of tin catalysts. *Polymer Chemistry* 13,1618-1647(2022).
123. D.S. Honeycutt, W.F. Charbonneau, A.J. North, S.L. Cobb, D. Lohmann and M.J. Miri, Effects of alkyl and phenyl-substituted 1,3-propanediols on the synthesis and properties of polyesters with 2,5-furandicarboxylic acid. *Polymer* 242,124584-124595(2022).
124. K.-K. Tremblay-Parrado, C. García-Astrain and L. Avérous, Click chemistry for the synthesis of biobased polymers and networks derived from vegetable oils. *Green Chem.* 23,4296-4327(2021).
125. L. Aliotta, M. Seggiani, A. Lazzeri, V. Gigante and P. Cinelli, A brief review of poly(butylene succinate) (PBS) and its main copolymers: Synthesis, blends, composites, biodegradability, and applications. *Polymers* 14,844-866(2022).
126. S. Taguchi and K. Matsumoto, Evolution of polyhydroxyalkanoate synthesizing systems toward a sustainable plastic industry. *Polymer J.* (Tokyo) 53,67-79(2021).
127. https://www.chemanalyst.com/industry-report/polyolefins-market-307, accessed June 18, 2022.
128. M. Baur, F. Lin, T.O. Morgen, L. Odenwald and S. Mecking, Polyethylene materials with in-chain ketones from nonalternating catalytic copolymerization. *Science* 374,604-607(2021).
129. Q. Liu and R.F. Jordan, Copolymerization of ethylene and vinyl fluoride by self-assembled multinuclear palladium catalysts. *Polymers* 12,1609-1616 (2020).
130. L.R. Parisi, D.M. Scheibel, S. Lin, E.M. Bennett, J.S. Lodge and M.J. Miri, Eugenol as renewable comonomer compared to 4-penten-1-ol in ethylene copolymerization using a palladium aryl sulfonate catalyst. *Polymer* 114,319 – 328(2017).
131. A.J. Peacock, The chemistry of polyethylene. *J. Macromol. Sci. Part C: Polymer Rev.* 41,285-323(2001).
132. G.A. Takacs, M.J. Miri and T. Kovach, Vacuum UV surface photo-oxidation of polymeric and other materials for improving adhesion: A critical review. *Rev. Adhesion Adhesives* 8,555–581(2020).
133. M. Vajpayee, M. Singh and L. Ledwani, Non-thermal plasma treatment of cellulosic biopolymer to enhance its surface property for various applications: A review. *Materials Today: Proceedings* 43,3250-3255(2021).
134. R. Bahrami , R, Zibaei , Z. Hashami , S. Hasanvand , F. Garavand , M. Rouhi, S.M. Jafari and R. Mohammadi, Modification and improvement of biodegradable packaging films by cold plasma; a critical review. *Crit. Rev. Food Sci. Nutrition* 62, 1936-1950(2022).
135. B. Dashtbozorg, X. Tao and H. Dong, Active-screen plasma surface multi-functionalisation of biopolymers and carbon-based materials – An overview. *Surf. Coat. Technol.* 442,128188-128201(2022).

136. F.J. Alonso-Montemayor, D. Navarro-Rodrıguez, M. Delgado-Aguilar, M.G. Neira-Velazquez, C.N. Aguilar, A.O. Castaneda-Facio, Y.K. Reyes-Acosta and R.I. Narro-Cespedes, Plasma-treated lignocellulosic fibers for polymer reinforcement. A review. *Cellulose* 29,659–683(2022).
137. C.-Y. Huang and W.-L. Lu, The effect of maleic anhydride/glycerol and plasma treatment on biodegradable LDPE/tapioca starch blends. *Material Science Forum* 426-432,3279-3284(2003).
138. D. Shao, Z. Jiang, X. Wang, J. Li and Y. Meng, Plasma induced grafting carboxymethyl cellulose on multiwalled carbon nanotubes for the removal of UO22+ from aqueous solution. *J. Phys. Chem. B* 113,860-864(2009).
139. Z. Honarvar, M. Farhoodi, M. R. Khani, A. Mohammadi, B. Shokri, R. Ferdowsi, and S. Shojaee-Aliabadi, Application of cold plasma to develop carboxymethyl cellulose-coated polypropylene films containing essential oil. *Carbohydrate Polym.* 176,1-10 (2017).
140. O.A. Laput, I.V. Vasenina, V.V. Botvin and I.A. Kurzina, Surface modification of polylactic acid by ion, electron beams and low-temperature plasma: a review. *J. Mater. Sci.* 57,2335-2361(2022).
141. G. Mann, L.P. Singh, P. Kumar, S. Singh and C. Prakash, On briefing the surface modifications of polylactic acid: A scope for betterment of biomedical structures. *J. Thermoplast. Composite Mater.* 34,977-1005(2019).
142. S.P. Davila, L.R. González, S. Chiussi, J. Serra and P. González, How to sterilize polylactic acid based medical devices? *Polymers* 13,2115-2132(2021).
143. S. Fu and P. Zhang, Surface modification of polylactic acid (PLA) and polyglycolic acid (PGA) monofilaments via the cold plasma method for acupoint catgut-embedding therapy applications. *Textile Res. J.* 89,3839-3849(2019).
144. Z. Mohammed, S. Jeelani and V. Rangari, Effect of low-temperature plasma treatment on surface modification of polycaprolactone pellets and thermal properties of extruded filaments. *JOM (The Journal of The Minerals, Metals & Materials Society)* 72,1523-1532(2020).
145. E. Vassallo, M. Aloisio, M. Pedroni, F. Ghezzi, P. Cerruti, and R. Donnini, Effect of low-pressure plasma treatment on the surface wettability of poly(butylene succinate) films. *Coatings* 12,220-232(2022).
146. Z. Wei, J. Gu, Y. Ye, M. Fang, J. Lang, D. Yang and Z. Pan, Biodegradable poly(butylene succinate) nanofibrous membrane treated with oxygen plasma for superhydrophilicity. *Surf. Coat. Technol.* 381,125147-125153(2020).
147. S. Sato, M. Ono, J. Yamauchi, S. Kanehashi, H. Ito, S. Matsumoto, Y. Iwai, H. Matsumoto and K. Nagai, Effects of irradiation with vacuum ultraviolet xenon lamp at 172 nm on water vapor transport through poly(lactic acid) membranes. *Desalination* 287,290-300(2012).
148. T. Saiga, S. Sato and K. Nagai, Water vapor solubility of poly(lactic acid) films modified the surface by vacuum ultraviolet irradiation. *J. Appl. Polym. Sci.* 132,42200/1-42200/8(2015).

5

Application-Related Optimization of Adhesion of Polymers Using Photochemical Surface Modification

Thomas Bahners[1]*, Jochen S. Gutmann[1,2] and Jörg Müssig[3]

[1]*Deutsches Textilforschungszentrum Nord-West gGmbH (DTNW), Krefeld, Germany*
[2]*Universität Duisburg-Essen, Physikalische Chemie and CENIDE, Essen, Germany*
[3]*The Biological Materials Group, Dept. Biomimetics, HSB – Hochschule Bremen City, University of Applied Sciences Bremen, Bremen, Germany*

Abstract
The present chapter describes the potential of UV-based photochemical surface modifications to optimize the adhesion of various systems to polymer substrates in reference to their specific field of application. This covers, e.g., coating, laminating, printing, adhesive bonding, composite materials, but also the development of innovative materials for biomedical applications. Following a short introduction to the fundamentals of photochemical processes – photooxidation as well as layer-forming processes along the "grafting to" and the "grafting from" concepts - and the general experimental designs in both laboratory and larger scales, individual sections cover three specific areas, which feature prominently in scientific literature. These sections refer to the "classic" problem of adhesion of coatings and matrix polymers to hydrophobic substrates, adhesion in composite materials made of biobased polymers, and the adhesion of cells and proteins to polymers used in tissue engineering and surgery.

Keywords: Photooxidation, photo-grafting, photo-polymerization, coatings, composites, bio-systems

**Corresponding author*: thomas.bahners@web.de

K.L. Mittal and A.N. Netravali (eds.) Polymer Surface Modification to Enhance Adhesion: Techniques and Applications, (155–198) © 2024 Scrivener Publishing LLC

5.1 Introduction

Photochemical concepts have been studied and applied for the surface modification of polymers for a long time with an increasing variety of modifications to a substrate's properties, including introducing new functionalities. The literature of the past 30 years refers to wettability, adhesion – as addressed in this review -, chemical and physical stabilities, surface electrical properties, and antibacterial properties.

In general, photochemical processes are based on radical activation and ensuing reactions with the surrounding atmosphere. Achievable effects include adding atoms, e.g., the introduction of oxygen (photooxidation), resulting in increased surface energy and the grafting of desired functional groups. Both have a potential for adhesion promotion; vinyl- or epoxy-functions could, for example, create a matrix-friendly surface for epoxy resins. An increase in hydrophilicity and adhesion of polyethylene (PE), polypropylene (PP), polyesters, and poly(ether ether ketone) (PEEK) following simple ultraviolet (UV) irradiation has been reported by McIntyre and Walzak [1], Mathieson and Bradley [2], and Gongjian *et al.* [3]. Various authors have reported effects on the wettability of polyolefins, polyamides as well as polyesters and resulting increase in adhesion of epoxy to these polymers following UV-induced grafting of acrylates, acrylic acid, or acrylamide [4-8]. Praschak *et al.* [9,10] as well as Bahners *et al.* [11] studied similar treatments of poly(ethylene terephthalate) (PET) films and fibers using monochromatic UV sources.

To put these early papers in context to general work on photochemical processes and identify further developments up to the present time, a search on Scopus [12] was done in preparation of the current review using the search word combination "UV AND photochemical AND adhesion". The investigation resulted in 239 entries for the time period 1973 to 2021, covering a variety of subject areas. Of the 239 publications, 179 related to the field of "Materials Science". One hundred papers related to the field of "Chemistry". The field of "Engineering" accounted for 85 contributions, and the keyword "Chemical Engineering" was stated in 73 papers. Seventy documents were assigned to "Physics and Astronomy", while 29 papers were assigned to the field of "Biochemistry, Genetics, and Molecular Biology". It is interesting to note that a Scopus search using the search word "photochemical" instead of "photo chemical" identified only 192 papers omitting several publications of relevance for the present chapter.

The first identified paper was published in 1973 by Vakula and Orlov entitled "The photo-aging of polyethylene and its effect on polyethylene adhesion to steel" in the journal *Polymer Science USSR* [13]. The development of publications in this area is shown in Figure 5.1 and can be summarized as follows:

- 1973 – 1993: 6 documents
- 1994 – 2003: 28 documents
- 2004 – 2013: 123 documents
- 2014 – 2021 (end of November): 80 documents

Using the software VosViewer [14,15], the construction and visualization of a co-occurrence network of terms, i.e., keywords and search terms, was carried out for the four time periods. The terms were extracted from the titles and summaries of the documents from Scopus literature research and checked for co-occurrence. The resulting network is shown exemplarily in Figure 5.2 for the period 2014-2021. In the visualization, each circle represents a term, with the size of the circle indicating the number of publications containing the corresponding term.

Comparing the specific co-occurrence networks of the four time periods reveals that the focus (most mentioned term) has shifted from *surface* in the first period to *adhesion*. The two terms remain the dominant terms in the following three periods, but the variety of topics increases.

To optimize the Scopus search for the focus of the present review, the search words "UV AND photo chemical AND adhesion AND polymer" were specified and individually combined with the terms (a) *coating, bonding* (b) *composite*, and (c) *cell* and *biofilm*. The search identified 119 related

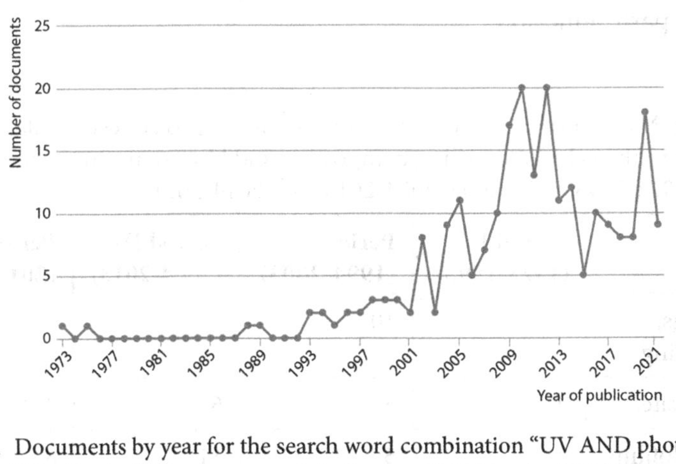

Figure 5.1 Documents by year for the search word combination "UV AND photo chemical AND adhesion"; Database search in Scopus, 2021 Elsevier B.V. [12].

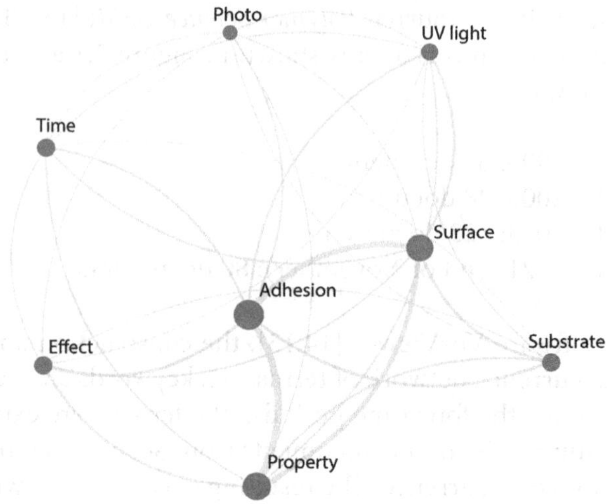

Figure 5.2 Co-occurrence network of keywords stated in papers published in the time period 2014 to 2021 according to an analysis using the software VosViewer [15].

papers, 44 of which were actually concerned with optimizing and modifying adhesion property by photochemical treatments. The remaining papers concentrated on UV-ageing, cross-linking of polymers, or modification of other polymer properties.

The 44 identified papers were further related to the four time periods mentioned above- which indicated the occurrence of new research fields related to the general term adhesion, as shown in Table 5.1 and Figure 5.3. As the analysis shows, most published studies of photochemical surface treatments in the 1990s were focused on improving coating, bonding, or printing paste adhesion.

Table 5.1 Number of identified publications referring to adhesion-related problems in the fields of coatings, composites, and biosystems in the four time periods 1973-1993, 1994-2003, 2004-2013, and 2014-2021.

	Period I (1973-1993)	**Period II (1994-2003)**	**Period III (2004-2013)**	**Period IV (2014-2021)**
Coatings, bonding	3	10	4	2
Composites	0	3	5	3
Cells, biofilms	0	5	11	8

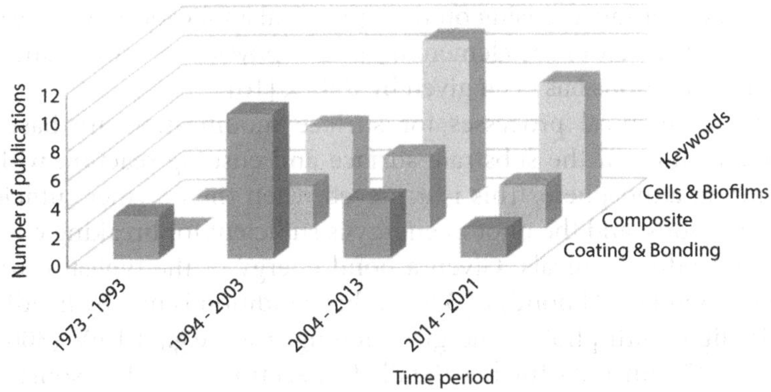

Figure 5.3 Number of identified publications referring to adhesion-related problems in the fields of coatings, composites, and biosystems in the four time periods 1973-1993, 1994-2003, 2004-2013, and 2014-2021; Database search in Scopus, 2021 Elsevier B.V. [12].

While the aforementioned topic lost prominence in literature, either due to successful industrial application or interest in competitive alternative technologies such as atmospheric plasma processing, photochemical surface treatments were increasingly studied in the context of emerging new materials and innovative applications of polymers. Most prominent examples are the increasing interest in bio-based materials, for example, composites, the design of biomaterials, and the expanding use of polymers for medical and surgical applications, namely implants and scaffolds for tissue engineering. Numerous researchers in these fields employ the more elaborate process design of photo-grafting and photo-induced polymerization of thin layers (photo-crosslinking) to adapt the surface properties of polymer substrates to the material combination at hand or the biological environment.

The present chapter is meant to provide an overview of the principles of the photochemical processes and illustrate the utilization of these processes to optimize adhesion properties in various fields of application of polymers on the background of the identified publications.

5.2 Photochemical Surface Modification

5.2.1 Fundamentals of the Process

The following section is meant to give the reader a short overview of the principal mechanisms of the major forms of photochemical processes for surface modification, focusing on their practical application. A more comprehensive overview of the elementary photo-physical processes and photochemical reactions has been given by Rabek [16].

All photochemical processes for surface modification are based on radical activation of the substrate surface and ensuing reaction with the surrounding atmosphere. This is achieved when the polymer sufficiently absorbs photons, and the photon energy is sufficient for breaking covalent bonds to produce radicals. Given a bond energy of the typical Carbon–Hydrogen bond (C-H bond) of 3.6 eV, this condition is met by irradiation with UV light with photon energies ranging from, e.g., 4.1 eV (300 nm) to 6.2 eV (200 nm). As further detailed in section 5.2.2, UV sources are available commercially as broadband lamps with varying emission characteristics and as monochromatic lamps in the form of excimer lamps. The use of monochromatic UV sources has the considerable advantage of allowing the choice of a wavelength, which is highly absorbed by the polymer, resulting in radical generation with a very high quantum yield. The monochromatic source is also free of any thermal emission. According to the polymers' absorption characteristics, a krypton-chloride ($KrCl^*$) excimer lamp emitting at 222 nm (cf. section 5.2.2) would be appropriate, for instance, for aromatic polymers such as PET. Aside from the absorption characteristics – and the related efficiency in radical generation – the polymer's internal structure and crystallinity are of further influence. According to the chemical composition of the PET macromolecule, backbone radicals have a very high probability of being generated. In the special case of polymer fibers, however, the high orientation of macromolecules and high crystallinity of at least 80 %, fast recombination of backbone radicals is a major factor, and the less probable side-chain radicals will be long-lived and available for the ensuing reactions.

If the reactive atmosphere is low- or non-absorbing, and the substrate is strongly absorbing, the actual radical generation and the ensuing reactions occur only at the substrate's very surface, forming a boundary between the atmosphere and the activated substrate. Basically, four different types of reactions are possible: (I) Recombination of radicals, (II) cross-linking of polymer chains, (III) addition of radicals from the reactive atmosphere, and (IV) addition of bi- or multi-functional molecules – typically allyls or acrylates - resulting in cross-linking between the functional groups in a "grafting from" process. If reactive media are strong absorbers themselves, the radiation is partly or fully absorbed, bulk radicals are generated, and *homo-polymerization* is initiated, resulting in a "grafting to" process (reaction type (V)). In view of adhesion, only reaction types (III), (IV), and (V)

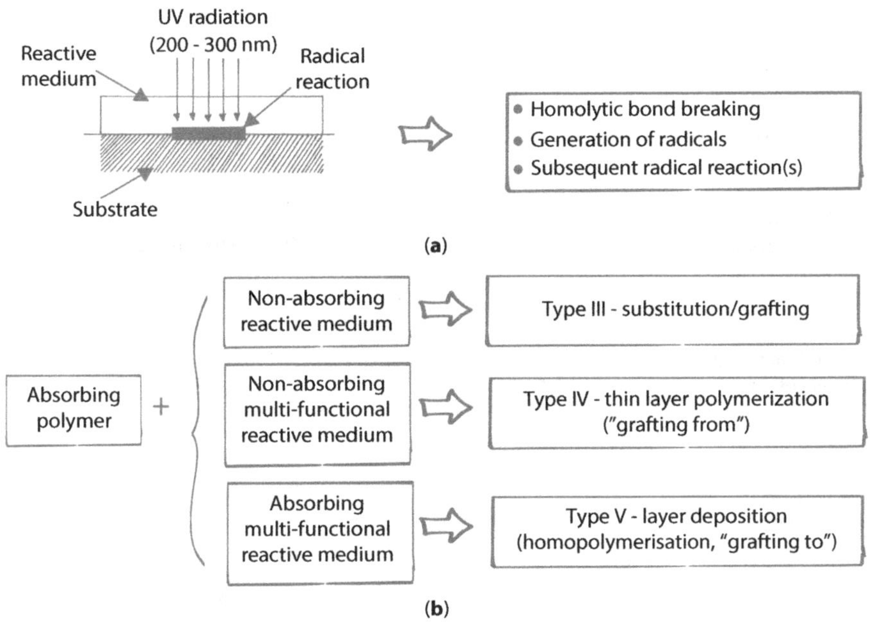

Figure 5.4 Fundamental reactions allowing photochemical modification of the surface chemistry of a polymer substrate, i.e., photo-oxidation, and photo-grafting (reaction type (III) and photo-induced layer polymerization (reaction types (IV) and (V)).

are candidates for the appropriate modification of surfaces properties (cf. Figure 5.4). These shall be discussed in the following sub-sections.

5.2.1.1 Photo-Addition or Photo-Grafting

Photo-addition or photo-grafting results from the UV irradiation of polymer substrate in a reactive atmosphere and subsequent addition according to reaction III. The simplest scenario of UV-induced modification of surface chemistry is the irradiation of the substrate in air, quite similar to the treatment of polymers by corona and atmospheric plasma. This generally leads to photooxidation, i.e., adding oxygen and generating a hydrophilic, hydroxyl functional group (OH-group). It is an idiosyncrasy of the photochemical process in contrast to plasma processes that the reactive "atmosphere" might be gaseous but also a liquid applied to the substrate before irradiation. The only condition is that the liquid medium is sufficiently transparent (non-absorbing) to the UV emission spectrum of the light source employed.

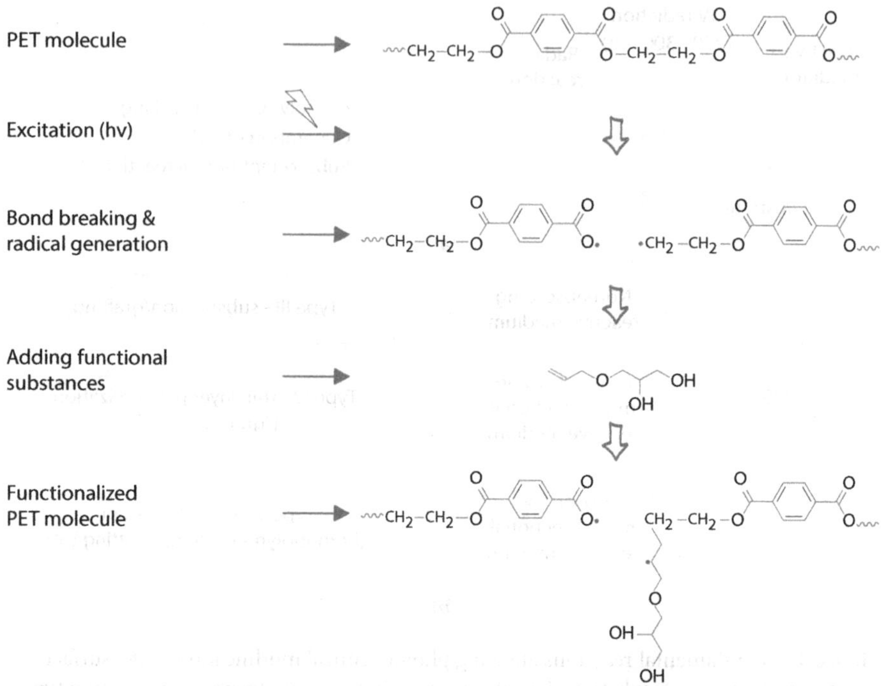

Figure 5.5 The principle of the surface modification of a polymer (here: PET) by addition of a functional DAP molecule via photo-grafting (reaction type (III)).

Following the same concept as reaction III, photo-grafting allows for far-reaching modifications to the surface chemistry of the substrate, as shown in Figure 5.5 for the example of diallylphthalate (DAP). The concept of photo-grafting of hydrophilic end groups is even more advantageous, as the modification is covalent and exhibits high stability/durability, which contrasts the known decrease of the effects achieved by, e.g., the corona treatment. A similar decrease – albeit far slower – has also been reported in early papers on photooxidation (see, e.g., [9]). The established functional end groups may affect interactions of the substrate with coating materials to great effect, as has been shown by many researchers [4–8]. The grafted end groups may also serve as moieties for ensuing chemical reactions.

5.2.1.2 *Layer Formation by Homo-Polymerization and Graft-Co-Polymerization*

If the reactive medium consists of a solution of polymerizable, multi-functional monomers, the polymerization of a thin layer is initiated by UV irradiation, which effectively masks the substrate surface and defines new chemical and physical surface properties. It has also been reported that bulk properties of the photo-polymerized layers can be exploited to introduce new surface functionalities (cf. [17, 18]).

Representatives of this class of multi-functional monomers reported for this type of surface modification are substances such as diallylphthalate (DAP), tetraallyloxyethane (TAE), pentaerythritoltriacrylate (PETA), 1,5-hexadiene, 1,7-octadiene, cyclohexane-1,4-dimethanoldivinylether (CHMV), or poly(ethylene glycol) dimethacrylate (PEGDMA), chemical structures of which are given in Figure 5.6.

Using the characteristic absorption properties of these monomers, graft-co-polymerization and homo-polymerization processes can be initiated in a given monomer system by appropriate choice of a light source, which could be either monochromatic light at various wavelengths or broadband UV. Most of the listed monomers absorb weakly at wavelengths above 250 nm (cf. Figure 5.7). One may conclude that the radiation penetrates the applied monomer solution in this case and generates radicals mainly at the substrate surface leading to co-polymerization initiated at the interface between the substrate and reactive medium ("grafting from", reaction type (IV)). This causes an actual layer growth on the surface with a rather homogeneous layer morphology. If irradiated at shorter wavelengths, e.g., between 200 and 250 nm, the absorbing end groups of monomers such as DAP or PETA, as shown in Figure 5.7, can be expected to react strongly with incident photons leading to bulk

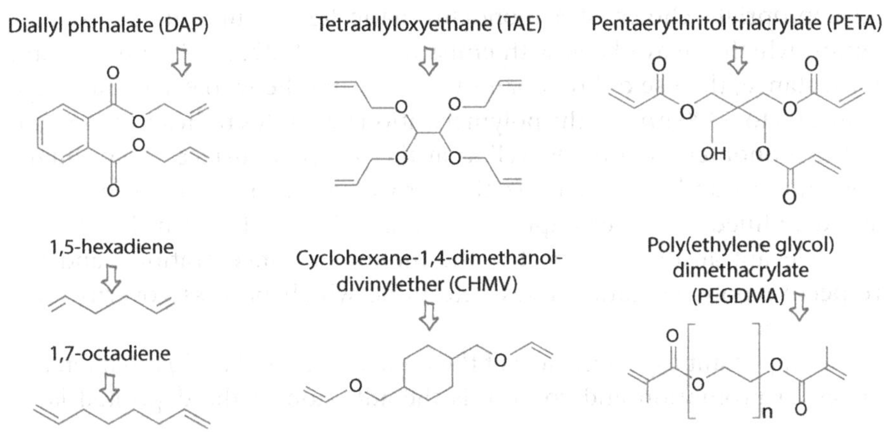

Figure 5.6 Chemical structures of several reactive monomers used in various studies making use of photo-polymerization for polymer surface modification.

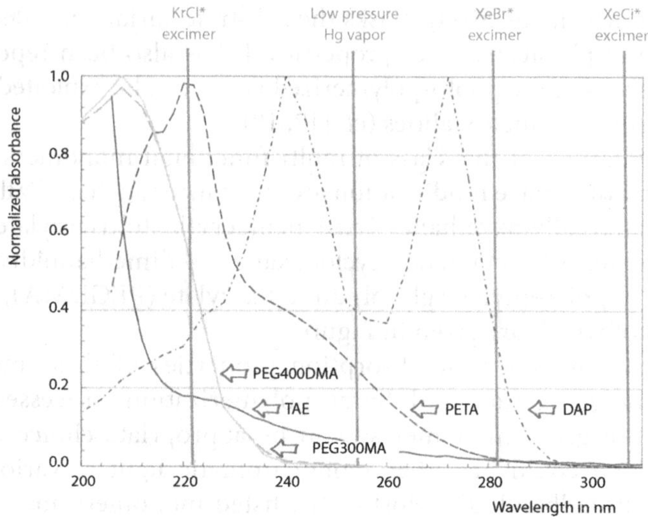

Figure 5.7 UV absorption spectra of the reactive monomers PETA, TAE, DAP, and the PEG methacrylates, PEG300MA and PEG400DMA, in ethanol (illustration adapted from [18]). All spectra are normalized to their absorption maximum. To indicate which UV source is absorbed strongly by the monomers and thus is able to initiate homo-polymerization, the wavelengths of relevant monochromatic UV lamps are also highlighted in the graph.

reactions in the monomer solution resulting in homo-polymerization. Layers are then deposited on the substrate surface ("grafting to", reaction type (V)).

While both types of polymerizing mechanisms may be initiated by an appropriate choice of a monochromatic light source using excimer lamps, which are available with emissions at 172, 222, 281, and 308 nm, for instance, the use of broadband UV covering the whole spectral range from 200 to >300 nm for the polymerization generates radicals in the bulk of the monomer solution as well as at the polymer surface. Accordingly, both homo- and co-polymerization processes are principally initiated, as exemplified by the examples of TAE and PETA shown in Figure 5.8. The relevant absorption spectra, the monomer concentrations, and the respective polymerization rates determine which process governs layer formation.

An important aspect in view of the topic considered in this chapter, i.e., adhesion promotion and control, is the adhesion of the deposited layer

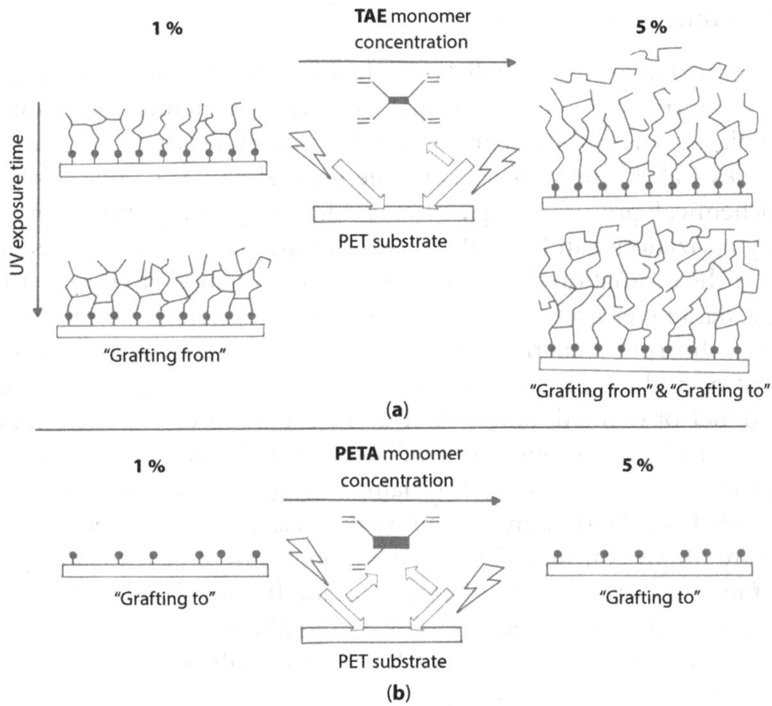

Figure 5.8 Schematic representation of the layer growth for TAE (a) and PETA (b) monomers depending on the monomer concentration and UV-exposure time. Dominating mechanism is indicated. Note that at high concentration of TAE (scheme (a)) both "grafting from" and "grafting to" occur simultaneously.

itself. The bonding of layers generated by graft-co-polymerization is always covalent, and its adhesion is accordingly strong. Homo-polymerized layers may be covalently bonded or just adhere to the substrate surface depending on the effective penetration of photons and whether radicals are generated at the substrate surface (i.e. boundary between substrate and deposited layer) or only in the bulk phase. Rather similar to plasma-polymerized layers, UV-induced homo-polymerization may, thus, produce highly cross-linked and mechanically strong networks, which nevertheless are easily detached from the substrate. Concerning porous substrates such as textile fabrics and non-wovens, the interpenetration of the photopolymerized layers between the macroscopic networks of the fibers is an important factor in the resulting overall adhesion property.

5.2.2 General Process Design

In the context of practical application, the first consideration must be made about the form of polymer substrate to be treated, namely three-dimensional plastics vis-à-vis planar, potentially moving products such as films or textiles (fabrics or webs). While the general concept of all discussed photochemical processes, e.g., photooxidation, photo-grafting, or photo-polymerization, can be applied to both types of substrates, one has to consider the irradiation dose (often also termed 'fluence") as the governing process parameter aside from UV lamp spectrum and material parameters, e.g., absorption characteristics. The irradiation dose is defined by the incident optical energy per unit area, and the optical energy is given by the product of emitted lamp power and irradiation or exposure time in most practical applications. The irradiated area is defined by masks directly above the substrate, windows in the lamp casing, or the distance from lamp to the substrate. With a given lamp layout (casing), most reported treatments are simply controlled by varying the exposure time with lamp power and lamp to substrate distance being constant. It immediately follows from this approach that the treatment is not really suitable to three-dimensional substrates unless complex robotics for lamp or substrate movement and intelligent lamp control are involved.

In contrast, the concept is easily applied and very favorable for planar substrates, whether static or moving. Important examples of the latter are films and textile fabrics, which are industrially processed as large-area material webs at considerable speed. The exposure time in these cases is simply defined by the processing speed and size of the irradiation zone, which can technically be determined by the mask or window used.

The design of the photooxidation process for moving webs is straightforward by adding a suitable UV source to an existing machine at the suitable processing step, e.g., directly before a coating step. It must be mentioned that depending on the state of the material or potential pre-treatments, a preceding cleaning step might be advisable. This aspect is, for instance, important in the context of textiles, which can carry significant amounts of residuals from preceding processing steps.

Because of the more complex grafting or polymerization processes, it is highly advantageous that, in contrast to similar plasma-based processes, the reactive medium or 'atmosphere' of the photochemical process can be gaseous as well as liquid, as has been pointed out before. The photochemical process then follows the concept to apply the reactive medium as a finish using simple padding, (screen) printing, or spraying. In addition, the treatment is executed under ambient conditions. However, flushing with inert gas might be required to avoid competing reactions with ambient oxygen for certain processes. This holds especially for a polymerizing process, where reactions with oxygen might terminate the cross-linking of monomers. Again, pre-cleaning might be necessary to remove residuals from preceding processing steps.

Accordingly, the general concept of photo-grafting and polymerization processes includes the following steps as presented in Figure 5.9. It includes (a) pre-cleaning depending on the state of the raw product ranging from simple rinsing to a two-step extraction in water and in the solvent of the monomer solution, (b) 'finishing' with the monomer solution, (c) irradiation, and (d) optional post-cleaning to remove residual monomers. The last step is especially advisable when porous substrates such as textile fabrics are treated. In these cases, it must be assumed that the UV radiation does not penetrate the macro-pores of the material and the unaffected monomers are present. An example of the technical concept of this process – in the example modification by photo-polymerization - is sketched in Figure 5.10.

The UV sources used in the vast majority of experiments described in this chapter are gas discharge lamps, where the nature of the gas fill defines the emission characteristics. Common broadband mercury vapor lamps are commercially available in sizes appropriate to cover the width of the material webs at a comparatively low cost. The specific UV emission characteristics of mercury vapor lamps are usually defined by doping the gas fill but have a main emission band from 200 to 320 nm. It is

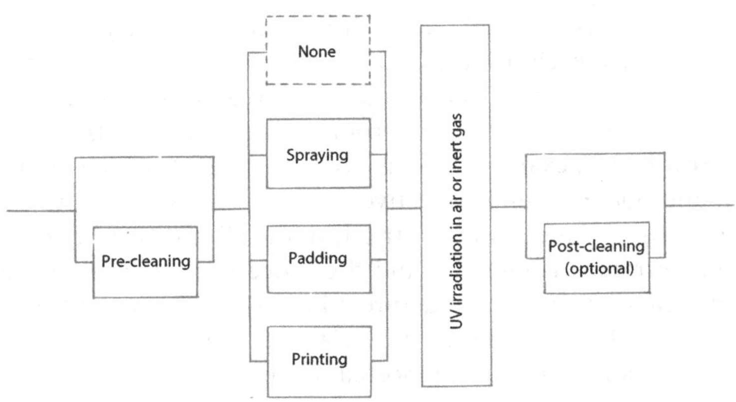

Figure 5.9 Schematic presentation of the sequence of steps involved in a photochemical treatment of polymer substrates, namely (a) optional pre-treatment, (b) definition/application of reactive atmosphere/medium, (c) irradiation, and (d) optional post-treatment.

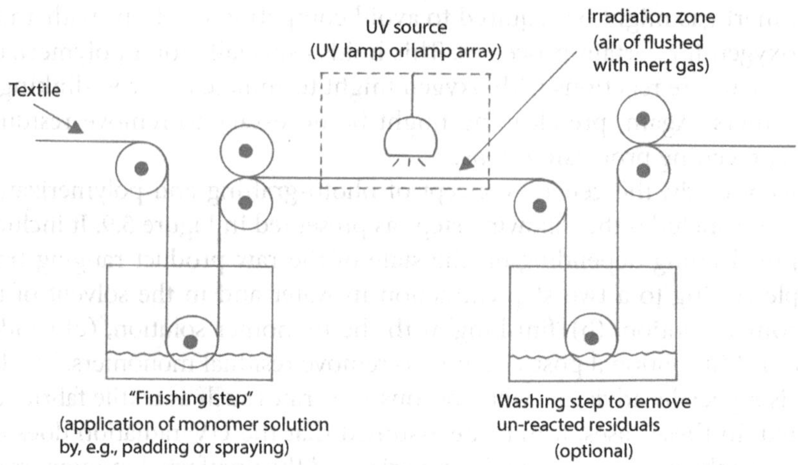

Figure 5.10 Concept for the photochemical surface modification of films or textile fabrics in an industrial inline process.

important to note that these lamps, aside from the main UV spectrum, generally emit over a wide spectral range, including certain visible and even infrared emissions. Therefore, despite using infrared filters during exposure to broadband lamps over long periods, notable sample heating occurs even above the glass transition temperatures of certain polymers. As an alternative, some of the experiments described here used monochromatic light sources of different designs. The so-called excimer lamps offer an interesting design – also gas discharge lamps – which makes use of excited dimer gases (= excimer) such as xenon (Xe_2*), krypton-chloride (KrCl*), and xenon-chloride (XeCl*). These gases are stable only for a short time in their excited states and emit monochromatic UV photons on their subsequent decay in the ground states. Excimer lamps emit in very narrow bands, examples being KrCl* and XeCl*, which emit at 222 ± 5 nm and 308 ± 5 nm, respectively. It is important to note that the emission of the excimer lamps is strictly in the UV range, i.e., no thermal radiation (infrared) is emitted, and the process is totally non-thermal. In recent years, several manufacturers have offered UV-emitting LEDs (light-emitting diode - LED) for material processing, but no application of such light sources has been reported in papers relevant to the topic of this chapter.

It must be noted that many of the cited studies for scientific purposes used pure monomer solutions without any cross-linkers (reaction starters). Several papers, however, describe the use of benzophenone (diphenylmethanone), 4-chlorobenzophenone or hydroxylcyclohexylacetophenone as photoinitiators, which would allow to shorten the reaction (irradiation) times by a factor of 10 or more (see, e.g., [19]). This has the considerable advantage of allowing the choice of a wavelength that is highly absorbed by the polymer, resulting in radical generation with a very high quantum yield and short processing time. Aside from processing speed and production efficiency, this aspect is also of importance because of minimization of thermal stress to the polymer substrate as discussed before, and – depending on the actual irradiation conditions – also in view of potential generation of ozone, which could create requirements in terms of exhaustion of the irradiation zone.

5.3 Using Photo-Addition and Photo-Grafting to Promote the Adhesion Property of Hydrophobic Substrates

Considering adhesion of coatings on films and textiles, bonding agents on films and plastics, or adhesion between fiber and matrix in composite materials, the majority of substrates in question are made of hydrophobic polymers such as PET, PP, polyamides, or aramids. Innovative materials such as bio-based polymers have gained importance only in recent years. The adhesion phenomena in these structures are directly related to surface free energy (SFE) and its components. The straightforward way to optimize adhesion is to increase SFE of the hydrophobic substrates, in general, by increasing the polar component of the SFE by appropriate modification of the surface chemistry. For further insight, we refer to a paper by Mittal [20], which gives a comprehensive discussion of the relationship between SFE and bond strength.

In the context of photochemical surface modification treatments of these hydrophobic substrates, photooxidation, i.e. addition of oxygen and subsequent generation of hydrophilic OH-groups following reaction type (III), is the simplest process concept to improve adhesion. In their 1994 paper, Zhang *et al.* proved the concept to improve the adhesion of metal layers on PMMA substrates using monochromatic lasers and lamps (excimer lamps) [21].

In the interest of high process efficiency, early papers describe the photooxidation of different polymers in ozone atmosphere – in contrast to air – in closed reaction chambers making use of the simultaneous effect of UV activation of the surface and production of reactive oxygen species in the atmosphere due to the radiation-induced decomposition of ozone. In 1995, McIntyre and Walzak [1] proposed the combined ultraviolet-ozone oxidation process (UVO) as an effective pre-treatment of 3-D polymer products made of PET and PP, which are difficult to process by the established corona treatment. While no quantitative values are given, the authors reported increased surface free energy and, accordingly, enhanced adhesion to paints, adhesives, and printing ink following treatments for 1 to 2 min. Mathieson and Bradley [2] studied the pre-treatment of polyethylene (PE) and poly(ether ether ketone) (PEEK) substrates and reported considerable oxidation and improved wettability for PE and PEEK and improved adhesion to epoxy systems. The surface modification of PE and polypropylene (PP) by UVO was investigated by Gongjian *et al.* [3]. XPS analysis showed that oxygen-containing functional groups such as carbonyl, carboxyl, and ether groups were produced on the sample surface. The authors reported that after UVO treatment at “suitable conditions”, the polyolefin samples’ hydrophilicity, adhesion property, and dyeability were improved significantly.

An important comment by Gongjian *et al.* [3] referred to the polymer damage, mainly due to intense ozone activity. Gongjian *et al.* stated that a short duration UVO treatment has a notable effect on sample properties. It is important to note that the simultaneous generation of ozone and subsequent decomposition to reactive oxygen species occurs principally in every photochemical process conducted in the presence of oxygen [18, 22]. This is especially notable where Vacuum Ultraviolet (VUV) sources such as, e.g., Xe_2^* excimer lamps (172 nm) are used. Various authors have described surface degradations similar to effects observed with air plasma treatments and also certain aging of the treatment effect indicated by a slow increase in the water contact angle measured after specific time intervals [23, 24].

It is easy to see that the UVO concept is most suitable for three-dimensional (3-D) objects and difficult to adapt to planar, potentially moving substrates such as films and textiles. Because of this Praschak *et al.* [9] and Netravali and Bahners [25] studied the “direct” photooxidation by UV irradiation in ambient air and its effect on the wettability and adhesion property of PET films and technical fabrics. These authors found that the water contact angle (advancing angle) on irradiated PET film surface decreased from 76° as measured on pristine film with increasing irradiation dose, i.e., incident energy per unit area and exposure time until it reached a saturation value of approximately 45°. X-ray photoelectron spectroscopy (XPS) studies as well as a cationic dye test showed an increase in the number of carboxylic carbons in relation to the other carbon types [9].

Netravali and Bahners described the effect of a similar surface modification on the adhesion of a condensed polyacrylate commonly used as pigment binding paste in pigment printing and pigment dyeing on PET film [25]. The film was irradiated using a KrCl* excimer lamp (222 nm) with varying irradiation doses. The variation was achieved by either varying the irradiation duration (a few seconds to 3 minutes) or the distance between lamp and substrate. Compared to the pristine PET film, the lamp treatment resulted in a fourfold increase in printing paste adhesion. The data also indicated that the effect saturates at higher irradiation doses, which is also of importance in view of the aspect of material damage due to ozone generation, as discussed before. Already in the experiments reported by Praschak *et al.* [9] as well as Netravali and Bahners [25], however, ozone reactions were considered to be negligible, as the short duration of the treatment was not sufficient for noticeable reactions of the PET, and the ozone concentration could be further kept minimal by permanent exhaustion of the irradiation zone.

As polyacrylate is also commonly used as a binding paste in pigment printing and pigment dyeing in textile finishing, it was of interest to study the potential of photooxidation for the pre-treatment of textiles [9,25]. The quality of pigment prints is commonly characterized by a standardized rubbing fastness test, during which the sample is rubbed against another (uncoated) fabric for a defined period of time and number of cycles. The abrasive removal of the pigment-containing paste ('print coat') is quantitatively described by the so-called rubbing fastness (also called 'crock' fastness) in grading from 0 (worst) to 5 (best). Measurements in the stated reports showed that the dry rubbing fastnesses of various samples increased from 1 to about 4 on UV-treated fabrics on average.

While the use of air or oxygen as the reactive atmosphere affects the adhesion between materials via hydrogen bonds and the concept is still applied to good effect in more recent studies (see [26, 27]), the concept of reaction III allows for far-reaching modifications to the surface chemistry of the substrate and the resulting effect on adhesion property, if the reactive "atmosphere" consists of molecules with functional end groups. The resulting "photo-grafting" establishes functional end groups on the substrate surface, which allow specific interactions between the modified substrate and coatings, adhesives, printing pastes or matrix polymers.

Yamada *et al.* [5], for instance, photo-chemically grafted hydrophilic methacrylic acid (MAA), acrylic acid (AA), methacrylamide (MAAm), and acrylamide (AAm) from monomer solutions onto polyethylene (PE) plates. Using contact angle measurements and XPS, the authors confirmed that contact angles became constant as the PE surface was fully covered with grafted chains, the amount of which was in the order MAA > AA > MAAm. The adhesive strength increased with the grafted amount for all the monomers used. Balart *et al.* [28] obtained similar results who photo-grafted methyl methacrylate (MMA) on PP. The treatment allowed increasing surface activity and also had a positive effect on PP-PP adhesive joints.

Ranby and Guo [6] devised a continuous photo-grafting process to establish acrylic acid (AA), and acrylamide (AAm) on PP fiber and film and commercial high-strength polyethylene (HSPE) yarn. The monomers were applied in solution with benzophenone (BP), 4-chlorobenzophenone (4-CBP), or hydroxylcyclohexylacetophenone (HHA) as photoinitiators, and the presoaked samples were irradiated using a high-pressure mercury lamp for just 5–20 s. XPS analysis confirmed that the treated surfaces were covered with a 2–8 nm thick layer of grafted polymer and the grafting efficiency was 70–80%. Grafting of the high-strength polyethylene yarn with AAm increased the adhesion to epoxy resin by a factor of 3–4 without affecting the mechanical properties of the fiber by no more than a few percent. In the case of a commercial HSPE yarn, grafting with AA increased the epoxy adhesion from 0.25 to 1.3 N/nm^2, i.e., by a factor of 5. Ranby and Guo [6] also studied the wettability of blown PP film and the adhesion of epoxy following photo-grafting of AAm. With increasing grafting, the water contact angle decreased from 90° to 20°. The adhesion to the epoxy resin increased from 0.35 to 1.7 N/mm^2 (by a factor of about 5) when the film surface was completely covered by the grafted polymer.

An interesting variant to the photo-grafting using UV lamp irradiation was studied by Netravali and co-workers [29-31], who used pulsed UV excimer lasers as light sources to improve the fiber/matrix adhesion in certain composite materials. As described elsewhere [25], irradiation with pulsed UV lasers was observed to generate microstructures on the fiber surface, which might already help to improve adhesion through mechanical contributions. The microstructures were generated by melting and recrystallization of thin surface layers. In addition, the introduction of polar groups to the surface can be expected to increase adhesion further if elementary processes following the intense energy deposition by the laser beam – e.g., ablation – do not prevent the grafting step.

In order to study the concept, Netravali and Song [29] irradiated ultra-high molecular weight PE (UHMWPE) fibers in a variety of environments, including air, ammonia, argon, and helium with XeCl (308 nm) excimer laser. A significant increase in the polar acid-base component of the surface free energy could be observed following laser irradiation in air and ammonia. There was also an excellent correlation between the acid-base component of the surface free energy and the interfacial shear strength (IFSS) between the treated fibers and epoxy resin. No additional effect from surface roughening occurred if fibers were irradiated in argon and helium atmospheres. The effect of the environment was extended by Zeng and Netravali [30], who used a diethylenetriamine (DETA) solution as the liquid environment to photo-graft UHMWPE fibers. In general, water contact angles for fibers irradiated in DETA were lower than

Table 5.2 Overview of processes and applications related to adhesion to hydrophobic substrates.

Process	Reactive medium	UV source	Substrate	Increase in adhesion to	Ref(s).
Photo-oxidation (UVO)	Ozone		PET, PP	Ink, paint, adhesives	[1]
			PE, PEEK	Epoxy	[2]
			PE, PP	(Unspecified)	[3]
Photo-oxidation	Ambient air	Excimer lamps (various)	PMMA	Metal layers	[21]
		Excimer lamp (172 nm)	PET film & textiles	Polyacrylate pastes	[9, 25]
			Polyurethane elastomers		[26]
			Pectin/ poly(vinyl alcohol) blends		[27]
Combined µ-roughening & oxidation	Air, ammonia	Excimer laser (308 nm)	UHMWPE fibers	IFSS in epoxy matrix	[29]
Photo-grafting	MAA, AA, MAAm, AAm		PE plates	(Unspecified)	[5]
	AA, AAm	Hg lamp	PP fiber & film, HSPE yarn	Epoxy	[6]
	MMA		PP	PP	[28]
Combined µ-roughening & photo-grafting	DETA	Excimer laser (248 or 308 nm)	UHMWPE & LC-PET (Vectran®) fibers	IFSS in epoxy matrix	[30, 31]

for those irradiated in air, and the IFSS was higher. As expected, the acid-base component of the SFE was significantly higher for fibers irradiated in DETA than those irradiated in air. Similar results were obtained from experiments using a krypton-fluoride (KrF) excimer laser (248 nm) [30]. In another study, Zeng and Netravali [31] used the excimer laser treatment to improve the adhesion of the fully aromatic liquid crystalline polyester Vectran® fibers to epoxy resin. These fibers have inert surface characteristics. In contrast to the UHMWPE fibers, the laser treatment in air had only little effect on surface chemistry, whereas some improvement in IFSS was apparently due to the laser-generated surface microroughness. If Vectran® fibers were irradiated in the presence of DETA significant increases in the acid-base component of the surface free energy and thus in IFSS were observed. In summary, the experiments of Netravali and co-workers give a strong indication of the versatile and effective modification of polymer surfaces by photo-grafting.

An overview of processes and applications in this area of adhesion-related problems is given in Table 5.2.

5.4 Enhancing Adhesion of Hydrophobic Materials on Hydrophilic Substrates – Biobased Composites as Case Study

In recent years, there has been a growing demand for sustainable materials such as matrix polymers and reinforcement fibers in the design of composite materials. Besides the well-known natural fibers like flax or hemp, industrially produced endless (continuous) fibers such as, e.g., polylactide (PLA) or regenerated cellulose (viscose) are interesting reinforcement fibers due to their more uniform mechanical properties. It has been shown that composites with viscose fibers embedded in a thermoset matrix have specific mechanical properties similar to glass fiber reinforced composites [32,33]. A major factor prohibiting wider use of viscose as reinforcement fiber is the poor adhesion of the highly hydrophilic viscose fiber to hydrophobic matrix polymers such as PP and other polyolefins [34]. A common approach is to modify the matrix polymer with bonding agents such as maleic anhydride to effect interfacial interaction and enhance fiber-matrix adhesion [35–38]. The nature of adhesion in these modified composite systems was studied by Felix and Gatenholm [35] by means of Fourier transform infrared spectroscopy (FTIR). It was found that the surface modifying agent, i.e. maleic anhydride, is covalently bonded to the fibers through esterification, which results in significant increase in adhesion [35].

Several reviews have described examples of modification techniques to establish functional groups on the surfaces of natural and synthetic cellulose fibers as an alternative way to modify the matrix polymer [39–41]. Among other concepts, UV treatments of cellulose reinforcement fibers have been reported (cf. [42]).

Gassan and Gutowski [43] compared the effects of corona and UV treatments on the properties of jute fiber-epoxy composites. The authors used a gas discharge lamp with a main UV emission at 254 nm in air and reported that the UV treatment, i.e., photooxidation of single fibers or yarns, led to a significantly higher effect on polarity than was achieved with corona treatment. It was stressed, however, that with increasing intensity of both types of treatments, yarn tenacity was decreased due to excessive surface oxidation, so that a compromise between an increase in fiber polarity and a decrease in fiber strength had to be sought to improve the overall mechanical properties of the jute/epoxy composites for both corona discharge and UV radiation. Zaman *et al.* [44] studied the effects of the photooxidation treatment on the mechanical properties of jute fiber-reinforced polyethylene/polypropylene composites. UV radiation at different intensities was applied both on matrices and jute. Mechanical properties of the composites were found to increase significantly following irradiation of the jute fibers as well as the matrix polymers. The results on UV-based treatments of plant cellulose reinforcement fibers (cf. [43–46]) were analyzed by Mohit and Selvan in a 2018 review which summarized the effects of a UV-treatment as follows [47]: "*The UV radiation can also be applied to change surface features for plant cellulose fiber laminates. The Ultra-violet analysis of sole jute yarn and fibrils of jute were examined which shows the greater achievement in polarity while comparing with corona plasma treatment. In composites, yarn tenacity and polarity are modified by extending time of treatment at consistent distance of sample or reduce the distance. The improvement in overall structural properties of plant cellulose fiber composites and reduced with increment in number of ultra-violet treatment due to decrease in fiber strength and increment in fiber surface polarity.*"

Aside from the direct irradiation in air – i.e., photooxidation – Khan *et al.* studied the UV-based grafting of jute yarn with methyl methacrylate (MMA), butyl acrylate (BA), and styrene (ST) and found enhanced yarn tenacity and water uptake, both of which can be expected to affect composite properties [48]. There is no explicit reference to properties of composite materials using the modified yarn for reinforcement in this paper, however. Recently, Masudur Rahman *et al.* studied photo-grafting of pineapple leaf fibers (PALFs) using a solution of 2-hydroxyethyl methacrylate (HEMA) and methyl methacrylate (MMA) in the presence of methanol and a photoinitiator [49]. The fibers were then used as reinforcement fibers in a biobased composite with polylactide (PLA) as the matrix polymer. Experimental results revealed that composites made of PALF grafted with 30% HEMA and 40% MMA exhibited the highest improvement in mechanical properties compared to the untreated PALF/PLA composite. The optimized solutions were further enhanced by adding 0.5–1.5% of urea, with the best mechanical features achieved using a 1% urea concentration. Detailed analysis also showed that incorporating photo-grafted PALF also enhanced the interfacial adhesion to the matrix.

In the given system of reinforcement fiber and matrix polymer, dispersion forces (London-van der Waals forces) can be expected to govern interaction at the fiber/matrix interface, and the work of adhesion to be determined by the surface free energies of the materials involved and the interfacial free energy. A vital issue introduced by the cellulose fiber is the very high amount of water adsorbed on the surface and in the pores of the cellulosic material. As could be shown by differential scanning calorimetry (DSC) and atomic force microscopy (AFM), this water content was quite evident even after longer thermal conditioning (cf. [50]). Due to the existence of the adsorbed water layer one has to consider the affinity of the matrix polymer to the water-rich polar surface of the viscose fibers as the major factor governing adhesion. Therefore, it must be concluded that the effects on polarity, which photooxidation processes might achieve, will always be counteracted by the significant sorption of water on and in the reinforcement fibers and will be of little significance for fiber/matrix adhesion.

In contrast, the deposition of a (thin) hydrophobic layer by means of photo-deposition, as described in section 5.2, would effectively mask the highly hydrophilic and water adsorbing fiber surface and thus prohibit water adsorption. Fiber/matrix adhesion would subsequently be based solely on interfacial forces between this thin layer and matrix polymer. The feasibility of this concept was shown as early as 1996 by Joly *et al.*, who partially masked cotton fibers with C_3H_7-, C_8H_{17}-, or $C_{18}H_{37}$-alkyl isocyanates and observed a significant decrease in hydrophilicity and in water sorption [51].

In a more recent study, Bahners *et al.* [50] evaluated the potential of hydrophobic organic thin layers to completely mask the hydrophilic fibers of viscose fiber reinforcement fabrics and increase adhesion in composites made from these fabrics and PP matrix polymer. Following the described concept, the surface modification was achieved by photo-polymerization of PETA or DAP in iso-propanol (cf. Figure 5.6) and irradiation by broadband UV. The photochemical modification led to a significant decrease in wettability of the otherwise highly hydrophilic and water adsorbing viscose fibers and increased their affinity towards non-polar substances as proven by inverse gas chromatography (IGC). Both effects were found to improve fiber/matrix adhesion, which was characterized by pull-out test and interfacial shear strength (IFSS). The poly-PETA deposition could increase IFSS by 30 %, which was still significantly lower than the increase of 130 % achieved using the maleic anhydride modified matrix polymer (MAPP). In the composite, however, the photochemically treated fibers yielded similar or even improved properties compared to the MAPP samples in terms of tensile and impact strengths. The latter is assumed to be related to the superior energy transfer by the thin organic layers forming a fiber/matrix interphase.

The last aspect must be seen in the context of modern design principles of composite materials, namely biomimetic concepts. Besides fiber/matrix adhesion, the ratios of Young's modulus and tensile strength of fiber and matrix polymers – or, more precisely their differences – are influential for the mechanical properties of a fiber-reinforced composite. Even if adhesion is strong, an abrupt change in mechanical properties at the fiber/matrix interface will result in reduced tenacity and insufficient damping of the composite.

It is an important observation that many composite plant structures exhibit graded transitions in mechanical properties at the fibers, fibrils, and matrix interfaces. An excellent example of a role model in biomimetics is bamboo. Bamboo is a very resilient and versatile material. The structure of the bamboo column corresponds to a composite material because the ground tissue is traversed by strong and stiff vascular bundles, which represent a fiber-like reinforcement in the axial direction. The distribution of the reinforcing vascular bundles varies over the cross section of the column. In addition, the cell walls of the vascular bundle themselves have a stiffness gradient [52]. Bamboo thus exhibits modulus gradients on several levels, which results in a perfect adaptation to flexural load.

It was shown earlier that the principles of the gradient structuring of plant cell walls could successfully be transferred to the modification of technical composite materials if appropriate coatings or thin layers are applied to the reinforcing fibers and serve as interphases or interlayers, creating graded transitions of mechanical properties, e.g., the elastic modulus, from fiber to matrix. Jones, for instance, reported the positive effect of polysiloxane layers on unmodified and plasma-treated carbon fibers on the mechanical properties of carbon fiber-epoxy composites [53].

Photochemical polymerization of vinyl monomers was employed by Kuttner *et al.* [54] to deposit brush-like polystyrene and poly(methyl methacrylate) layers on glass fiber surfaces. Analysis by Kuttner *et al.* confirmed that the "grafting-from" mechanism yields an excellent adhesion of the growing interphase layer on the fiber surface and a graded transition of mechanical properties by means of "stiffness mapping". Kuttner *et al.* summarized the essence of their findings by stating that "*Composite materials, which are essentially multiphase materials, rely on synergetic effects by combination of materials of different mechanical properties. The chemical formulation of (polymeric) interphases can be used to adjust and control the mechanical performance of the composite as they can serve as a compatibility agent, as well as mediating mechanical gradient at the same time. Apart from simple matchmaking, the interphase can be used to control the interfacial adhesion and therewith the energy absorption capability of composites.*" [54].

Given this background, Müssig *et al.* [55] evaluated the potential of the organic poly-PETA thin-layers described above (cf. also [50]) to mimic natural structures and to increase both adhesion and damping behavior of cellulose fiber-reinforced PP composites at the same time. As indicated by the results on impact behavior reported in their earlier paper [50], Müssig and co-workers found a significant effect on fatigue strength under cyclic load. After 1,000,000 loading and de-loading cycles, the residual strength

Table 5.3 Overview of processes and selected applications related to improving fiber/matrix adhesion in bio-based composites made of hydrophilic reinforcement fibers.

Process	Reactive medium	UV source	Reinforcement fiber	Matrix	Ref(s).
Photo-oxidation	Ambient air	Lamp (254 nm)	Jute	Epoxy	[43]
		Broadband	Jute	PP, PE	[44]
		Broadband	Wood fiber	PP	[46]
Photo-grafting	MMA, butyl acrylate, styrene		Jute	-	[48]
	HEMA + MMA		Pineapple leaf fiber	PLA	[49]
Photo-polymerization	PETA, DAP	Broadband	Viscose	PP	[50, 55]

of the photo-chemically modified composites was 60 to 70 % of the original strength. In contrast, the composite strength dropped to <60 % in the case of the composites made of pure and unmodified fiber and matrix materials and even less (50 %) in the case of the conventional system using maleic anhydride grafted polypropylene (MAPP) as matrix polymer, which remarkably proved to be inferior to the unmodified samples [54].

An overview of processes and selected applications in this area of fiber/matrix adhesion in bio-based composites are given in Table 5.3.

5.5 Biosystems: Cell and Protein Adhesion, Antifouling Surfaces

The control of cell growth and biofilm formation is a major aspect in designing materials serving as implants or biomedical microdevices (BioMEMS). In contrast to the coated fibers and composite materials discussed in the preceding sections of this chapter, cell growth and adhesion of biosystems may be wanted or unwanted depending on the application of materials and medical products. On the one hand, it is generally accepted that protein adsorption following material implantation is an initial critical step that ultimately may lead to deleterious fibrosis and fibrin deposition [56, 57] or may lead to clogging of artificial blood vessels. On the other hand, cell attachment and proliferation are essential factors for surgical reconstruction, where implant materials serve as scaffolds for rapid cell growth (see, e.g., [58]). It is interesting to note that given this background, the majority of recent papers on this topic focus on the *minimization of adhesion* of proteins and cells, contrary to the goal to improve adhesion addressed in the paper reviewed in the previous sections.

The governing factors for cell growth and adhesion are the adsorption and adhesion of proteins, e.g., fibrinogen, on the surfaces in question. A good overview of the relevant mechanisms determining protein adsorption and adhesion is given by Zhang [59]. Proteins are large and amphiphilic molecules and are intrinsically surface-active. As detailed by Zhang, the adsorption mechanism is based on four fundamental interactions, (I) ionic or electrostatic interactions, (II) hydrophobic interactions, (III) hydrogen bonding, and (IV) van der Waals forces (cf. Figure 5.11). In the case of most technical polymers, hydrophobic interactions are assumed to be the governing factor. The hydrophobic interaction is defined as the interaction of non-polar groups in aqueous media. Non-polar surface groups – 'patches' [59] – are known to exist on proteins and interact with a

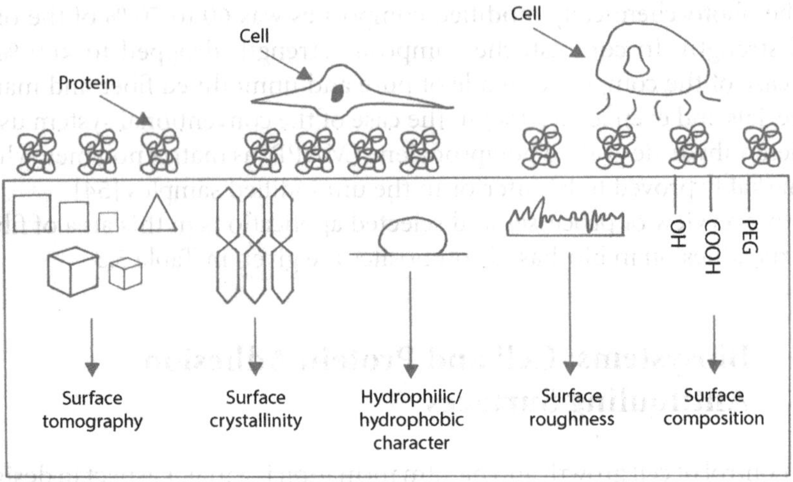

Figure 5.11 Principal factors governing biosystem-substrate interaction; illustration adapted from a scheme proposed by Zhang [59].

non-polar substrate surface, a process entropically driven by the release of water from the hydrophobic surface domains.

Accordingly, the precise definition of a hydrophilic or hydrophobic character of a surface is advantageous to control protein and cell adhesion. Low adhesion is generally expected from super-hydrophobic surfaces, but, according to the mechanisms sketched above, also from easier to prepare surfaces with high surface free energy. This was experimentally proven and quantified by Xu and Siedlecki [60], who measured the adhesion force of bovine serum albumin (BSA) to surfaces of varying SFE. They found that the adhesion force decreased dramatically if SFE exceeded 45 mJ/m^2. More specific requirements with respect to the chemical functionality of protein- and cell-resistant surface layers have been linked to their kosmotropic properties [61]. 'Kosmotropes' or 'kosmotropic' solutes stabilize proteins in an aqueous solution.

Gumpenberger *et al.* proposed a photochemical concept to *enhance* the adhesion and proliferation of human endothelial cells (HECs) [62] and muscle cells [63] on hydrophobic polytetrafluoroethylene (PTFE). The polymer surfaces were modified for several minutes by irradiation with Xe2*-excimer lamp radiation (172 nm) in an ammonia atmosphere. The photochemical modification resulted in a significant increase in the number of adhered cells and in the formation of a confluent cell layer after 3–8 days of incubation. It is important to note that the results were comparable to those obtained on standard cell cultivation petri dishes made of polystyrene, which served as a benchmark. In their subsequent study, Gumpenberger and co-workers [64] combined their photochemical concept – i.e., irradiation in NH_3 atmosphere - with a subsequent grafting with amino acid alanine from aqueous solution and found increased adhesion of rat aortic smooth muscle cells (SMCs) and mouse fibroblasts (3T3 cells) as well as human endothelial cells to the modified PTFE. The authors concluded that PTFE modified following the described method could be a promising material for the fabrication of artificial vascular protheses and implants or the cultivation of skin substitutes.

In contrast to the improvement of cell adhesion to polymer surfaces, photochemical surface treatments have found increasing interest in recent years as an avenue to minimize protein and cell adhesion to substrates used for medical application, for example, to increase blood compatibility or tissue engineering. Partly, the modified surfaces are micro-patterned for spatial control of cell attachment. Most published studies make use of the concepts of photo-grafting or photo-initiated cross-linking.

For instance, Bae *et al.* [65] designed heparinized micropatterned surfaces for the spatial control of the attachment of human mesenchymal stem cells (hMSCs). A substrate polymer synthesized from poly(styrene-co-vinylbenzyl N,N-diethyl-dithiocarbamate) (poly(ST-co-VBDC)) was coated with poly(ethylene glycol) methacrylate (PEGMA), and subsequently UV irradiated, which led to a polymerized poly-PEGMA layer. After chemically immobilizing heparin onto this PEGMA-grafted surface, surface micropatterning was achieved by additional photo-polymerization of PEGMA using a photomask. Even after several hours of incubation the surface exhibited the desired effect, as hMSCs adhered only to the heparinized areas of the surface, while no cell adhesion was detected on the hydrophilic poly-PEGMA domains. This type of surface design gives a good example of an additional advantage of the photochemical process that can be used for patterning. As described in detail by Bahners *et al.* [66], patterning may be achieved by masked irradiation, as applied by Bae *et al.* [65], or by, e.g., using ink-jet or sieve printing for patterned application of the reactive medium. Similar photo-crosslinked micropatterned polyelectrolyte multilayer films for the spatial control of cellular adhesion were also described by Chien *et al.* [67]. Another concept was proposed by Pidhatika *et al.* [68], who presented the fabrication of substrates with dual-functional surfaces, which contain bioactive spots on a biopassive background. The fabrication employed a two-step process based on photo-initiated cross-linking. Poly(N,N-dimethylacrylamide) (PDMAA) solution containing 5 mol% of benzophenone was coated on various organic substrates and UV irradiated in a highly controlled process to *partially* cross-link the PDMAA to prepare the samples. The addition of laminin as bioactive ligand to the partially cross-linked layer followed by further UV irradiation enabled local C,H-insertion cross-linking (termed CHiC process by Pidhatika *et al.* [68]) between the polymer and the bioactive ligands providing the desired dual-functional surface.

A paper by Li *et al.* [69] described a method to graft poly(N-vinyl pyrrolidone) (PNVP) onto biomedical poly(styrene-b-(ethylene-co-butylene)-b-styrene) (SEBS) elastomer substrate by UV-induced graft polymerization following a plasma surface pre-treatment. The authors reported significantly suppressed protein adsorption and platelet adhesion. Employing the same photo-grafting concept, Kawasaki and Iwasaki [70] reported improved blood compatibility of poly(ether ether ketone) (PEEK) following covalent grafting of methacryloyl-functionalized phospholipid polymers using a narrow-band light source (350 ± 50 nm). The authors stated that the method effectively reduced plasma protein adsorption regardless of the type of protein, resulting in a significant improvement in blood compatibility of the modified surfaces.

Zhirong *et al.* [71] studied the effect of sugar-based trisiloxane surfactants on protein adhesion to polyurethane (PU) substrate in a two-step process. Photo-grafting was employed as the first step to establish 2-carboxy-ethyl acrylate on the PU substrate. In the second step, either N-siloxanegluconamides (Si(3)NGA) or N-siloxanelactobionamides (Si(3)NLA)) was introduced by conjugating the amino groups and acid groups. Zhirong *et al.* found that the modification with Si(3)NGA obviously suppressed protein adsorption and platelet adhesion, while the Si(3)NLA-modified substrate exhibited higher protein adsorption and a tendency for blood coagulation [71].

PEEK is an interesting polymer to replace metals in orthopedic implants. In order to minimize protein adhesion, Nakano *et al.* [72] employed the photo-polymerization process to establish layers of zwitterionic polymers on PEEK substrates. In their study, aqueous solutions of 2-(methacryloyloxy)ethyl phosphorylcholine (MPC) and the zwitterionic monomer 2-[2-(methacryloyloxy)ethyldimethylammonium] ethyl benzophenoxy phosphate (MBPP) were applied in varying monomer ratios to the samples, which were subsequently UV irradiated for 180 min. The formation of grafted layers of zwitterionic polymers was confirmed analytically (XPS) and by apparent surface properties as water contact angle and significantly reduced surface friction, which was attributed to preserving the hydration state of grafted polymers under compressive stress. Bacterial adhesion was completely suppressed.

In contrast to the aforementioned researchers, Tanaka and Iwasaki [73] employed the simple photooxidation process of a photo-reactive phospholipid polymer, which was spin-coated on silicon wafers to create non-fouling surfaces. Poly[2-methacryloyloxyethyl phosphorylcholine (MPC)-ran-N-methacryloyl-(l)-tyrosinemethylester (MAT)] (P(MPC/MAT)) was synthesized by conventional radical polymerization and subsequently coated on the wafers from an aqueous solution to prepare the substrates. Following photooxidation by 254nm UV irradiation, active species such as catechol and quinone were alternately generated in the co-polymer. The authors were able to show that the general properties of the modified wafers were mainly governed by the oxidized MAT units in P(MPC/MAT), which were advantageous not only for polymer adhesion to a solid surface but also for protein conjugation with the adhered polymers and the amount of immobilized protein.

Besides the overall aspect of cell adhesion (or non-adhesion), cell orientation, migration and proliferation are the main parameters for tissue and organ formation and thus are central factors in the design of materials such as polymer implants. It is well known that patterns of adhesive and non-adhesive domains are instrumental in controlling cell growth on these materials. A concept based on photochemical surface modification was proposed by Matsuda and Sugawara in 1996 [74]. The authors prepared patterned structures of cell adhesive and non-adhesive domains, namely stripes, on polystyrene tissue culture dishes to stimulate cellular migratory and orientation responses. For the modification, the dishes were coated with photo-reactive poly(N,N-dimethylacrylamide-co-3-azidostyrene) co-polymer, masked with patterned Pyrex glass, and subsequently UV irradiated for 2 min (UV dose 2.2 mW/cm^2, no further information supplied). After washing to remove the non-irradiated coating, stripes of covalently bonded nonionic hydrophilic material remained on the dishes. The authors could show that endothelial cells proliferated only within the non-irradiated (hydrophobic) channels between the hydrophilic stripes.

Following a totally different concept, Mayuri *et al.* [75] used UV radiation to cross-link fibroporous mats made of poly(ethylene-co-vinyl alcohol) (EVAL) intended to be used as leukodepletion filters with the hydrophilic monomer 2-hydroxyethyl acrylate (HEA) in order to modify the EVAL surface chemistry. EVAL is considered a suitable material for blood-contacting applications. The modified fibroporous mats were prepared by fiber formation in an electrical field ("electro-spinning") from a solution of EVAL and HEA with benzophenone as a photoinitiator and subsequent UV irradiation to induce cross-linking. *In vitro* hemocompatibility tests revealed that the modified EVAL surface exhibited significantly lower hemolytic activity, protein adsorption, and platelet adhesion than neat EVAL. Mayuri *et al.* found a significant reduction in the adhesion of red blood cells, while white blood cell consumption remained above 90%. This indicates that the photochemical cross-linking of EVAL and HEA substantially improves the hemocompatibility of the EVAL mats and selectivity if used as leukodepletion filters.

It was mentioned earlier that unwanted cell growth might also cause bio-fouling on technical products exposed to biologically active elements. Relevant product classes include filter membranes for water purification, products used in medical engineering or the pharmaceutical industry, outdoor textiles as, e.g., used for roof construction in modern architecture, and geotextiles. Again, protein adsorption and adhesion are the governing factors, as the formation of a biofilm usually can be divided into four basic steps [76], (I) adsorption of proteins and anchoring of cells to adsorbed protein, e.g., fibrinogen, (II) networking of cells to form monolayers, in most cases by means of extracellular polymeric substances, (III) formation of microcolonies, and (IV) biofilm formation through a network of monolayers and microcolonies, as schematically shown in Figure 5.12.

In tissue engineering as well as in the design of BioMicroElectro-Mechanical systems (BioMEMS) such as DNA chips, it is well known that poly(ethylene glycol)s (PEGs) deposited as a coating on a surface has the potential to depress protein adsorption and cell growth (see for instance

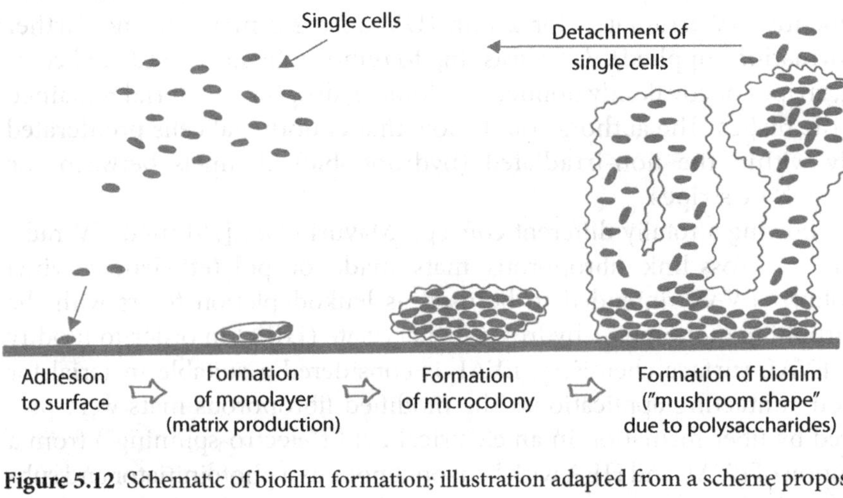

Figure 5.12 Schematic of biofilm formation; illustration adapted from a scheme proposed by Vasudevan [77].

[78–80]). PEGs are hydrophilic. Thus, they minimize hydrophobic protein-surface interactions and are also non-toxic. An essential aspect in designing a PEG-based surface modification of a technical product is the method chosen for the PEG immobilization. For silicon substrates to be used in BioMEMS, various techniques for PEG immobilization have been proposed, e.g., by physical adsorption [81] or by covalent immobilization [82]. In view of the latter, photochemical process concepts have been increasingly studied and have proven favorable in terms of ease of the process, effect, and durability, as exemplified in the following section.

In a study by DeFife *et al.* [83], photochemical immobilization technology was utilized to covalently couple various polymers to silicone rubber either at multiple points along a polymer backbone or at the endpoint of an amphiphilic chain. PEGs and a synthetic polypeptide were synthesized with 4-benzoylbenzoic acid (BBA) as the photo-reactive moiety, applied to the substrate, and subsequently UV irradiated (no detailed information on the actual irradiation step was provided). Overall, the photochemical immobilization of hydrophilic polymers effectively prohibited the adsorption of proteins. All coating variants suppressed the adsorption of fibrinogen and immunoglobulin G, and most also inhibited fibroblast growth by 90–99% compared to uncoated silicone rubber. The most passivating coating variant in the study by DeFife *et al.* was an endpoint immobilized polypeptide that reduced protein adsorption, inhibited fibroblast growth by 90%, elicited low cytokine secretion from monocytes, and reduced fibrous encapsulation by 33% [83].

Ahmed *et al.* [84] described a complex two-step process to establish PEG films on tissue culture polystyrene (TCPS) as well as glass slides. In the first step, layers of 4-benzoyl-[2,2] paracyclophane (Benzoyl-PPX) were deposited on the substrates using CVD polymerization as a photo-reactive platform, onto which PEG was conjugated via UV illumination. In the second step, solutions of PEG200, PEG2000, and PEG20000 of various concentrations were placed on 96-well Benzoyl-PPX plates. The plates were then illuminated at 365 nm for exposure times ranging from 1 to 60 min. The authors showed that the deposited PEG films effectively reduced protein adsorption and cell attachment. The low-fouling properties of the PEG films were positively correlated with the molecular weight and concentration of PEG. The authors found that a PEG film, thicker than 16 nm and with a water contact angle of 30°, is a prerequisite for effective inhibition of cell attachment. It should be noted that the stated water contact angle is in good correlation with the results from the earlier mentioned study by Xu and Siedlecki [60]. Ahmed *et al.* [84] further found that poly(N-isopropyl acrylamide), PNIPAAm, could also be tethered on the Benzoyl-PPX via UV illumination and possessed thermo-responsive properties. Intact cell sheets could be released from the PNIPAAm film by decreasing culture temperature. In summary, the results of these authors indicate that Benzoyl-PPX is an excellent photo-reactive platform for the conjugation of functional polymers in a photochemical process for the modulation of cell attachment.

The direct photo-induced immobilization of PEGs on textiles in a one-step process was shown by Praschak *et al.* [10] for the example of PEG200 and by Bahners *et al.* for PEG600DMA [85]. The latter study [85] confirmed the principal effect of UV cross-linked poly-PEGs on protein and bacteria adsorption on textiles. In these experiments, fabrics made of PET and PP were irradiated in the presence of PEG600DMA, and tested for the adsorption of enzyme catalase and of *Escherichia coli* (*E. coli*). According to qualitative analysis using ninhydrin as marker dye, the enzyme adsorption could be reduced significantly. Additionally, the adsorption of the *E. coli* on PET fabric treated in the presence of PEG600DMA was reduced to 36 ± 15 % compared to the untreated fabric.

A comprehensive study by Bahners *et al.* compared the deposition characteristics, layer morphology, and protein adsorption of mono- and di-methacrylated PEGs of comparatively short chain lengths [86]. The chemical structures of methacrylated PEGs are given in Figure 5.13. The experiments showed that the photo-induced polymerization of di-methacrylated PEGs resulted in the deposition of thick layers that effectively masked the fabric's texture and pore system, which would prohibit these fabrics for

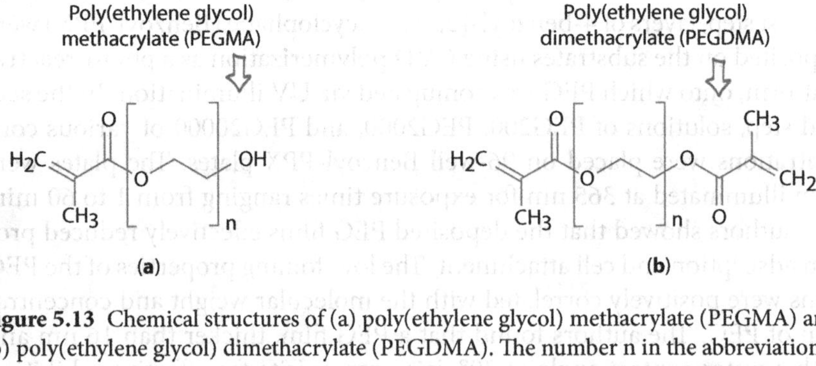

Figure 5.13 Chemical structures of (a) poly(ethylene glycol) methacrylate (PEGMA) and (b) poly(ethylene glycol) dimethacrylate (PEGDMA). The number n in the abbreviations of PEG derivatives, e.g., PEG300MA or PEG400DMA, always indicates the average molar mass of the PEG in g/mol, whereas the number n in the structure formula gives the average number of PEG monomer units, e.g., n = 8 for PEG400DMA and n = 14 for PEG600DMA.

filtration purposes. Significantly thinner and non-obtrusive layers were obtained from the mono-methacrylated PEG300MA with maximum add-on of the order of 1 mass-%. In the context of bio-fouling, studies of protein adsorption indicated that the highest reduction in adsorption of bovine serum albumin (BSA) could be observed for samples prepared using PEG300MA under conditions where the drop penetration time was at its minimum [86].

The striking effect on biofilm formation and adhesion is exemplified by the photographs and SEM micrographs in Figure 5.14, taken from untreated and photo-chemically modified PET fabrics after three months of use in water filtration in a large outdoor water tank (photographs and SEM-micrographs taken from [87]). The outdoor conditions allowed

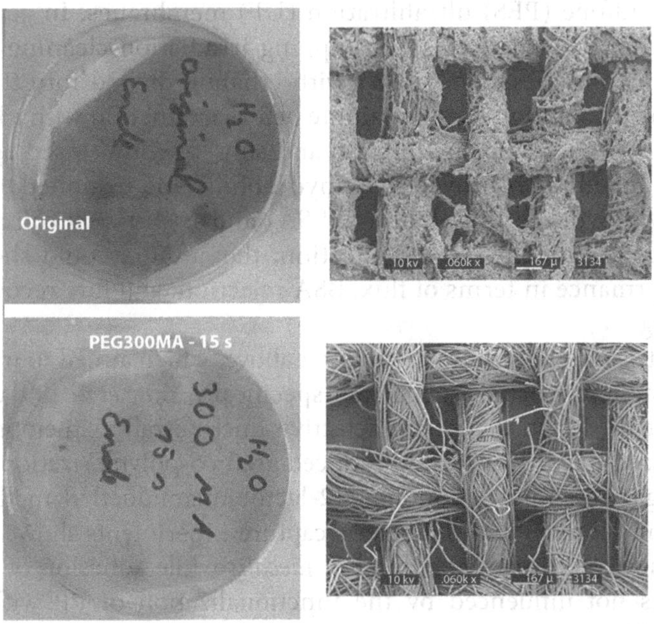

Figure 5.14 The effect on biofilm formation and adhesion (from [18]): Photographs (left) and SEM micrographs (right) of PET fabric samples after 14 weeks storage in the biologically active aqueous environment. The samples were rinsed with distilled water after removal from the water tank. The "original" sample (top) refers to the untreated fabric, which was only cleaned from production residuals. The modified sample (bottom) was irradiated in the presence of PEG300MA with broadband UV; the irradiation time was 15 s.

biological activity producing chlorophytae, zygnemae, and algae-consuming microorganisms in the water. On removal, all samples showed distinct biofouling. Samples modified with PEG300MA not only appeared to be slightly brighter, but the residual biofilm was also more or less completely removed from the modified sample [87].

A different concept to photo-chemically creating low bio-adhesion surfaces was reported by Lin *et al.* [88]. These authors report using a newly synthesized molecule, 4-azidotetrafluorobenzoyl tetraaniline (ATFB-TANI), to couple tetraaniline (TANI) to other materials via perfluorophenylazide photochemistry. TANI is the smallest conjugated repeat unit of polyaniline and has similar hydrophilic and antibacterial properties but can be readily dissolved in common organic solvents for facile processing. Besides several carbon-based materials, the authors also studied the modification of large area PET films for biomedical applications and polyethersulfone (PES) ultrafiltration (UF) membranes. In general, the modification followed a cleaning-dipping-irradiation-cleaning protocol, i.e., PES membranes were rinsed with ethanol, dipped into the ATFB-TANI ethanol solution for one minute before irradiation with a handheld 254 nm UV source for one minute, and subsequently rinsed again. The modified membranes proved to be hydrophilic and exhibited extraordinarily low bovine serum albumin (BSA) and *Escherichia coli* adhesion. In view of the membranes' application, the authors could show superior performance in terms of flux, BSA rejection, and flux recovery after biofouling.

In contrast to the studies mentioned above, which aimed to increase or suppress adhesion of biosystems non-specifically, Yang *et al.* [89] described a concept to design a bacteria-selective microfiltration membrane. The authors used the process of UV-induced graft co-polymerization to establish a sugar-containing monomer (2-lactobionamidoethyl methacrylate, LAMA) on PP membranes. Bacteria capture experiments showed that the membranes could selectively bind *E. faecalis* while adhesion of *S. maltophilia* was not influenced by the functionalization of PP with grafted poly(LAMA).

An overview of processes and selected applications in the field of surface modification for the control of biosystem adhesion to polymers is given in Table 5.4.

Table 5.4 Overview of processes and selected applications related to biosystem adhesion (refer to the text for material abbreviations). It shall be noted that, with few exceptions, broadband UV lamps were used as light sources in the studies.

Process	Substrate	Reactive medium	Effect	Ref(s).
Photo-oxidation	PTFE	Ammonia	Enhanced adhesion and proliferation of HEC and muscle cells	[62, 63]
	Si wafer	P(MPC/MAT)	Protein conjugation	[73]
Photo-oxidation, chem. grafting	PTFE	NH_3, amino acid alanine	Increased adhesion of SMC, 3T3 and HEC cells	[64]
Photo-grafting	PEEK	Functionalized phospholipid polymers	Reduced plasma protein adsorption	[70]
	Silicone rubber	PEG/BBA	Suppressed protein adsorption	[83]
Photo-grafting, chem. grafting	PU	2-Carboxy-ethyl acrylate, Si(3)NGA	Suppressed protein adsorption and platelet adhesion	[71]
Photo-grafting (patterned)	Polystyrene	Poly(N,N-dimethylacrylamide-co-3-azidostyrene	Cell adhesive and non-adhesive domains	[74]
Photo-polymerization (patterned)	Poly(ST-co-VBDC)	PEGMA	Suppressed adhesion of hMSCs	[65]
	Various	PDMAA + laminin	Local bioactivity	[67]

(*Continued*)

Table 5.4 Overview of processes and selected applications related to biosystem adhesion (refer to the text for material abbreviations). It shall be noted that, with few exceptions, broadband UV lamps were used as light sources in the studies. (*Continued*)

Process	Substrate	Reactive medium	Effect	Ref(s).
Photo-polymerization	PNVP	SEBS elastomer	Suppressed protein adsorption and platelet adhesion	[69]
	PEEK	Zwitterionic polymers	Suppressed bacterial adhesion	[72]
	PET	PEGMA, PEGDMA	Reduced protein adsorption, anti-fouling	[85–87]
	PES	ATFB-TANI	Reduced BSA and *e-coli* adhesion, anti-fouling	[88]
	PP	LAMA	Selective bacteria attachment	[89]

5.6 Summary

The photochemical modification of a polymer surface is generally initiated by activating the surface following absorption of energetic photons, homolytic bond breakage, and generation of radicals, which form the basis for subsequent chemical reactions at the very surface of the irradiated material. UV radiation is most favorable for this kind of surface modification given the high photon energy, which in the spectral range below 300 nm exceeds the binding energy of C-H-bonds. Depending on the environment – "atmosphere" – in which the irradiation occurs, the fundamental processes of photooxidation, photo-grafting of functional end groups, or photo-polymerization of highly cross-linked thin layers are possible. All stated processes can be used to adapt the chemical property of the surface and, accordingly, the related adhesion properties to specific cases of application. Based on the literature discussed in detail in this chapter, the following scheme helps to relate the main photochemical process designs to adhesion control in major areas of technical products, here characterized by the nature of the polymer substrate and adhesion-related problem:

	Coating, printing, adhesive bonding	Composite	Bio-system (reduction of adhesion)
Hydrophobic polymer	Increase in SFE by Photo-oxidation or Photo-grafting	Increase in SFE by Photo-oxidation or Photo-grafting	Photo-grafting of functional groups or polymer
		Interlayer deposition by photo-polymerization	Deposition of water-rich network by photo-polymerization
Hydrophilic polymer, bio-based, natural		Masking of surface by photo-polymerization	
		Interlayer deposition by photo-polymerization	

In many examples of "classic" problems of adhesion, e.g., coating polymers and pastes, adhesives, printing pastes and inks on hydrophobic polymers, the UV irradiation in ambient air is an attractive alternative to corona or atmospheric plasma treatment. The photochemical process is even more advantageous if the concept of photo-grafting of hydrophilic end groups is applied, as the modification is covalent and exhibits high stability/durability, which is in contrast to the known decrease of the effects achieved by, e.g., the corona treatment. A similar decrease – albeit far slower – was also reported in early papers on photooxidation. Also, in the case of the photochemical process, the "atmosphere" can be a liquid, e.g., monomer solution, which is easy to apply by conventional processing techniques without the need for special reactors.

The photo-polymerization concept can also be used to deposit thin layers of totally differing chemical properties on the surface of a polymer substrate. Here, multi-functional monomers are used, which can also be applied as solutions by conventional methods. Depending on the UV absorption properties of the monomers, the layer deposition follows the "grafting-to" or the "grafting-from" mechanism, the latter producing a very thin and covalently bonded layer. The resulting complete masking of the substrate not only allows to adapt, for instance, a basically hydrophilic polymer to hydrophobic coating material and improve the substrate-coating adhesion; it is also possible to exploit the bulk properties of the deposited thin layers, which act as *interphase* or *interlayer* in the substrate-coating system. This concept has been successfully applied in composite material design with excellent damping behavior. The photo-polymerization technique was also evaluated with the aim to establish hydrophilic networks of PEGs and other similar compounds, to control the adhesion of bio-systems. The cross-linked PEGs create a water-rich environment, which prohibits the adsorption of proteins and cells to the basically hydrophobic polymer substrate and consequently *reduces* adhesion. A related application of the concept can be found in medical materials and products such as implants or artificial blood vessels, where unwanted cell growth and blood clogging must be minimized or even totally suppressed. There are also interesting examples of technical polymer products used for (outdoor) architecture or water filtration.

Acknowledgement

Our special thanks goes to Johanna Malika Müssig for her creative drawings and illustrations.

References

1. S. N. McIntyre and M. J. Walzak, New UV/ozone treatment improves adhesiveness of polymer surfaces, *Modern Plastics* 72, 79-83 (March 1995).
2 I. Mathieson and R. H. Bradley, Improved adhesion to polymers by UV/ozone surface oxidation, *Int. J. Adhesion Adhesives* 16, 29-31 (1996).
3. B. Gongjian, W. Yunxuan and H. Xingzhou, Surface modification of polyolefine by UV light/ozone treatment, *J. Appl. Polym. Sci.* 50, 2397-2402 (1996).
4. S. Edge, S. Walker, W. J. Feast and W. F. Pacynko, Surface modification of polyethylene by photochemical grafting with 2-hydroxyethylmethacrylate, *J. Appl. Polym. Sci.* 47, 1075-1082 (1993).

5. K. Yamada, H. Tsutaya, S. Tatekawa and M. Hirata, Hydrophilic and adhesive properties of polyethylene plates grafted with hydrophilic monomers, *J. Appl. Polym. Sci.* 46, 1065-1085 (1992).
6. B. Ranby and F. Z. Guo, "Surface-photografting": new applications to synthetic fibers, *Polym. Adv. Technol.* 5, 829-836 (1994).
7. G. G. Bogoeva, N. Y. Pimonenko and G. Petrov, Photo-induced acrylamide graft polymerization onto polyamide-6, *Textile Res. J.* 63, 51-57 (1993).
8. E. Uchida, Y. Uyama and Y. Ikada, Grafting of water-soluble chains onto a polymer surface, *Langmuir* 10, 481-485 (1994).
9. D. Praschak, T. Bahners and E. Schollmeyer, PET-surface-modification by treatment with monochromatic excimer-UV-lamps, *Appl. Phys.* A66, 69-75 (1998).
10. D. Praschak, T. Bahners and E. Schollmeyer, Excimer UV lamp irradiation induced grafting on synthetic polymers, *Appl. Phys.* A 71, 577-581 (2000).
11. T. Bahners, T. Textor and E. Schollmeyer, Photon-based processes for the surface modification of synthetic fibers, in: *Polymer Surface Modification: Relevance to Adhesion, Vol 3*, K.L. Mittal (Ed.), pp.97-124, CRC Press, Boca Raton, FL (2004).
12. *Scopus® - Expertly curated abstract & citation database* <https://www.scopus.com> (date of execution 2021-11-30).
13. V. L. Vakula and E.B. Orlov, The photo-ageing of polyethylene and its effect on polyethylene adhesion to steel, *Polym. Sci. USSR*, 15, 3061-3067 (1973).
14. N.J. Van Eck and L. Waltman, Visualising bibliometric networks, in: *Measuring Scholarly Impact: Methods and Practice*, Y. Ding, R. Rousseau and D. Wolfram (Eds.), pp. 285–320. Springer (2014).
15. N.J. Van Eck and L. Waltman, *VosViewer Manual.* (2021) <https://www.vosviewer.com/download>
16. J. F. Rabek, *Mechanisms of Photophysical Processes and Photochemical Reactions in Polymers*, J. Wiley & Sons, New York (1987).
17. T. Bahners and J.S. Gutmann, Using bulk properties of photo-polymerized thin layers for fiber modification, *Surface Innovations* 4, 14-22 (2016).
18. T. Bahners and J.S. Gutmann, UV-based photo-chemical surface modification of textile fabrics, in: *Textile Finishing – Recent Developments and Future Trends*, K.L. Mittal and T. Bahners (Eds.), pp. 261-296, Wiley-Scrivener, Beverly, MA (2017).
19. K. Opwis, A. Wego, T. Bahners and E. Schollmeyer, Permanent flame retardant finishing of textile materials by a photochemical immobilization of vinyl phosphonic acid, *Polym. Degrad. Stab.* 96, 393-395 (2011).
20. K.L. Mittal, The role of the interface in adhesion phenomena, *Polym. Eng. Sci.* 17, 467-473 (1977).
21. J.-Y. Zhang, H. Esrom, U. Kogelschatz and G. Emig, Modification of polymers with UV excimer radiation from lasers and lamps, *J. Adhesion Sci. Technol.* 8, 1179-1210 (1994).

22. M.J. Walzak, S. Flynn, R. Foerch, J.M. Hill, E. Karbashewski, A. Lin and M. Strobel, UV and ozone treatment of polypropylene and poyl(ethylene terephthalate), in: *Polymer Surface Modification: Relevance to Adhesion*, K.L. Mittal (Ed.), pp.253–272, CRC Press, Boca Raton, FL (1996).
23. K. Gotoh, Wettability and surface free energies of polymeric materials exposed to excimer ultraviolet light and particle deposition onto their surfaces in water, in: *Polymer Surface Modification: Relevance to Adhesion*, Vol. 3, K.L. Mittal (Ed.), pp.125-137, CRC Press, Boca Raton, FL (2004).
24. B. Viallet and E. Daran, Adhesion improvement of epoxy resist to a benzocyclobutene layer: Application to nanoimprint lithography, in: *Polymer Surface Modification: Relevance to Adhesion*, Vol. 4, K.L. Mittal (Ed.), pp.231–240, CRC Press, Boca Raton, FL (2007).
25. A. Netravali and T. Bahners, Adhesion promotion in fibers and textiles using photonic surface modifications, *J. Adhesion Sci. Technol.* 24, 45–75 (2010).
26. K.M. Zia, I.A. Bhatti, M. Barikani, M. Zuber and I. Shahazad, Surface characteristics of UV-irradiated polyurethane elastomers extended with α, ω-alkane diols, *Appl. Surf. Sci.* 254, 6754-6761 (2008).
27. J. Kowalonek, H. Kaczmarek and A. Dbrowska, Air plasma or UV-irradiation applied to surface modification of pectin/poly(vinyl alcohol) blends, *Appl. Surf. Sci.* 257, 325-331 (2010).
28. J. Balart, V. Fombuena, J.M. España, L. Sánchez-Nácher and R. Balart, Improvement of adhesion properties of polypropylene substrates by methyl methacrylate UV photografting surface treatment, *Materials & Design* 33, 1-10 (2012).
29. A. N. Netravali and Q. Song, Laser surface modification of ultra-high strength polyethylene fibers: Correlation between acid-base interactions and adhesion to epoxies, in: *Acid–Base Interactions: Relevance to Adhesion Science and Technology*, Vol. 2, K. L. Mittal (Ed.), pp. 525–537. CRC Press, Boca Raton, FL (2000).
30. J. Zeng and A. N. Netravali, KrF excimer laser modification of ultrahigh molecular weight polyethylene fibers for improved adhesion to epoxy resins, in: *Polymer Surface Modification: Relevance to Adhesion*, Vol. 3, K.L.Mittal (Ed.), pp. 159–182. CRC Press, Boca Raton, FL (2004).
31. J. Zeng and A. N. Netravali, XeCl excimer laser treatment of Vectran® fibers, *J. Adhesion Sci. Technol.* 20, 387-409 (2006).
32. A. Mader, E. Volkmann, R. Einsiedel and J. Müssig, Impact and flexural properties of unidirectional man-made cellulose reinforced thermoset composites, *J. Biobased. Mater. Bioenergy* 6, 481–492 (2012).
33. A Mader, A. Kondor, T. Schmid, R. Einsiedel and J. Müssig, Surface properties and fiber-matrix adhesion of man-made cellulose epoxy composites—influence on impact properties, *Composites Sci. Technol.* 123, 163–170 (2016).
34. C. Albano, J. Gonzalez, M. Ichazo and D. Kaiser, Thermal stability of blends of polyolefins and sisal fiber, *Polym. Degrad. Stab.* 66, 179–190 (1999).

35. J. Felix and P. Gatenholm, The nature of adhesion in composites of modified cellulose fibers and polypropylene, *J. Appl. Polym. Sci.* 42, 609-620 (1991).
36. T. Huber and J. Müssig, Fibre matrix adhesion of natural fibres cotton, flax and hemp in polymeric matrices analyzed with the single fibre fragmentation test, *Composite Interfaces* 15, 335–349 (2008).
37. P. Joseph, M. Rabello, L. Mattoso, K. Jospeh and S. Thomas, Environmental effects on the degradation behaviour of sisal fiber reinforced polypropylene composites, *Composites Sci. Technol.* 62, 1357–1372 (2002).
38. J. Müssig and N. Graupner, Test methods for fibre/matrix adhesion in cellulose fibre-reinforced thermoplastic composite materials: A critical review, *Rev. Adhesion Adhesives* 8, 68-129 (2020).
39. C. Jose, M. S. Thomas, B. Deepa, L. A. Pothan and S. Thomas, Adhesion and surface issues in biocomposites and bionanocomposites, *Rev. Adhesion Adhesives* 2, 173–225 (2014).
40. K. Li, R. Qiu and W. Liu, Improvement of interfacial adhesion in natural plant fiber-reinforced unsaturated polyester composites: A critical review, *Rev. Adhesion Adhesives* 3, 98–120 (2015).
41. J. Müssig and N. Graupner, Characterisation of fibre/matrix adhesion in biobased fibre-reinforced thermoplastic composites, in: *Textile Finishing: Recent Developments and Future Trends*, K.L. Mittal and T. Bahners (Eds.), pp 485–556, Wiley-Scrivener, Beverly, MA (2017).
42. M. Belgacem and A. Gandini, The surface modification of cellulose fibers for use as reinforcing elements in composite materials, *Composite Interfaces* 12, 41–75 (2005).
43. J. Gassan and V.S. Gutowski, Effects of corona discharge and UV treatment on the properties of jute-fibre expoxy composites, *Composites Sci. Technol.* 60, 2857–2863 (2000).
44. H.U. Zaman, M.A. Khan and R.A. Khan, Improvement of mechanical properties of jute fibers-polyethylene/polypropylene composites: effect of green dye and UV radiation, *Polym. Plast. Technol. Eng.* 48, 1130–1138 (2009).
45. Y. Zhou, M. Fan and L. Chen, Interface and bonding mechanisms of plant fiber composites: an overview, *Composites B* 101, 31–45 (2016).
46. R. Selden, B. Nystorm and R. Langstorm, UV aging of poly(propylene)/wood-fiber composites, *Polym. Composites* 25, 543–553 (2004).
47. H. Mohit and V. Arul Mozhi Selvan, A comprehensive review on surface modification, structure interface and bonding mechanism of plant cellulose fiber reinforced polymer based composites, *Composite Interfaces* 25, 629-667 (2018).
48. M.A. Khan, M.M. Rahman and K.S. Akhunzada, Grafting of different monomers onto jute yarn by *in situ* UV-radiation method: effect of additives, *Polym. Plast. Technol. Eng.* 41, 677–689 (2002).
49. A.N.M. Masudur Rahman, S. Alimuzzaman, and R.A. Khan, Performance evaluation of PLA based biocomposites reinforced with photografted PALF, *Tekstilec*, 64, 230-246 (2021).

50. T. Bahners, M. Kelch, B. Gebert, X.L. Osorio Barajas, T.C. Schmidt, J.S. Gutmann and J. Müssig, Improvement of fibre-matrix adhesion in cellulose/polyolefin composite materials by means of photo-chemical fibre surface modification, *Cellulose* 25, 2451-2471 (2018).
51. C. Joly, R. Gauthier and M. Escoubes, Partial masking of cellulosic fiber hydrophilicity for composite applications. Water sorption by chemically modified fibers, *J. Appl. Polym. Sci.* 61, 57–69 (1996).
52. X. Wang, H. Ren, B. Zhang, B. Fei and I. Burgert, Cell wall structure and formation of maturing fibres of moso bamboo (*Phyllostachys pubescens*) increase buckling resistance, *J. R. Soc. Interface* 9, 988-996 (2012).
53. F.R. Jones, A review of interphase formation and design in fibre-reinforced composites, *J. Adhesion Sci. Technol.* 24, 171-202 (2010).
54. C. Kuttner, M. Tebbe, H. Schlaad, I. Burgert and A. Fery, Photochemical synthesis of polymeric fiber coatings and their embedding in matrix material: morphology and nanomechanical properties at the fiber–matrix interface, *ACS Appl. Mater. Interfaces* 4, 3484–3492 (2012).
55. J. Müssig, M. Kelch, B. Gebert, J. Hohe, M. Luke and T. Bahners, Improvement of the fatigue behaviour of cellulose/polyolefin composites using photo-chemical fibre surface modification bio-inspired by natural role models, *Cellulose* 27, 5815–5827 (2020).
56. J.M. Anderson, Mechanisms of inflammation and infection with implanted devices, *Cardiovasc Pathol* 2, 33-41 (1993).
57. T.A. Horbett, Principles underlying the role of adsorbed plasma proteins in blood interactions with foreign materials, *Cardiovasc Pathol* 2, 137-148 (1993).
58. S. Salehi, M. Czugala, M. Fathi, S.H. Javanmard, T.Bahners, J.S. Gutmann, B.B. Singer and T.A. Fuchsluger, Poly (glycerol sebacate)-poly (ε-caprolactone) blend nanofibrous scaffold as intrinsic bio- and immunocompatible system for corneal repair, *Acta Biomater.* 50, 370-380 (2017).
59. M. Zhang, Biocompatibility of materials, in *Biomaterials and Tissue Engineering*, D. Shi (Ed.), pp. 83-137, Springer, Berlin (2004).
60. L.C. Xu and C.A. Siedlecki, Effects of surface wettability and contact time on protein adhesion to biomaterial surfaces, *Biomaterials* 28, 3273–3283 (2007).
61. R.S. Kane, P. Deschatelets, and G.M. Whitesides, Kosmotropes form the basis of protein-resistant surfaces, *Langmuir* 19, 2388-2391 (2003).
62. T.Gumpenberger, J.Heitz, D.Bäuerle, H.Kahr, I.Graz, C.Romanin, V.Svorcik and F.Leisch, Adhesion and proliferation of human endothelial cells on photochemically modified polytetrafluoroethylene, *Biomaterials* 24, 5139-5144 (2003).
63. V. Svorcik, K. Walachova, J. Heitz, T. Gumpenberger and L. Bacakova, Muscle cell adhesion on polytetrafluorethylene modified by UV irradiation, *J. Mater. Sci. Lett.* 20, 1941-1942 (2002).
64. J. Heitz, V. Svorcik, L. Bacakova, K. Rockova, E. Ratajova, T. Gumpenberger, D. Bauerle, B. Dvorankova, H. Kahr, I. Graz and C. Romanin, Cell adhesion

on polytetrafluoroethylene modified by UV-irradiation in an ammonia atmosphere, *J. Biomed. Mater. Res. A* 67, 130-137 (2003).

65. J.W. Bae, J.H. Choi, T.E. Kim, K.D. Park, J.Y. Kim, Y.D. Park and K. Sun, Heparinized micropatterned surfaces for the spatial control of human mesenchymal stem cells, *J. Bioact. Compat. Polym.* 24, 493-506 (2009).
66. T. Bahners, T. Mayer-Gall, W. Mölter-Siemens and J.S. Gutmann, Patterned functionalization of textiles using UV-based techniques for surface modification – patterned wetting behavior, in: *Advances in Contact Angle, Wettability and Adhesion, Vol. 3*, K.L. Mittal (Ed.), pp147-166, Wiley-Scrivener, Beverly, MA (2018).
67. H.W. Chien, T.Y. Chang and W.B. Tsai, Spatial control of cellular adhesion using photo-crosslinked micropatterned polyelectrolyte multilayer films, *Biomaterials* 30, 2209-2218 (2009).
68. B. Pidhatika, N. Zhao, M. Zinggeler and J. Rühe, Surface-attached dual-functional hydrogel for controlled cell adhesion based on poly(N,N-dimethylacrylamide), *J. Polym. Res.* 26, 69 (2019).
69. X. Li, S. Luan, H. Yang, H. Shi, J., Zhao, J. Jin, J. Yin and P. Stagnaro, Surface modification of poly(styrene-b-(ethylene-co-butylene)-b-styrene) elastomer via photo-initiated graft polymerization of poly(ethylene glycol), *Colloids Surfaces B* 93, 127-134 (2012).
70. Y. Kawasaki and Y. Iwasaki, Surface modification of poly(ether ether ketone) with methacryloyl- functionalized phospholipid polymers via self-initiation graft polymerization, *J. Biomater. Sci. Polym. Ed.* 25, 895-906 (2014).
71. X. Zhirong, D. Binbin, Y. Shunjie, D. Shanshan, Z. Chunyu, S. Miao and G. Yuan, Surface modification of polyurethane via covalent immobilization of sugar-based trisiloxane surfactants, *Designed Monomers Polymers*, 18, 284-294 (2015).
72. H. Nakano, Y. Noguchi, S. Kakinoki, M. Yamakawa, I. Osaka and Y. Iwasaki, Highly durable lubricity of photo-cross-linked zwitterionic polymer brushes supported by poly(ether ether ketone) substrate, *ACS Appl. Biomaterials* 3, 1071-1078 (2020).
73. M. Tanaka and Y. Iwasaki, Photo-assisted generation of phospholipid polymer substrates for regiospecific protein conjugation and control of cell adhesion, *Acta Biomater.* 40, 54-61 (2016).
74. T. Matsuda and T. Sugawara, Control of cell adhesion, migration, and orientation on photochemically microprocessed surfaces, *J. Biomed. Mater. Res.* 32, 165-173 (1996).
75. P.V. Mayuri, A. Bhatt, R. Joseph and P. Ramesh, Effect of photografting 2-hydroxyethyl acrylate on the hemocompatibility of electrospun poly(ethylene-co-vinyl alcohol) fibroporous mats, *Mater. Sci. Eng. C* 60, 19-29 (2016).
76. A. Carré and K.L. Mittal (Eds.), *Surface and Interfacial Aspects of Cell Adhesion*, CRC Press, Boca Raton, FL (2011).
77. R. Vasudevan, Biofilms: microbial cities of scientific significance. *J. Microbiol. Exp.* 1, 84-98 (2014).

78. S.I. Jeon, J.H. Lee, J.D. Andrade and P.G. de Gennes, Protein—surface interactions in the presence of polyethylene oxide: I. Simplified theory, *J. Colloid Interface Sci.* 142, 149-158 (1991).
79. L.D. Unsworth, H. Sheardown, and J.L. Brash, Protein resistance of surfaces prepared by sorption of end-thiolated poly(ethylene glycol) to gold: Effect of surface chain density, *Langmuir* 21, 1036-1041 (2005).
80. C. Zhou, V.K. Khlestkin, D. Braeken, K. De Keersmaecker, W. Laureyn, Y. Engelborghs, and G. Borghs, Solvent-controlled organization of self-assembled polymeric monolayers on gold: An easy approach for the construction of protein resistant surfaces, *Langmuir* 21, 5988-5996 (2005).
81. M. Zhang and M. Ferrari, Reduction of albumin adsorption onto silicon surfaces by Tween 20, *Biotechnol. Bioeng.* 56, 618–625 (1997).
82. S. Sharma, R.W. Johnson, and T.A. Desai, Evaluation of the stability of nonfouling ultrathin poly(ethylene glycol) films for silicon-based microdevices, *Langmuir* 20, 348-356 (2004).
83. K.M. DeFife, M.S. Shive, K.M. Hagen, D.L. Clapper and J.M. Anderson, Effects of photochemically immobilized polymer coatings on protein adsorption, cell adhesion, and the foreign body reaction to silicone rubber, *J. Biomed. Mater. Res.* 44, 298–307 (1999).
84. I.N. Ahmed, R. Chang, M.-C. Keng, H.-C. Chien, H.-Y. Chen and W.-B. Tsai, Immobilization of functional polymers on poly(4-benzoyl-pxylylene-co-pxylylene) films via photochemical conjugation for modulation of cell adhesion, *Colloids Surfaces B* 174, 360–366 (2019).
85. T. Bahners, M. Milster, K. Opwis, A. Wego and E. Schollmeyer, Photochemical surface modification for the control of cell growth on textile substrates, in: *Polymer Surface Modification: Relevance to Adhesion Vol 5*, K.L. Mittal (Ed.), pp.107-124, CRC Press, Boca Raton, FL (2009).
86. T. Bahners, K. Klingelhöller, M. Ulbricht, A. Wego and E. Schollmeyer, Photo-chemical surface modification for the control of protein adsorption on textile substrates, *J. Adhesion Sci. Technol.* 25, 2219-2238 (2011).
87. T. Bahners, A. Wego and J.S. Gutmann, Photochemische Oberflächenmodifizierung textiler Materialien mit Polyethylenglykolen zur Verminderung der mikrobiellen Adhäsion, *Melliand Textilber.* 93, 50-51 (2012).
88. C.-W. Lin, S. Aguilar, E. Rao, W.H. Mak, X. Huang, N. He, D. Chen, D. Jun, P.A. Curson, B.T. McVerry, E.M.V. Hoek, S.-C. Huang and R.B. Kaner, Direct grafting of tetraaniline via perfluorophenylazide photochemistry to create antifouling, low bio-adhesion surfaces, *Chem. Sci.* 10, 4445–4457 (2019).
89. Q. Yang, M. Strathmann, A. Rumpf, G. Schaule and M. Ulbricht, Grafted glycopolymer-based receptor mimics on polymer support for selective adhesion of bacteria, *ACS Appl. Mater. Interfaces* 2, 3555-3562 (2010).

6

UV/Ozone Surface Treatment of Polymers to Enhance Their Adhesion

Johannes A. Poulis* **and Adriaan Kwakernaak**

Adhesion Institute Delft University of Technology, Faculty of Aerospace Engineering, Delft, The Netherlands

Abstract

The surface treatment of plastics as well as metals or ceramics includes a thorough surface cleaning as an essential step prior to adhesive bonding and coating processes. Besides this, surface activation of polymers is often needed because their surface free energy is too low for durable adhesion of a coating or adhesive. In this chapter various types of UV/Ozone sources with different light spectra as well as the influences of spectra and ozone concentration are investigated and compared. Also the surface wetting and adhesive bond strength as a result of UV/Ozone, atmospheric plasma, or corona treatments on thermoset, thermoplastic, and rubber materials are presented.

UV/Ozone treatment was found to show an excellent cleaning performance on all kinds of materials, and especially as a very useful technique for surface functionalisation of polymers, resulting in durable adhesion both for adhesives as well as coatings. This chapter is a condensed overview of over 30 years of experiments done with UV/Ozone treatments at The Delft University of Technology.

Keywords: UV-light, ozone, atmospheric plasma, corona, contact angle, surface free energy, rubber, CFRP

6.1 Introduction

6.1.1 Adhesive Bonding of Polymers

The generally accepted main criteria for surface treatment for successful adhesive bonding of polymers are: a well cleaned surface, wetting capability

**Corresponding author*: j.a.poulis@tudelft.nl

K.L. Mittal and A.N. Netravali (eds.) Polymer Surface Modification to Enhance Adhesion: Techniques and Applications, (199–272) © 2024 Scrivener Publishing LLC

(high surface free energy), homogeneity, roughness, and stability (no weak boundary layers). Moreover, spreading of the adhesive is the minimum requirement to allow molecular interactions and physical adsorption. For this reason, polymers in general (and polyolefin in particular) are difficult to bond because of the absence of polar or reactive groups on their surfaces and their low surface free energy. Surface treatment (without affecting the bulk properties of the polymers) to increase the surface free energy is thus an essential and critical part of the adhesive bonding, sealing or coating processes and to impart reactive chemical groups on polymer surfaces (surface modification, functionalisation) to enhance their adhesion characteristic [1, 2].

Throughout the last century a large number of polymer surface modification methods have been developed and commercialised (Table 6.1) all with their specific pros and cons. They can be divided into three main groups: physicochemical (e.g. laser, flame, plasma, corona, and UV/Ozone treatments), chemical only (e.g. pickling, etching, ozone gas exposure), or mechanical (e.g. grit blasting, grinding) surface treatments. Physical and chemical methods are mainly applied to increase the surface free energy and microroughness, while mechanical treatments can only create a fresh clean and chemically active surface, by removing the upper surface layers, creating an increased surface area to allow more interactions on a molecular scale and optimize the topography to make mechanical interlocking possible [2, 3].

6.1.2 UV-C Sensitivity of Polymers

Surfaces of organic materials, and especially those of non-pure (commercial) polymer and rubber blends, are in contrast to metals, far more dynamic with the additional probability of a greater surface variation which might affect the adhesive bond strength. The atomic bonds of organic materials (such as rubbers and polymer blends) are sensitive to UV-C light, especially in combination with reactive gas species like ozone, causing visually a loss of shine or gloss (etching) and often colour change (yellowing) as a result of cross-linking, chain scissions and oxidation [3]. The formation of functional groups during such processes will lead to an improvement in adhesion property. However, the sensitivity to UV-C light can be (strongly) reduced when small weight percentages of UV- scavengers and/or filler materials such as carbon black are added, which absorb the photon energy [4, 5]. Even these polymer blends were often found to be susceptible to UV-C light in combination with ozone gas and could be successfully treated for improved adhesive bonding.

Table 6.1 UV/Ozone surface treatment of polymers compared to other surface treatment techniques. +++: excellent, ++: very good, +: good, +/- average, -: disappointing, --: bad, ---: very bad, n.a.: not applicable.

	Surface treatment						
	Physicochemical				**Chemical**	**Mechanical**	
	UV/Ozone	**Atm. Plasma**	**Vacuum plasma**	**Corona**		**Blasting**	**Grinding**
Removing small traces of surface contamination	+++	-	+++	-	+++	---	---
Surface activation	+++	++	+++	+	++	n.a.	n.a.
Complex substrate shape	--	++	+++	++	++	++	++
Activation speed	-	+++	--	+	+	n.a.	n.a.
Low investment costs	+++	+/-	---	+/-	--	+/-	+++
Low running costs	+++	++	+/-	++	--	+	+
Reproducibility	+++	++	+++	++	+++	++	++
Environmentally friendly	+++	+++	++	+++	--	-	-

Adhesive bonds of industrially blended polymer products that will definitely fail at the interface, even after an initial successful UV/Ozone surface treatment, are those which include a (low molecular weight) release agent which diffuses from the polymer bulk to the surface as a function of time, leading to adhesive bond failure. Another group of materials which are impossible to pre-treat by UV/Ozone for improved adhesive bonding are the fluorpolymers, Teflon® being the best known, and will definitely show adhesion failure.

Environmental regulations and the development of new materials as substrates and the aim for improved (environmental) performance are a constant driving force to improve or fine-tune existing chemical and physical surface treatment processes for an optimum adhesive bonding. To minimize liquid waste during such processes, dry surface modification techniques combining oxygen plasma and an ultraviolet light source are often employed. They form oxygenated functional groups on the polymer surface, making it more hydrophilic and thus more amenable to adhesion. One of these methods employed at low pressure is known as vacuum UV (VUV) with wavelengths ranging from 100 to 200 nm [3].

This chapter, however, mainly focuses on dry physicochemical surface treatment processes during which UV-light is generated by a low cost low pressure mercury or an Excimer lamp in a laboratory environment as an alternative to the more sophisticated and more well-known chemical surface treatments as mentioned above. They were applied in combination with ozone gas to improve wetting of polymer surfaces by increasing the surface free energy.

6.1.3 UV/Ozone Treatment: Advantages

One of the major advantages of the UV/Ozone cleaning/surface activation process is that after a simple cleaning by either soap and water or a solvent (and removal of the inactive oxide layers in the case of metal treatment) no further chemicals are needed before a successful adhesive bonding, printing, or coating can take place. This makes the UV/Ozone process an environmentally-friendly alternative treatment method for adhesive bonding of inert polymer materials such as Ethylene-Vinyl Acetates (EVAs) as used in the shoe industry, and will further help to abandon the use of hazardous chemicals [6, 7].

Though not widely reported in literature, the use of UV/Ozone equipment for surface cleaning of non-polymeric surfaces is found to give excellent results [8], even better when compared to other far more popular methods (Table 6.1). This technique has been described as very effective for durable adhesive bonding of plastics and rubbers, as well as for metals, specifically in combination with sol-gel and water-based primers [9–13].

A major advantage of the UV/Ozone treatment, as described below, is the fact that it is very reproducible, as a result of the limited number of variables involved, the most important ones being the distance between surface and UV-source; the source power, spectrum and its age; and the concentration of ozone gas used. The equipment itself is easy to use (simply on/off) and is low cost [14]. Finally, the process can take place at room temperature and is environmentally-friendly, leaving no hazardous (solid or liquid) waste, as the small amount of organic surface contamination leaves the process as volatile molecules such as CO_2, H_2O, NO_2, and N_2.

6.1.4 UV/Ozone Treatment: Disadvantages

The biggest disadvantage of the UV/Ozone cleaning/ surface functionalisation technique is that non-flat or 3D-shaped substrates are difficult to treat. However, McIntyre and Walzak explored the possibility to make use of separate UV-light and ozone generators to treat 3D shaped objects [15].

Another disadvantage of the technique is that the treatment is rather time consuming, though the efficiency (treatment speed) can be improved when the concentration of ozone gas and/ or the UV intensity/cm^2 is increased [16, 17]. For example, by adding polished aluminium reflectors (Alzak) to focus the light on the surface of the object.

Another minor disadvantage might be that often a cleaning step of the surface by either a solvent or soap solution should be done prior to the UV/Ozone treatment to improve reproducibility. Otherwise, one is simply oxidizing surface contaminants which will result in an improved surface wetting, but might very well lead to unexpected bond failures.

The next sections are an assembly of experimental data based on 30 years research on surface cleaning and activation mainly done with UV/ Ozone. Mostly done on hydrophobic polymers but also on metals to improve adhesive bonding. Examples include a wide variety of thermoplastic and thermoset materials such as polypropylene (PP) polyethylene (PE), Carbon Fibre Reinforced Polymers (CFRPs), as well as rubbers,

6.2 Historical Development of UV/Ozone Surface Treatment

As early as 1972 Vig and LeBus described cleaning of silicon surfaces and stripping of photoresist by UV-light in combination with ozone gas generated by a single source for electronic applications [16]. The same approach was used by Zafonte and Chiu for the removal of organics and photoresist residues on silicon wafers during production [18], or by Tabe for the removal of carbon from silicon wafers [19].

UV/Ozone treatment has been applied for the surface modification of thermoplastic screening systems based on microarray techniques which play a vital role in the fast advances of medical diagnostics as a part of life sciences [20]. Another researched application was the surface oxidation of batches of natural wool fibres, which appeared to cause a reaction of the proteinaceous carbon, leading to an increase of carbonyl functionality [21]. Other applications are the adhesion improvement of carbon fibres and carbon nanotubes to the matrix material leading to improved mechanical performance [22, 23]. UV/Ozone treatment was also applied on thin film solar cell surfaces to improve the wettability of the ink and therefore the surface coverage prior to inkjet printing by an indium-thiourea (In_2S_3) compound dissolved in a mixture of water and ethanol [24]. The use of low-pressure mercury UV-lamps as a direct source of germicidal disinfection is widespread, and can be found as a disinfection technique for public spas, hot tubs, waste or drinking water [25, 26] and also in the germicidal disinfection of recirculated air streams and surfaces [27]. Specific UV-lamps are known (and used) effectively for the photochemical generation of ozone gas, as the absorption of UV radiation by oxygen has a maximum efficiency at approximately 160 nm [28], which is about 15 nm below the lowest wavelength of the mercury spectrum.

6.3 Parameters Controlling the UV/Ozone Surface Treatment Process

6.3.1 Ultraviolet (UV) Light Spectrum

UV-light is a small fraction of wavelengths that belong to the visible part of the electromagnetic spectrum which runs from X-rays (wavelength 10^{-2}m), visible light (380 – 750 nm) (1nm = 10^{-9}m), radio waves up to television transmission wavelengths (10 – 10^{-1}m) [29].

The UV spectrum is divided into three main parts:

- UV-A: 400 – 315 nm;
- UV-B: 315 – 280 nm;
- UV-C: 280 – 100 nm (and lower) wavelengths.

The UV-A part of the light spectrum is part of the sunlight that drives photochemical processes, while UV-C is dangerous for human beings, as it is strongly absorbed by deoxyribonucleic acid (DNA) and ribonucleic acid (RNA) and as a result can cause skin cancer, and for this reason it is effective in germicidal disinfection processes [30].

6.3.2 UV-Light Generation

There are several types of UV- light sources such as:

- Fluorescent lamps: Mercury-vapour filled low, medium and high pressure fluorescent discharge lamps. An electric current in the gas excites mercury-vapour, which produces shortwave ultraviolet light. Typically 30 to 40% of the electrical power input is transferred into UV-light. The ozone generating UV-C lamps have an emission spectrum with a radiance of 90 to 98% at two wavelengths, with the main resonance line at 253.7 nm, which is close to maximum of the germicidal curve, and a much smaller one at 184.9 nm (Figure 6.1). These low-pressure mercury arc discharge types (10^{-3} Torr) are often used as a disinfection light source [31] but are also found to be well suited for pre-treating polymer surfaces. The optimum operating temperature of fluorescent lamps when operated in the open is 20 – 30 °C should not reach more than 40 °C. A rapid growth in light technology showed the development of amalgam fluorescent lamps, which use hard alloys of mercury with metals that allow

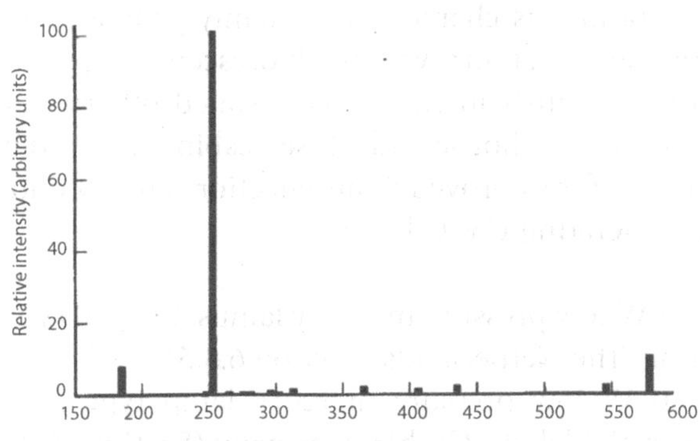

Figure 6.1 Low pressure mercury lamp UV-C spectral energy distribution [14].

higher light output. This implies, however, a slightly longer warm-up time before use [32].

Besides this, the user should be aware of the limited lifetime of fluorescent UV-lamps, as their housing is made of different quartz materials. Normal window glass allows mainly UV-A and visible light [33], while synthetic and natural quartz glass housings are available for germicidal disinfection purposes and added ozone production, respectively. However, the transparency of these quartz housings shows the steepest radiation reduction at 253.7 nm in the first 100 hours. The overall radiation power depreciation at 253.7 nm is about 25% after 8000 hours [34].

- Excimer lamps: These are Radiofrequency (RF) electric discharge light sources, which radiate quasi-monochromatic light, operating over a wide range of narrow emission bands with wavelengths from 351 down to 108 nm. Light is generated by a silent discharge between two electrodes. Due to an internal cooling system, the generated light is considered as "cold" and typically shows a high power spectral density of UV radiation [35].
- Light Emitting Diodes (LEDs): The electrical efficiency of LEDs is far higher than those of all other light generating sources, and the output shows narrow bandwidths down or near to 255 nm, good enough to be applied in germicidal technology [36].

6.3.3 UV-Light Sources Used

All presented data in this chapter were mainly generated using different types of low pressure mercury vapour fluorescent lamps and setups. The basic design of these instruments (unless stated otherwise) is shown in Figure 6.2. It shows an in-house built closed cabinet covered by aluminium sheets on the inside for improved light reflection, and it contained any of 3 types of ozone generating UV-C lamps:

- Three 4 W low pressure mercury lamps, type HUV5, Philips Lighting, The Netherlands (Section 6.3.5);
- One 80 W low pressure mercury lamp type NNIQ 120, Heraeus Noblelight GmbH, Germany (Sections 6.3.6–6.3.8, 6.4.5, 6.3.3, 6.3.5, 6.3.6, 6.6.4);

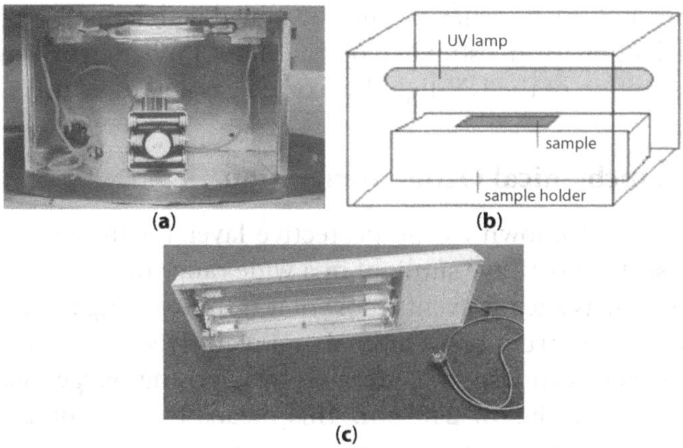

Figure 6.2 (a) Photograph of an early UV/Ozone chamber used for the experiments described in section 6.3.3 with three stainless steel cylindrical samples on a mechanical lift [14]. (b) schematic representation of the UV/Ozone exposure chamber used for the described measurements. (c) photograph of one of the most often used setups of three combined low pressure mercury lamps for the discussed experiments.

- Three 30 W low pressure mercury lamps type UVN 30 with a spacing of 50 mm between the lamp centres, and a radiation flux at 253.7 nm of 10 W, measured after 100 hours [34], UV-Technik Speziallampen GmbH, Germany, (Sections 6.5.1, 6.5.6, 6.6.2, 6.3.1, 6.3.2, 6.3.4, 6.3.7-6.3.11).

The specimens were always placed at a distance of max. 20 mm from the UV lamp (unless stated otherwise). All tests were performed at laboratory conditions.

Some of the experiments described below were performed either with an Excimer 222 nm or an Excimer 172 nm monochromatic light source. Just as lasers they have proved viable in the modification of polymer surfaces such as fluorocarbons [35]. The Excimer prototype light sources used in the described experiments below were supplied by Heraeus Noblelight GmbH, Germany. This first-generation Excimers had an efficiency of 10% of the electrical power input, which equals 1.5 W/cm lamp length at 172 nm. In comparison, a low-pressure mercury UV-source has an electrical input power of 2 W/cm and with its efficiency of 30 - 40% it generates about 0.06 W/cm at 184.9 nm. The light-photon bombardment per unit surface area of the Excimer lamp thus is about 25 times higher.

The cabinet of the Excimer 172 nm was filled with laboratory air with a pressure reduction to an "estimated optimum" of 1 mbar, in order to reduce the full absorption of the UV-light by oxygen gas and moisture.

6.3.4 Photochemical Ozone Generation

Ozone gas is well known as the protective layer for life on Earth, as it strongly absorbs shortwave sunlight in a wide range of lower wavelengths (Figure 6.3) and as such prevents most of it from reaching the Earth's surface, preventing destruction of animal and plant DNA [37]. However, on earth Ozone (O_3) can also be generated by exposing oxygen gas (O_2) to shortwave UV-light below 240 nm. This means that all three discussed UV- sources (low pressure mercury, Excimer 172 nm (Xe gas fill) and Excimer 222 nm (KrCl gas fill) will generate ozone from oxygen gas.

However, the ozone generation varies depending on the design and lamp power from 6-12 g/kWh^{-1} for low pressure mercury lamps to 60-100 g/kWh^{-1} for the 172 nm Xenon Excimers. For a 40 W low pressure mercury lamp the amount of ozone generated would be around 0.3-0.5 gh^{-1}. Hence, near to the lamp the concentration might be significantly higher [28].

Ozone is an unstable molecule with a "half-life" of about 705 minutes (24 °C, 45% RH) up to 5 days (based on the literature) [e.g. 28]. It has a strong oxidizing nature, a typical smell and is dangerous to human beings with a Maximum Allowable Concentration (MAC) value of 0.1-0.3 Parts

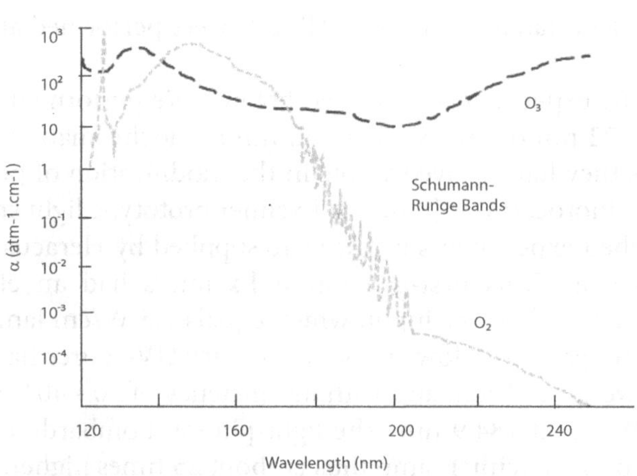

Figure 6.3 Wavelength dependent absorption (α) of oxygen and ozone gas. Please note the exponential vertical scale.

Per Million (PPM) [16, 34, 38–40]. However, the values for maximum allowed concentration and exposure time differ internationally.

6.3.5 Relation between Ozone Generation and UV-Source Power

The photochemical ozone generation by two HUV 4 W low pressure mercury lamps (Philips Lighting, The Netherlands) with essentially both 184.9 and 253.7 nm (Figure 6.1) was measured by an ozone analyser (model 8810, Monito Labs, Envico, Zoeterwoude, The Netherlands) as a function of medical oxygen gas input in a laboratory built UV/Ozone cabinet of 1.5 litres (Figure 6.2a). The results showed that an increase in the source power as well as an increase in oxygen gas flow leads (asymptotically) to an increased ozone concentration (Figure 6.4). The lower the UV-source power, the faster the maximum ozone concentration is obtained, though at a lower concentration than would have been obtained at a higher power [14].

Note that the light absorption of both oxygen and ozone gas covers the whole UV-C spectrum (Figure 6.3). This means that the breakdown of ozone into oxygen by the absorption of UV-light has to be taken into account. The calculated relative light intensity as a function of the distance from the UV-source is shown in Figures 6.5a and b. The UV-light is strongly absorbed by molecular oxygen below roughly 200 nm (Figure 6.5a) and by ozone molecules above 200 nm (Figure 6.5b). As a result, these graphs show the importance of the minimal distance between the

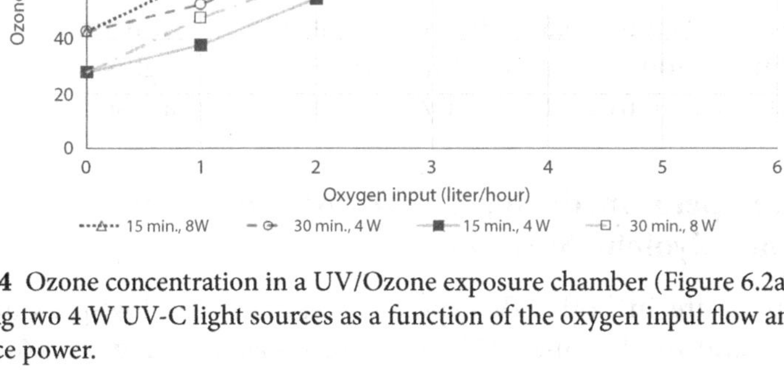

Figure 6.4 Ozone concentration in a UV/Ozone exposure chamber (Figure 6.2a) containing two 4 W UV-C light sources as a function of the oxygen input flow and the UV-source power.

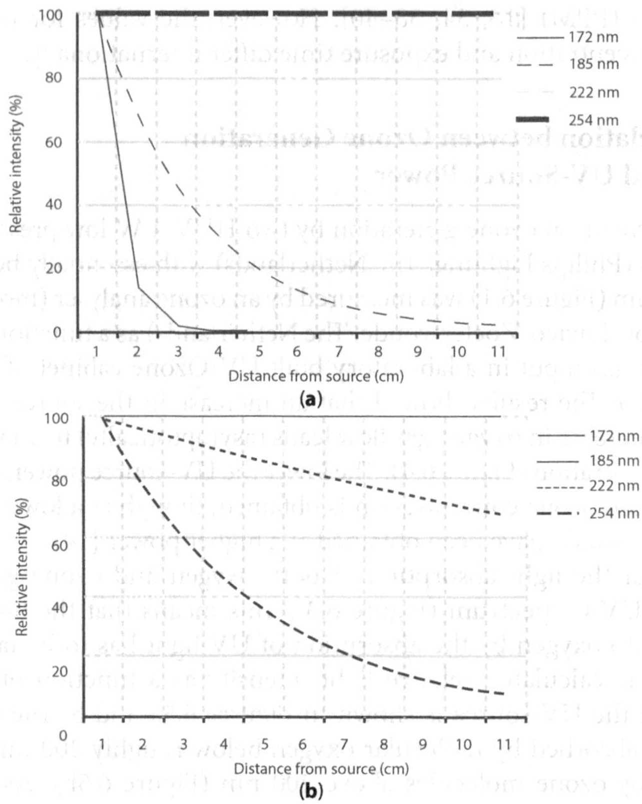

Figure 6.5 Relative intensity of UV-light at 172, 185, 222 and 254 nm as a function of distance from the source due to absorption of oxygen (a) and ozone gas (b). (a) UV-light absorption by oxygen. (b) UV-light absorption by ozone.

UV-source (max. around 2 cm, Figure 6.5a), or a gas pressure reduction is needed, when an Excimer 172 nm is used. The same graph shows that 222 nm and 253.7 nm are hardly affected, and thus the resulting data are overlapping. A similar phenomenon occurs with the absorption of ozone gas molecules (Figure 6.5b): the UV-light at 253.7 nm has the highest absorption of ozone gas from all investigated wavelengths, while 172 nm and 184.9 nm are hardly affected by ozone and are, as a result, overlapping.

6.3.6 Temperature during UV/Ozone Treatment on Polyolefin Surfaces

The temperature in both UV-exposure cabinets, the low-pressure mercury lamp and the Excimer 222 nm were measured (Figure 6.6), as this

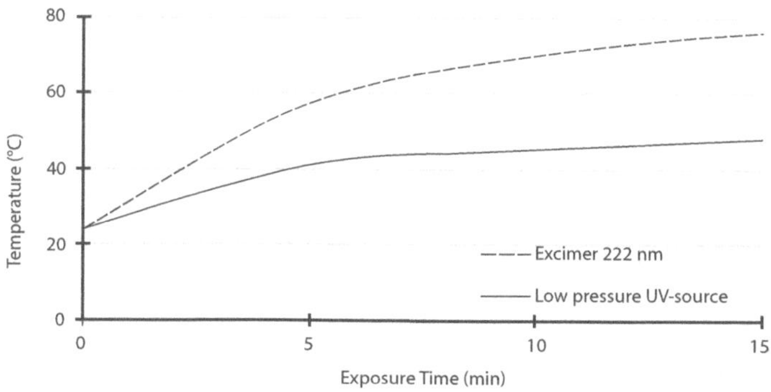

Figure 6.6 Typical sample temperatures as a function of treatment time on exposure to two different UV-sources.

temperature might influence the reaction velocity of the chemical processes, the lamp efficiency, as well as the stability of the sample materials. The measurements showed that the temperature of the low-pressure mercury lamp was rather stable and near the optimal performance temperature of fluorescence lamps of 40 °C, while the sample temperature during the surface treatment by the Excimer 222 nm was much higher, increasing with time and reaching over 70 °C after 15 minutes.

6.3.7 Influence of Wavelength(s) and Gas Fill on Polyolefin Surfaces

Polypropylene (PP) is a thermoplastic, partly crystalline, material that, like polyethylene (PE), belongs to the group of polyolefins [17] with rather similar properties, though it is slightly harder and more heat-resistant. Its non-polarity (hydrophobic nature) and low surface free energy of about 30 mJ/m^2 (Figures 6.7a, b) makes it difficult to bond without increasing its wetting performance [41, 42].

The surface free energy and bond strength were measured on pure Polypropylene samples (PP, Vink Kunststoffen, The Netherlands) in order to find the optimal combination of wavelengths and ozone gas for the surface activation. Three different types of UV-sources were used: a low-pressure mercury UV-lamp, an Excimer 172 nm and an Excimer 222 nm.

The treatment chamber of the Excimer 172 nm was filled (pressure: 1 mbar) with nitrogen, laboratory air, or laboratory air saturated with water. The surface free energy of the PP surface was measured by test inks (Arcotest, Mönsheim, Germany) as a function of the exposure time

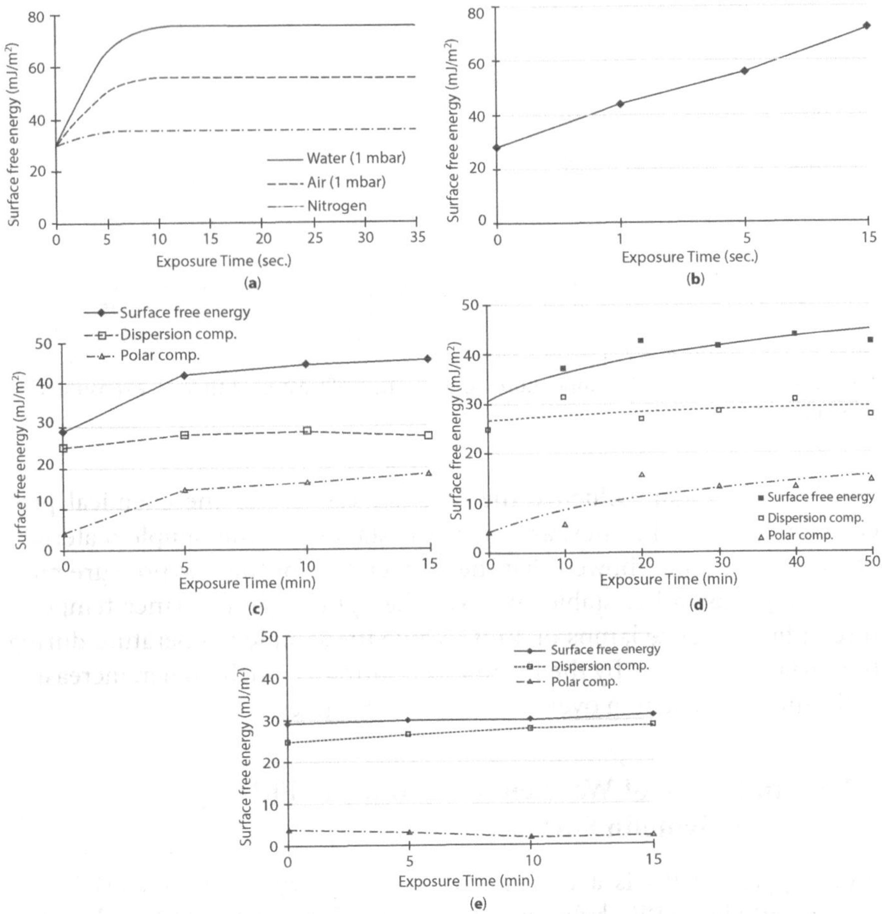

Figure 6.7 (a) Trend lines showing the surface free energy of PP as a result of exposure to the Excimer 172 nm in different gas atmospheres. (b) Surface free energy of PP exposed to the Excimer 172 nm as a function of exposure time. (c) Surface free energy of PE due to exposure to the (whole) spectrum of a low-pressure mercury UV-source. (d) The same as (c), but with a 185 nm UV-filter, and (e) the surface free energy on exposure to the Excimer 222 nm.

(Figure 6.7b). The surface free energy of the PP remained at the same low value when only nitrogen gas was present in the reaction chamber, meaning that no polar groups were generated on the surface. But introducing oxygen (laboratory air), with or without water, showed a strong increase of polar groups on the surface, which was the highest when water molecules were present. The water molecules tend to break apart into reactive radical species like OH* which can react with the polymer surface. Hence, these experiments show that oxygen radicals need to be present for surface oxidation which are increased in the presence of water vapour.

Figures 6.7a and b show that exposure of PP to 172 nm resulted in a maximum surface free energy of 72 mJ/m^2 when exposed to moist air. This value clearly surpasses that obtained from the other two gas environments: just laboratory air and definitely that of nitrogen gas. Note that the activation time is in the order of seconds instead of minutes (like that of the low-pressure mercury source).

Further experiments to determine the effects of the combinations and separate UV wavelengths were carried out on pure polyethylene (PE, Vink Kunststoffen, The Netherlands). For the Excimer 222 nm and the low pressure mercury UV- source the surface oxidation was determined by water contact angle (*) measurements as a function of the exposure time.

(*) All contact angle measurements for wetting and surface free energy experiments were done at room temperature at a RH of 50% using either a KSV CAM 200 (Chem. Instruments, USA), or a Krüss G2 (Krüss, Germany) computer controlled drop shape analysis system with drop volumes of 3 - 5 μl. All reported data are averages over at least three droplets, placed at different positions on the specimens.

An Action Research Corporation (ARC) 01720 filter (peak wavelength of 184 nm, bandwidth 24 nm and a peak transmission of 24.8%) was placed 5 mm above the polymer surface to block all other wavelengths (thus ozone gas as generated by the UV-lamp could still reach the surface). This arrangement made it possible to measure the effect of just 184.9 nm together with ozone gas without exposing the polyethylene (PE) to 253.7 nm. As a result PE could be exposed to the low-pressure mercury lamp both with (Figure 6.7d) and without the 185 nm filter (Figure 6.7c). The results show an increase in the total surface free energy, independent of the filter used. This leads to the observation, that 184.9 nm plays a negligible direct role in the surface modification, though needed for the creation of the oxygen radicals which is in agreement with the results found by MacManus *et al.* [43].

However, as measured by test inks, the Excimer 172 nm (Figures 6.7a, b) showed the largest increase of surface free energy (during the shortest exposure time) for the three tested UV-sources, when used in a moist environment [44]. In contrast, the Excimer 222 nm hardly affected the surface free energy, and was unable to increase the surface polarity (Figure 6.7b). The effect is also visible when the same tests were done on PP instead of PE. The above measurements show the necessity of specific wavelengths in combination with ozone gas for a fast photo-oxidation process to create polar groups on the surface of the polymer.

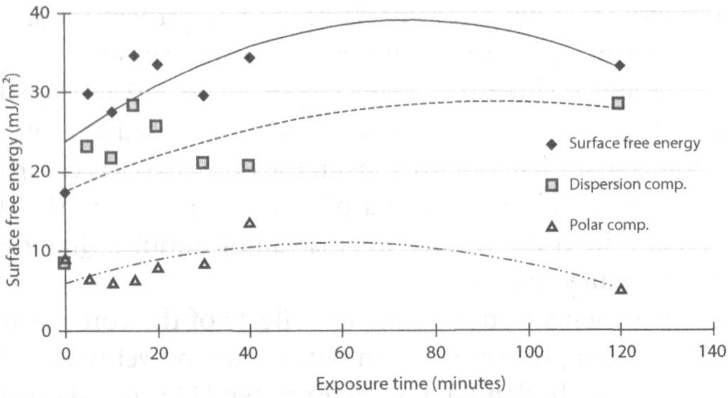

Figure 6.8 Measurements with added trend lines to show the course of a prolonged surface treatment by a low pressure mercury UV-source on the surface free energy of PP.

Figure 6.8 shows the occurrence of an optimum value of the surface free energy as a function of treatment time, though the highest adhesive bond strength appears to occur at a treatment time that is slightly less. Increasing the treatment time beyond the optimum value of the surface free energy will lead to an even further decrease of the adhesive bond strength, due to the creation of Low Molecular Weight Oxidized Molecules (LMWOMs) [45, 46].

6.3.8 Influence of Ozone Gas and UV-Light Spectrum on Polymer Surface Wetting

The influence of combinations of the UV-light spectrum and ozone gas concentration on the wetting of polymer surfaces was investigated to give a better insight into the relation between these two process variables to improve surface wetting.

The data shown below are the experimental results of a low-pressure UV-source with a transmission starting at 253.7 nm from a TUV-glass lamp sleeve instead of a fused quartz sleeve as used in all other discussed experiments in this chapter. Test samples were made of pure polypropylene (PP) (Vink Kunststoffen, The Netherlands) and only pre-cleaned with ethanol as a reference before exposure, shown as (254 nm) 0 min. in Figure 6.9.

The influence of the ozone concentration during the exposure was already found by some researchers [14, 16, 17, 47], though no experiments have been reported where the relations between the generated

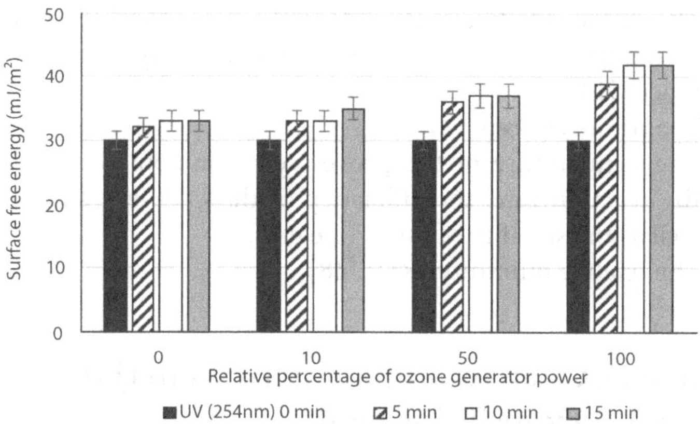

Figure 6.9 Surface free energy of PP measured by test inks. Pre-cleaned by ethanol as a function of various UV-light (>253.7 nm) exposure times and ozone concentrations.

separate wavelengths and the ozone concentration are shown. The ozone in this experiment was generated by an Ozone Technology OT20 (Ozone Technic, Turkey) ozone generator (20 g/m^3 at 100% power range). The surface free energy test results on the PP samples were determined by test inks (Arcotest, Mönsheim, Germany).

Exposure to ozone gas alone did not show any increase of the surface free energy within the standard deviation (STD). Though increasing the ozone concentration together with a longer UV exposure time showed a small (but a clear) increase of the average surface free energy.

The above results show that the efficiency of surface activation by ozone and/or 253.7 nm without 184.9 nm is negligible and low. The surface free energy due to 253.7 nm only increased when the ozone concentration was increased, but it was still far lower than when the combined wavelengths (183.9 and 253.7 nm) were generated by a single low pressure mercury UV/Ozone lamp (Figure 6.8).

6.3.9 Main Process Variables: Overview

The number of variables related to the UV/Ozone cleaning and surface functionalisation process are limited. The main ones are:

- The type, power and age of the UV- light source (generated wavelengths and intensities);

- The distance between the UV- light source and the treated object (the efficiency decreases exponentially with the distance) [14];
- The ozone concentration.
- The temperature of the process environment (influencing the reaction rate, the efficiency of the UV-light source, as well as the stability of the ozone gas);
- The relative humidity (RH) [48]

6.4 Surface Changes of Polymeric Materials by UV/Ozone Treatment

6.4.1 Polymer Bonds

The bond energies of several atomic bonds which are of importance during the UV/Ozone process are listed in Table 6.2. For comparison, the photon energies at different wavelengths are given in Table 6.3.

The generated photons from UV-light of e.g. 253.7 nm are clearly much less energetic than those from 172 and 184.9 nm but still gives them the ability to generate chain scissions in the carbon-carbon polymer bonds with bonding energies as shown in Table 6.2, because the photon energy (Table 6.3) should at least be equal to the bond energy (Table 6.2) to break these bonds.

Table 6.2 Bond energies of several atomic bonds of importance during the UV/Ozone process.

Type	H_2C-H	HO-H	HOO-H	OO-H	H_3C-CH_3	O=C=O	N≡N	O=O
Bond energy (J)	6.6 E-19	7.5 E-19	5.6 E-19	5.6 E-19	5.6 E-19	8.1 E-19	14.3 E-19	7.6 E-19
Bond energy (kJ/mole)	438	498	369	374	376	532	945	498

Table 6.3 Relevant spectral lines during the UV/Ozone process in relation to the transmitted energy/tube length and the number of photons/cm², together with their photon energies.

Wavelength (nm)	Energy/cm tube length	No. of photons/ cm² per second	Photon energy (J)	Photon energy (eV)
254	2.7 W *(low pressure UV-lamp)*	3 E18	7.7 E-19	4.8
222	12.5 W *(first generation Excimer)*	14 E18	8.9 E-19	5.6
185	0.3 W *(low pressure UV-lamp)*	3 E18	10 E-19	6.8
172	50 W/cm *(new generation Excimer)*	41 E18	12 E-19	7.2

6.4.2 Surface Cleaning by UV/Ozone: Increasing Hydrophilicity and Surface Free Energy

The lower wavelength of the mercury spectrum which is transmitted by a low pressure mercury lamp with a natural quartz housing is 184.9 nm (with an energy of 155 kcal/mol) and is thus capable to break various polymer bonds like: H_3C-H (102 kcal/mol), CH_{2-}CH-H (105 kcal/mol) and CH_3-CH_3 (84 kcal/mol). This results in the phenomenon that during exposure to UV/Ozone from a low pressure mercury UV-source, the polymer chains and organic contaminant molecular bonds are broken (chain scissions) by the photochemical process of the highly energetic light quanta of 253.7 nm and 184.9 nm [25, 26].

Besides this, a kind of plasma arises, where reactive oxygen radicals O (^{3}P) from oxygen molecules (O_2), as shown schematically in Figure 6.10a, might react with (unstable) ozone molecules (O_3), which can be dissociated (photo-decomposed) into molecular oxygen (O_2) and an oxygen radical O(^{1}D) by 253.7 nm.

This radical O(^{1}D) is a reactive atomic species, which, as a result, can oxidize the surface and create functional polar surface groups such as OH, OOH, and C=O, which increase the wettability [15, 48]. Meanwhile,

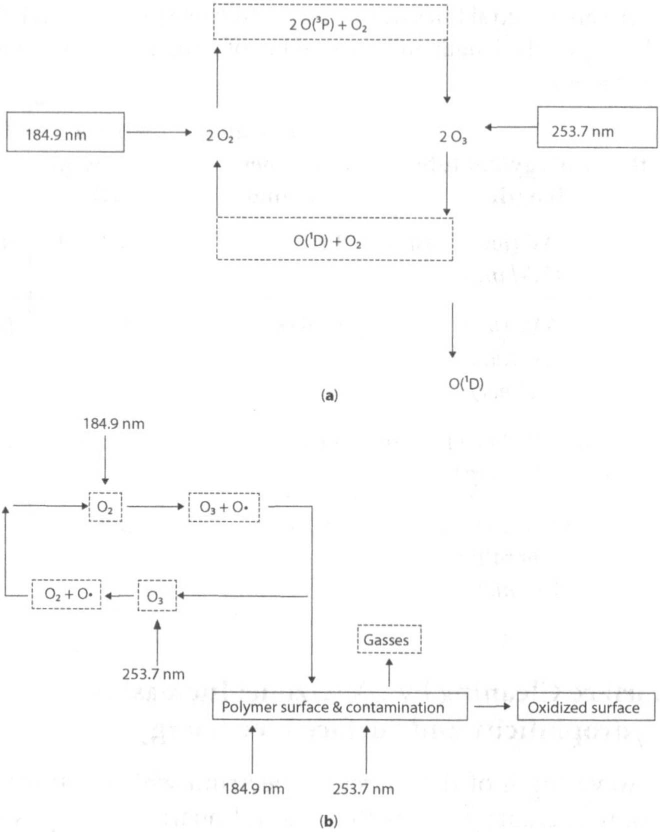

Figure 6.10 Simplified schematic models of (a) ozone generation and destruction by low pressure mercury UV-sources and (b) the UV/Ozone surface cleaning and activation process.

traces of oxidized surface contaminants leave the process in a volatile state (Figure 6.10b) [15, 16].

This continuous process of forming and destroying ozone gas and the generation of oxygen radicals is in fact a kind of reactive cold plasma. As a result, polymers are oxidized in this aggressive environment of UVC-light, ozone gas and oxygen radicals. On a nano-scale polymer surfaces will either become nano-roughened [49] or nano-polished [45], depending on the type of polymer. Very thin layers of organic contamination will thus be eroded, and will finally disappear from the surface.

In literature only few authors like Vig and LeBus [16] investigated the importance of the combination of UV-light and ozone gas on the cleaning performance (the decomposition of organic contaminants) on materials such as metals and silicon wafers, which increases the surface wetting by hydrophilic liquids like water.

Indeed, UV/Ozone proved to be an excellent cleaning method for the thorough removal of the last traces of organic contaminants from metal surfaces [19]. Furthermore, previous research showed that UV/Ozone treatment improved the (initial) adhesive bond strength and durability between a metal and epoxy, especially when used in combination with sol-gel and water-based primers [19, 20, 45, 50, 51].

As already mentioned above, UV/Ozone was found to be capable of removing the last traces of organic surface contamination and is very often able to increase the hydrophilicity of the surface as a result of surface oxidation which is the minimum requirement for good wetting. This was the precursor for high quality initial and durable adhesive bonding of many polymers.

UV/Ozone treatment of polymers for a combined cleaning and surface functionalisation by low pressure UV-sources generally takes between 5 and 15 minutes. Depending on the polymer material and process variables such as the power of the UV-source and its distance from the substrate, it often results in an increased polarity and an increased wettability of polar liquids [52]. This is a large advantage for the interfacial (or adhesion) strength of an adhesive, sealant, coating, paint or printing ink as well as adhesion of metallized polymers and of micro-organisms or cells. The adhesive bond strength of polymer surfaces is often found to increase compared to the initial (untreated) value as a result of the larger number of polar and non-polar interactions resulting in a larger density of near-surface cross-links [53]. Hence, the photonic degradation and oxidation process strengthens the outermost surface layer and increases the interaction necessary for improved adhesion. Besides, the durability of the adhesive bond in harsh environmental conditions often increases as a result of these increased molecular interactions across the interface.

6.4.3 Photo-Degradation: Surface Roughness and Morphology Changes

Light that strikes a plastic surface may either be reflected from the surface or absorbed. The last process may remove a hydrogen atom from the polymer forming an unstable radical which starts the photo-degradation process [54]. As a result, plastic weathering involves changes in the physical, mechanical, and chemical properties of polymers, particularly at the surface, leading to changes in gloss, morphology and roughness [55, 56]. Thus it is essential to know the physical state of the polymer and its fundamental mechanism of oxidative degradation in the absence of UV-light in order to understand what is happening in the presence of ultraviolet radiation and oxygen [57].

6.4.4 UV-Light Treatment Depth in Polymer Surfaces

An important criterion is the depth to which the surface is oxidized. Preferably, only the outermost layer is oxidized, thus keeping its bulk mechanical properties unchanged. Experiments on polypropylene (PP) and poly(ethylene terephthalate) (PET) have shown that the oxidation depth of flame treatment is the lowest, followed by corona and vacuum plasma. UV/Ozone treatment has a far deeper penetration depth [58]. Atmospheric plasma is somewhat less efficient and will penetrate less deep into the bulk mass than the vacuum plasma type, thus taking slightly longer to obtain the same result as the vacuum plasma surface treatment [59].

UV/Ozone treatment is found to penetrate relatively deep into the bulk of the polymer in comparison with the more familiar surface treatments such as flame, corona and plasma treatments on e.g. PP and PET films. Hill and co-workers argue that the surface modification by a UV/Ozone treatment as well as by ozone treatment alone extends quite deeply (1 μm), well beyond a few monolayers. This should be considerably less for corona- or plasma-treated samples. Flame treatment appears to be the "shallowest" meaning that the oxygen incorporated by the treatment is mostly concentrated near the outermost layer of the polymer surface (see also sections 6.5.3 and 6.5.4) [49].

6.4.5 Surface Relaxation of HDPE

The surface oxidation of a polymer as a result of a physical treatment will (partly) disappear as a function of time due to reorientation and migration of the (functional) groups on the surface. The three graphs of Figure 6.11 representing measurements on HDPE show that the decrease in surface wetting by water was found to be dependent on the surface treatment method and was found to be far higher after the atmospheric plasma (**) than after either corona (***) or UV/Ozone treatment.

(**) All atmospheric plasma experiments were done in a laboratory environment using a Tigres Plasma-Blaster (Tigres GmbH, Rellingen, Germany) equipped with three separate plasma guns. (300 W power supply, gas pressure 4 bar, distance 20 mm from the object).

(***) All corona experiments were done in a laboratory environment using a TIGRES Corona instrument (Tigres GmbH, Rellingen, Germany) equipped

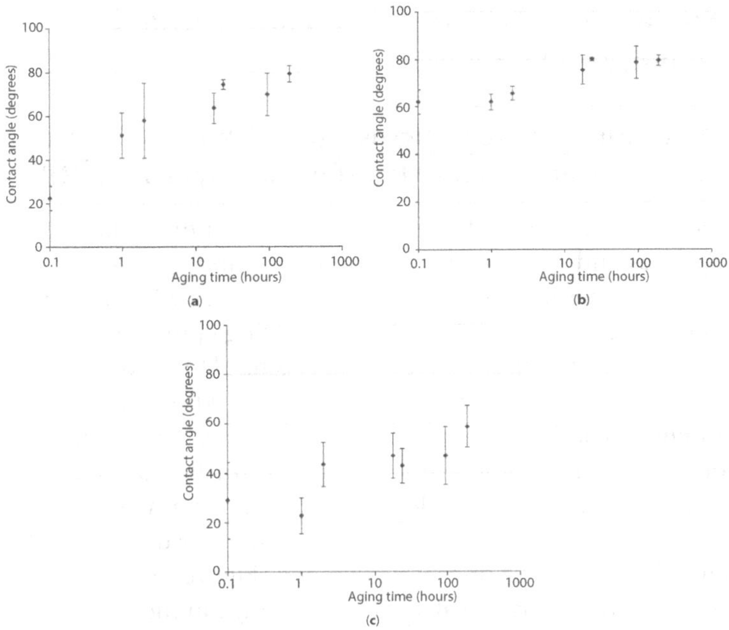

Figure 6.11 Aging of a HDPE surface determined by water contact angle as a function of exposure time in a laboratory environment after treatment by: Corona (a), UV/Ozone (b), and atmospheric plasma (c).

with a 300 W power supply, gas pressure: 4 bar, torch distance 10 mm from the object.

All oxidized polymer surfaces will show this effect, depending on a number of factors such as crystallinity, purity and molecular weight. Higher crystallinity (HDPE: 60% - 80%) will increase the stiffness of the polymer chains, leading to a less likely or slower reorganisation of the functional groups on the surface to return to the bulk. Especially, increased crosslinking will increase the lifetime of functionalities at the surface. Reorganization to the bulk mainly takes place in the amorphous part of the polymer and by the low molecular weight oxidized molecules, which have more freedom of movement.

Kim and co-workers found that the surface relaxation is also dependent on the type of gas (mixture) used. Plasma treatment was found to lead to an optimal balance between a high oxidation level and survival time of the oxidized groups at the surface at a gas mixture of argon and oxygen in the ratio 9:1 due to an increased amount of crosslinks and thus stiffness of the polymer chains [60].

6.5 Surface Analysis of UV/Ozone Treated Polymeric Surfaces

6.5.1 Scanning Electron Microscopy (SEM) on UV/Ozone Treated Carbon Fibre Reinforced Polymer (CFRP)

The measurable changes that UV/Ozone treatment might have on the surface properties and morphology of a polymer, and specifically CFRP, are discussed below in more detail. The CFRP used was a HexPly 8552 unidirectional prepreg with a toughened amine curing epoxy resin in combination with AS4 carbon fibres (Hexcel Composites, Duxford, UK).

SEM micrographs (instrument: JEOL JSM-7500F) were taken and are shown in Figures 6.12a-c, in order to discover if any changes in the surface morphology would become visible as a result of the UV/Ozone surface treatment. A single CFRP sample (sized 2.5 x 2.5 cm, with a thickness of 10 plies) was subjected to various UV/Ozone exposures, using a folded aluminium foil mask, and images were made (Figures 6.12a-c).

The SEM micrographs revealed no clear change in the surface morphology, as expected, and thus Atomic Force Microscopy (AFM) scans with a far higher resolution were made (section 6.5.2).

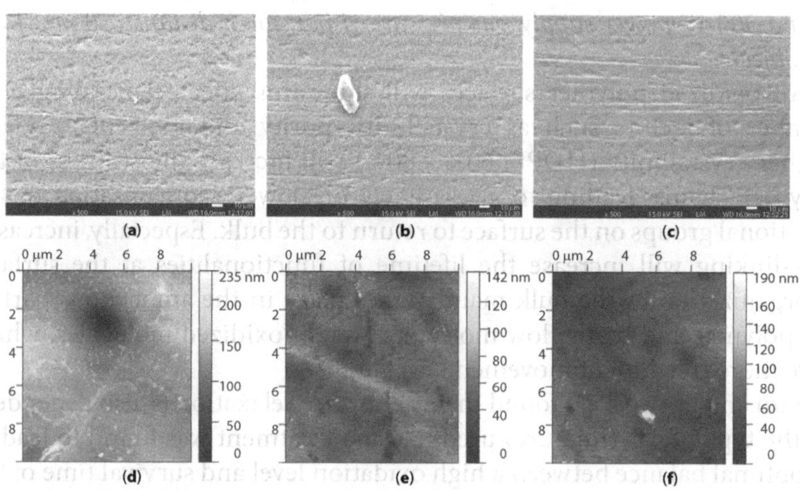

Figure 6.12 SEM (a, b, c) and AFM (d, e, f) micrographs of (a, d) the CFRP surface acetone cleaned only (reference), (b, e) after 6 minutes of UV/Ozone (a dust particle is visible in the SEM micrograph) and (c, f) 10 minutes of UV/Ozone treatment. a: (Acetone cleaned (reference), 500x, b: UV/Ozone 6 min., 500x, c: UV/Ozone 10 min., 500x.

6.5.2 Atomic Force Microscopy (AFM) on UV/Ozone Treated CFRP

AFM imaging was done to measure the morphology of the surface on a nanoscale level by a Dimension Edge Scanning Probe Microscope (Bruker, Germany) with a silicon probe TESPA (Bruker). The sample for the AFM measurements was 20 x 25 mm (with 10 plies thickness). It was visually divided into sections made by an Edding 780 silver paint marker pencil, and during the surface treatment partly covered with a mask of folded aluminium foil to allow for 0, 6 and 10 minutes of UV/Ozone treatments on a single sample. Each presented value is the result of at least three measurements.

The AFM micrographs are presented in Figures 6.12d-f and the test data can be found in Figure 6.13a/b, showing two of the CFRP surface parameters the root mean square roughness (R_q) and the arithmetical mean height (R_a). Each value was averaged from three measurements on 10 x 10 μm scanned surfaces.

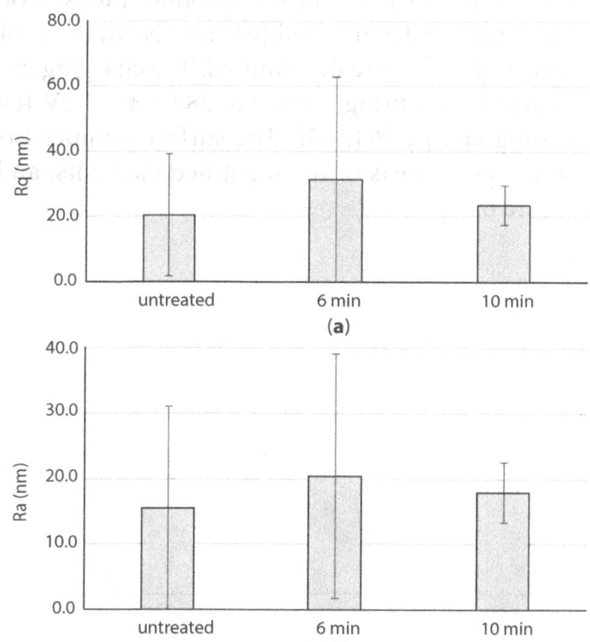

Figure 6.13 Averaged AFM data, showing: (a) The root mean square average height (Rq,) and (b) the arithmetical mean height (Ra) of the CFRP surface after acetone cleaned only ("untreated") as well as after 6 and 10 minutes of UV/Ozone treatment.

Neither (SEM) nor AFM scans showed any convincing evidence for a change in surface roughness of the CFRP as a function of UV/Ozone exposure time up to 10 minutes. These results are in contrast to the reported "less structured"/"smoothed" appearance of the same material as a result of a treatment by vacuum UV that Arikan *et al.* had found [45], possibly as a result of the large standard deviation in our AFM data.

6.5.3 X-Ray Photoelectron Spectroscopy (XPS) on UV/Ozone Treated CFRP

Three CFRP samples (sized: 50 x 30 mm and 10 plies thickness) were treated and analysed within 4 hours to observe the surface composition after a 7 minutes UV/Ozone treatment. For this purpose, a K-Alpha X-ray photoelectron spectrometer (XPS) (Thermo Fisher, The Netherlands) was employed at ambient temperature using an Al Kα X-ray source. The cathode was set to 12 kV and the beam current to 3 mA, leading to a nominal spot size of 400 μm². The spectra were evaluated with Avantage software (version 5.9922) of Thermo Fisher Scientific. At the beginning, the satellite peak subtraction was performed. The background intensity of the spectra was described with the Shirley method [61–64]. Next, the positions of the photoelectron spectral lines were determined. The charging up of the specimens was corrected by adopting a value of 284.8 ± 0.1 eV for the adventitious C 1s binding energy [64, 65]. The surface composition was then calculated from the peak areas of the identified elements, and the results are shown in Figure 6.14.

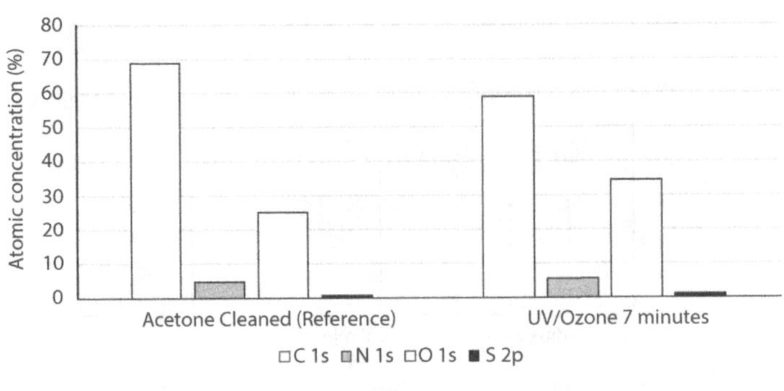

Figure 6.14 XPS data showing atomic concentration percentages of the elements carbon, nitrogen, oxygen and sulphur of the CFRP before and after surface treatment by UV/Ozone.

The O 1s/C 1s ratio increased from 0.37 to 0.52 as a result of the 7 minutes UV/Ozone treatment, clearly indicating an increased oxygen content at the CFRP surface. The XPS results thus showed an increased amount of oxygen on the surface of the UV/O_3 treated sample compared to the non-treated one due to oxidation. Washing the 7 minutes UV/Ozone exposed surface with acetone reduced the O/C value to 0.42 which was in line with expectation [45]. Etching of the above sample surfaces in the XPS instrument to a depth of 2 nm and 4 nm showed a strongly reduced oxygen content with O/C values less than 0.1, independent of the washing procedure. Hence, the oxygen content at this depth was far less than would be expected, based on the results obtained from other researchers [e.g. 45, 65].

6.5.4 Attenuated Total Reflection Fourier Transform Infrared Spectroscopy (ATR-FTIR) Investigation on UV/Ozone Treated CFRP

The change in surface oxidation of the CFRP due to exposure to a low pressure mercury UV-source and ozone gas was studied by FTIR, as this method detects changes in the polymer far deeper into the bulk than XPS (respectively: 1000 – 5000 nm and 2 - 5 nm) [66, 67].

Making use of a mask of aluminium foil, different exposure times could be obtained on a single sample sized 50 x 35 mm. To take into account any surface heterogeneity, a minimum of three spectra in different areas were recorded per sample and analysed on a Perkin Elmer Spectrum 100 FTIR spectrometer equipped with a universal attenuated total reflectance (ATR) ZnSe crystal sample accessory. Data were collected from 16 scans with a resolution of 4 cm^{-1} and are presented in Figure 6.15. No notable changes

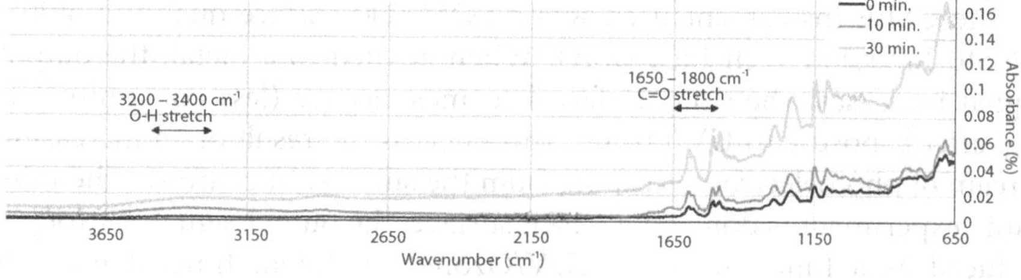

Figure 6.15 ATR-FTIR absorption spectra of CFRP surfaces at three stages of surface activation. The reference sample (0 min.) was acetone cleaned only, after which the sample was treated by UV/Ozone for 10 and 30 minutes.

were observed, therefore the ATR-FTIR spectra were not normalized, and only one spectrum for each measurement was chosen for representation.

The main changes in the chemical composition as a result of the increased UV/Ozone exposure time of the surface layer by oxidation were found by the formation of alcohol and carbonyl groups with strong stretching peaks between 3200 - 3400 and 1620 - 1870 cm^{-1} respectively as can be found in Figure 6.15. This increase in oxidation is in line with the results found by XPS (section 6.5.3) and the contact angle measurements (section 6.5.6. Furthermore, they are in agreement with those found by other researchers [46, 68].

Washing the samples with acetone directly after the UV/Ozone treatment did not show a measurable difference in the FTIR scan results. This implies that the washing procedure did not remove a measurable amount of some Low Molecular Weight Oxidized Molecules (LMWOMs) as an effect of chain scissions, from the surface to a depth of 0.5-2 μm, considering the scan depth of the infrared technique.

6.5.5 Optically Stimulated Electron Emission (OSEE) Investigation on UV/Ozone Treated CFRP

An accurate quantitative method to determine the surface properties is to make use of the photoelectric effect, described by Einstein who showed that electrons could be released from a surface when the absorbed energy of a photon exceeded the energy by which it was attracted [69]. The intensity of electrons emitted from such a surface is found to be related to the surface condition (e.g. the level of oxidation) and the amount (thickness) of contaminant layer present. The intensity of emitted electrons can be measured by an OSEE instrument, which exposes the surface to a narrow beam of UV- light and measures the number of emitted electrons as a result. This technique has found an important application in the adhesive bonding technology, especially in space related objects [70–72].

The OSEE instrument used was a SQM 200 surface quality monitor (Photo Emission Tech Inc., USA). When applied to a metal, the output signal increases when the surface becomes cleaner (less contaminated) e.g. after exposure to UV/Ozone. This effect is the result of an increasing stream of photoelectrons escaping from the surface. In contrast, when the same experiment is done on a CFRP surface, the output signal is strongly reduced as a function of the (UV/Ozone) treatment time, due to an increased static charge which prevents electrons from escaping the surface (Figure 6.16).

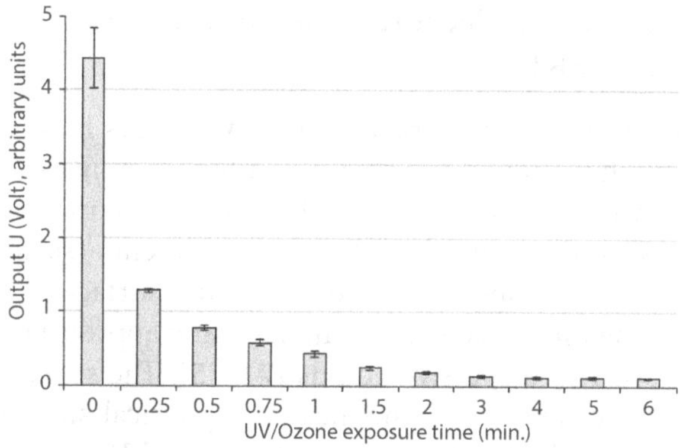

Figure 6.16 OSEE output voltage as a function of an incremental exposure time to UV/Ozone.

However, wiping/washing the surface with a polar liquid like acetone or water showed a (repeatable) steep rise of the OSEE initial signal as shown in Figure 6.17. The noticeable slight overall rise of the recovered OSEE signal is the result of a permanent change of the oxidation state and thus the electrical conductance of the CFRP surface.

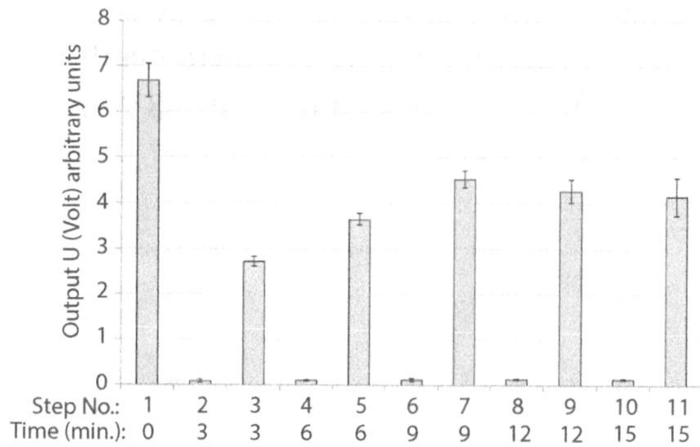

Figure 6.17 OSEE output voltage as a function of the UV/Ozone exposure time with an increment of 3 minutes/step and intermediate cleaning by acetone. Step1 is "As is" or no UV/Ozone exposure. Step 2 is after 3 minutes of UV/Ozone exposure. Step 3 is after 3 minutes UV/Ozone exposure including acetone wash. Step 4 is after 6 minutes UV/Ozone exposure, etc.

6.5.6 Contact Angle Measurements on UV/Ozone Treated CFRP

Contact angle measurements are sensitive to changes in surface chemistry and topography. Though the changes in topography as a result of the UV/Ozone treatment were not detected, as they were much smaller than 100 nm, as observed during the AFM measurements (section 6.5.2 [49, 73]). There are various approaches to determine surface free energies of polymers from the measured contact angles using appropriate test liquids, ranging from polar to dispersive liquids [74, 75]. The surface free energy values as a result of combined chemical and physical surface treatments (Figure 6.18) were calculated using the Owens and Wendt [76] approach from the static contact angle measurement data (droplet volume: 5 μl). They were measured on CFRP samples, sized 25 x 20 mm, with a range of liquids: water, formamide, diiodomethane, and glycerol.

The results of 10 minutes of UV/Ozone treatment showed a large increase of the polar component of the surface free energy compared to the just acetone cleaned surface. This is in agreement with the increase of the oxygen content of the surface as found with both XPS (section 6.5.3) and ATR- FTIR measurements (section 6.5.4).

Washing the samples with acetone directly after the UV/Ozone treatment did not show a measurable difference in the polar component of the total surface free energy. This implies that the washing procedure did not remove a measurable amount of polar groups generated by chain scissions during the UV/Ozone process, considering its treatment depth as 1 μm [58] and the scan depth of the ATR- FTIR technique as being only 0.5 – 1 nm [77].

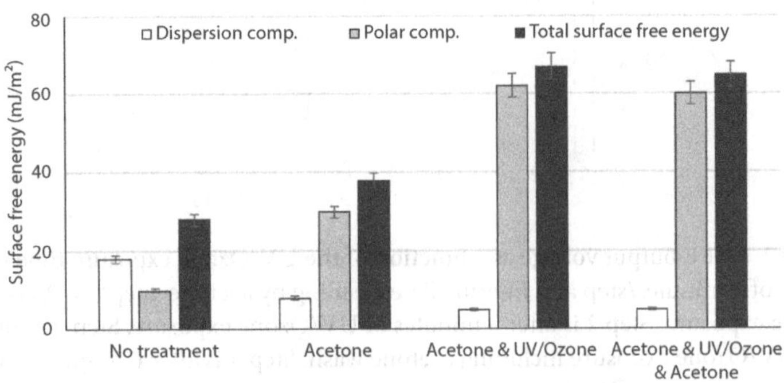

Figure 6.18 Surface free energy of CFRP together with its dispersion and polar components, when untreated, acetone cleaned only, acetone & 10 minutes UV/Ozone treated, as well as an added acetone cleaning step after the UV/ozone treatment.

Thus, while UV/Ozone treatment penetrates rather deep into the surface (about 1 μm), but contact angle measurements are sensitive to a depth of only 0.5-1 nm [77] and as a result far less sensitive to changes in the bulk mass. It might be concluded from the above results together with the mechanical measurement results (section 6.6.2.2) that in the uppermost layers only a few chain scissions occurred. These upper layers were thus rather uninfluenced by the washing procedure. In other words, the washed out LMWOMs must come from deeper polymer layers.

6.6 UV/Ozone Treatment of Polymers: Improved Wetting and Adhesion

6.6.1 Introduction: UV/Ozone Treatment of Polymers

The removal of photoresist polymers by UV-light and ozone gas (depolymerisation) was already described by Bolon and Kunz as early as 1972 [78], while its cleaning capability was extensively described by Vig and LeBus [16] in 1976. Still the application of UV/Ozone as a treatment for improved adhesion between polymer surfaces like polypropylene (PP) and adhesives was discussed by only a few researchers [43, 47]. On a molecular level, UV/Ozone treatment of polymers was discussed to explain the importance of high surface free energy (or increased surface wetting) for successful adhesive bonding e.g. by Mittal [52] or van der Leeden and Frens [79] and more extensively by Sham *et al.* for the functionalization of carbon nanotubes to improve the mechanical performance of an epoxy matrix by modifying the surface properties of nanoscale reinforcements such as nanotubes and graphite [80].

The application of this technique in the sealing of disposable thermoplastics for microfluidics was described in literature in 2009 [81]. Some researchers make use of added oxidizing agents during the UV/Ozone treatment such as chlorine dioxide (ClO_2) [82] or water which influences the atmosphere in the reaction chamber and as a result the ozone production [44].

The advantage of the UV/Ozone technique itself is that even in its most basic form it can be regarded as easy to use and highly reproducible for extremely high quality surface cleaning [44] to increase the surface polarity of hydrophobic polymers, including difficult-to-bond materials such as Ethylene-Propylene-Diene Monomer (EPDM) rubber or Ethylene-Vinyl-Acetate (EVA). The UV/Ozone technique was found by the authors to be capable of cleaning (metal and polymer) surfaces down to the very last layers of contamination, even thin layers of Teflon© or silicone oil, metal drilling fluids or fingerprints. Hence, bringing environmentally-friendly high-quality cleaning for adhesive bonding within reach for commercial applications like e.g. in the shoe industry [6].

Several commercial UV/Ozone applications were explored [81, 83, 84] and sometimes even successfully commercialized like the cleaning of flat surfaces in laboratory environments. However, despite the technological successes, the technique never became as popular as corona or (atmospheric) plasma treatments in the field of commercially-based adhesion improvement. As high quality and reproducible surface treatment is becoming more and more important, especially with the increasing use of adhesive bonding in multi-material applications such as in the microelectronics, automotive, and aerospace industries where lightweight structures of Carbon Fibre Reinforced Polymers (CFRPs) are often adhesively bonded to metal parts such as titanium. UV/Ozone surface cleaning and activation was found to show valuable results as shown in various scientific publications [8, 9, 15, 17, 50, 70, 84].

6.6.2 UV/Ozone Treatment of Thermoset Materials

6.6.2.1 Introduction: UV/Ozone Treatment of CFRP

CFRP material combines high strength-to-weight ratio and stiffness which is a key issue in the aerospace industry in order to minimize the environmental impact of transportation. To achieve this, weight reduction of aircraft without compromising the structural integrity is a key issue. Within the scope of energy reduction falls the Hybrid Laminar Flow Control (HFLC) technology which reduces turbulence and increases laminar flow on the wing's skin of the leading edge. Effectively, this construction consists of a thin porous titanium grade CP 40 sheet which is adhesively bonded to a Carbon Fibre Reinforced Polymer (CFRP) inner structure (Figure 6.19) [85, 86].

The key to success of high quality durable adhesive bonding of such a structure strongly relies on the surface treatment of both substrates prior to the bonding process. Thus, it is crucial to promote the interfacial adhesion between the adherends and adhesive to obtain the highest bond strength and durability. With a properly selected surface treatment the adhesive bond should never fail at the adhesive-substrates interfaces, but cohesively within either the adhesive or adherends.

The surface treatments of both CFRP and titanium have been extensively studied, and are summarized in a review paper [87]. Various titanium

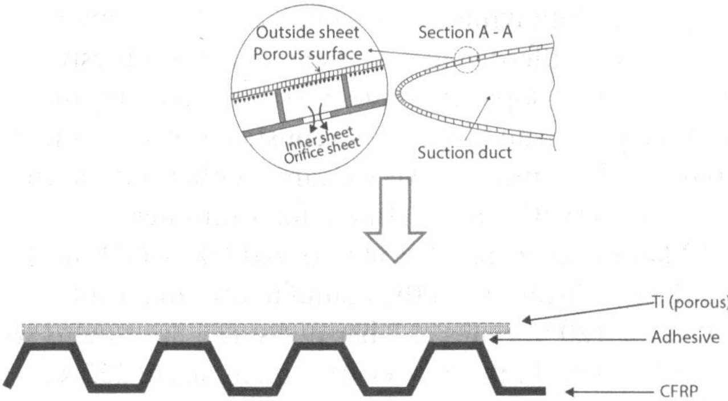

Figure 6.19 Sketch of Hybrid Laminar Flow Control (HFLC) (top) and a simplified 2D-version (bottom) showing adhesive bonding between titanium CP 40 top sheet and CFRP (inner structure of airplane wing) [86].

alloys have been used to study the effects of surface treatments. However, up to now, the application of the UV/Ozone technique as an integral part of the surface treatment for CFRP and titanium for structural adhesive bonding was reported by only a few researchers [e.g. 50, 68, 88].

The study of the UV/Ozone surface treatment of CFRP was extended to determine in more detail its effect on the epoxy matrix of the CFRP from a sub-micro to a macro-level. These sub-micro results showed that the observed changes occurred mainly on the outermost layer of the CFRP surface as a result of the UV/Ozone treatment. Additionally, mechanical test results are discussed when CFRP samples were tested in peel on a macro-scale (sections 6.6.2.2 and 6.6.2.3) and specifically when bonded to titanium CP 40 on a large test panel (resembling a realistic loading case of Figure 6.19) that was tested in mechanical fatigue loading (section 6.6.2.3).

6.6.2.2 *Mechanical Tests on UV/Ozone Treated CFRP*

Peel tests are the best mechanical test method to determine the weakest point of an adhesive bond [89]. Floating roller peel tests (ASTM D 3167-10) [90] are an excellent method to test adhesively bonded flexible to rigid multi-layer materials such as CFRP. The CFRP adherend was a laminate with a layup of either two plies (0°/90°) for the flexible part or ten plies (0°/ 90°/0°/90°/0°) for the rigid part, with a total thickness of approximately 0.27 mm and 1.55 mm respectively. During the fabrication process the surfaces of the laminates were in contact with a Fluorinated

Ethylene Propylene Copolymer release film (FEP Copolymer A 4000 clear red, Airtech Europe, Differdange, Luxembourg). The adhesive used in this study was the (red) knit supported carrier structural epoxy film AF163-2K designed for honeycomb and solid panel constructions, obtained from 3M, The Netherlands. The laminates were cleaned either with acetone or acetone with an added UV/Ozone treatment for 7 minutes.

The CFRP laminates were fabricated (sized 600 x 600 mm) and cured according to the specifications of the manufacturer in an autoclave (curing cycle of 180 °C for 120 minutes at 7 bar pressure). Specimens were cut by a diamond cutting wheel into smaller specimens sized: 255 x 25 and 305 x 25 mm for floating roller peel testing.

Mechanical test results were obtained on a Zwick 1455 test machine equipped with a 1kN load cell. The results (Figure 6.20) showed that the average peel strength values of these two groups of samples hardly differed. However, the acetone cleaned only samples show an adhesion (interfacial) failure while the UV/Ozone treated surfaces show an interlaminar (cohesive) failure of the CFRP, which indicates an increased adhesive bond strength across the interface. This is in line with previous observations by other researchers such as e.g Wang *et al.* [68] and Teixeira de Freitas *et al.* [91]. The change in failure type from adhesion failure between the adhesive and the CFRP to interlaminar failure is also in line with the above results, showing that an increased polarity of the CFRP surface layer results in stronger interactions across the CFRP-adhesive interface which were found to be, altogether, eventually stronger than those within the CFRP laminate [92–94].

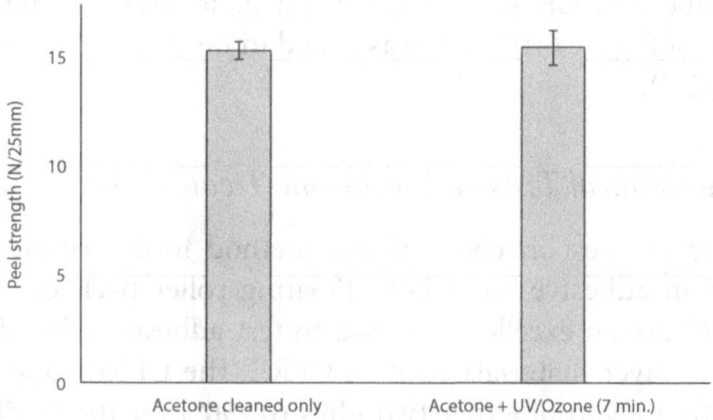

Figure 6.20 Floating roller peel test results for acetone cleaned only and those with a 7 minutes added UV/Ozone treatment of CFRP to CFRP bonded samples.

Washing the samples with acetone directly after the UV/Ozone treatment did not show a measurable difference in the mechanical test results (not shown in Figure 6.20) nor in the failure mode. This implies that the washing procedure and the removal of the LMWOM had no measurable effect on the bond strength as found in previous research [45].

6.6.2.3 Mechanical Fatigue Loading of UV/Ozone Treated CFRP

Two representative panels, based on the drawing shown in Figure 6.19, with a small artificially created defect between the titanium CP 40 sheet and one of the CFRP stringers, were subjected to in-plane cyclic loading. This loading involved an increasing in-plane compressive load, unloading, and then reloading to a higher load level until fracture occurred. The failure found was mostly adhesion failure at the titanium surface located in the vicinity of the artificial defect and cohesive failure within the adhesive layer further away from the artificially made defect [95]. This failure type is in line with previously found results discussed in section 6.6.2.2.

6.6.2.4 Adhesive Bonding of UV/Ozone Treated CFRP to Aluminium

UV/Ozone treatment was successfully applied for the surface preparation of aluminium bonded to CFRP when a study was done to replace Bisphenol A by an eco-epoxide adhesive synthesized from bio-renewable raw material (tannic acid—TA) and used for bonding lightweight materials (aluminium (Al) and a carbon fibre reinforced polymer (CFRP) as discussed by Tomić *et al.* [96].

6.6.2.5 UV/Ozone Treatment and Testing of Aerospace Primers

Generally, the surface treatment of primers before a top-coat layer is applied is done in a standard sequence that consists of degreasing (by solvent wipe), surface roughening to remove a thin and inactive surface layer and, again, a solvent wipe to remove dust particles [97]. Surface modification using UV/Ozone was explored by Haack *et al.* as an approach for robust inter-coat adhesion in multi-layered automotive coating systems. This treatment was used to reduce the variability in adhesion performance linked to changes in its surface chemistry resulting from migration of additives [98]. This approach was again investigated in the laboratories of the Adhesion Institute of the Delft University of Technology, amongst other alternative treatments, to reduce the amount of dust particles which are released during the surface roughening of adhesive bond primers on the airplane's fuselage onto which a paint primer and a topcoat are applied in the painting process.

In standard practice, the surface treatment of adhesive bond primers is: solvent cleaning with methyl ethyl ketone (MEK) followed by light abrasion with Scotch-Brite® (3M) and again cleaning with MEK to remove dust particles. This is a time consuming and unhealthy surface treatment process due to the formation of dust particles. Thus, a reference treatment and three physical surface treatments:

- Atmospheric plasma (10 – 120 seconds);
- Corona (10 – 40 seconds), or
- UV/Ozone (5 – 15 min.).

were done on bond primer protected chromic acid anodized aluminium test panels.

The test results were compared with the standard conventional three-step process: solvent cleaning, primer abrasion, and again solvent cleaning. After the surface treatment of the bond primer layer, a paint primer and paint top-coat were applied in a professional spray coating facility. The sequence of layers of the chromic acid anodized aluminium panels and those of the paint layers is shown in Figure 6.21. The test panels were coated with one of the following adhesive bond primers:

- BR® 127 corrosion inhibiting primer (Solvay);
- BR® 6700 NC a non-chromate water-based epoxy primer (Solvay);
- Redux 101 phenolic primer (Hexcel).

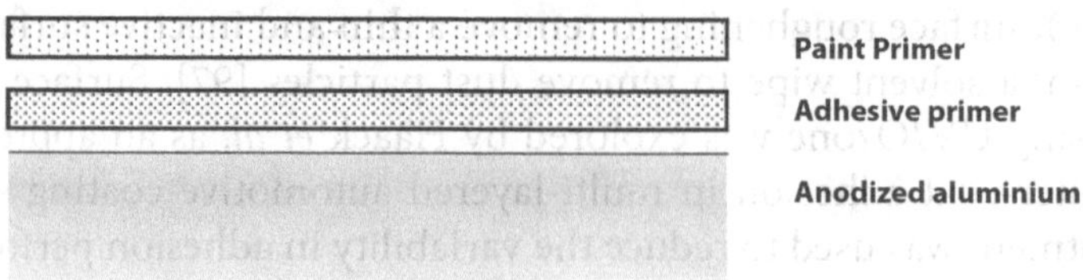

Figure 6.21 Schematic representation of chromic acid anodized aluminium with the adhesive primer and paint layers (not to scale).

The surface cleaning/activation effect on the three adhesive bond primers was studied by static water contact angle measurements. The test results (Figure 6.22) showed clearly the best overall wetting performance for UV/Ozone compared to atmospheric plasma and corona treatments.

After surface treatment of the three bond primers by UV/Ozone, atmospheric plasma, or corona, a paint primer (Akzo 37035 A), and a paint top-coat (Aerodur Finish C21/100) were applied. Subsequently cross-cut tests [99] and were done. The panels (Figures 6.23a and b) were cured according to the material specifications, and stored in the laboratory at room temperature for at least 1 week before cross-cut tests were carried out in laboratory conditions. The adhesion between the layers was evaluated in the initial dry state as well as after 14 days of tap water exposure at room temperature (RT). After removal from the water and drying of the panels further cross-cut tests were done (tape: Scotch 250, 3M, The Netherlands). Some of the visual test results on the test panels for the combinations of the adhesive primer with the paint primer and topcoat are presented in the upper sections of Figures 6.23a and b.

All three adhesive bond primers showed a rather similar behaviour as the prepared panels (either with paint primer or paint primer with top-coat) and showed no flaking during the initial (dry) state, as can be seen from the cross-cut test results. However, after exposure to tap water for 14 days at RT and subsequently drying, partial flaking, blister formation and detachment of the topcoat from the paint primer occurred, which

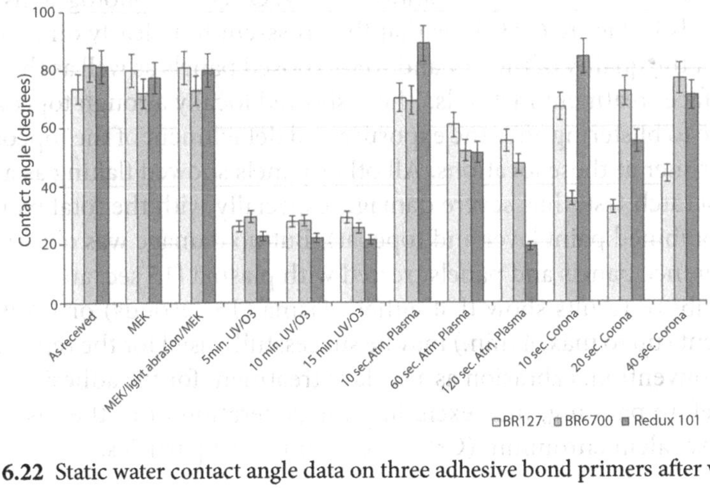

Figure 6.22 Static water contact angle data on three adhesive bond primers after various surface treatments.

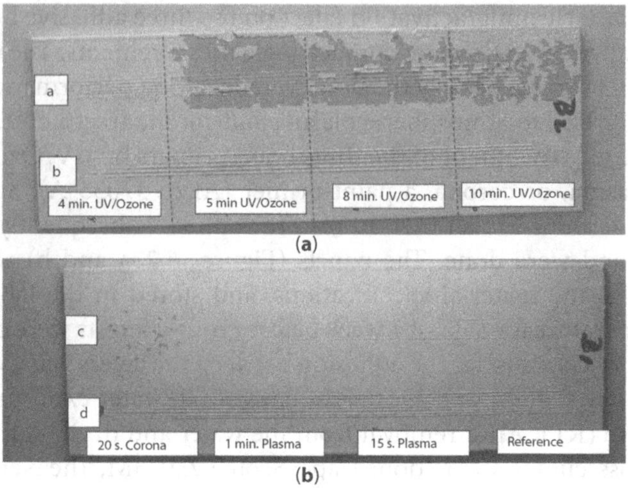

Figure 6.23 Cross-cut tape test results on chromic acid anodized BR127 primed together with paint primer and top coated panels after various surface treatments. The top section (a or c) of each panel is the 2 weeks water exposed part, and the lower section (b or d): non-exposed. The following surface treatments were done: a: 4, 5, 8 and 10 minutes UV/Ozone treated and painted, water exposed. b: 4, 5, 8 and 10 minutes UV/Ozone treated and painted, dry. c: 20 sec. corona, 60 sec. and 15 sec. plasma treated, the reference, water exposed. d: 20 sec. corona, 60 sec. and 15 sec. plasma treated, the reference, dry.

were obviously caused by a sub-optimal pre-treatment. This also occurred for the panels treated by UV/Ozone with treatment times longer than 4 min., possibly due to the generation of LMWOMs. The photographs of the test panels in Figure 6.23 show that the cross-cut test clearly differentiates between the quality of the dry and wet-exposed panels as well as the different surface treatment methods. These showed locally a rough topcoat surface due to blistering after the exposure and detachment of the topcoat and paint primer at these locations. All other panels showed flaking as a result of the scratch test, thus severe damage, especially with the total paint system (combined paint layer and topcoat). But no damage was observed for the reference panels and panels treated with plasma (15 sec. and 1 min.).

The above results show that either plasma (15 seconds) or UV/Ozone treatment (up to max. 4 min.) may be successfully used for the replacement of the conventional abrasion as a surface treatment for the adhesive primer layer before painting, thus excluding the generation of in the case of BR® 127, hexavalent chromium (Cr^{+6}), containing dust particles.

6.6.2.6 *Mechanical Tests on UV/Ozone Treated Epoxy Coated Magnets*

Static magnets of electro-motors were coated with an epoxy or a metallic coating to prevent corrosion. These surfaces were pretreated either by atmospheric plasma or UV/Ozone in order to improve the bonding to an adhesive. The physical surface treatments improved the surface wetting considerably compared to cleaning with a solvent wipe (methyl ethyl ketone, MEK) only, as can be seen in Figure 6.24.

Compared to the solvent wipe, a strongly increased surface wetting after UV/Ozone and atmospheric plasma treatments due to an increased surface polarity became visible, but no large difference in wetting between the two physical activation methods was evident. Generally, the best results were obtained either after 10 min. UV/Ozone or 60 sec. atmospheric plasma treatment. The resulting increase in polarity due to these physical surface treatments was expected to improve the adhesive bond strength between the coatings and a two-component room curing structural epoxy with 1 w% added glass beads of ø 150 µm or a no-mix acrylate adhesive. The adhesive bonds were pre-cured at room temperature for 1 day and then cured for 1 hour at 80 °C and 100 °C. Both shear (ASTM D1002 [100]) and butt joint tests (EN 15870 [101]) were performed on a 250 kN Zwick static test machine with a 1kN load cell. The lap shear and butt joint test results of the no-mix acrylate adhesive are shown in Figure 6.25. The lap shear test

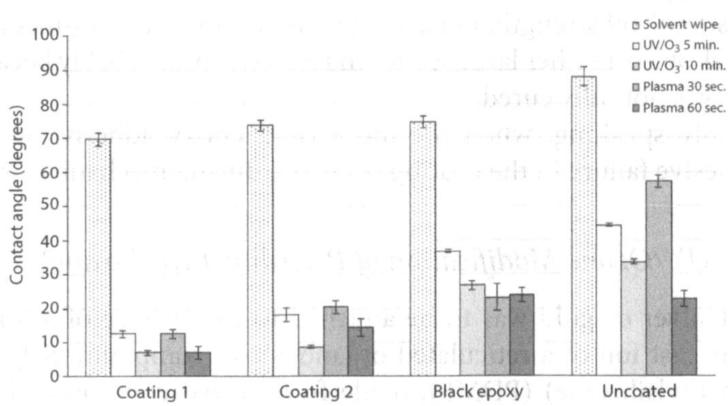

Figure 6.24 Static contact angles of water on magnet coatings after 5 different surface treatments.

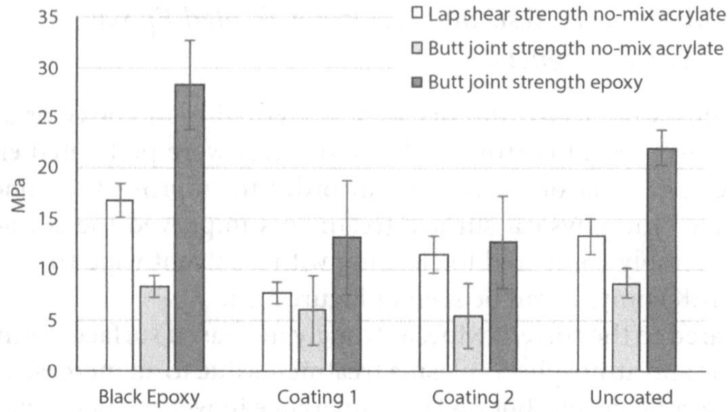

Figure 6.25 Lap shear strength together with butt joint test results for a no-mix acrylate and an epoxy adhesive on magnet coatings.

results of the epoxy adhesive bonds are not shown, but they were all at least twice as high as those found for the no-mix adhesive.

Fracture analysis of the lap shear and butt joint test samples for the no-mix adhesive showed that 90% of adhesion failure occurred on the coating surface. Magnets bonded by the epoxy adhesive showed 50% cohesive failure within the epoxy adhesive and 50% within the coating. The black colored epoxy showed about 90% cohesive failure of the coating itself, possibly due to too much filler material, increasing its brittleness.

The butt-joint test results were found to be the most discriminating for the failure type for both adhesives, with the black epoxy coating showing the largest pull-off strength. On average, the results of the no-mix acrylate adhesive showed a rather large scatter in results, most probably because the adhesive was not fully cured.

Generally speaking, when the more rigid epoxy adhesive was used, more cohesive failure in the coating occurred during mechanical testing.

6.6.2.7 UV/Ozone Modification of Poly(dimethylsiloxane)

Colloidal silver or gold was to be attached to the surface of a thin optically clear cast foil of a reticulated organosilicon compound Sylgard-184 poly(dimethylsiloxane) (PDMS), made from a two-component kit from Dow Corning. The viscoelastic PDMS, when cast as a foil, was to be placed on top of a valuable piece of art such as a drawing, photograph, or letter [e.g. 102, 103] for improved Raman analysis. The colloidal material functions to amplify the Raman signal and suppress the fluorescence of the object. After analysis, the PDMS foil can easily be removed without leaving any traces.

The surface of the PDMS is highly hydrophobic and thus it has to be made more hydrophilic to allow adhesion of the colloidal particles. This can be done by exposing it to a so-called "piranha" solution, during which silicon oxide with hydrophilic (mainly -OH) groups are created on the treated surface [104]. However, this method was found to be too risky for the specific purpose (analysis of valuable pieces of art), non-uniform and also non-reproducible. A far simpler, safer and more reproducible method of surface oxidation was expected from exposure to an ozone generating low-pressure mercury lamp. In this process the macromolecules undergo chain scissions, involving both the main chain backbone and side groups, after which the UV-modified specimen interact with the molecular oxygen and ozone [105, 106]. The oxidised PDMS surface is then chemically reacted with (3-aminopropyl) triethoxysilane (APTES) and via this process amino groups are bonded to the PDMS which can then react with the colloidal material.

The "As received" cast PDMS surface (thickness ~ 1 mm) showed an average (water) contact angle of 98 degrees. But a surface treatment for 90 minutes by UV/Ozone resulted in a spectacular increase of the surface wetting with an average contact angle of 8 degrees (Figure 6.26). However, immediately after the treatment a "hydrophobic recovery" started and during such occurrence the chemical composition of the treated PDMS

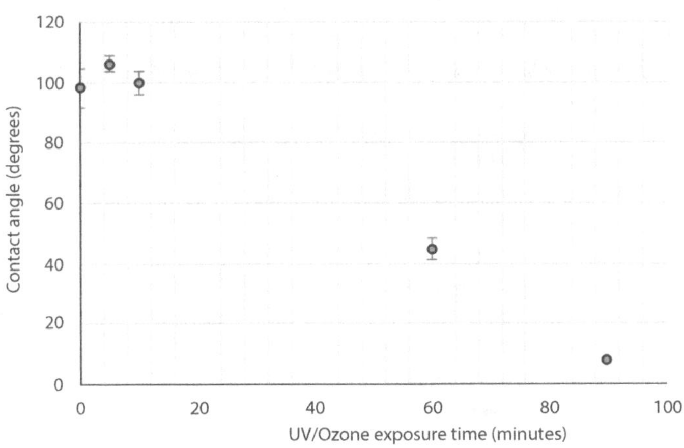

Figure 6.26 Static contact angle (water, degrees) on PDMS surface as a function of UV/Ozone exposure time.

surface changed over time due to the uncontrollable diffusion to the surface of non-crosslinked silicone oligomers through sub-microscopic surface cracks [105, 107, 108]. This cracking was not visible after the UV/Ozone treatment as shown by Owen and Smith [108]. Thus it is assumed that simply diffusion of hydrophobic molecules occurred. In contrast, the initial visible waviness of the surface as a result of casting was clearly visible in the micrograph (Figure 6.27a), but was removed by the UV/Ozone treatment. Indeed, after 60 and 90 minutes of UV/Ozone treatment (Figures 6.27b and 6.27c respectively) all visible signs of this waviness had disappeared.

Immediately after the UV/Ozone treatment the surface started to turn back to a hydrophobic state, which already became clearly visible within minutes and this increase of hydrophobicity continued (Figure 6.28). After 5 days exposure to a laboratory environment (20 °C, 55% RH) the same surface was exposed again to the UV/Ozone treatment for 90 minutes, after which it showed a complete wetting by water, though the trend to return to hydrophobic state started again directly after the treatment.

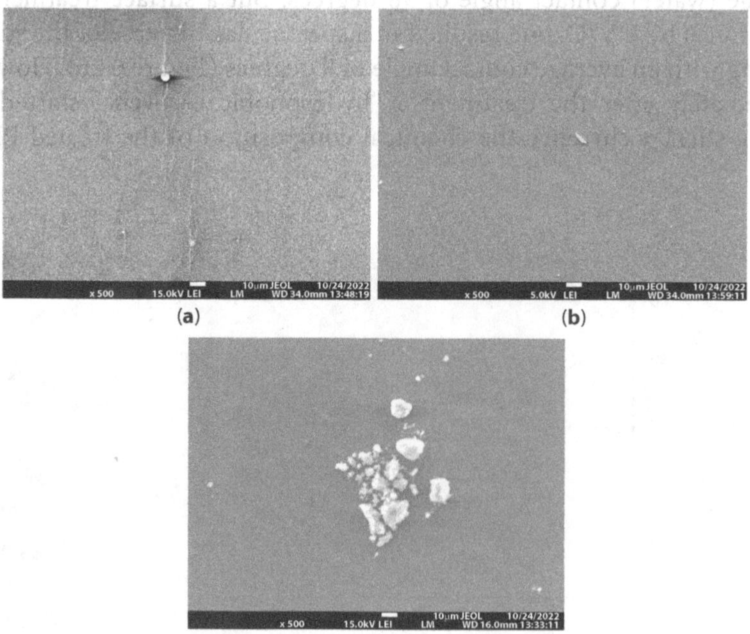

Figure 6.27 SEM micrographs of a cast PDMS surface after (a) 0 minute, (b) 60 minutes and (c) 90 minutes exposure to UV/Ozone.

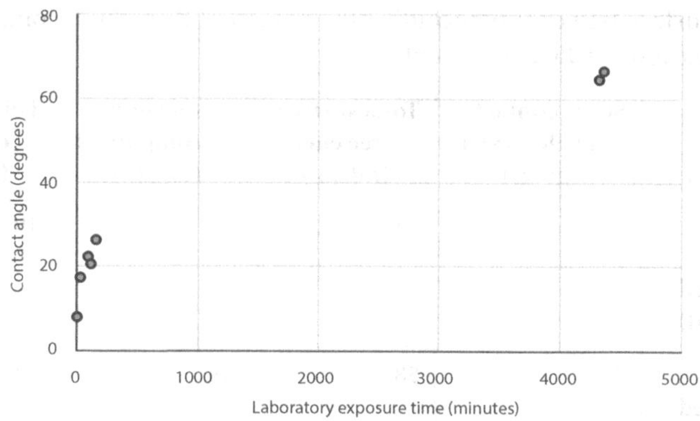

Figure 6.28 Static contact angle (water, degrees) as a function of the exposure of PDMS to laboratory air after an initial UV/Ozone treatment for 90 minutes.

6.6.3 UV/Ozone Treatment of Thermoplastics

6.6.3.1 Adhesive Bonding of POM to Aluminium

Small cylindrical stand-offs (ø 15 mm, thickness 5 mm) made of Polyoxymethylene (POM), an engineering thermoplastic, were bonded with a two-component paste adhesive to aluminium (2024 T3). These are used as cable guides in spacecraft constructions. However, the POM's surface free energy was found to be too low to obtain a reliable adhesive bond without adhesion failure on the POM surface to withstand at least 7 up to 33 temperature cycles between -100 °C and +100 °C.

Hence, different surface treatments were carried out to improve the bond strength. The POM surface was cleaned with isopropanol as a pre-cleaning step, roughened (emery paper P1000) and again cleaned with isopropanol (reference treatment, see Table 6.4). Then the samples were successfully treated for 5 to 20 seconds in 85% phosphoric acid at 50 °C, or exposed for 15 minutes to UV/Ozone, or exposed for 15 seconds to Corona. The adhesive bonding to aluminium was done either by the above-mentioned two-component room curing epoxy or a two-component silicone adhesive. After temperature cycles the bonded joints were tested in shear on a 20kN Zwick 1455 test machine with a 1kN load cell.

The average adhesive bond strength clearly improved by the UV/Ozone treatment (Figure 6.29), but the fracture areas still showed 100% adhesion failure at the POM surface. This failure is the result of a too large stress at the POM - adhesive interface during the temperature cycles as a result of the difference in thermal expansion between the POM and aluminium

Table 6.4 Surface free energy including the polar and dispersion components of POM after various surface treatments.

Type of treatment	Static contact angle (water, degrees)	Total surface free energy (mJ/m²)	Dispersion component (mJ/m²)	Polar component (mJ/m²)
Reference (Degreased, roughened, degreased)	61	43	33	10
Degreased, roughened, degreased UV/Ozone (15 minutes)	38	58	24	34
Degreased, roughened, degreased, phosphoric acid	52	49	29	20
Degreased, roughened, degreased, corona (15 seconds)	42	56	28	28

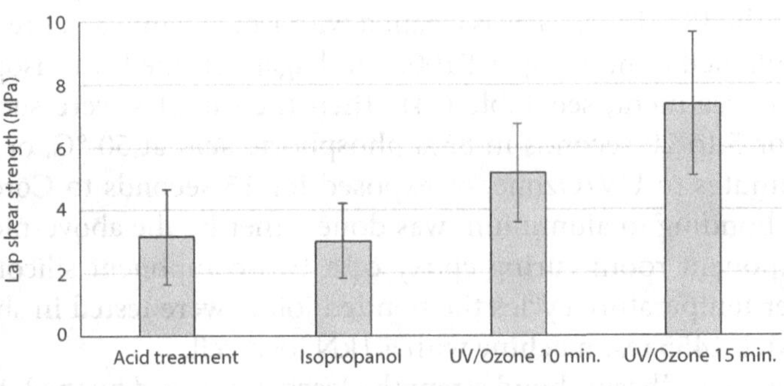

Figure 6.29 Average lap shear strength of POM bonded to aluminium with a 2- component room temp. curing epoxy (EC2216), as a function of surface treatment after 33 temperature cycles (from -100 °C to +100 °C).

substrate. Part of the solution was found by increasing the thickness of the adhesive bondline from 0.1 mm to 0.8 mm by adding 2 w% ø 0.8 mm glass beads to the mixed adhesive. An alternative solution might be to apply a more flexible adhesive and/or an increased density of polar groups on the POM surface.

6.6.3.2 Adhesive Bonding of Polyethylene (PE) to Stainless Steel

An instrument that was constructed from PE and stainless steel showed an adhesion bond failure at the surface of the PE. The part had to operate in a saturated saltwater environment at a temperature of -15 °C. The adhesive bond had to withstand this environment for up to 6 weeks. Therefore, it was decided to increase the hydrophilicity of the PE by a surface treatment.

Samples of both PE and stainless steel were pre-cleaned with ethanol before they were treated by UV/Ozone for 5 to 15 minutes which resulted in a strongly improved wetting performance for both materials (Table 6.5). Alternatively, the so-called DELO-SACO process was applied to the stainless steel [109].

After various surface treatments, the parts (PE to stainless steel) were bonded by a 2-component polyurethane adhesive with an adhesive bondline thickness of 0.2 mm, using 0.1w% glass beads. Samples were tested in shear (ASTM D1002) [100] and test results are shown in Figure 6.30.

Bond failure still occurred at the PE surface, despite the improved surface wetting as a result of any of the UV/Ozone treatments of the PE, and as a result, an increased bond strength, meaning that the combination of surface treatment and adhesive was not sufficient enough for this application. The data presented in Figure 6.30 show that the effect of surface treatment

Table 6.5 Contact angle of water on PE and stainless steel as a function of UV/Ozone treatment.

Type of treatment	Contact angle on PE (water, degrees)	Contact angle on steel (water, degrees)
No treatment	90	72
Ethanol	85	70
5 min. UV/Ozone	57	27
10 min. UV/Ozone	56	20
15 min. UV/Ozone	50	24

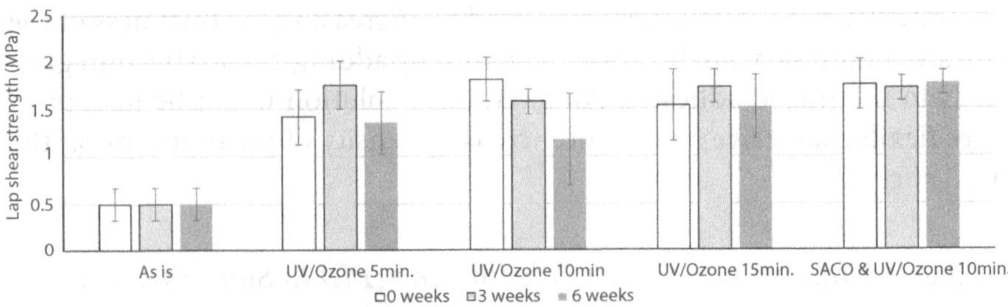

Figure 6.30 Lap shear strength test results of adhesively bonded PE to stainless steel after 0 to 6 weeks of aging in salt water at -15 °C as a function of various surface treatments.

by UV/Ozone leads to a strong increase in the adhesive bond strength, including after salt water exposure. Besides, the average bond strength after the DELO-SACO surface treatment with an added UV/Ozone treatment for 10 minutes showed a stable lap shear strength with time and an average bond strength in the same range as the UV/Ozone treatment alone, but with a slightly lower standard deviation.

It should be noted that the average bond strength reaches a maximum value, though not significant, at 10 min. of UV/Ozone treatment after which it reduces again due to the formation of Low Molecular Weight Oxidized Molecules (LMWOMs), further described in section 6.6.3.8.

6.6.3.3 *Adhesive Bonding and Aging of HDPE*

Only a few researchers have compared the effects of different physical surface treatments on surface wetting, adhesive bond strength, and relaxation of low energy surfaces [58, 110–112]. Hydrophobic materials with their non-polar nature such as polypropylene (PP), thermoplastic polyolefin (TPO) and High Density Polyethylene (HDPE) are generally difficult to bond adhesively, and therefore need some kind of surface treatment prior to the adhesive bonding process [109]. Sometimes a specific primer (a chlorinated polyolefin in an organic solvent) is used to increase the surface polarity for an improved interaction with the adhesive [113–115].

The physical surface treatment of high density polyethylene (HDPE) used as a construction element of a fuel container was investigated. The treatment had to provide an increased surface wetting to improve the lap shear strength performance after adhesive bonding. The HDPE material was pre-treated with a commercial cleaning solvent mixture called PFQD [116] as a pre-cleaning, after which any of the following physical surface treatments was used:

- UV/Ozone treatment: using three 80 W Heraeus Noble Light (NNIQ120) low pressure UV-lamps at a distance of 20 mm from the surface and a treatment time between 5 and 30 minutes;
- Corona treatment: at a distance of 10 mm from the surface and treatment times between 15 and 75 seconds;
- Atmospheric plasma treatment: at a distance of 10 mm from the torch and treatment times between 5 and 20 seconds.

The resulting effects of these treatments on the wetting by water of the HDPE surface as a function of the treatment time can be found in Figure 6.31.

The adhesive bonding was done using a mechanically relatively weak methyl methacrylate (MMA) adhesive specifically suited for bonding thermoplastic adherends, although acrylic adhesive types exist that might give a much higher strength (up to 7 MPa).

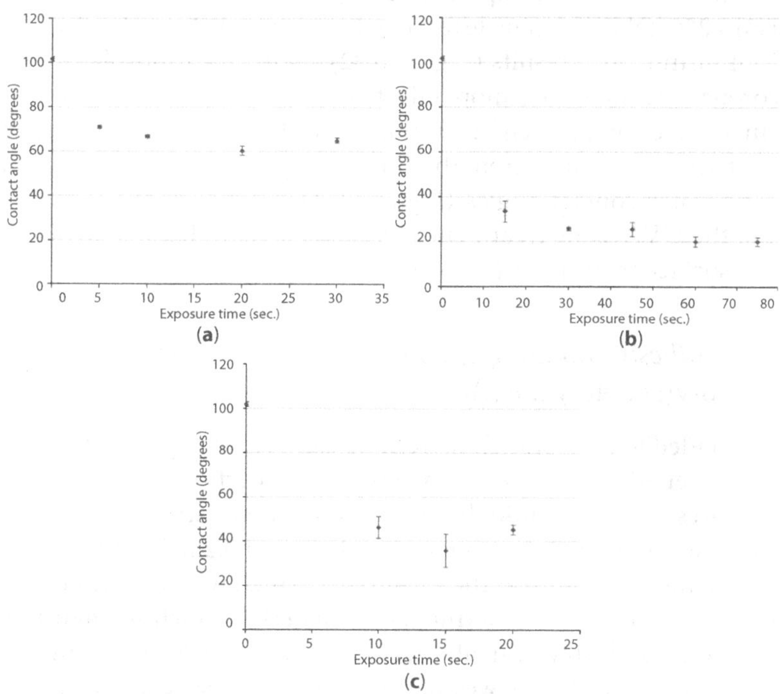

Figure 6.31 Static contact angle of water on HDPE as a function of the surface treatment time by UV/Ozone, corona or atmospheric plasma treatment. Please note the different time scales on the horizontal axis. (a) UV/Ozone. (b) Corona. (c) Atm. Plasma.

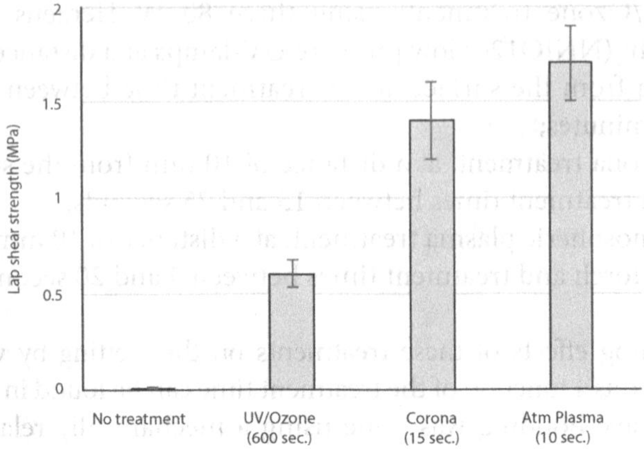

Figure 6.32 Lap shear strength of HDPE bonded to HDPE by a two-component epoxy as a function of various surface treatments.

Destructive mechanical lap shear tests were performed by ASTM D1002 [100] on a 250 kN Zwick tensile test machine with a 1kN load cell at a speed of 0.13 mm/min. Test results (Figure 6.32) show that atmospheric plasma and corona treatments are more effective than UV/Ozone and result in a large increase of bond strength compared to the untreated surface, which showed hardly any bond strength. This is in agreement with the results of the static water contact angles (Figures 6.31a-c), where the wetting as a result of the UV/Ozone treatment was also less than that by the other two physical surface treatments (corona and atm. plasma).

6.6.3.4 Adhesive Bonding of Polyethylene (PE) to Acrylonitrile Styrene Acrylate (ASA)

PE was bonded to acrylonitrile styrene acrylate (ASA) by a modified silane (MS) Polymer adhesive as a part of a pipe construction for use at elevated temperatures. However, initially the adhesive bond strength (with a silicone hot-melt adhesive) was found to be less than 0.2 MPa at 65 °C, showing 100% adhesion failure at the PE surface, which was insufficient.

ASA is an amorphous thermoplastic material, which is often used in the automotive industry as an alternative to acrylonitrile butadiene styrene (ABS) as it has a far better weathering resistance, especially to UV light. PE is one of the most often used common plastics today with many applications in packaging (bags, films, containers), but it is also a non-polar material, showing poor wetting characteristic and a low surface free energy

(~30 mJ/m^2) and is thus difficult to bond without (physical or chemical) treatment or the use of specific polyolefin primers [117, 118].

Various surface treatments were investigated in order to increase the initial bond strength as well as the durability of the adhesive bond. Thus, the following surface treatments were chosen:

- Solvent cleaning, roughening (Scotch-Brite®);
- Making use of a coupling agent/primer;
- Activation by UV/Ozone or atm. plasma.

An MS polymer was chosen as an alternative to the original hot-melt adhesive, which provided the opportunity to make use of a dedicated polyolefin primer system on the PE (Figure 6.33).

The optimum average lap shear strength was found by testing the adhesive bond in shear (ASTM D1002). Just roughening the surface did not lead to any improvement in the bond strength, as it still lacked polarity for adequate adhesive bonding. Nor the use of atmospheric plasma or a primer showed convincing improvement in the max. shear strength of the bonded samples. However, the surface treatment that led to a highly improved adhesive bond strength and 100% cohesive failure within the adhesive was found to be a 5 to 10 minutes UV/Ozone treatment (Figure 6.33).

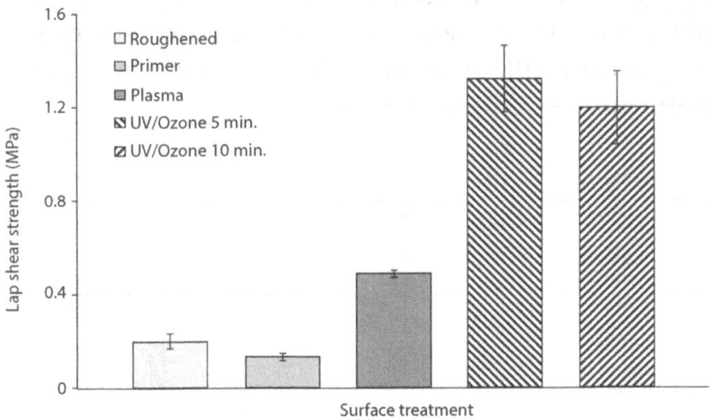

Figure 6.33 Comparison of different types of surface treatment methods on the lap shear strength of PE bonded to PE by an MS polymer adhesive.

6.6.3.5 *Adhesive Bonding of Polypropylene (PP)*

The surface functionalisation by photo-oxidation and the resulting increase of the adhesive bond strength to hydrophobic materials such as pure PP and PE was investigated by adhesive bonding with a one-component MS polymer mixed with 1 w% glass beads with a diameter of 0.125 mm. Bonded samples were cured for at least a week at RT. The test results from the lap shear strength measurements (ASTM D1002 [100]) are shown in Figures 6.34a and b, respectively.

All samples failed 100% at the interface when untreated. The fracture surfaces of the PP lap shear joints treated by Excimer 172 nm moved gradually from 100% interfacial failure without surface treatment to 100% cohesive failure within the adhesive layer after 15 seconds of photo-oxidation. After the surface treatment either by the Excimer 222 nm or the low-pressure mercury lamp, the failure mode changed to 100% cohesive failure in the adhesive layer. But the highest bond strength with both materials was obtained when the low-pressure mercury lamp was used. For the surface treatment of PP, the 222 nm Excimer is clearly more effective than the 172 nm. Though for PE the 172 nm seems to give better results, which are reached after an exposure time of only a few seconds.

6.6.3.6 *Surface Treatment of Nylon (Polyamide 6)*

Polyamide 6 (nylon) is an important structural material used in numerous industrial sectors. Surface treatment prior to adhesive bonding had been successfully demonstrated before using various methods like ozone gas only [119], plasma [120], or corona [121] treatments, but not yet by UV/Ozone treatment.

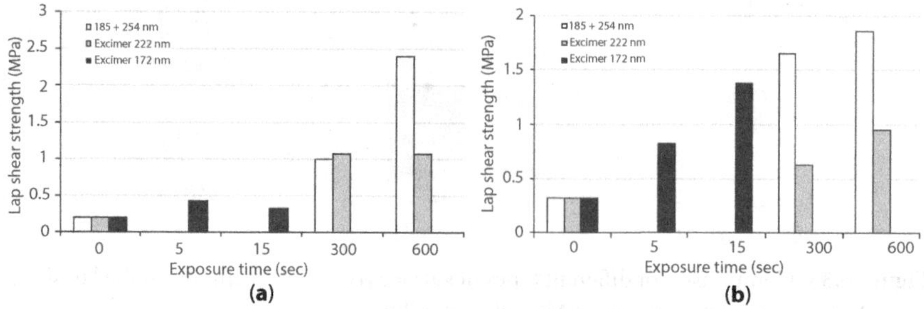

Figure 6.34 Lap shear strength as a result of the exposure time of PP (a) and PE (b) to: a low-pressure mercury UV-source, an Excimer 222 nm and an Excimer 172 nm. The bonded samples were made of PP to PP and PE to PE joints.

A flexible hollow nylon cylindrical tube was found to have occasionally adhesion bond failure when bonded to a stiff cylindrical housing of polycarbonate by an urethane/methacrylate UV-curing adhesive blend. Nylon is known for its absorption of water, which is, in practice, frequently found to be the reason for a premature adhesion failure in bonded structures. However, pre-drying the parts was found to be impossible at the production site. For this reason only a small set of samples was pre-dried in a stove in the Adhesion Institute of the Delft University of Technology laboratory for 2 hours at 60 °C in order to investigate the sensitivity of the chosen adhesive to absorbed moisture. The mechanical data did not show any improvement in the adhesive bond strength nor in the failure type, and pre-drying was therefore not included in the adhesive bonding process.

All parts were physically pre-treated either by UV/Ozone or corona in order to obtain the largest possible number of polar groups on the surface to obtain more molecular interactions between the nylon surface and adhesive.

Initially, the dispersion component contributed about three times as much to the surface free energy than the polar component (Figure 6.35). This contribution decreased as a function of the treatment time as a result of decreasing material density. In contrast, the value of the polar contribution more than doubled after either of the two physical surface treatments. After 5 seconds corona treatment the nylon parts show an increase in the polar component together with an increase in the total surface free energy. The contribution of the average polar component to the surface free energy is even larger than the one obtained for the UV/Ozone treated samples, though not significant.

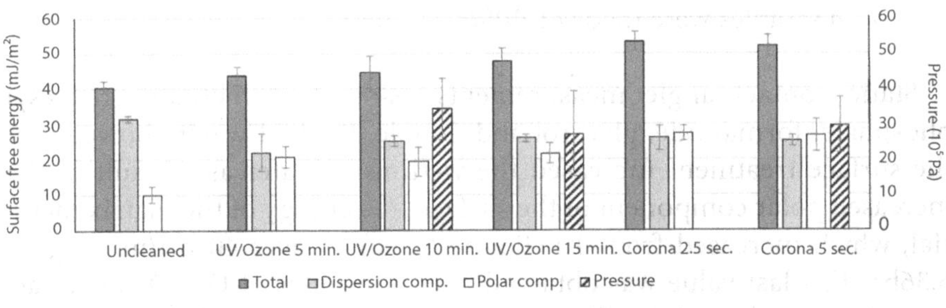

Figure 6.35 Bar graphs representing the surface free energy of nylon including its dispersion and polar components. The diagonally hatched bar graphs show the mechanical test results (due to an applied liquid pressure) after adhesive bonding to a polycarbonate housing.

10 and 15 minutes of UV/Ozone and 5 seconds of corona were used as surface treatments before testing the adhesive bond strength. The mechanical strength was tested by applying a hydrostatic pressure of 1 bar/second to the nylon tube which resulted in a peel force on the adhesive bond. The average adhesive bond strength (measured as a pressure until failure) was found to result into the highest value after 10 - 15 minutes of UV/Ozone treatment (Figure 6.35). But, within the standard deviation, no difference is visible between the results. Although the adhesive bond strength increased, the failure type remained 100% adhesion failure at the nylon surface as a result of the highest stress peaks which naturally occur near the surface of the most flexible adherend of the adhesive bond.

6.6.3.7 UV/Ozone Treatment of Poly(phenylene sulphide) (PPS)

Some parts of the Airbus A340-600 aircraft are made from (PPS), a thermoplastic matrix reinforced with either carbon or glass fibres. The parts made of these hybrid materials are difficult to bond adhesively, due to the high chemical resistance of PPS and its low surface free energy. Some research has been published regarding its adhesion behaviour after low pressure plasma treatment [122]. Hence, an alternative physical surface treatment such as UV/Ozone was investigated in order to increase the surface free energy. Therefore, 20 samples were selected for further testing:

- 8 samples were made of PPS/carbon fibre (2 were roughened by sandpaper);
- 8 samples were made from PPS/carbon fibre/Astrostrike® (an aerospace lightning strike damage and shielding protection material) which included a nonwoven copper mesh, two of which were roughened with sandpaper;
- 4 samples were made of PPS/glass fibres.

Static contact angle measurements using diodomethane, tricresyl phosphate, formamide, glycerol and double distilled water showed that the surface treatment increased the wetting of water as a result of the increased polar component of the surface free energy of the matrix material, which increased from 10 (Figure 6.36a) to over 40 mJ/m^2 (Figure 6.36b). This last value was obtained after 5 minutes of UV/Ozone treatment, independent of the filler material (Figure 6.37). No further increase of the polar component could be found after about 5 minutes UV/Ozone exposure.

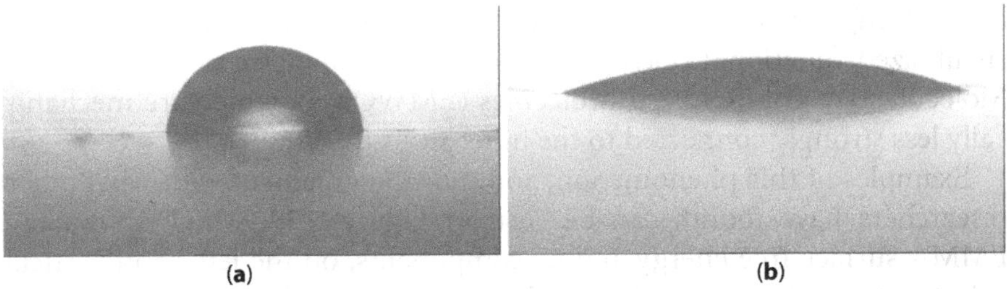

Figure 6.36 Double distilled water droplets (volume 3 microliters) on PPS, (a) without treatment, and (b) after 5 minutes of UV/Ozone treatment.

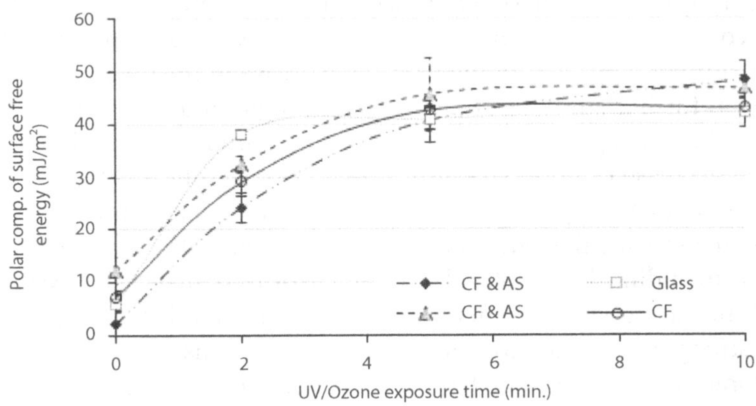

Figure 6.37 Trend lines connecting the measurements of polar component of the surface free energy of PPS as a function of the UV/Ozone treatment time. Abbreviations of filler materials: CF: Carbon fibres, AS: Astrostrike® lightning protection.

The gain in the polar component of the surface free energy had the implication that adhesive bonding by a two-part structural paste adhesive curing at RT became now successful. No significant difference could be found in the surface free energy in relation to the used filler materials. However, the dispersion component of the surface free energy slightly lowered during the surface treatment due to the effect of reduced polymer density, which compensated for the polar increase.

6.6.3.8 *UV/Ozone Treatment of Poly(methyl methacrylate) (PMMA)*

Generally, physically treated polymer surfaces show a maximum in the wetting performance as a function of treatment time. However, just slightly before this maximum wetting value is obtained, a reduction of the adhesive bond strength is often measurable. This is the result of the creation of small sized functional polymer groups on the outer surface, called Low Molecular Weight Oxidized Molecules (LMWOMs), which are mechanically less strongly connected to the inner mass [6, 45].

Examples of this phenomenon, and thus in agreement with what other researchers have found, can be seen in Figure 6.38, which shows the PMMA surface free energy and its components, on the left-hand vertical axis, together with the mechanical lap shear test results, on the right hand side vertical axis, as a function of three different UV/Ozone exposures: 0, 5 and 10 minutes in a single graph. The average polar component is shown to increase up to an exposure time of 10 minutes. The failure type changed from 100% adhesion failure on the non-treated PMMA surface to 100% cohesive failure within the silicone adhesive after 5 minutes of surface treatment. This was in line with the increase of the polar component. However, further study of the lap shear strength and failure type clearly showed that an exposure time of more than 5 minutes resulted in a slight average decrease of bond strength (though not significant) together with an increase of the polar component. This is a general observation that can be made with the physical surface treatments: The density of polar groups, including low molecular weight oxidized material (LMWOM), is still increasing which coincides with the start of a decreasing adhesive bond strength as a function of treatment time. This is the effect of the deterioration of the surface layer due to polymer chain scissions by the

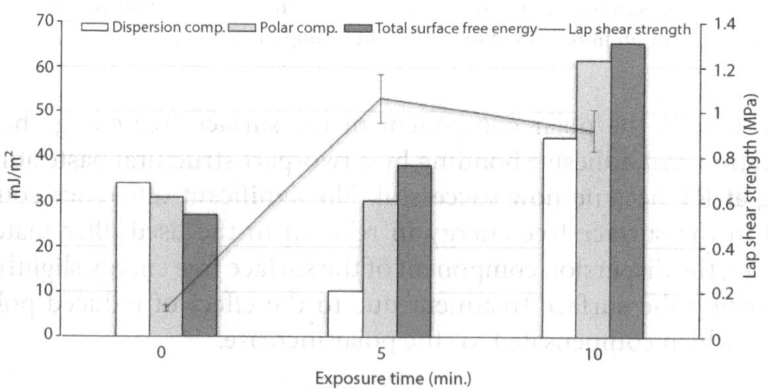

Figure 6.38 Total surface free energy (mJ/m^2) of PMMA, including the dispersion and polar components (left hand side vertical axis) shown as a function of the UV/Ozone treatment time. The lap shear strength of PMMA bonded to PMMA by a silicone adhesive is shown on the vertical axis (MPa) on right-hand side of the graph.

UV-light and oxidation, resulting in the formation of LMWOM which are less strong attached to the bulk mass, eventually weakening the adhesive bond strength.

6.6.3.9 *Adhesive Bonding of Flexible Polymeric Solar Cells*

A key issue in the development of solar powered vehicles is the adhesive bonding and framing of solar cells by an adhesive tape with gap filling ability and the capability to deform. The flexible solar cells were encapsulated in fluorine-based plastic foils, poly (ethene-co-tetrafluoroethene) (ETFE, Tefzel®), with a surface free energy ~30 mJ/m^2 or poly (ethene-co-chlorotrifluoroethene) (ECTFE, Halar®) as an alternative (semi-crystalline) encapsulating material, which shows high corrosion resistance and strength in a wide temperature range.

A large number of surface treatments are available, though practically rather limited for the above materials and application, as fluoropolymers are resistant to most oxidizing agents and solvents and solar cells are too sensitive to be treated by corona or flame treatments. Still, the resulting surface topography and chemistry should be optimised to the characteristics that best fit to those of the chosen adhesive. For this reason, several passive chemical cleaning agents and physical and mechanical treatments of the encapsulating foils were compared before being bonded, as a decisive factor in the successful bonding of low surface free energy substrates. Any such treatment should always start with degreasing the substrate using a cleaning agent. The preferred cleaning agents for fluorocarbons are chlorinated alcohols or ketone solvents. The solvent can be applied by wiping the surface in one direction using a special paper or a clean cloth. But the highest efficiency of degreasing can be obtained by ultrasonic agitation of the substrate submerged in a detergent solution.

In this case, alternative physical treatments were investigated, and for this reason all specimens were degreased with a commercial solvent mixture called PFQD [116] using a paper or cloth before the physical treatment took place. Only abraded specimens (D-Ab) were degreased again in order to remove dust particles.

Figure 6.39 displays the average static contact angle of double distilled water including the standard deviation indicating that ECTFE showed an overall better wetting (lower contact angle) than ETFE. This is due to the chloride group in the molecule's backbone giving rise to a higher surface polarity. The standard deviation is high for DP-60 and D-Ab-P60 both

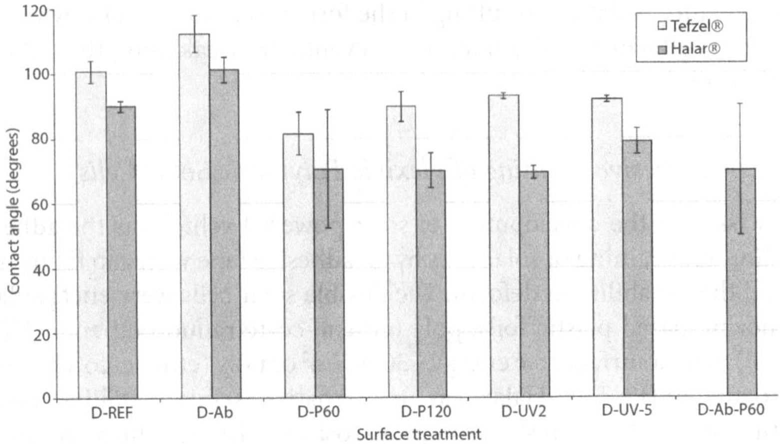

Figure 6.39 Static contact angle measurements (double distilled water) on poly(ethene-co-tetrafluoroethene) (ETFE, Tefzel®) and poly(ethene-co-chlorotrifluoroethene) (ECTFE, Halar®) surfaces as a function of the surface treatment. D-: degreased by solvent only, Ab: abrasion, P60, P120: respectively 60 and 120 seconds atmospheric plasma, UV2, UV5: respectively 2 and 5 minutes UV/Ozone treatment, Ab-P60: abrasion followed by 60 seconds atmospheric plasma treatment.

involving 60 seconds of open air plasma treatment, as can be found in the bar graph. The reason for this phenomenon is the difference of the direction of the focussed plasma beam torches angles (-45°, 0°, -45°) relative to the surface. Furthermore, it can be concluded that:

- Halar® showed a better average wetting performance (lower contact angle with water) than Tefzel®. This difference resides in the chlorine atoms in ECTFE, which are substituted by a fluorine atom in the structure of ETFE;
- There was no noticeable wetting difference between 60 sec. plasma and 120 sec. plasma treatments. However, some de-colourisation of the surface occurred when exposed to a 120 sec. plasma treatment;
- There was a large deviation in the test results of the 60 sec. plasma treatments, due to the sensitivity to the nozzle angle used;
- The lowest, and most consistent, contact angle was established on Halar® foil by degreasing, abrading and 60 sec. plasma treatment. The resulting contact angle was 54°;

- Worst wetting, or highest contact angle, was obtained after abrading the Tefzel® foil, as abrading only improves the wetting for surfaces with a contact angle below 90°;
- UV/Ozone treatment led to a good wetting performance, and resulted in contact angles roughly comparable to atmospheric plasma;
- The UV/Ozone treatment results showed a slightly lower standard deviation than the ones obtained with atmospheric plasma surface treatments and/or abrasion.

6.6.3.10 Treatment of ABS for Adhesive Bonding

Acrylonitrile butadiene styrene (ABS) is a lightweight, hard and impact resistant low-cost amorphous polymer. As a result, it is often used as a housing material for consumer products, especially in the automotive industry. However, some kind of surface treatment is often needed to obtain a reliable adhesive bond [123].

A poor peel adhesion with an adhesion failure of 100% was found when ABS parts were bonded with a polyurethane adhesive without surface treatment, except degreasing. For this reason mechanical and two physical (atmospheric plasma and UV/Ozone) surface treatments were investigated in combination with different types of adhesives.

After pre-cleaning with PFQD, a commercial degreaser [116], the mechanical abrasion was done using grade 500 emery paper which visually showed a clear increase of the surface roughness. This step was followed by cleaning (again) with PFQD to remove the generated dust particles.

During the second set of experiments the physical surface treatments were done. After solvent cleaning with PFQD to remove most of the organic contaminations, either atmospheric plasma or UV/Ozone surface treatments were carried out in order to improve the wetting performance.

Static contact angle measurements using water were done to determine the best wetting performance. Plasma treatment was found to improve the surface wettability but to a lesser extent than the 4 minutes of UV/Ozone treatment (Figure 6.40). Hence, a PFQD [116] pre-cleaning together with 4 min. UV/Ozone treatment showed the best wetting results with an average contact angle of 28 degrees.

After a sequence of surface treatments, which always started with a pre-cleaning with PFQD (Figure 6.40), the substrates were bonded with four different adhesives (2 different types of methacrylate and two different one-component polyurethanes).

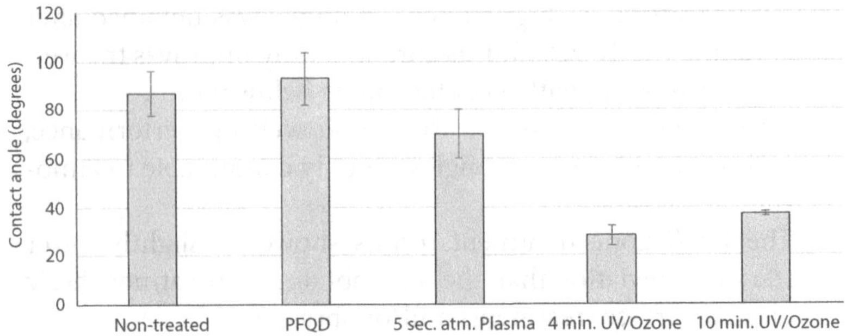

Figure 6.40 Static contact angle of water on an ABS surface "as is" and after solvent pre-cleaning with PFQD followed by atmospheric plasma or UV/Ozone surface treatments.

After the adhesives were fully cured at RT mechanical tests were done according to ASTM D D1876 [124] on a 20kN Zwick test machine using a 1kN load cell (Figure 6.41).

All alternative treatments showed an increase in the peel strength, compared to the original test results (non-treated or "as is"), especially when abrasively treated (Figure 6.42). In the "As is" & PU (1) the failure mode was found to be 30% – 70% interfacial on the ABS surface, while adding abrasion changed the failure mode to 90% cohesive failure within the adhesive layer.

The use of the methacrylate adhesive changed the failure mode to a cohesive failure within the ABS material, meaning that the ABS was weaker than the adhesive bond, and at the cost of an increasing standard deviation when UV-light is used in combination with mechanical abrasion.

The main results of this research, which show an increased adhesive performance when abraded and/or physically treated by e.g. atmospheric plasma treatment, are in agreement with those found by Frascio *et al.* [123].

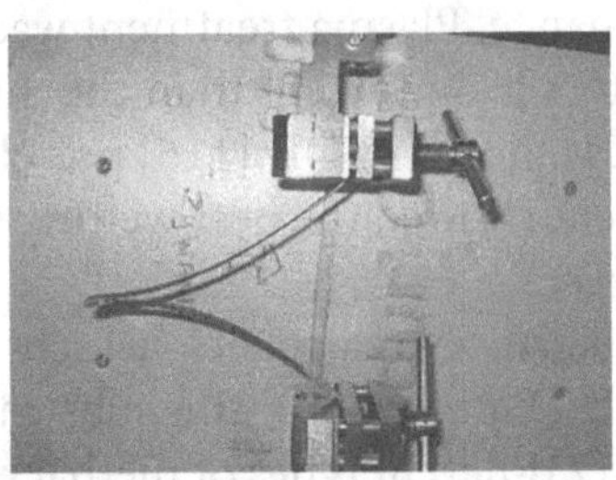

Figure 6.41 Typical example of a peel test on one of the ABS samples tested according to ASTM D1867.

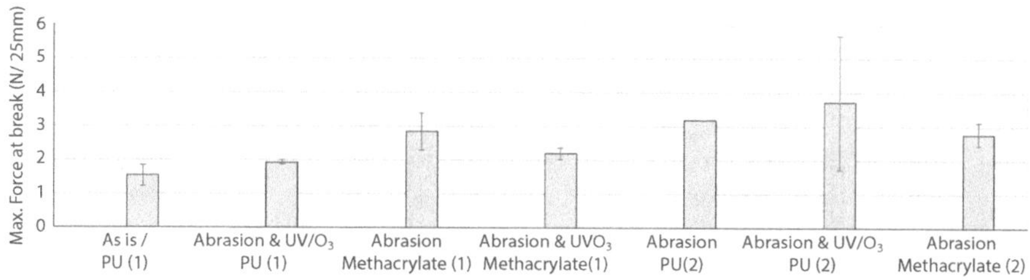

Figure 6.42 Maximum peel force at break for ABS bonded to ABS as a function of various surface treatments and adhesives. Peel strength by ASTM D1867. Not shown on the x-axis is the effect of only cleaning with PFQD, carried out prior to the presented surface treatments.

6.6.3.11 Adhesive Bonding of Styrene-Acrylonitrile (SAN) to a Thermoplastic Elastomer (TPE)

UV/Ozone treatment was used as one of the physical surface treatments to improve the adhesive bond strength of a silicone sealant to an injection moulded medical product that consisted of a glass-filled styrene-acrylonitrile (SAN) resin, a colourless and transparent thermoplastic, bonded to either a thermoplastic elastomer TPE (an unknown mixture of a plastic and a rubber) or to synthetic quartz glass.

Various surface treatments were carried out on the SAN, TPE and the quartz glass surfaces in order to improve their wettability and to obtain a higher initial bond strength. The adhesive bonding was done with a light-curing epoxy adhesive.

Figure 6.43 shows the results of the static contact angle measurements with water. A large standard deviation with unknown origin is observed in the obtained data which shows that neither degreasing nor plasma or corona treatment were able to increase the wetting performance of SAN material. However, the 10 min. UV/Ozone treatment did decrease the average contact angle (with water) and showed the best wetting results.

In contrast to the results found for SAN, treating TPE by UV/Ozone hardly influenced the contact angle, most possibly because the reduction in the density of the polymer is larger than the increase in the polar component, as it seemed that hardly any oxidation of the TPE surface had occurred.

The observed effect of the 5 min. UV/Ozone surface treatment on quartz led to spontaneous surface wetting, indicating a (near to) zero contact angle, a result which was often observed on glass and quartz materials by the authors.

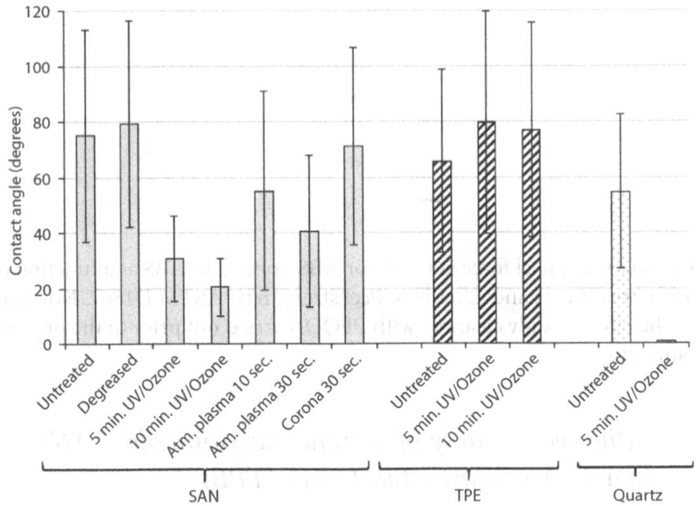

Figure 6.43 Test results of the static contact angle measurements (water) on SAN, TPE and Quartz after various surface treatments.

6.6.4 UV/Ozone Treatment of Rubbers

6.6.4.1 UV/Ozone Treatment of SBS Rubber

SBS rubber is a block copolymer and a special type of rubber that can be classified as a thermoplastic elastomer; it behaves like a rubbery material at room temperature without being crosslinked like most other rubbers, though at higher temperatures it can be moulded into a specific shape. This characteristic makes it useful for applications like tire treads and the soles of shoes where material durability is an important aspect. However, the material is non-polar, and as such difficult to bond without a specific surface treatment. As chemical treatments generate large amounts of hazardous waste, or require expensive instrumentation [e.g 125] UV/Ozone treatments were investigated as an alternative [47].

As regards the Excimer 222 nm, a small but steady increase of polar group density on the surface as a function of treatment time up to 15 minutes was formed after which the static contact angle with water became almost zero (Figure 6.44a). But the effect flattens out when the maximum density of polar groups is reached as the density of the non-polar component is steadily decreasing as a function of exposure time due to a decrease of the molecular density.

The low pressure mercury lamp showed a steady increase of both the polar and dispersion components. Though the molecular density started to decrease after about 5 minutes of exposure (Figure 6.44b).

After 15 minutes, both surface treatments resulted in an increase of the total surface free energy as a function of treatment time which was up to three times as high as the initial value.

The highest average lap shear strength of the SBS rubber was obtained by the longest exposure time to both the Excimer 222 nm and the low pressure mercury lamp (Figure 6.45). It is thus worth mentioning that for SBS-rubber the effect of the Excimer 222 nm on the adhesive bond strength is roughly the same as that of the low pressure UV-s ource. The adhesive bond strength is strongly increased after a surface treatment with either UV/Ozone equipment. The type of bond failure after both UV/Ozone treatments, independent of the treatment time, moved from 100% adhesion failure at the interface to about 50% cohesive failure in the adhesive layer. These results are in agreement with the results found by Romero-Sánchez *et al.*, who studied the influence of different loadings of fillers in relation to the UV/Ozone treatment time and the adhesion to polyurethane adhesive [7].

6.6.4.2 *Surface Treatment of Ethylene Propylene Diene Monomer (EPDM) Rubber to Optimize the Adhesion of a Coating*

EPDM is an elastomer, resilient up to 600% elongation, extremely durable and for these reasons finds its use in commercial products such as roof covers (housing) and as (coated) door seals in automotive and refrigerator constructions. The investigated EPDM rubber had to be coated for cosmetic reasons for the application as an automotive car door seal.

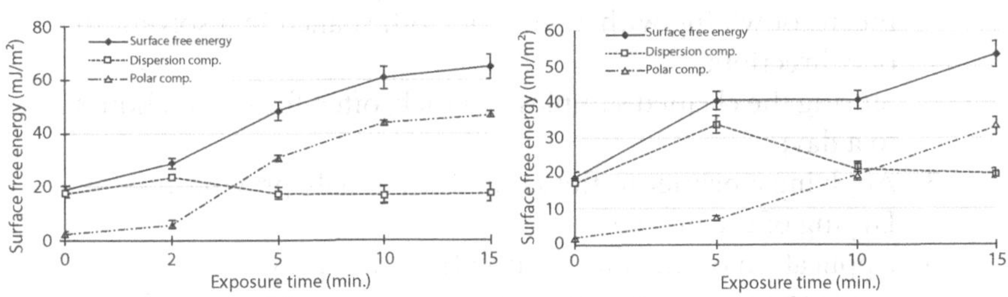

Figure 6.44 Surface free energy of SBS-rubber as a function of treatment time by (a) a 222 nm Excimer and (b) a low-pressure mercury lamp.

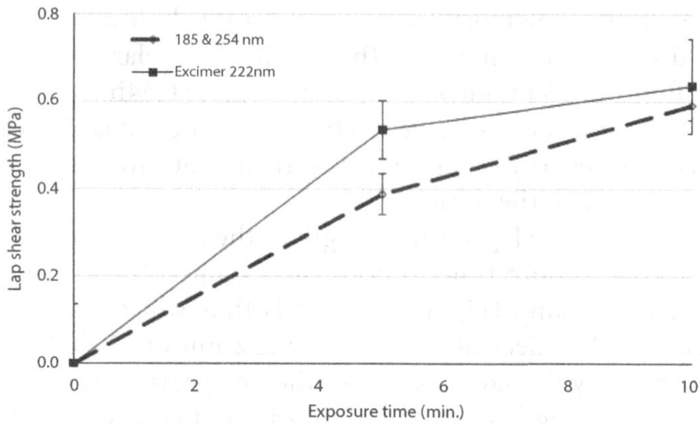

Figure 6.45 Bond strength of SBS rubber bonded to SBS rubber after the surface treatment by an Excimer 222 nm or a low pressure mercury light source.

Therefore, an alternative surface treatment was investigated for the generally applied labour-intensive and unhealthy solvent–based cleaning procedure. Various surface modification methods for EPDM have been published in literature, most of them are based on plasma treatments either with [126], or without [127–129] organosilicon precursors, but also UV/Ozone surface activation as an adhesion promoting treatment was described by Hamdi and Poulis [130].

6.6.4.3 *EPDM Rubber Pre-Treated by a Low Pressure UV-Source*

An industrially often used sequence of surface preparation steps before applying a coating on rubber materials is:

1. Removal of dust from the rubber samples and degreasing by means of wiping with a piece of cloth soaked in a solvent in one direction;
2. Storing the cleaned samples in a rack, often for several hours to a day;
3. Applying a primer to the samples (manual-spraying);
4. Drying of the primer;
5. Application of the final coating (manual spraying);
6. Drying of the coating in an oven.

UV/Ozone treatment was investigated in order to reduce the number of production steps and thus a simplified production that was less labour intensive. Therefore, contact angle measurements, using different liquids, were done on an EPDM-rubber surface after treatment by a low-pressure mercury UV-source in order to determine its effect on the surface free energy. The results are presented in Table 6.6.

These measurements clearly reveal an increase of polar surface groups and a decrease of the dispersion component as a function of treatment time by the low pressure UV-source. The effect on the adhesion of the coatings applied to the UV/Ozone treated and untreated samples is shown in Table 6.7. Note that a pre-cleaning step by means of a solvent (naphtha) was done to the non-UV/Ozone treated samples. After UV/Ozone treatment for 0, 5 and 10 minutes a solvent-based primer was applied to 50% of the samples after which an electrostatic paint coating was applied.

The adhesion of the paint coating was tested by measuring the pull-off force of the coating from the substrate by the cross-cut test (ISO 2409 [99]). Six parallel scratches were made in two perpendicular directions by means of a certified tool consisting of a set of sharp knives. Subsequently a tape was applied and pulled off within 5 minutes. The more the squared coating patches remained on the surface, the better the level of adhesion.

The final coating was thus applied to both primed and un-primed samples. The untreated samples showed no adhesion whatsoever to either the primer or combination of primer and/or coating due to too low surface free energy as a result of lack of EPDM polar surface groups. The naphtha-only cleaned samples were used as a reference which showed a better performance but still showed unsatisfactory adhesion test results. Test results of all UV/Ozone treated EPDM samples showed a strongly

Table 6.6 Surface free energy of EPDM-rubber as a function of surface treatment time using a low pressure UV-source. Please note that the measured value of the solvent (naphtha) cleaned rubber is only temporary due to adsorbed solvent.

	Total surface free energy	**Dipersion component**	**Polar component**
Type of treatment	**(mJ/m²)**	**(mJ/m²)**	**(mJ/m²)**
Not treated	29.7	24.9	4.8
Solvent treated	46.4	24.5	21.9
5 minutes UV/Ozone	44.7	22.5	22.2
10 minutes UV/Ozone	46.7	20.7	26.0

Table 6.7 Surface free energy and coating adhesion to the EPDM-rubber as a function of surface treatment by a low pressure UV-source. Each presented result is the averaged value of 4 different samples.

Type of treatment	Total surface free energy (mJ/m^2)	Adhesion of coating with primer	Adhesion of coating without primer
Not treated	29.7	n.a.	n.a.
Solvent cleaned	44	50%	n.a.
5 minutes UV/ Ozone	44.7	good	good
10 minutes UV/ Ozone	46.7	good	good

improved adhesion performance: the coating showed no more adhesion failure during the tape test [99]. Thus it can be concluded that the UV/Ozone treatment not only gave satisfactory results from the point of view of the quality aspect, but also offered the opportunity to reduce the number of production steps.

6.7 Prospects

Ongoing investigations lead to new technological developments in physical surface treatments for plastics. This is stimulated by the increasing use of water-based printing inks, adhesives and coatings which require polymer surfaces to be treated to increase their surface free energy. Different methods such as chemical, mechanical or physical treatments like UV/Ozone, corona, flame, atm. plasma, etc. might improve their wetting property resulting in a higher initial adhesion strength and long-term stability in structural applications [131, 132]. However, the developments and commercial applications of equipment that produces UV-C light and ozone gas for surface cleaning and specific functionalisation are still far behind the many available "alternatives".

This is a pity, as despite its successful treatment results, low investment costs, ease of use, surface cleaning capability (both for polymers and metals), up to the removal of organic surface contaminants [133] and the environmental friendliness of the UV/Ozone technique [134–136], it is expected that this surface treatment of polymers will stay behind its "alternatives". Most probably due to its rather low process speed as well as the higher commercial profits that can be made in the sales of other surface cleaning and activation techniques. An increase of the application of Excimer sources might be the game changer though.

6.8 Summary

Test results presented here show that the exposure to the wavelengths 253.7 nm and 184.9 nm, as generated by an ozone generating low-pressure mercury UV-light source, is generally very effective as a surface treatment for improving the adhesion, and as a result adhesive bond strength and durability of polymer (and metal) surfaces. However, changes in the sample distance to the UV-C source, wavelength used, or the ozone gas concentration as process parameters influence strongly the efficiency of the surface cleaning and activation for improved adhesive bonding.

The UV/Ozone process is, in general, very well suited for increasing the density of polar groups (and oxygen) on a polymer surface, showing an increased wetting and an increase in the adhesive bond strength as a function of treatment time. Though a too long treatment time might lead to the generation of low molecular weight oxidized molecules (LMOWMs), which may reduce the bond strength.

More than 30 years of experimental work has shown that UV/Ozone treatment by low pressure mercury light sources is very effective for increasing the initial as well as the long-term adhesive bond strength where other physical surface treatments such as corona and atmospheric plasma hardly showed any improvement. In some cases, especially when used in combination with metals, it succeeds to give increased bond strength [131] by reproducible removal of thin layers of hydrophobic contaminating materials, such as many types of greases, including finger grease, Teflon® and silicone oils, which is a characteristic that was hardly found with the other investigated physical surface treatments such as atmospheric plasma and corona.

Acknowledgements

The authors are very grateful to the following individuals for their contributions: A. Kooijman, B. Boshuizen, C. Del Grosso, S. Teixeira de Freitas, W. Wandong, D. Zarouchas, I. Vieira, F. Badan and H. Nevel. Staff members: B. Out, J. Vervloed, P. de Regt, J. Hofstede, C. de Ruijter, R. Steenbreker,

F. de Wit, R. J. Lemmen, D. Borger, M. van der Leeden, I. Cringus-Fundeanu and P. Beentjes. Management assistants: M. Oosterloo, W. Veldhoven, P. Peers and G. van der Windt. Board members: G. Frens, B. Vogelesang, A. Vlot and R. Benedictus. Faculty colleagues: R. Alderliesten, F. Oostrum, B. Grashof, D. Mainali, and all laboratory staff and workshop technicians.

References

1. M. Strobel, C. S. Lyons and K. L. Mittal (Eds.), *Plasma Surface Modification of Polymers: Relevance to Adhesion*, CRC Press, Boca Raton, FL (1994).
2. K.L. Mittal and T. Bahners (Eds.), *Laser Surface Modification and Adhesion*, Wiley–Scrivener, Beverly, MA (2015).
3. G.A. Takacs, M.J. Miri and T Kovach, Vacuum UV surface photo-oxidation of polymeric and other materials for improving adhesion: A critical review, *Rev. Adhesion Adhesives*, 8, 555-581 (2020).
4. C.A. Del Grosso, J.A. Poulis and E.R. de la Rie, The photo-stability of acrylic tri-block copolymer blends for the consolidation of cultural heritage, *Polym. Degradation Stability*, 159, 31-42 (2019).
5. G.A. El-Hiti, D.S. Ahmed, E. Yousif, O. S.A. Al-Khazrajy, M. Abdallh and S.A. Alanazi, Modification of polymers through the addition of ultraviolet absorbers to reduce aging effect of accelerated and natural irradiation, *Polymers*, 14, 20 (2022).
6. M. D. Landete-Ruiz and J. M. Martin-Martinez, Surface modification of EVA copolymer by UV treatment, *Int. J. Adhesion Adhesives*. 25, 139-145 (2005).
7. M.D. Romero-Sánchez, M.M Pastor-Blas, J. M. Martín-Matínez and M.J. Walzak, UV treatment of synthetic styrene-butadiene-styrene rubber, *J. Adhesion Sci. Technol.*,17, 25-45 (2003).
8. J.R. Vig, UV/Ozone cleaning of surfaces, in: *Treatise on Clean Surface Technology*, K.L. Mittal (Ed), pp. 1-26, Springer, Boston MA (1987).
9. J.A. Poulis, J.C. Cool and E.H.P. Logtenberg, UV/Ozone cleaning, a convenient alternative for high quality bonding preparation, *Int. J. Adhesion Adhesives*. 13, 89-96 (1993).
10. A.R. Rodríguez, L.A. Boshuizen, B.C. Rans and J.A. Poulis, The influence of grit blasting and UV/Ozone treatments on Ti-Ti adhesive bonds and their durability after sol-gel and primer application, *Int. J. Adhesion Adhesives*, 104, 1-10 (2021).
11. A.R. Rodríguez, L.A. Boshuizen, B.C. Rans and J.A. Poulis, Effect of surface morphology on the Ti-Ti adhesive bond performance of Ti6Al4V parts fabricated by selective laser melting, *Int. J. Adhesion Adhesives*, 110, 1-10 (2021).
12. N. Pichler, W. Wang, J.A. Poulis and E. Ghafoori, Surface preparations and durability of iron-based shape memory alloy adhesively-bonded joints, *Int. J. Adhesion Adhesives* (2023). Accepted, but not yet published

13. K. Jousten, *Handbook of Vacuum Technology*, pp. 463-474, John Wiley & Sons (2008).
14. J.A. Poulis, Small cylindrical adhesive bonds, PhD thesis, Delft University of Technology, Delft, The Netherlands (1993).
15. N. S. McIntyre and M.J. Walzak, New UV/ozone treatment improves adhesiveness of polymer surfaces, *Modern Plastics Int.* 25, 69-75 (1995).
16. J.R. Vig and J. LeBus, UV/Ozone cleaning of surfaces, *IEEE Trans. Parts Hybrids Packaging*, 12, 365-370 (1976).
17. Ann., Bright-Euram3, FP4, European Craft project, No. *BE-ST-2032,*UV/Ozone surface treatment of polymers for high quality adhesion in the footwear and coating industry, Grant agreement ID: BRST965055 (1996-1998).
18. L. Zafonte and R. Chiu, UV/ozone cleaning for organics removal on silicon wafers, *Proc. SPIE*, 470, 164-177 (1984).
19. M. Tabe, UV ozone cleaning of silicon substrates in silicon molecular beam epitaxy, *Appl. Phys. Lett.* 45, 1073-1075 (1984).
20. G.A. Diaz-Quijada, R. Peytavi, A. Nantel, E. Roy, M.G. Bergeron, M.M. Dumoulin and T. Veres, Surface modification of thermoplastics - towards the plastic biochip for high throughput screening devices, *Lab Chip* 7, 856-862 (2007).
21. R.H. Bradley, I.L. Clackson and D.E. Sykcs, UV/Ozone modification of wool fibre surfaces, *Appl. Surf. Sci.* 72, 143-147 (1993).
22. K. Sato, A. Ijuin, Y Imai and Y Hotta, A feasibility study on direct measurements of interaction forces between carbon fiber surfaces and other substances, *Polym. Composites*, 41, 5209-5215 (2020).
23. M.A. Downey and L.T. Drzal, Toughening of carbon fiber-reinforced epoxy polymer composites utilizing fiber surface treatment and sizing, *Composites Part A*, 90, 687-698 (2016).
24. A. Debot, V. B. Chu, D. Adeleye, J. Guillot, D. Arl, M. Melchiorre and P.J. Dale, Inkjet-printed indium sulfide buffer layer for $Cu(In,Ga)(S,Se)_2$ thin film solar cells, *Thin Solid Films* 745, 139096-139101 (2022).
25. R.A. Crandall, The use of ultraviolet light in the treatment of water in public spas and hot tubs, *J. Environmental Health*, 49, 16-23 (1986).
26. J. Wang, J. Shen, D. Ye, X. Yan, Y. Zhang, W. Yang, X. Li, J. Wang, L. Zhang and L. Pan, Disinfection technology of hospital wastes and wastewater: Suggestions for disinfection strategy during corona virus disease 2019 (COVID-19) pandemic in China, *Environmental Pollution*, 262 114665-114674 (2020).
27. M. Raeiszadeh and B. Adeli, A critical review on ultraviolet disinfection systems against COVID-19 outbreak: applicability, validation, and safety considerations, *ACS Photonics* 7, 2941-2951 (2020).
28. H. Claus, Ozone generation by ultraviolet lamps, *Photochem Photobiol* 97, 471-476 (2021).
29. Ann., Desinfektion mit UV-Strahlung, Bestell Nr. 1817, Stand 5/59, Philips Licht, Philips GmbH, Hamburg, Germany.

30. G. Knight, Monitoring of ultraviolet light sources for water disinfection, Conference Record of the 2004 IEEE Industry Applications Conference, 39th IAS Annual Meeting, Vol.2, 1016-1018 (2004).
31. C.C.E. Meulemans, The basic principles of UV–disinfection of water, *Ozone: Sci. Eng.* 9, 299-313 (1987).
32. M. van der Meer, F. van Lierop and D. Sokolov, The analysis of modern low pressure amalgam lamp characteristics, https://www.semanticscholar.org/paper/The-analysis-of-modern-low-pressure-amalgam-lamp-Meer-Lierop/8484103e410249bb329c44e6868aa71f6ab48f1a (2015).
33. C. Tuchinda, S. Srivannaboon and H.W. Lim, Photoprotection by window glass, automobile glass, and sunglasses, *J. Am. Acad. Dermatol.*, 54, 845-854 (2006).
34. UV-Technik Special Lampen GmbH, Wümbach, Germany (2015).
35. U. Kogelschatz, Dielectric-barrier discharges: Their history, discharge physics, and industrial applications, *Plasma Chem. Plasma Proc.*, 23, 1-46 (2003).
36. A. Fujioka, K. Asada, H. Yamada, T. Ohtsuka, T. Ogawa, T. Kosugi, D. Kishikawa and T. Mukai, High-output-power 255/280/310 nm deep ultraviolet light-emitting diodes and their lifetime characteristics *Semicond. Sci. Technol.*, 24, 084005- 084010 (2014).
37. F. Anwar, F.N. Chaudhry, S. Nazeer, N. Zaman and S. Azam, Causes of ozone layer depletion and its effects on human: Review, *Atmospheric Climate Sci.* 6, 129-134 (2016).
38. J.D. McClurkin and D.E. Maier, Half-life of ozone as a function of air conditions and movement, Proceedings 10th Int. Working Conf. on Stored Product Protection, 425 (2010).
39. R. G. Rice and N. Aharon, Introduction, in: *Handbook of Ozone Technology and Applications*, Vol. 1, pp. 1-9, R. G. Rice and N. Aharon (Eds), Ann Arbor Science Publishers, Ann Arbor, MI (1982).
40. Ann., WHO, Environmental Health Criteria 7; Photochemical Oxidants, World Health Organization, Geneva (1979).
41. C. Mühlhan and H. Nowack, Plasma pretreatment of polypropylene for improved adhesive bonding, *Surf. Coatings Technol.*, 1-3, 1107-1111 (1998).
42. A. Popelka, I. Novak and I. Krupa, Polyolefin adhesion modifications, in: *Polyolefin Compounds and Materials, Fundamentals and Industrial Applications,* M. Al-Ali AlMa'adeed and I. Krupa, (Eds), pp. 201-230, Springer Series on Polymer and Composite Materials, Heidelberg (2016).
43. L.F. MacManus, M.J. Walzak and N.S. McIntyre, Study of ultraviolet light and ozone surface modification of polypropylene, *J. Polym. Sci, Part A: Polym. Chem.*, 37, 2489-2501 (2000).
44. R. Ono, Y. Nakagawa, Y. Tokumitsu, H. Matsumoto and T. Oda, Effect of humidity on the production of ozone and other radicals by low-pressure mercury lamps, *J. Photochem. Photobiol.* A 274, 13-19 (2014).
45. E. Arikan, J. Holtmannspötter, F. Zimmer, T. Hofmann and H-J Gudladt, The role of chemical surface modification for structural adhesive bonding

on polymers - washability of chemical functionalization without reducing adhesion, *Int. J. Adhesion Adhesives*, 95, 1-12 (2019).
46. S. Sapieha, J. Cerny, J. E. Klemberg-Sapieha and L. Martinu, Corona versus low pressure plasma treatment: Effect on surface properties and adhesion of polymers, *J. Adhesion*, 42, 91-102 (1993).
47. M.J. Walzak, S. Flynn, R. Foerch, J.M. Hill, E. Karbashewski, A. Lin and M. Strobel, UV and ozone treatment of polypropylene and poly(ethylene terephthalate), *J. Adhesion Sci. Technol.* 9, 1229-1248 (1995).
48. M.M. Makogon, Yu. N. Ponomarev and B.A. Tikhomirov, The problem of water vapour absorption in the UV spectral range, *Atmospheric Oceanic Optics*, 26, 45-49 (2013).
49. J.M. Hill, E. Karbashewski, A. Lin, M. Strobel and M.J. Walzak, Effects of aging and washing on UV and ozone-treated poly(ethylene terephthalate) and polypropylene, *J. Adhesion Sci. Technol.* 9, 1575-1591 (1995).
50. W. Wang, S.T. de Freitas, J.A. Poulis and D. Zarouchas, A review of experimental and theoretical fracture characterization of bi-material bonded joints, *Composites Part B*: 206, 108537-108551 (2021).
51. S. Bok, G-H. Lim and B. Lim, UV/ozone treatment for adhesion improvement of copper/epoxy interface, *J. Ind. Eng. Chem.* 46, 199-202 (2017).
52. K.L. Mittal, The role of the interface in adhesion phenomena, *Polym. Eng. Sci.* 17, 467-473 (1977)
53. E.M. Petrie, *Handbook of Adhesives and Sealants,* 2nd edition, pp. 267-270, McGraw Hill, New York (2007).
54. L.W. McKeen, Introduction to the weathering of plastics, in: *The Effect of UV-Light and Weather on Plastics and Elastomers*, PDL Handbook Series, third edition, S. Ebnesajjad (Ed.), pp.17-41, William Andrew Publishing, Norwich (2013).
55. G.A. El-Hiti, D.S. Ahmed, E. Yousif, O.S.A. Al-Khazrajy, M. Abdallh and S.A. Alanazi, Modifications of polymers through the addition of ultraviolet absorbers to reduce the aging effect of accelerated and natural irradiation, *Polymers*, 14, 20 (2022).
56. T. Ojeda, A. Freitas, K. Birck, E. Dalmolin, R. Jacques, F. Bento and F. Camargo, Degradability of linear polyolefins under natural weathering. *Polym. Degrad. Stab.*, 96, 703–707 (2011).
57. J.G. Neevel, The identification of van Gogh's inks for drawing and writing, in: *Accelerated Aging: Photochemical and Thermal Aspects, Research in Conservation*, R.L. Feller (Ed.), pp. 23, 129-141, J. P. Getty Trust (1994).
58. M. Strobel, M.J. Walzak, J.M. Hill, A. Lin, E. Karbashewski and C.S. Lyons, A comparison of gas-phase methods of modifying polymer surfaces, *J. Adhesion Sci. Technol.* 9, 365-383 (1995).
59. H.S. Choi, V.V. Rybkin, V.A. Titov, T.G. Shikova and T.A. Ageeva, Comparative actions of a low pressure oxygen plasma and an atmospheric pressure glow discharge on the surface modification of polypropylene, *Surf. Coatings Technol.* 200, 4479-4488 (2006).

60. B.K. Kim, K.S. Kim, C.E. Park and C.M. Ryu, Improvement of wettability and reduction of aging effect by plasma treatment of low-density polyethylene with argon and oxygen, *J. Adhesion Sci. Technol.* 16, 509-521 (2002).
61. D. Briggs and M.P. Seah, (Eds.), *Practical Surface Analysis – Vol. 1: Auger and X-ray Photoelectron Spectroscopy*, 2nd ed., John Wiley & Sons (1990).
62. Ann., Operators Multipak Software Manual – version 6.0, Physical Electronics Inc. (1998).
63. American Society for Testing and Materials standard practice for checking the operating characteristics of X-ray photoelectron spectrometers (E902-82), *Surf. Interf. Anal.*, 10, 55-57 (1987).
64. D.A. Shirley, High-resolution X-Ray photoemission spectrum of the valence bands of gold, *Phys. Rev. B*, 5, 4709-4714 (1972).
65. NIST Standard Reference Database 20, Version 3.5 (Web Version, 2007), http://srdata.nist.gov/xps/
66. S. Oswald, X-ray photoelectron spectroscopy in analysis of surfaces, in: *Encyclopedia of Analytical Chemistry*, R.A. Meyers (Ed.), pp. 1-49, John Wiley & Sons Ltd, online library (2013).
67 J. Oelichmann, Surface and depth-profile analysis using FTIR spectroscopy, *Zeitschrift Analytische Chemie*, 333, 353-359 (1989).
68. W. Wang, J.A. Poulis, S. Teixeira De Freitas and D. Zarouchas, Surface pre-treaments on CFRP and titanium for manufacturing adhesively bonded bi-material joints, Proc. 18th European Conf. on Composite Materials, Athens,1-8 (2018).
69. A. Einstein, On a heuristic point of view about the creation and conversion of light, *Annalen der Physik*, 17, 132-148 (1905).
70. M. Romand, F.Gaillard, M. Charbonnier, N.S. Prakash, L. Deshayes and I. Linossier, Adhesion science and surface analysis: Typical examples, *J. Adhesion*, 55, 1-16 (1995).
71. R. Ledesma, W. Yost, F. Palmieri, D. Cataldo and J. Connel, Correlation of optically stimulated electron emission with failure mode of adhesively bonded epoxy composites, *Int. J. Adhesion Adhesives* 84, 257-264 (2018).
72. G.D. Davis, Contamination of surfaces: Origin, detection and effect on adhesion, *Surf. Interf. Anal.* 20, 368 – 372 (1993).
73. J. Wang, Y. Wu, Y. Cao, G. Li and Y. Liao, Influence of surface roughness on contact angle hysteresis and spreading work, *Colloid Polym. Sci.* 298, 1107-1112 (2020).
74. F.M. Etzler, Determination of the surface free energy of solids: A critical review, *Rev. Adhesion Adhesives*, 1, 3-45 (2013).
75. F.M. Etzler, Characterization of surface free energies and surface chemistry of solids, in: *Contact Angle, Wettability and Adhesion*, Vol. 3, K.L. Mittal (Ed.) pp. 219-264, CRC Press, Boca Raton, FL (2003).
76. D.K. Owens and R.C. Wendt, Estimation of the surface free energy of polymers, *J. Appl. Polym. Sci.* 13, 1741-1747 (1969).

77. C.W. Karl, A.E. Krauklis, A. Lang and U. Giese, Characterization of rough PTFE surfaces by the modified Wilhelmy balance technique, *Polymers* 12, 1528 (2020).
78. D.A. Bolon and C.O. Kunz, Ultraviolet depolymerisation of photoresist polymers, *Polym. Eng. Sci.* 12, 109-111 (1972).
79. M.C. van der Leeden and G. Frens, Surface properties of plastic materials in relation to their adhering performance, *Advanced Eng. Mater.* 4, 280-289 (2002).
80. M.L. Sham, J. Li, P. Ma and J.K. Kim, Cleaning and functionalization of polymer surfaces and nanoscale carbon fillers by UV/Ozone treatment: a review, *J. Composite Mater.* 43, 1537-1564 (2009).
81. C-W. Tsao and D.L. De Voe, Bonding of thermoplastic polymer microfluidics, *Microfluidics Nanofluidics* 6, 1-16 (2009).
82. Y. Jia, H. Asahra, Y. Hsu, T-A. Asoh and H. Uyama, Surface modification of polycarbonate using the light-activated chlorine dioxide radical, *Appl. Surf. Sci.* 530, 147202 (2020).
83. Y. Hong, Z. He, N.S. Lennhoff, D.A. Banach and J. Kanicki, Transparent flexible plastic substrates for organic light –emitting devices, *J. Electronic Mater.* 33, 312-320 (2004).
84. A.J. Yáñez-Pacios and J.M. Martín-Martínez, Surface modification and adhesion of wood-plastic composite (WPC) treated with UV/ozone, *Composite Interfaces*. 25, 127-149 (2018).
85. K.S.G. Krishman, O. Obertram and O. Seibel, Review of hybrid laminar flow control systems, *Progr. Aerospace Sci.* 93, 24-52 (2017).
86. Ann., European Clean Sky project "Titanium Composite Adhesive Joints (TICOAJO)", Clean Sky 2 Joint Undertaking under the European Union's Horizon 2020 research and innovation program under grant agreement No 7377856.
87. P. Molitor, V. Barron and T. Young, Surface treatment of Titanium for adhesive bonding to polymer composites: a review, *Int. J. Adhesion Adhesives* 21, 129-136 (2001).
88. B Kolesnikov, L. Herbeck and A. Fink, CFRP/titanium hybrid material for improving composite bolted joints, *Composite Structures* 83, 368-380 (2008).
89. S.T. de Freitas and J. Sinke, Test method to assess interface adhesion in composite bonding, *Appl. Adhesion Sci* 3, 1-13 (2015).
90. Standard test method for floating roller peel resistance of adhesives, ASTM D 3167-10 (2017).
91. S. Teixeira de Freitas, D. Zarouchas and J.A. Poulis, The use of acoustic emission and composite peel tests to detect weak adhesion in composite structures, *J. Adhesion* 94, 743-766 (2018).
92. E.M. Petrie, *Handbook of Adhesives and Sealants*, 2nd edition, pp. 232-234, McGraw Hill, New York (2007).
93. A.J. Kinloch, *Adhesion and Adhesives: Science and Technology*, pp. 103-105, Chapman and Hall, New York (1990).

94. G.D. Davis, Surface treatments of selected materials, in: *Handbook of Adhesion Technology*, Vol. 1, L.F.M. da Silva, A. Öchsner and R.D. Adams (Eds), pp.171-173, Springer-Verlag, Berlin, Heidelberg (2011).
95. M. Saeedifar, M.N. Saleh, P. Nijhuis, S.T. de Freitas and D. Zarouchas, Damage assessment of a titanium skin adhesively bonded to carbon fiber-reinforced plastic omega stringers using acoustic emission, *J. Structural Health Monitoring* 2, 407-423 (2021).
96. N.Z. Tomić, M.N. Saleh, S.T. de Freitas, A. Živković, M. Vuksanović, J.A. Poulis and A. Marinković, Enhanced interface adhesion by novel eco-epoxy adhesives based on the modified tannic acid on Al and CFRP adherends, *Polymers* 12, 1541 (2020).
97. G. Bockmair and K. Kranzeder, Surface protection for aircraft maintenance by means of zinc-rich primers. *Advanced Mater. Res.* 138, 41–46 (2010).
98. L.P. Haack, A.M. Straccia. J.W Holubka, A. Bhurke, M. Xie and L. T. Drzal. Chemistry of surface modification with UV/Ozone for improved intercoat adhesion in multi-layered coating systems, *Surf. Interf. Anal.* 29, 829-836 (2000).
99. Paints and varnishes - cross-cut test, ISO 2409 (1992).
100. Standard test method for apparent shear strength of single-lap-joint adhesively bonded metal specimens by tension loading (metal-to-metal), ASTM D-1002-10 (2019).
101. Adhesives - Determination of tensile strength of butt joints, EN 15870 (2009).
102. H. Neevel, Logwood writing inks: History, production, forensics, and use, *Restaurator. Int. J. Preservation of Library and Archival Mater.*, 42, 169-191 (2021).
103. J.G. Neevel, The identification of van Gogh's inks for drawing and writing, in: *The Identification of van Gogh's Inks*, M. Vellekoop, M. Geldof, E. Hendriks, L. Jansen and A. de Tagle (Eds), pp. 420-436, Yale University Press, New Haven and London (2013).
104. K. Efimenko, W.E. Wallace and J. Genzer, Surface modification of Sylgard-184 Poly(dimethyl siloxane) networks by ultraviolet and ultraviolet/Ozone treatment, *J. Colloid Interface Sci.*, 254, 306-315 (2002).
105. K-S Koh, J. Chin, J. Chia and C-L Chiang, Quantitative studies on PDMS-PDMS interface bonding with piranha solution and it swelling effect, *Micromachines*, 3, 427-441 (2012).
106. Y Berdichevsky, J. Khandurina, A. Guttman and Y.-H Lo, UV/ozone modification of poly(dimethylsiloxane) microfluidic channels, *Sensors Actuators B* 97, 402-408 (2004).
107. H. Hillborg and U.W. Gedde, Hydrophobicity changes in silicone rubbers, *IEEE Trans. Dielectrics Electrical Insulation*, 6, 703-717 (1999).
108. M.J. Owen and P.J. Smith, Plasma treatment of polydimethylsiloxane, *J. Adhesion Sci Technol.* 8, 1063-1075 (1994).

109. J. Friedrich, *Metal-Polymer Systems: Interface Design and Chemical Bonding*, pp. 173-237, Wiley-VCH Verlag (2018).
110. R. Oostrom, T.J. Ahmed, J.A. Poulis and H.E.N. Bersee, Adhesion performance of UHMWPE after different surface modification techniques, *Medical. Eng. Phys.* 28, 323-330 (2006).
111. L. Mathieson and R.H. Bradley, Improved adhesion to polymers by UV/ozone surface oxidation, *Int. J. Adhesion Adhesives*, 16, 29-31 (1996).
112. A.M. Guedes Pinto, A.G. Magalhães, F.G. da Silva and A.P. Monterio Baptista, Shear strength of adhesively bonded polyolefins with minimal surface preparation, *Int. J. Adhesion Adhesives*. 28, 452-456 (2008).
113. H. Tsubaki, S. Inokuchi and T. Ogawa, Joined body and method of producing joined body, US Patent 8974902B2, assigned to Nitto Denko Corporation, Osaka, Japan (2015).
114. E. Orgilés-Calpena, F. Arán-Aís, A M. Torró-Palau and M.A. Martínez Sánchez, Adhesives in the footwear industry: A critical review, *Rev. Adhesion Adhesives*, 7, 69-91 (2019).
115. K.A. Williams, L.T. Germinario and R. Eagan, Adhesion and gasoline resistance performance properties of chlorinated and non-chlorinated polyolefin adhesion promoters and blends for TPO substrates, Proc. Int. Coatings for Plastics Symposium, Troy, MI, pp. 1-15 (2003).
116. PFQD, quick drying residue –free solvent, PT Technologies, Ireland.
117. E.M. Petrie, *Handbook of Adhesives and Sealants*, 2nd edition, pp. 890-892, McGraw Hill, New York (2007).
118. C.Q. Sun, D. Zhang and L.C. Wadsworth, Corona treatment of polyolefin films-A review, *Adv. Polym. Technol.*, 18, 171-180 (1999).
119. M. Kłonica, J. Kuczmaszewski, M. Kwiatkowski and J. Ozonek, Polyamide 6 surface treatment layer following ozone treatment, *Int. J. Adhesion Adhesives*, 64, 179-187 (2016).
120. E.M. Liston, L. Martinu and R. Wertheimer, Plasma surface modification of polymers for improved adhesion, *J. Adhesion Sci. Technol.*, 7, 1091-1127 (1993).
121. P. Blais, D.J Carlsson and D.M. Wiles, Effects of corona treatment on composite formation. Adhesion between incompatible polymers, *J. Appl. Polym. Sci.* 15, 129-143 (1971).
122. N. Anagreh, L. Dorn and C. Bilke-Krause, Low-pressure plasma pretreatment of polyphenylene sulphide (PPS) surfaces for adhesive bonding, *Int. J. Adhesion. Adhesives*, 28, 16-22 (2008).
123. M. Frascio, C. Mandolfino, F. Moroni, M. Jilich, A. Lagazzo, M. Pizzorni, L. Bergonzi, C. Morano, M. Alfano and M. Avalle, Appraisal of surface preparation in adhesive bonding of additive manufactured substrates, *Int. J. Adhesion Adhesives*. 106, 102802 (2021).
124. Standard test method for peel resistance of adhesives, (T-Peel test), ASTM D 1876-08 (2016).

125. R. Kapica, J. Tyckowski, J. Balcerzak, M. Makowski, J. Sielski and E. Worwa, Enhancing adhesive joints between commercial rubber (SBS) and polyurethane by low-pressure plasma surface modification, *Int. J. Adhesion Adhesives* 95, 102415 (2019).
126. C.W. Karl, W. Rahimi, S. Kubowicz, A. Lang, H. Geisler and U. Giese, Surface modification of ethylene propylene diene terpolymer rubber by plasma polymerization using organosilicon precursors, *ACS Appl. Polym. Mater* 2, 3789-3796 (2020).
127. M.D. Romero-Sánchez and J.M. Martín-Martínez, UV-Ozone surface treatment of SBS rubbers containing fillers: Influence of the filler nature on the extent of surface modification and adhesion, *J. Adhesion Sci. Technol.* 22, 147-168 (2008).
128. J.H. Moraes, A.S. da Silva Sobrinho, H.S. Maciel, J.C.N Dutra, M. Massi, S.A.C. Mello and W.H. Schreiner, Surface improvement of EPDM rubber by plasma treatment, *J. Phys. D: Appl. Phys.* 40, 7747-7752 (2007).
129. K.F. Grythe and F.K. Hansen, Surface modification of EPDM rubber by plasma treatment, *Langmuir* 22, 6109-6124 (2006).
130. M. Hamdi and J.A. Poulis, Effect of UV/Ozone treatment on the wettability and adhesion of polymeric systems, *J. Adhesion*, 97, 651-671 (2021).
131. E.M. Petrie, *Handbook of Adhesives and Sealants*, 2nd edition, pp. 62-64, McGraw Hill, New York (2007).
132. A.C. Marques, A. Mocanu, N.Z. Tomić, S. Balos, E. Stammen, E. Lundevall, S.T. Abrahami, R. Günther, J.M.M. de Kok and S. Teixera de Freitas, Review on adhesives and surface treatments for structural applications, *Materials* 13, 1-43 (2020).
133. M.A Moyano and J.M Martín-Martínez, Surface treatment with UV/ozone to improve adhesion of vulcanized rubber formulated with an excess of processing oil, *Int. J. Adhesion Adhesives.* 55, 106 –113 (2014).
134. M. de Boo, Solvents replaced by UV irradiation and ozone, *Outlook*, 2, 17-20 (1993).
135. J. Seo, W.S. Chang and T-S Kim, Adhesion improvement of graphene/copper interface using UV/ozone treatments, *Thin Solid Films* 584, 170-175 (2015).
136. R. Kohli, Applications of UV/Ozone cleaning technique for removal of surface contaminants, in: *Developments in Surface Contamination and Cleaning*, Vol. 11, R. Kohli and K.L. Mittal (Eds.), Chapter 9, Elsevier, Oxford, UK (2019).

7

Adhesion Enhancement of Polymer Surfaces by Ion Beam Treatment

Endu Sekhar Srinadhu[1]*, Dinesh P. R. Thanu[3], Srilakshmi Putta[5], Mingrui Zhao[4], Bishwambhar Sengupta[6], Lakshmi Phani Arabandi[7], Jatinder Kumar[2], Radhey Shyam[1], Vinay H. Keswani[8] and Manish Keswani[3]

[1]*Department of Physics and Astronomy, Clemson University, Clemson, SC, USA*
[2]*Department of Physics and Astronomy, University of Kansas, Lawrence, KS, USA*
[3]*Materials Science and Engineering, University of Arizona, Tucson, AZ, USA*
[4]*Applied Materials, Santa Clara, CA, USA*
[5]*Department of Electronics and Physics, GITAM Institute of Science, GITAM University, Vishakhapatnam, India*
[6]*University of Washington, Seattle, WA, USA*
[7]*School of Medicine, Avalon University, Willemstad, Curacao*
[8]*Department of Information Technology, G.H.Raisoni College of Engineering, Nagpur, India*

Abstract
Polymers are widely used in different industries ranging from microelectronics, medical, to space. However, polymer materials are seldom used in their pristine state and need selective surface treatment to induce a specific response which is a challenging and complex task. Adhesion enhancement of polymers is one of the major requirements that can be achieved with ion beam technology at low cost. Surface enhancement involves keeping the bulk properties of materials unchanged and modifying only the surface properties to achieve optimum results.

In this chapter, we illustrate the use of ion beam technology to modify the surface properties of polymers for potential biomedical and microelectronics applications. This chapter focuses on effects on the adhesion characteristics of different polymeric materials with various optimizable parameters such as type of ion used, ion energy regime (low to medium to high) and the ion fluence range with respect to singly and multi-charged ion beams.

**Corresponding author*: esrinad@g.clemson.edu

K.L. Mittal and A.N. Netravali (eds.) Polymer Surface Modification to Enhance Adhesion: Techniques and Applications, (273–328) © 2024 Scrivener Publishing LLC

Keywords: Polymers, multi-charged ion beams, ion beam treatment, biomedical applications, microelectronics applications, adhesion

7.1 Introduction

Polymers are characterized by their excellent electronic, thermal, physical, and optical properties. Even though polymers have generated significant interest in the research community but are seldom used in their natural state. Most polymers require pretreatment for potential industrial applications. Nowadays almost every industrial application relies on some sort of surface treatment as a preliminary step. Surface treatment of polymers has evolved over the years due to their excellent properties and low cost. Surface treatment of polymers with traditional methods such as chemical treatment or heat treatment will induce chemical reactions on the surface resulting in changes of surface physical and chemical properties. Alternate methods, such as ion sputtering with noble gases like argon, would eliminate the risk of chemical reactions on the surface, but these methods involve the risk of ion implantation and there is not much freedom to choose the selectivity, whereas ion beam treatment methods have proven to be one of the reliable methods for the surface enhancement. Ion beam methods have been widely used in the microelectronics industry for implanting, doping, adhesion improvements, etc.

7.1.1 Ion Beam - Surface Interactions: Background

When a solid target is treated with ion irradiation, it could induce a significant modification of the microstructure and subsequent surface properties. Over decades, a tremendous amount of research has been conducted to understand the mechanism of ion-target interaction upon impact and the subsequent surface modifications. These modifications usually occur from the combined effects of electronic friction, roughening, and amorphization. In general, the dominant phenomena in ion beam treatment induced surface modification are energy dependent, ranging from quantum mechanical diffraction effects at a few meV energies to track formation via atomic displacements at MeV energies [1]. Understanding ion-solid interactions is very important in the microelectronics industry as ion-implantation is widely used in semiconductor device fabrication. Following the continuing trend for scaling down device dimensions, a more comprehensive understanding of ion-solid interactions is required for the new devices consisting of very shallow and laterally constrained target regions.

Ion beam induced interactions with material surfaces, especially polymer type materials, can be broadly classified into four widely known mechanisms: ablation, activation, deposition, and grafting. Ablation involves mechanical removal of surface contaminants, which are comprised of weak C–H bonds, by energetic electron and ion bombardment. Argon sputtering is often used for this process. Surface activation is achieved by removal of C-C and C-H surface bonds and creation of new functional groups on the surface using plasma gases such as O_2, H_2, N_2, etc. Ion beam is efficient enough to break weak surface bonds present in the polymers and replace them with highly reactive groups to increase adhesion strength. Cross-linking is a phenomenon by which the ion beam interactions set up chemical links between the molecular chains of polymers. Mostly inert gases are used for this purpose. Finally, deposition involves formation of a thin polymer layer on top of the surface of the substrate through polymerization of the process gas. The new layer can significantly alter the physical and chemical characteristics of the polymer [2]. Treatment of polymers by ion irradiation is a common practice profoundly utilized in several industrial applications such as optical, semiconductor, solar, etc. In the next section, the energetics of ion-solid interactions is briefly discussed along with the most widely used simulation techniques.

7.1.2 Ion Beam - Surface Interactions: Kinetics

The interactions of ion beams with the polymers have been discussed in the previous section. In this section the kinetics of ion beam-surface interactions will be discussed. The interactions of ions with a target material are typically described in terms of the kinetic energy of the incident ions as shown in Figure 7.1. For incident energies less than thermal energy (below ~25 meV), incident ions are trapped at the target surface through physisorption or chemisorption. In the energy range 1 eV to 1 keV, surface trapping is less effective and there is a high probability of scattering of the ions from the surface. When the ion energy goes above 1 keV, more ions

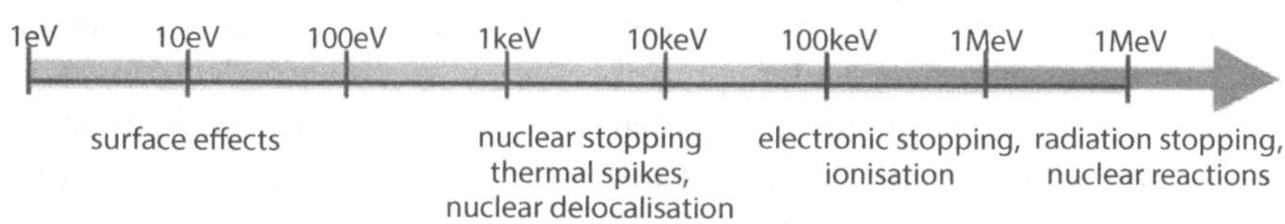

Figure 7.1 Schematic showing the primary interactions of an ion with a solid target as a function of the ion kinetic energy [3].

are implanted into the target. The dominating interactions in this energy range involve Coulombic collisions and elastic collisions between ions and target atoms as the ions gradually lose energy and get fully implanted in the target. Additionally, the energy transfer to the target surface lattice atoms induces oscillations that can heat parts of the target surface to a very high temperature (thermal spike). Above 100 keV, the main interactions are between the ions and the target electrons, and this becomes the dominant interaction in the MeV range. Finally, at energies above several MeV, energy loss occurs by radiative processes.

During the impact of an ion with a material, the ion will lose kinetic energy and the energy loss can be described using the stopping power $S(E)$, which is defined as the energy loss ΔE of the ions per unit distance traveled in the material Δx [3],

$$S(E) = -\frac{\Delta E}{\Delta x} \tag{7.1}$$

The stopping power refers to macroscopic systems and has units of [keV/cm]. This concept is extended to the microscopic level using the atomic density of the target material, ρ, as in

$$S(E) = -\frac{1}{\rho}\frac{\Delta E}{\Delta x} = \frac{1}{\rho}S(E) \tag{7.2}$$

The units of microscopic stopping power, $S(E)$, are [MeV·cm^2/g]

At lower energies, the ions can be stopped or slowed by two distinct processes: electronic and nuclear stopping. The sum of these two processes is the total stopping power, $S(E)$, which is defined as

$$S(E) = S_e(E) + S_n(E) \tag{7.3}$$

Nuclear stopping power ($S_n(E)$) is the momentum transfer between the ion and the target nuclei. Electronic stopping power ($S_e(E)$) is generated from inelastic collisions between electrons in the target material and the ion. Since ions experience a large number of collisions with electrons, and as the ion traverses the medium, its charge state will also change frequently, and it would be difficult to predict and describe all possible interactions for all possible ion charge states. As a comparison, the electronic stopping power is given as a simple function of energy, which is an average taken over all energy loss processes in different charge states. Theoretically, it can be determined with very high accuracy in the energy range of a few hundred keV or above, and the best known being the Bethe-Bloch formula [3, 4]. However, it could be difficult to determine the electronic stopping power as the energy level goes below 100 keV. Hence several semi-empirical stopping power formulas have been developed. The most commonly used model today is the one given by Ziegler, Biersack, and Littmark (ZBL stopping) [4, 5].

7.1.3 Computer Simulations of Ion Beam - Solid Interactions

Transport of ions in matter (TRIM) and stopping range of ions in matter (SRIM) are two widely used computer simulation packages for the theoretical estimation of ion implantation effects in solids, including polymers, such as ion penetration depths, distributions of scattered atoms, electrons and transmitted ions. TRIM and SRIM softwares are based on the Monte Carlo method and simulation results mostly agree with the experimental results [3].

7.2 Ion Beam Treatment of Polymers

Polymers have attracted much interest in a variety of applications such as electrical, telecommunications, optical lenses, automotive lighting, biosensors, solar cells, and nuclear track detectors, etc. because of their lightweight, durability, high optical transmittance, high deformation temperature, design flexibility, and cost-effectiveness [5–14]. Polymer-type materials are generally composed of long chains or networks of chemically bonded monomers. Monomers are representative units and several of these monomers join to form polymers. Ion beam irradiation of monomers can significantly alter the surface and mechanical and structural properties. As a result, there has been increasing interest in the modification of polymer materials to improve their structural, chemical, optical, and mechanical properties.

7.2.1 Principle of Technique

Many techniques such as Wet Chemical Etching, Electrical, Plasma, UV, γ-rays, X-rays, Electron beams, Swift Heavy ions have been extensively studied to modify the optical, thermal, structural, and chemical properties of polymers [14–23]. First, ion beams were developed and proven to be an efficient technique for surface modification of metals, semiconductors, and ceramics. Ion beams are effectively proven to modify polymer surface properties because the bonding energy of polymer molecules is much lower than that of metals, semiconductors, and ceramics. However, the interaction between the ions and the substrate involves high deposited energy and the interaction timescale is shorter than the elementary chemical event, thus it is expected to initiate non-conventional chemistry in polymers [24, 25]. Thus interaction of ion beams with polymers leads to complex chemical reactions like chain scission, cross-linking, formation of reactive atoms and free radicals, etc., which result in a change in structure as well as properties of polymers. Since the intra- molecular bonding in polymers is generally weak, it is easy to initiate chain scission or breakage of bonds with ion beams. Chain scission initiates the breakage of molecular bonds and the overall shortening of the polymer chain. Cross-linking results in the formation of chemical bonds between two adjacent polymer molecules. Chain scission decreases the molecular weight and cross-linking increases the molecular weight [26–29]. Studies have shown that ion beam irradiation could improve the adhesion, conductivity, wettability, solubility, etc. of polymers [30–38].

7.2.2 Types of Ion Beams and Interactions

In this chapter, we primarily discuss the surface modification of polymers using two widely known ion beam types, singly-charged and multi-charged ion beams.

The interaction of singly-charged ion beams (SCIs) with polymers leads to complex chemical reactions like chain scission, cross-linking, formation of reactive atoms and free radicals, etc., which result in the change in structure as well as properties of polymers. Chain scission initiates the breakage of molecular bond and overall shortening of polymer chain. Cross-linking results in formation of chemical bonds between two adjacent polymer molecules. In contrast to singly-charged ions where the kinetic energy of the projectile constitutes the important parameter, potential energy plays a key role in the highly charged ions (HCIs) interaction.

Unlike SCIs, highly charged ions (HCIs) possess large potential energies, which is sum of the ionization energies of the removed electrons. This potential energy is liberated during the impact of the ions with the solid surfaces and is being deposited into nanometer sized region on the surface. The deposited potential energy induces radiation effects such as formation of craters, hillocks, pits etc. and can facilitate chemical reactions that require high temperatures [39–50]. This makes HCIs a promising technique for surface modification.

7.2.3 Impacts and Outcome of Polymer Surface Modification

In most applications, polymer materials are selected and developed because of their excellent bulk properties, while the inherent poor surface properties create challenges. There are extensive studies on plasma treatment methods and techniques have been carried out to improve polymer surface properties [51, 58]. Generally, plasmas are used to incorporate chemical reactive functionalities like chemical groups onto the surface which activates the non-reactive substrate. Variety of gases such as Ar, O_2, N_2, NH_3 or CF_4 can be used for the plasma treatment depending on the polymer surface characteristics and applications [51]. Additionally, processing parameters such as power, pressure, flow rate, gas composition and treatment time could be utilized to fine-tune surface properties while not affecting bulk conditions. For example, in polypropylene (PP) film modification, Ar plasma treatment creates more active sites on the surface through chain scission resulting from energetic ion bombardment. In comparison, oxygen plasma modifies polymer surface through both physical ion bombardment and surface reactions triggered by reactive species from the plasma [52]. It was also found that long-term plasma exposure could result in inner chain scissions, whereas short time treatment would limit chain scissions in the outer surface [52, 53]. XPS results have shown that after Ar/O_2 plasma treatment there is an enrichment of oxygen-containing groups (e.g., C-O, C=O, O-C=O, etc.) on the surface, indicating the active sites can bind to other atoms/molecules especially oxygen in air when exposed to the atmosphere [52, 53]. These ion beam-induced polar functional groups, responsible for the surface hydrophilicity, and the resulting degree of crosslinking of polar groups on the surface are the control factors for the surface wettability [53]. Further, the introduction of carboxyl/carbonyl and other functional groups allows for improved adhesion at the polymer surface [54].

Another important aspect of plasma treated polymers is the degradation of surface functionalities due to reaction or reorientation, which leads to hydrophobic recovery with aging of the treated surface [51, 54–56]. On one hand, the plasma treated polymers have increased surface free energy and reorientation of polar chemical groups into the bulk of the material occurs, which follows the tendency of the material to attain thermodynamic equilibrium [51–54]. On the other hand, polar chemical groups are susceptible to reaction with oxygen in the atmosphere environment [54]. An example of aging effect is shown in Figure 7.2, where poly(ethylene terephthalate) (PET) substrate was treated by O_2 and N_2 plasmas with identical dose amount. It was observed that although both cases showed initial

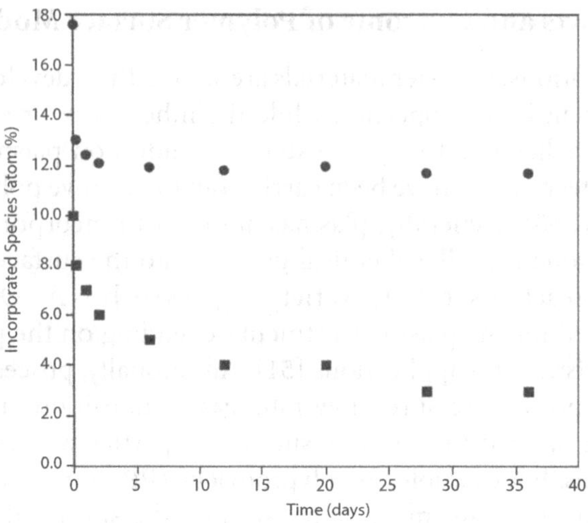

Figure 7.2 Aging data for a PET surface modified in a N_2 (•) and an O_2 (▪) plasma. Adapted from [57].

rapid decrease in the incorporated species, N_2 plasma treated surface was more stable in the long-term while O_2 plasma treated surface depicted depletion of oxygen species. It was suspected that oxygen plasma induced more chain scission on the target surface and resulted in an increased number of oxidized species with lower molecular weight. In addition, the degree of crosslinking in the near-surface region may also be different and it could affect the mobility of low-molecular weight species to diffuse into the bulk [55]. Studies have shown that aging effect could be reduced by increasing crosslinking through Ar plasma treatment, and by minimizing air exposure and maintaining at low ion temperatures [56–58].

7.3 Analysis Techniques to Analyze Post Ion Beam Treatment

Ion beam-treated polymers exhibit a change in color depending on the fluence (ions/area). Discoloration can be observed for all types of irradiated polymers, whereas the impact is highly evident for polymers with high carbon compounds (more carbon-containing species). On visual inspection, the discoloration can be observed only on the top surface layer to only a few angstroms. This is due to the low penetration depth of the incident ion beam.

There are several analytical techniques available for the characterization of polymers. Techniques such as Raman Spectroscopy, Ultraviolet-Visible (UV) Spectroscopy, Nuclear Magnetic Resonance (NMR), Fourier-Transform Infrared Spectroscopy (FTIR), etc. are utilized for chemical analysis of polymers. For the structural analysis of polymers, techniques such as X-ray Diffraction (XRD), Scanning Electron Microscopy (SEM), Transmission Electron Microscopy (TEM), Atomic Force Microscopy (AFM) are widely used. Differential Scanning Calorimetry (DSC), Thermal-Mechanical Analysis (TMA), Dynamic Mechanical Analysis (DMA) are used for thermal analysis of polymers. Some of the widely used techniques are discussed in the following sections.

7.3.1 X-Ray Diffraction

XRD is an analytical technique, where a sample is bombarded with X-rays to produce diffraction patterns which are recorded and analyzed to study the nature of the crystal. Two primary types of XRD analysis techniques are X-ray powder diffraction and single-crystal XRD. The underlying principles of the experiments are the same in both powder diffraction and single-crystal diffraction. When two parallel monochromatic X-ray beams of wavelength λ are incident on successive planes of a crystal at an angle θ, constructive interference of the reflected beams from two successive planes occurs only when the path difference between the two diffracted beams fulfills the Bragg condition:

$$2d\sin\theta = n\lambda \tag{7.4}$$

where n is an integer, λ is the wavelength of the X-ray beam and d is the distance between the planes.

As X-rays have a wavelength of the same order of magnitude as the lattice constants of crystals, the positions of the diffraction peaks and their intensities can be used to determine the structure, phase, etc., of a sample.

Diffraction patterns are typically plotted as the intensity of the diffracted X-rays vs. the angle 2θ. Peaks will appear in the diffraction pattern at 2θ values when constructive interference is at a maximum, i.e., when the Bragg Law is satisfied. By measuring the 2θ value for each diffraction peak, the d-spacing (which is essentially the distance between the diffracting planes) for each diffraction peak can be calculated. The peak identities can be confirmed or matched from the XRD database.

7.3.2 Scanning Electron Microscopy (SEM)

SEM is essentially a high magnification microscopy, which is a widely used technique to investigate the morphology of the prepared samples. In general, a Scanning Electron Microscope is comprised of an electron gun, a condenser, an aperture, an X-Y coil, an objective lens, and a detector. The electron beam is generated through thermionic or field emission. The generated electron beam is attracted towards the anode and is condensed by a condenser lens system. By using an objective lens system, the condensed electron beam is focused as a very fine point on the sample. The primary electrons of the electron beam strike the specimen surface and undergo inelastic scattering by the atoms in the sample. The scattered electrons are detected by a secondary or backscatter detector and converted to a voltage and amplified. This amplified voltage is applied to the grid of the cathode ray tube (CRT) to amplify the intensity of the spot of light. Topography of the image consists of thousands of spots of varying intensities. The quality of the image depends on the electron beam voltage and current. The latest models of SEM are capable of imaging dielectric samples without coating with a conducting layer.

7.3.3 Scanning Tunneling Microscopy (STM)

Scanning tunneling microscopes capture images of materials with atomic precision and can be used to manipulate individual molecules or atoms. Researchers have been using these instruments for many years to explore the world of nanoscopic phenomena. Recently, scanning tunneling microscope works without any moving parts and is almost vibration-free at extremely low temperatures as low as 30 millikelvin. The instrument can help researchers unlock the exceptional properties of quantum materials, which are crucial for the development of quantum computers and sensors [59]. Scanning Tunneling Microscope is a non-contact technique, and hence, artifacts due to interactions of nanoparticles with the tip usually do not occur, which is among the most significant advantages compared to electron beam-based techniques. For example, Scanning Tunneling Microscope is a sensitive technique in which impurities can affect the graphene image. In Scanning Tunneling Microscope, interactions of high-energy electrons with the sample may lead to changes and/or damage of the sample, which in turn may give rise to topological artifacts. Thus, STM is limited to imaging conductive or semiconductive substrates [60].

7.3.4 Fourier Transform Infrared Spectroscopy

When radiation (or) light interacts with matter, different processes such as reflection, scattering, absorbance, etc., can take place. FTIR-Spectroscopy technique is based on light or radiation absorption due to molecular vibrations. In this technique, infrared (IR) radiation is incident on the sample, and the resultant vibrational frequencies are analyzed to understand the chemical structure. Since every material is a unique configuration of atoms, no two compounds will have the same IR spectra. From the spectrum collected, two useful conclusions can be made:

(a) Vibrational peak frequency can be used to identify the chemical bonds
(b) The intensity of the peak corresponds to the quantity of material present.

The vibrational transition frequency can be described as:

$$\gamma = \frac{1}{2\pi c}\sqrt{\frac{k}{\frac{M_X M_Y}{M_X + M_Y}}} \tag{7.5}$$

Where γ is the vibrational or stretching frequency, c is the velocity of light, k is force constant and M_X,M_Y are masses of the two atoms involved in the vibrational bond stretching.

IR technique is also widely used to determine the chemical structure of the unknown polymers especially in the case of organic molecules by analyzing the specific absorption IR frequencies. The range of IR frequencies that can be detected depends on the spectrometer efficiency but, in general, the scan range is of the order 500-5000 cm^{-1}.

7.3.5 Raman Spectroscopy

When radiation (or) light interacts with matter, different processes such as reflection, scattering, absorbance, etc. can occur. Raman spectroscopy technique is based on the light that is scattered due to vibrational transitions in the molecules. According to the Raman Effect proposed by Sir C. V. Raman, "The scattered light varies depending on the symmetric vibrational modes of the scattered molecules". IR and Raman techniques can be considered as complementary where IR technique detects the asymmetric modes and the Raman technique detects the symmetric modes. Since the

Raman technique is most responsive to the symmetrical stretching type of transitions, which are most prominently observed in C-C bonds, it is a viable technique to study the structure of polymers since most of the polymers are made of carbon chains.

Raman spectroscopy is not easy to adapt for in-line, high-volume manufacturing (HVM) requiring regular calibrations. It is also prone to being impacted by environmental conditions such as temperature fluctuations or acoustic and vibrational noise. These issues can be effectively mitigated by using the NOVA ELIPSON™ - an instrument that offers rapid, ultra-stable, self-calibrated and fully automatized Rs(resolution, response and repeatability) based metrology [59].

7.3.6 UV Spectroscopy

When radiation (or) light interacts with matter, different processes such as reflection, scattering, absorbance, etc. can occur. UV-Vis spectroscopy techniques involve the measurement of the light beam reflected after it passes through a sample or after reflection from a sample surface.

$$Transmitted\ radiaton(T) = \frac{I}{I_0} \tag{7.6}$$

$$Absorbed\ radiation = -log\left(\frac{I}{I_0}\right) \tag{7.7}$$

Where I is the transmitted light (output) and I_o is the incident light (input)

UV-visible absorption and emission spectroscopy is a well-known method that is commonly used in polymer structure analysis since the lifetimes of excited states transitions of polymers are within the detection limits of electronic spectroscopy techniques such as UV. As the polymer undergoes excited states transitions such as $\pi \rightarrow \pi^*$ upon irradiation of light, the chromophores of the polymer, which are determined by the chemical structure, can be detected with UV analysis.

UV-visible absorbance spectroscopy is a powerful tool for detecting noble metal nanoparticles, because the localized surface plasmon resonance (LSPR) of metal nanoparticles allows for highly selective absorption of photons. UV-visible absorbance spectroscopy can also be used to detect various factors that affect the localized surface plasmon resonance of noble metal nanoparticles [61].

7.3.7 X-Ray Photoelectron Spectroscopy (XPS)

XPS, also known as Electron Spectroscopy for Chemical Analysis (ESCA), uses soft X-rays to produce photoelectrons from the surface atoms in a solid sample. XPS analyzes the kinetic energy of the emitted electrons and generates the corresponding spectrum to further identify the elements and their chemical states.

XPS is based on the photoelectric effect. Each atom has core electrons with characteristic binding energies that are conceptually equivalent to the ionization energy of that electron. When an X-ray beam strikes a sample surface, the core electron of atom absorbs the energy from the X-ray photon, and if the photon energy, *h*u, is large enough, the core electron will escape from the atom of the target surface. This emitted electron with certain kinetic energy (KE) is known as the photoelectron. As a result, the binding energy of the core electron can be calculated by the Einstein relationship:

$$KE = h\upsilon - BE - \varphi \tag{7.8}$$

where $h\upsilon$ is the X-ray photon energy (for K_α x-rays, E_{photon} = 1486.6 eV), BE is the binding energy (unknown), KE is the kinetic energy of photoelectron (measured by the energy analyzer), and φ is the work function depending on the analyzer, about 4-5 eV.

7.3.8 Atomic Force Microscopy (AFM)

AFM technique is different from optical microscopy techniques. Unlike optical microscopy techniques, AFM does not use a light source to illuminate the sample or eyepiece and lenses. AFM technique utilizes a mechanical device known as a cantilever, which measures changes in the forces when atoms or molecules come close to each other. The cantilever is fixed to the rigid support on one end and the tip on the other end is used to detect the changes in the atomic or molecular forces. The basic working principle of AFM depends on the local attractive or repulsive force between the cantilever tip and the sample, where the tip will deflect towards or away from the sample depending on whether the force is attractive or repulsive. A small laser source that is attached to the back of the cantilever reflects the laser beam on the photodetector to track the deflections of the cantilever. The major advantage of AFM or scanning probe microscopy (SPM) is that the samples do not need to be conductive and therefore need not be coated. AFM happens to be a very useful technique for surface analysis of materials that are dielectric or nonconducting such as polymers, living cells, DNA, proteins, etc. Apart from the surface structure, information on surface roughness and adhesion can also be obtained.

7.3.9 Wettability Measurements

Wettability studies usually involve the measurement of water contact angles. The contact angle is the angle where water, air, and solid meet and the value defines the surface wettability. Small contact angles (<90°) correspond to high wettability where the water is spread over a large area of the surface, while large contact angles (>90°) correspond to low wettability where the surface tends to repel water or minimum contact between the water and surface exists. Superhydrophobic surfaces, where the contact angles are very high (>150°), correspond to no contact between the water and surface. Thomas Young in 1805 defined the contact angle of a liquid drop on an ideal solid surface in terms of three interfacial tensions

$$cos\theta_Y = \frac{\gamma_{sv-\gamma_{sl}}}{\gamma_{lv}} \tag{7.9}$$

where, γ_{sv}, γ_{sl}, and γ_{lv} represent the solid-vapor, solid-liquid and liquid-vapor interfacial tensions, respectively, and θ_Y is Young's contact angle.

The contact angle is measured using an apparatus known as a contact angle goniometer. The experimental procedure involves cleaning the sample in an ultrasonic bath in acetone/ethanol medium and placing the sample on the stage. After cleaning the syringe it is filled with the required liquid (generally water) and the knob is opened to let the liquid drop fall on the surface after the sample stage is properly adjusted. The attached charge-coupled device (CCD) camera records the shape and angles at both ends. The software records the data and calculates contact angle values after each experimental measurement. It is a general practice to measure the contact angles at five different locations on the sample surface and report the averaged value.

In summary various spectroscopic techniques available for polymer analysis are represented in Table 7.1.

In recent years, there has been a growing interest in the modification of polymers using ion beam-based techniques such as ion implantation, ion sputtering, etc., for utilization in industrial applications such as nuclear reactors, biomedical applications, thin film growth, surface modification, nanostructures growth, etc. As a consequence, there has been a demanding need to understand the ion beam-surface interaction mechanism and the major factors affecting polymer surface properties. Bottom-up grown single-crystalline silicon nanowires (SiNWs) are highly intriguing to build nanoscale probes, for instance for atomic force microscopy (AFM), due to their mechanical robustness and high aspect ratio geometry. Several strategies to build such nanowire-equipped probes have been explored but their fabrication is still elaborate, time-consuming and relies partly on single-crystalline substrates. New strategy to fabricate AFM probes that are equipped with single-Si nanowire scanning tips [62–64] have been explored by research community recently.

In the following sections, some of the major ion beam processing applications of polymers will be discussed in detail.

Table 7.1 Spectroscopic methods for polymer analysis. Adapted from [65].

Transition(s)	Spectroscopic technique	Polymer structural analysis
Electronic	UV–Vis	Chromophore composition
Absorption, fluorescence, phosphorescence	Laser	Conformation, excited state behavior, polymer mobility
Vibrational	IR, Raman	Chemical structure, tacticity, conformation, chemical modification
Magnetic resonance	ESR, NMR	Chemical structure, tacticity, conformation, polymer mobility (NMR); radical, triplet state structure and behavior (ESR)

Note: ESR (electron spin resonance)

7.4 Polymer Surface Modifications for Biomedical Applications

Polymers come from either natural or synthetic sources and can be easily shaped into structures that range from simple foils to various complex shapes. Their application in medicine is typically limited by their mechanical properties like tensile strength and Young modulus, which are subpar in comparison with other materials, such as metals or ceramic composites [65, 66]. However, with the advancement in research, polymers offer a wide spectrum of physical-chemical properties due to the large variety in their chemical composition, which makes them candidates for biomedical applications [67–72]. Also, they are very flexible and capable of withstanding large deformations due to the large degree of freedom in the movement of the individual polymer chains. They are also semicrystalline materials, and the degree of crystallinity influences their stress-strain behavior, which can range from brittle to highly elastic [73]. While polymers are becoming the preferred materials in the biomedical industry, they are seldom used in a pristine state. Surface treatment by suitable techniques is often required to achieve properties necessary for specific applications [74–76]. There are multiple methods of surface treatment, each with their own advantages and disadvantages, such as UV lamp treatment, etching, grafting, plasma and laser treatment, metallization, ion sputtering, etc. An appropriate treatment can alter both the physical and chemical properties of the surface of a polymer in a way that makes it conducive for a variety of biological compounds, or, on the contrary, makes the polymer exhibit cytotoxic or antibacterial properties, thus making the polymer attractive for biomedical applications. Appropriate surface modification can alter the morphology of the polymer surface, giving rise to various nanoscale structures which serve as anchor points for specific proteins of the cell membrane, thus improving the adhesion of the cell to the treated surface [77]. The adhesion of cells to a substrate is a two-step process [78]. At first, the cell adheres to the surface due to non-covalent interactions (van der Waals, hydrogen bond, electrostatic, polar and ionic interactions) between specific molecules in the cell membrane and polar groups on the surface of the substrate. The second step is adhesion of molecules, such as fibronectin, vitronectin, collagen and laminin, from an extracellular matrix controlled by receptors. The cells attach themselves, via integrin receptors located in the cytoplasmic membrane, to certain sequences of amino acids [79, 80]. The most common sequence of amino acids to which the cell attaches itself is the triplet Arg-Gly-Asp [81]. Due to the complex mechanism of cell adhesion and proliferation, the interaction of the cell with the polymer is dependent on all the physico-chemical and morphological properties of the polymer surface as mentioned earlier, which gives the option to modulate the surface appropriately for compatibility improvement.

A traditional way of altering the biocompatibility of a polymer surface is ion implantation, which can increase both the adhesion and the subsequent proliferation of cells [82]. High-energy ions (in the range 10^{13} to 10^{15} #/cm^2) are separated and accelerated in an electric field before being implanted into the surface of the substrate and thus energy is quickly dissipated and causes changes in the structure of the polymer by breaking down macromolecular chains. Breaking of the polymer chains leads to an increased concentration of highly reactive radicals in the surface layer. This leads not only to an increased reactivity of the surface but also to crosslinking of the broken macromolecular chains, formation of double bonds and release of gaseous degradation products [83, 84]. The structures formed in the polymer by ion implantation are similar to those formed by pyrolysis, but they are found only in the surface of the polymer rather than the bulk of the material [85]. For biomedical applications, the ions used are either those of noble gasses, which are assumed to be biologically inert, or ions of elements such as oxygen or nitrogen, which can be commonly found in biomolecules [86, 87]. For specific applications, ions of cytotoxic elements such as silver can be used to create surface with no cell adhesion [88]. The process of ion implantation has several modulators. The depth of penetration is affected typically by both the mass and energy of the ions. Low energies in the range of tens keV are often utilized in this field since it is assumed that cells only interact with the uppermost layer of the material and thus shallow penetration depths are sufficient [89]. Alternatively, for the adhesion and proliferation of mammalian cells, good anchoring to the substrate is necessary. Here appropriate physico-chemical properties and morphology of the surface are needed. Pristine polymers are usually hydrophobic or negatively charged, which makes cell adhesion very difficult. Ion implantation allows modulation of surface properties of a polymer, including its conductivity and biocompatibility. The increase in biocompatibility can be attributed to an increased polarity of the modified surface and the presence of oxygen-rich groups, as well as to an increase in its roughness, which leads to an increase in surface area [90]. The adhesion of cells to a substrate is mediated by bonding via electrostatic forces and also via specific molecules of the extracellular matrix [91]. Lower ion doses lead to the formation of oxygen-containing groups and an increase in free volume, thus facilitating further penetration into the modified layer, creating a slightly porous structure. On the contrary, higher doses lead to more prominent crosslinking and the formation of two- and three-dimensional structures, accompanied by a decrease of free volume [92]. It is, therefore, possible to modulate the depth of penetration not only by the energy of the implanted ions but also by their density. Another key modulator is the size of the individual atoms. For example, small atoms such as fluorine are not incorporated into the modified layer and are released with other gaseous byproducts, while larger atoms such as iodine are incorporated well and remain in the polymer substrate [93]. The implanted atoms are bound either to the highly reactive free radical sites or to the newly formed double bonds [94]. Given equal energy, lighter ions have a higher mean distance range of penetration into the polymer in comparison to heavier ones, which are implanted closer to the surface of the modified substrate. Since ions always penetrate deeper than the uppermost surface layer regardless of mass and energy, it is a safe assumption that the adhesion of cells is dependent not only on the very surface of the polymer but also on the physico-chemical properties of a certain depth range of the surface layers [95].

Ion implantation also significantly changes the morphology of the treated surface. Argon and oxygen ions, for example, increase the roughness of a surface proportionally to the dose, and lead to the formation of a fibrous structure. However, in some cases, the earlier mentioned cross-linking and the decrease of free volume caused by higher doses of ions can lead to a decrease in surface roughness in comparison to the effect of lower ion doses [96]. The morphology of a surface is a key aspect when it comes to cell adhesion. Cells have been known to align themselves along grooves formed during surface treatment [97]. Increase of surface roughness on polymers often leads to an improved cell adhesion compared to its unmodified state. In general, the effect of surface roughness and morphology is dependent on both the polymer substrate and the type of seeded cells [98]. The electrical conductivity increase caused by ion implantation is attributed not only to bond breaking of macromolecule chains by the implanted ions and the subsequent formation of free radicals and conjugated double bonds, but also to the presence of conductive dopants in the treated surface layers. Carbonization and graphitization, which also occurs during ion implantation given sufficient energy and ion doses, also influences the electrical conductivity of the surface [99]. Interactions between the surface charge and the charge on the seeded cells are an important factor for proper adhesion. Specific sites with a negative charge on the cellular membrane are utilized in both recognition and adhesion processes when a cell gets into contact with a foreign surface. Whether a slightly negative or a slightly positive charge will facilitate adhesion depends on the type of the seeded cells. Furthermore, slightly charged groups influence the adsorption of specific proteins of the cellular membrane responsible for proper adhesion [100]. The changes in the wettability of the polymer substrate caused by ion implantation, affected by the modulation of chemical composition, roughness and morphology of the surface layers also have a noticeable effect on its biocompatibility [101–104].

7.4.1 Poly(lactic acid) (PLA)

Biodegradable polymer materials have vast industrial applications due to their ability to decompose into carbon dioxide and water under ambient conditions. However, biodegradable plastics, such as PLA, tend to decompose with photodeterioration when exposed to ultraviolet radiation despite its high transparency. To address this issue, Takano *et al.* [105], utilized ion beams as a processing tool to modify the surface properties while maintaining high transmittance to the visible rays as well as improving the surface wettability and hardness.

The biodegradable PLA polymer samples were bombarded with Ar^+ and He^+ ions with an ion energy of 2–18 keV and fluence of 3.13×10^{15} and 6.25×10^{15} ions/ cm^2. C1s and O1s XPS spectra of PLA polymers irradiated with Ar^+ and He^+ ions are shown in Figure 7.3. Pristine PLA polymer consists of C=O and C-O bonds at 298.1eV and 287.5 eV respectively. The observed shift in the C1s spectrum is due to the breakage of C=O and C-O bonds due to Ar^+ irradiation. The disappearance of O1s peaks in XPS spectra taken post Ar^+ irradiation indicates that the surface chemical structure has been changed to graphitic structure [105–107].

Another major effort is to induce surface modification that would facilitate the transmission of visible x-rays and intercept or block ultraviolet radiation. The UV radiation from the solar rays can induce bond breakage in the polymeric chains that leads to a decrease in the surface hardness. Thus, the lifetime of a biodegradable polymer is determined by its ability to withstand solar radiation without deterioration. From Figure 7.4, it is evident that the transmittance of visible radiation decreased with the increase in the ion energy along with the gradual decrease in the transmittance of UV radiation in the energy range from 2-10 keV. It can be understood that the ion beam irradiation modified the surface maintaining the transmittance to visible radiation while blocking most of the UV radiation. In summary, He^+ beam irradiation increased the hardness of the biodegradable polymer compared to Ar^+ due to modification of a thicker surface layer by helium ion beam; at energies below 10 keV the irradiated polymer samples exhibited higher protection from UV radiation while allowing the transmittance to visible rays due to modified surface layer [106, 107].

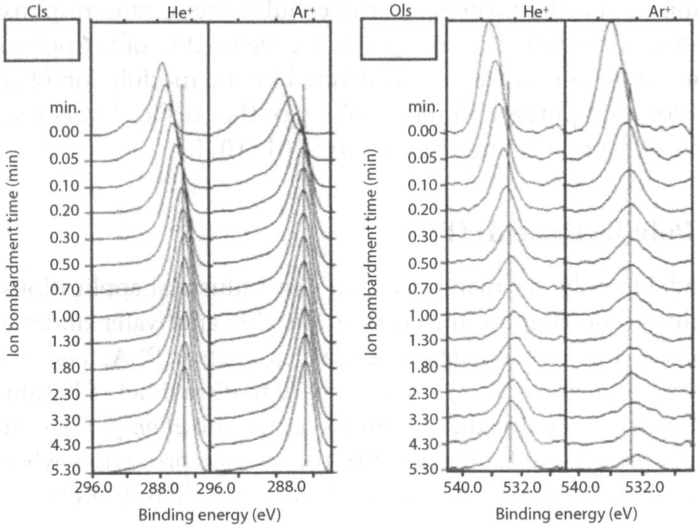

Figure 7.3 XPS spectra of biodegradable polymer (PLA) bombarded by He⁺ and Ar⁺. Adapted from [107].

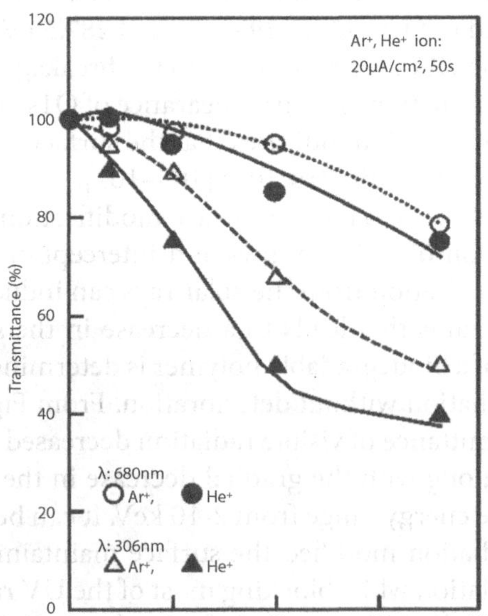

Figure 7.4 Transmittance of biodegradable polymer (PLA) irradiated with Ar⁺ and He⁺ ion beams/680 nm wavelength corresponds to visible radiation, and 306 nm corresponds to UV radiation. Adapted from [106].

7.4.2 Poly(L-lactic acid) (PLLA)

PLLA is widely used in the medical and pharmaceutical industries due to its bioabsorption capability. PLLA is hydrolyzed into nontoxic byproducts such as carbon dioxide and water by the human body's metabolic activity; however, the lack of reactive groups to promote cell attachment on the surface has limited its application. Yotoriyama *et al.* [108] irradiated PLLA films with Ar^+ ion-beams at an energy of 50 keV with fluences of 1×10^{14} and 1×10^{15} ions/cm^2 to induce new functional groups on the surface that enhanced cell attachment.

The XPS analysis of PLLA films irradiated with Ar^+ ion-beams at an energy of 50 keV with fluences of 1×10^{14} and 1×10^{15} ions/cm^2 showed a decrease in the intensity of C=O and C-O bonds, whereas an increase in the intensity of C-C bonds was observed. This observed phenomenon is attributed to the C=O and C-O bond scission due to Ar^+ irradiation and crosslinking of byproducts to form C-C bonds. XPS analysis shows clear evidence of a decrease in the oxygen concentration and increase in the carbon concentration thus revealing the modification of PLLA surface post ion beam treatment [106, 108].

The fluorescence antibody technique was employed to investigate the fibronectin (FN) adsorption on the ion irradiated surfaces. From Figure 7.5, the fluorescence intensity of ion-treated surfaces after adsorbed FN labeling with fluorescein isothiocyanate (FITC) shows an increase in the intensity compared to the non-treated surface. FN adsorption has increased drastically with ion irradiation with the maximum being for the sample treated with 1×10^{14} ions/cm^2.

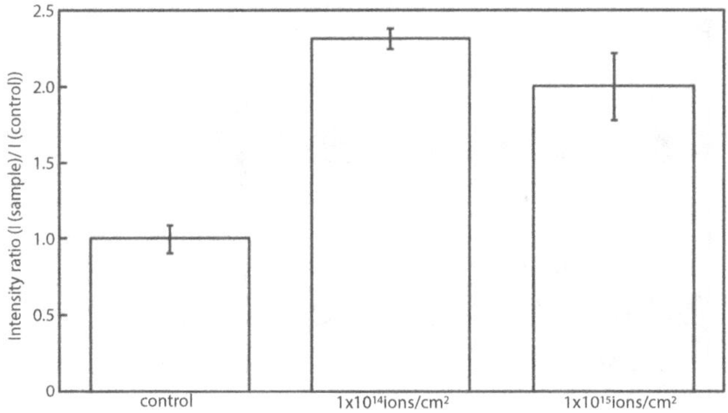

Figure 7.5 Fluorescence intensity of pristine(control) and ion irradiated PLLA films measured by a fluorometer. Adapted from [106].

In summary, the ion beam treatment has resulted in surface modification along with the increase in the adsorption for cell adhesion [106, 108].

7.4.3 Poly(L-lactide) (PLA), Poly(D, L-Lactide-co-glycolide) (PDLG) and Poly(L-lactide-co-caprolactone) (PLC)

Biodegradable polymers such as PLA, PDLG, and PLC are widely used in biomedical and tissue engineering applications. These polymers can hydrolyze into nontoxic byproducts such as carbon dioxide and water, however, their low surface energy and lack of reactive groups to interact with the cellular environment have limited their application. Sokullu [107] has explored the possibility of modifying the surface chemistry and wettability by implanting 60 keV Au^{2+} and 30 keV C^{+} ions over a range of fluences into the biodegradable polymers. XPS results revealed breakage of the C-O bonds on the surface and subsequent modification in the surface chemistry due to high ion irradiation [107, 109].

The hydrophobic polymers need to be modified to a hydrophilic state to improve their surface wettability, and this can be achieved by increasing the surface carboxyl and C-O-C groups. From Figure 7.6, all three polymers exhibited improvement in surface wettability (i.e., lower contact angle) due to ion implant. For the case of PLC samples, the water contact angle

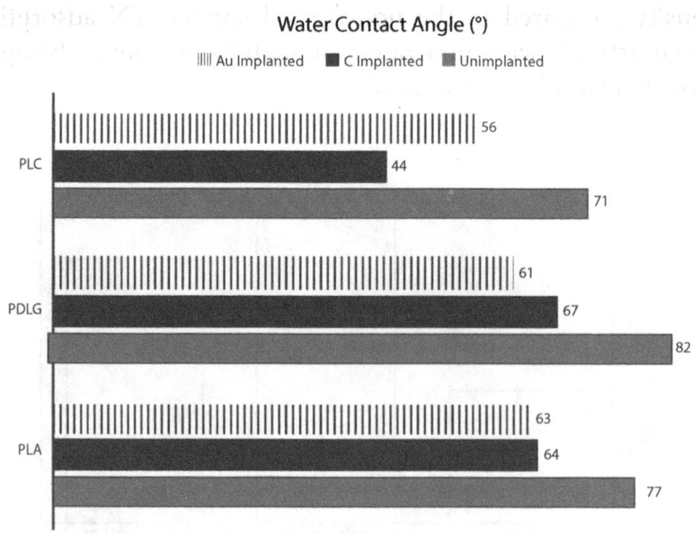

Figure 7.6 Mean water contact angles for pristine(unimplanted) and Au, C implanted PLA, PDLG and PLC films. Adapted from [109].

decreased from 71° to 44° with C implant and the contact angle decreased from 82° to 61° for PDLG polymer with Au implant and from 77° to 63° for PLA polymer with Au implant [107, 109].

The researchers above also found that polymers implanted at both low and high fluences showed the formation of fewer oxygen-containing groups. In comparison, when intermediate implantation fluence was applied, more oxygen-containing groups were generated. Researchers concluded that this phenomenon could be from the migration of species along the chains and associated atmospheric effects. These results suggest that ion implantation at optimum fluence affects the behavior of biodegradable polymeric materials in highly oxidative environments [107, 109].

In summary, it is shown that ion beams of several species and different fluences can be used for the modification of cell structures, wettability, and scaffold systems in biodegradable polymers. Ion modification alters the surface layer which increases the conductivity and biocompatibility. Lower ion doses lead to the formation of oxygen-containing groups on the surface with a slightly porous structure, thus facilitating further penetration into the modified layer. On the contrary, higher doses lead to chain scission and crosslinking with the formation of two- and three-dimensional structures on the surface, which are harder to penetrate by the subsequent ions [106–110].

In another study, PLA and poly(L-lactic-co-glycolic acid) (PLGA) (70/30) formed porous structures and were used as scaffolds for cell seeding, where hydrophobicity and low surface energy are roadblocks as they impact cell migration, attachment and growth on the polymer surface [111, 112]. Tests were carried out by treating PLA and PLGA films with anhydrous ammonia plasma for 120s under vacuum, and post-treatment results showed contact angles of these two polymer films were reduced from 78.0° and 70.1° to 21.5° and 15.9°, respectively. Further XPS analysis indicated polar amine groups and positively charged groups were introduced and incorporated onto the polymer surface, which results in improved hydrophilicity and enhanced surface energy. As a result, plasma treated surface showed increased cell seeding efficiency (>99%) and greatly reduced cell loss, suggesting improved cell affinity compared to the regular chemical method of using ethanol prewetting [112].

Another important application of polycaprolactone scaffold is in bone tissue engineering, where increased wettability and surface activation are critical [112]. The conventional chemical immersion induced surface modification (e.g., NaOH) is not favored due to its toxicity and extremely long treatment period up to 48 hours [113]. In comparison, Ar or air plasma treatment that only interacts with polymer surface without damaging the bulk properties is superior [114]. A detailed study was presented by Yang *et al.* in which electrospun poly(ε-caprolactone) (PCL) scaffold underwent Ar plasma treatment, followed by extensive soaking in simulated body fluid (SBF) [112]. By Ar plasma treatment, contaminants were eliminated and the weak chemical bonds on the targeted surfaces were replaced by highly reactive carboxyl groups, which resulted in a increase in surface energy and number of bonding sites for the following calcium and phosphate ion grafting. Immersion of the treated PCL in SBF for sufficiently long time caused mineralization of bone-like CaP onto polymer surface, which is similar to the natural bone structure [113].

7.4.4 Poly(dimethylsiloxane) (PDMS)

Tissue engineering has recently gained significant importance in the fields of both engineering and life sciences. The ability to create various surface topographies on the substrate is highly desirable since the interaction of the tissue cells within the surface topographical features can determine the cell behavior. To achieve specific cell function, which is a critical aspect of scaffold design, it is highly important to be able to understand the tissue interactions with the neighboring environment and to engineer the desired topographical patterns.

PDMS is often used for fabrication of microfluidic devices because of its low cost, biocompatibility, oxygen permeability and optical transparency [62, 114]. However, the inherent hydrophobicity of PDMS creates challenges in certain applications. Moreover, its very low glass transition temperature (<-120°C), and fast hydrophobic recovery (or aging effect) after surface hydrophilization are other major challenges for PDMS based microfluidic devices [114]. Tan *et al.* recently developed a technique that utilizes a second and extended oxygen plasma treatment (>300s) for bonded PDMS devices and showed that the hydrophilicity could be maintained up to several weeks [114]. It is worth mentioning that the amount of ionized gas diffusion into the inner wall surface was limited because the tiny fluidic access holes are the only entrance from bonded PDMS device. Unlike regular oxygen plasma treated PDMS surface that forms a wettable but brittle silica-like layer, this type of treatment results in a crack-free film that can retain the hydrophilicity for a few hours on exposure to the ambient air. These researchers also found that storage of the treated PDMS devices in water under vacuum could further prolong the hydrophilicity for more than a week, which may be explained by the fact that the high surface tension of water inhibited the reorientation of silanol groups on the inner wall surface.

Ion beam techniques have proven to be highly efficient for this purpose and for a wide variety of applications. PDMS substrates were prepared by the chemical route as explained [115, 116]. The PDMS substrates were irradiated with Ar^+ ion beams normal to the surface with irradiation times in the range of 0 to 20 min.

The AFM images of the unirradiated and irradiated PDMS substrates are shown in Figure 7.7. Periodic ripples or patterns formed in the direction perpendicular to the direction of the strain are shown in Figures 7.7 (a)-(d).

The conditions of Ar beam irradiation are shown in Table 7.2. Ar ion beam irradiation parameters significantly impact the surface topography. The wavelength and amplitude increased as the irradiation time increased. The surface topography can be tailored using ion beam parameters.

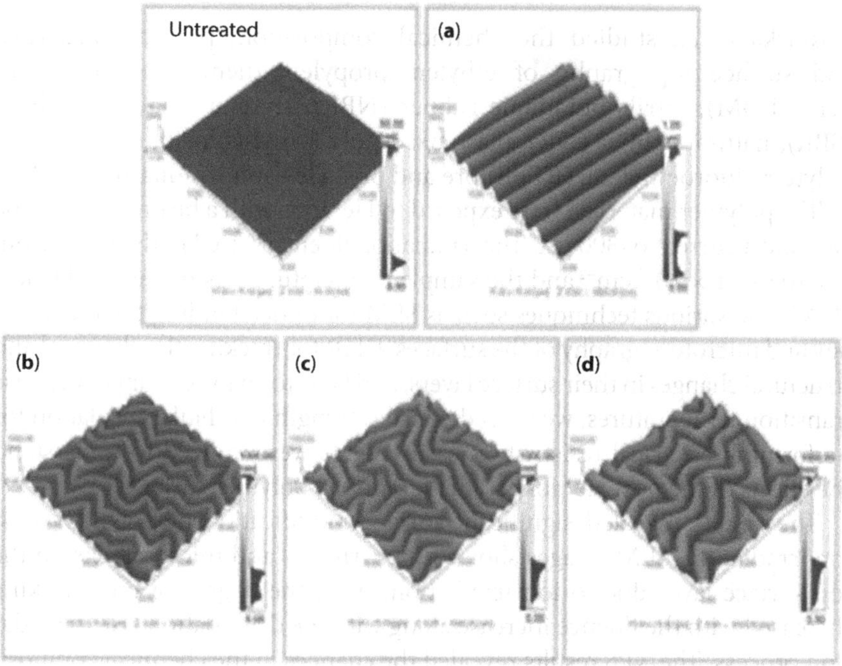

Figure 7.7 AFM images of the unirradiated and irradiated PDMS substrates. Adapted from [117].

Table 7.2 The conditions of Ar ion beam irradiation of PDMS samples. Adapted from [117].

Pre-strain (%)	20	0	0	0
Ar flow rate (sccm)	8	12	12	12
Anode voltage (kV)	1	1	1	1
Bias voltage (V)	-600	-600	-600	-600
Irradiation time (min)	15	0.5	5	20

Moon and coworkers [118, 119] have performed a similar study on PDMS substrates using Ga ions and concluded that the morphology of the wrinkles depends on the ion fluence, and wrinkle pattern can be controlled by adjusting the substrate orientation and ion velocity. They have also created other patterns on PDMS substrates such as nano-tunnels, and nanofiber-like tilted microcapillaries.

7.4.5 He^+ Ion Irradiation of Selected Polymeric Materials

Kosińska *et al.* studied the chemical composition, physical structure, and surface topography of ethylene-propylene-diene monomer rubber (EPDM), nitrile-butadiene rubber (NBR), styrene-butadiene rubber (SBR), natural rubber (NR), plasticized poly (vinyl chloride) (PVC), and polytetrafluoroethylene (PTFE) pre and post He+ ion irradiation [120].

The polymer materials were exposed to He^+ ions with a beam energy of 60 keV and a current of 300 μA. The irradiation fluence varied in the range from 1×10^{15} to 1×10^{17} cm^{-2} and the sample temperature was maintained below 80 °C. The various techniques such as SEM for understanding the morphology and microtopography of the surfaces, FTIR for tracking the chemical and structural changes in their surface layers, and DSC to analyze changes in phase transition temperatures, were used Tensile strength and Fluke Insulation for performing electrical resistivity measurements were utilized to understand the effects of the post He^+ ion irradiation on the polymer surfaces.

The results revealed significant surface modification of polymers post ion irradiation. SEM images showed the wrinkles and micro-cracks on the top surface layer due to ion irradiation. Also, the degree of cross-linking increased with the fluence increase along with the degradation due to oxidation process. The XPS results revealed the changes in the chemical compositions of the polymer surfaces as a result of irradiation as shown in Figure 7.8. The appearance of carbonyl bonds and the vibrations of C=O double bonds on the surface indicates oxidation of the material during irradiation. The presence of —OH groups in the materials after irradiation indicates degradation of the material. A significant decrease in the hydrogen concentration about 5 times lower than the starting material indicates the change in the surface composition. The removal of hydrogen known as chain scission and the oxidation process which is known as cross-liking occur simultaneously as a result of ion irradiation. The reduction in the hydrogen concentration causes a significant stress which, in turn, leads to the formation of cracks as observed from the SEM images. The impact of high-energy ion beam is observed to be only limited to its surface not extending beyond few micrometers, which helps the polymer materials to maintain their elastoplasticity.

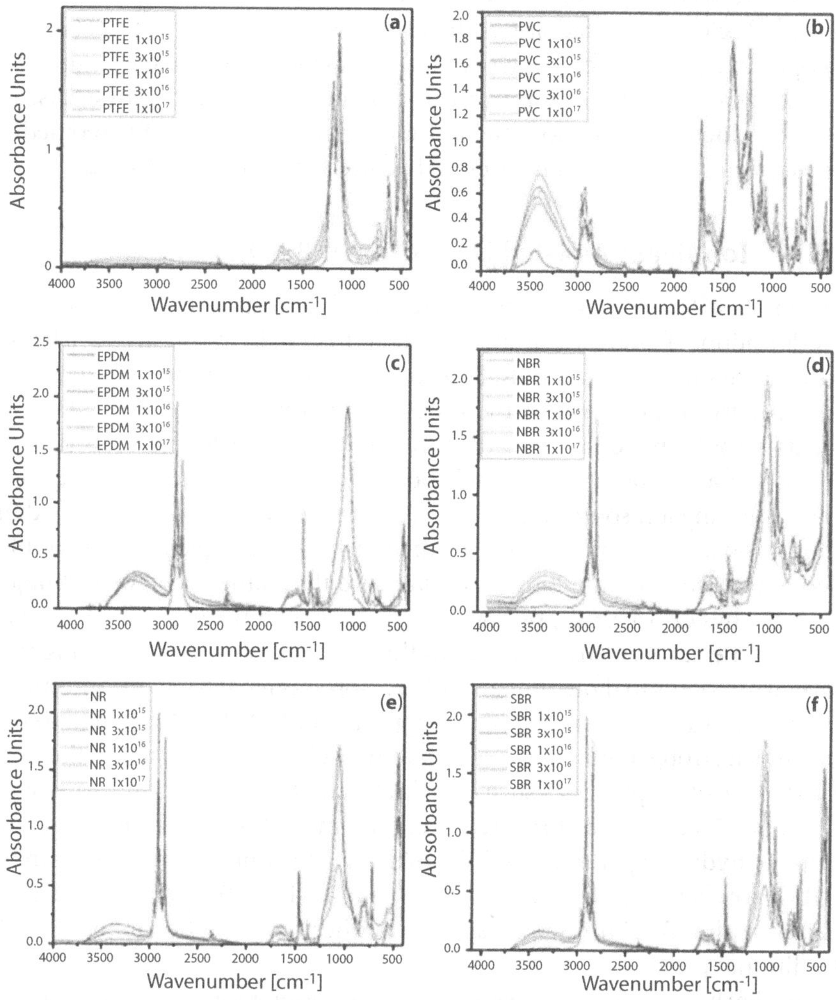

Figure 7.8 Influence of He+ ion bombardment on FT-IR spectra of the polymer materials studied: (a) PTFE, (b) PVC, (c) EPDM, (d) NBR, (e) NR, and (f) SBR. Adapted from [120].

The other properties such as wear resistance and hardness of the surface layers increased significantly due to ion irradiation. Apart from all the significant enhancements in the surface properties of the polymer materials due to ion irradiation, a decrease in the electrical resistivity has also been observed which makes them unlikely materials for applications such as a cover material for electrical cables exposed to ionizing radiation [120].

Ion beams have proven to be a valuable technique to pattern the various morphologies which require both physicists and biologists to collaborate to optimize patterns for the biocompatible devices. This tissue engineering field has gained significant interest from materials scientists, physicists, and medical doctors to develop highly functionalized scaffolds for biomedical applications. Significant advances in this field are expected in the near future.

7.4.6 Ion Beam Assisted Deposition (IBAD)

Ion implant technology has shown remarkable positive outcomes with the modification of both metal and polymer materials, especially to alter the surface region of the materials. Similarly, several other techniques have been explored to further improve the functionality. In this section, the improvements in the biomaterials using the IBAD technique are discussed. IBAD setup consists of a vacuum pump, sample holder to mount the sample, water coolant, sputtering ion source for the deposition of thin film, and low to medium energy ion sources for ion bombardment. IBAD technique involves simultaneous deposition of thin films using standard thin film deposition techniques such as physical vapor deposition (PVD) along with the ion-beam bombardment in the vacuum environment. In this process, ion energies ranging from several hundreds to thousands are utilized. Ion species, ion energy, and current density are varied depending on the application. IBAD coatings tend to possess much stronger adhesion strength to the substrate compared to the traditional deposition processes such as PVD, thermal deposition, or plasma spray.

Figure 7.9 shows the force required to remove the film using the scratch test on hydroxyapatite-coated Ti–6Al–4V specimens prepared using A (IBSD-ion beam sputtered deposition) and B (IBAD) techniques. It is evident that the IBAD films possess higher adhesion strength compared to the traditional process.

Similarly, several other coatings were developed for biomedical applications using the IBAD technique. To list a few,

1. Silver coatings on the catheters and other medical devices act as infection resistant. Silver coatings prevent the attack of infection/bacteria on the surface of the catheters and therefore help to provide long and safe life to the catheters [120].
2. IBAD ceramic coatings are widely used in drug vials, and polymer bags are used to collect blood and other biological fluids. For this purpose, IBAD ceramic coatings such as Alumina are applied to the polymer material to prevent oxygen and water penetration [115].

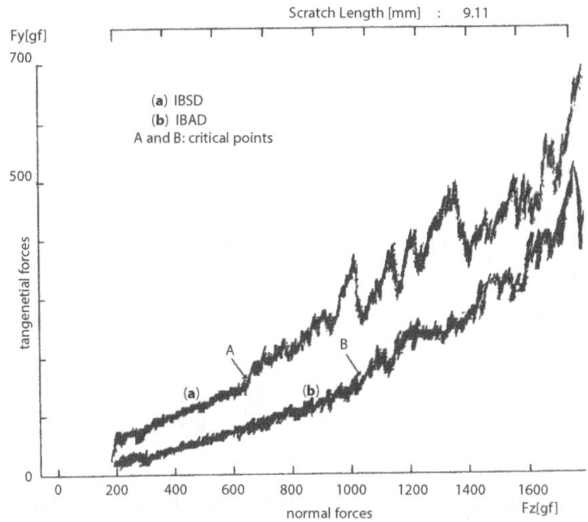

Figure 7.9 Arrow A (critical load for traditional deposition process such as ion beam sputtering deposition (IBSD) is 660 gf and B indicates the point where coating removal first occurs (critical load with ion beam assisted deposition (IBAD) process). Fy and Fz are the forces in the tangential and normal directions, respectively; measured in the units of gf(gram-force). Adapted from [115].

7.4.7 Ion Beam Texturing (IBT)

Surface morphology and adhesion strength are the two most important properties that determine biomaterial functionality. IBT technique is mainly intended to modify the surface morphology uniquely. Unlike the deposition techniques discussed in the above section, the IBT technique involves the removal of surface atoms to modify the surface by the bombardment of high-energy ions in the energy range from a few hundreds to thousand eV. This technique is extremely helpful to create the micro-to macro-features on a biomaterial. For example, various patterns can be etched on the surface of the biomaterials with the help of meshes. Figure 7.10 shows the modified silicone surface post ion beam treatment. The mesh is placed on the surface and two different patterns (a) square-shaped and (b) circular patterns were etched with Ar^+ beams of 1 keV. The area covered with the mask is protected from the ion etching and the rest of the non-covered or exposed area is etched. This technique is extremely useful in the field of tissue engineering to create several kinds of patterns on the surface of the cell structure scaffold materials.

In summary, ion beams techniques are extremely useful in modifying or optimizing the surface properties without alteration of the bulk

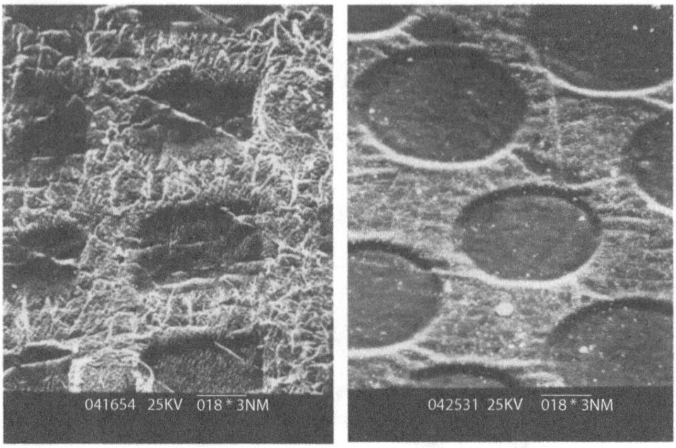

(a) (b)

Figure 7.10 SEM morphology of the silicone rubber modified by the IBT process. The etched shapes of square (a) and round (b) meshes are seen in the SEM images. Adapted from [115].

properties. Reliability and reproducibility are two key aspects to determine if the technique can be vastly applied and ion beam techniques have proven to be highly consistent, reliable, and cleaner techniques compared to the traditional approaches. The ion beam techniques help to optimize the surface properties and to increase the adhesion strength to the modified surface.

7.5 Polymer Surface Modification for Microelectronics Applications

Over the past several decades, plasma treatment has become a very useful method for surface modification and deposition of various materials in microelectronic applications [24, 121–123]. In integrated circuit packaging, plasma is employed to prepare surfaces for die attach, wirebonding mold/encapsulation and also to address interfacial delamination due to contamination. Plasma based contamination removal and surface activation processes are known to improve the quality, reliability and yield of microelectronic packages in advanced packaging. In many optical packaging (optoelectronic) devices, plasma-enhanced contamination removal is used to prepare surfaces prior to die attach and wirebonding. Substrate materials and the adhesives employed for attachment often do not possess the needed properties to allow for good adhesion and generally require surface modification for enhanced reliability. Plasma surface modification involves the interaction of the plasma species with a solid surface. The plasma process results in a physical and/or chemical modification of the molecular layers of the surface, without modifying the properties of the bulk. Typical materials employed in the microelectronics and optoelectronics industries include polymers, ceramics, glass and metals. The effectiveness of the plasma on these complex surfaces is determined by the plasma source, the configuration of the plasma system and the plasma operating parameters. Surface modification processes can be classified into four categories: Contamination Removal; Surface Activation; Etching; and Cross- linking. Selection of a specific process is determined by the physical and chemical compositions of the material to be processed as well as by the subsequent process required. Surface modification is often sensitive to time, environmental exposure such as heat and moisture where the surface may lose its plasma-induced physical and chemical properties. Several automated in-line plasma systems which allow surface modification process to be performed individually are currently available in the market.

Surface contamination removal or modification typically involves the use of the physical and/or chemical energy of the plasma. This process employs ablation, where the positive ions bombard the surface. The ablation process can dislodge contamination from the surface, and can roughen the surface on an atomic scale, as revealed by atomic force microscopy [121]. The chemical process is widely employed to remove residual materials, typically less than a few micronmeters, such as organic films, and oxidized surfaces. The chemical process employs either reduction or oxidation chemistry via the gas-phase radicals. Specific microelectronic and optical package reliability issues are due to poor wirebond pull strength, delamination at an interface and voiding. Wirebond pad contamination can be a by-product from previous processing steps such as die attach epoxy bleed or environmental exposure. An oxygen-based plasma will take advantage of the oxygen radicals to chemically react with the epoxy, producing volatile gas-phase by-products that can be pumped from the vacuum chamber. The success of an oxygen plasma for removal of die bond epoxy bleed has been widely demonstrated in literature and is frequently used in industries [124]. In cases where oxidation is of concern, a physical process can be employed to prepare the bond pad surfaces. An argon-based plasma treatment of ball grid array (BGA) strips has been shown to improve the wirebond pull strength by as much as 24.3 percent. Metal oxidation can act as a physical barrier for both wirebonding and solder reflow in microelectronic packaging. A combined physical and chemical process using argon and hydrogen can reduce the metal oxides. Reduction of copper oxide to copper is achieved in a hydrogen plasma via the reaction of the hydrogen radicals with the metal oxide as shown in Equation (7.10) below

$$CuO + 2H\bullet \rightarrow Cu + H_2O \tag{7.10}$$

Even in the absence of a contamination source, ablation will help to roughen the surface and provide a larger surface area for wirebonding, resulting in improved wirebond uniformity [123].

Plasma surface activation employs gases, such as oxygen, nitrogen, hydrogen, and ammonia, which, when exposed to the plasma, will dissociate and react with the surface, creating different chemical functional groups on the surface. The different functional groups can aid to modify the surface from hydrophobic to hydrophilic. The new functional groups have strong chemical bonds with the bulk material and have the capability to further bond adhesives to promote better adhesion. The functional groups also increase the surface area available for the adhesive and thus will distribute the load over a larger area resulting in improved adhesion strength. Gas selection and surface type determine the functional groups that will be incorporated on the surface. In microelectronic applications, plasma surface activation prior to die attach provides better contact, improved heat transfer and minimal voiding thereby enhancing the use condition of the end product.

The purpose of the mold/encapsulant materials for semiconductor applications is to provide adequate mechanical strength, adhesion to various package components, good corrosion and chemical resistance, matched CTE to the materials of interfaces with high thermal conductivity and high moisture resistance in the temperature range used. The ability to form good adhesion with package components and to remain bonded is of paramount importance, since delamination along the interfaces is a major concern. Plasma treatment has been demonstrated to improve the bond strength at the plastic encapsulants, and gold-plated copper-lead frame interface via an enhanced chemical compatibility with the molding compound [121]. Research works have also shown that treatment of nickel surfaces with water-based plasma improves the adhesion at mold-nickel interface. The etching process utilizes gases that dissociate in the plasma, creating a mixture of highly reactive species and the benefit of this chemical plasma is its chemical selectivity. The process chemistry can be optimized to make sure that a particular material can be selectively etched in the presence of other materials. For example, the dissociation of carbon tetrafluoride (CF_4) and oxygen in appropriate concentrations produces highly reactive oxy, oxyfluoro, and fluoro radicals that rapidly break carbon-carbon bonds in numerous materials. Plasma etching has many applications specific to semiconductor and optoelectronic processing, including photoresist removal, thin film etching, and polymer etching. In optoelectronic manufacturing, plasma etching has been employed to produce stripped fibers through the controlled removal of the urethane acrylate buffer coating. Plasma-induced cross-linking employs inert gases such as argon or helium to remove some atomic species from the surface, and generates reactive surface radicals. These radicals react with the surface forming chemical bonds, which results in cross-linked surface. This approach is employed on polymeric substrates, such as those used for plastic ball grid array (PBGA) packages. Argon plasma effectively sputters nanometers of material from the sample surface, thus roughening the surface on the nanometer scale. The cross-linking improves the adhesion of metal layers to the plasma-treated polymer laminate.

Polymer materials are widely used in the microelectronics industry due to their excellent electronic, thermal, physical and optical properties. They have attracted much interest in a variety of applications such as electrical, telecommunications, optical lenses, automotive lighting, bio-sensors, solar cells and nuclear track detectors, etc. [125–128]. As a result, there has been increasing interest in surface modification of polycarbonate to improve its structural, chemical, optical and mechanical properties [129–132]. We will discuss the surface modification of polymer materials using singly-charged and multi-charged ion beams in the next sections.

7.5.1 Bisphenol A Polycarbonate (PC)

Choi *et al.* [134], studied the role of oxygen flow in the modification of PC surface with Ar^+ irradiation in keV range. From Figure 7.11, the water contact angle decreases from 78° to 53° after $5 \times 10^{16}/cm^2$ Ar^+ ion irradiation. However, with the introduction of oxygen gas in the chamber, the contact angle decreased to 22° and 12° with 2 and 4 sccm of oxygen flow. With a further increase in oxygen gas flow, the contact angle did not decrease, so 4 sccm of oxygen flow was determined to be optimal for the best results. From XPS results, the decrease in the C-C bond intensity and increase in the C-O and C=O bonds intensities at 4 sccm of oxygen flow are observed. The decrease in the C-C bond intensity is attributed to the bond scission by Ar^+ ion and the increase in C=O is attributed to the cross-linking on the

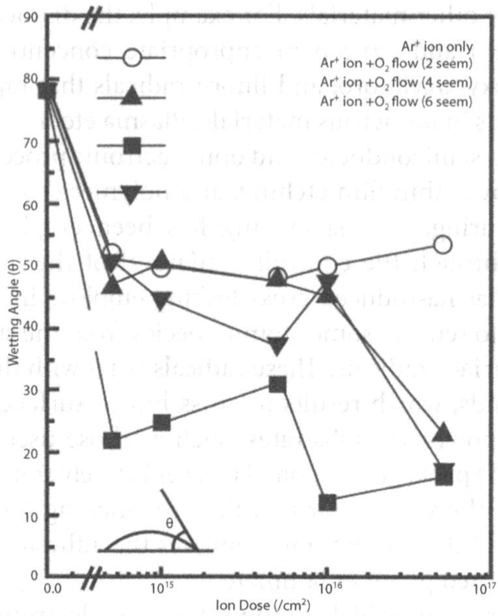

Figure 7.11 Water contact (wetting) angle of PC with varying Ar^+ dose and oxygen flow rate. Adapted from [134].

surface. Also, the increase in the C=O directly corresponds to the decrease in the contact angle or increase in the wettability due to the hydrophilic nature of C=O bond [134].

In summary, a two-step reaction mechanism was proposed to explain the enhanced wettability results observed in presence of oxygen gas flow with Ar^+ irradiation. Initially, the creation of an unstable polymer chain by ion irradiation and as a consequence a chemical reaction between highly unstable reactive radicals and introduced reactive oxygen gas which leads to the formation of new C-O and C=O bonds [133, 134].

The initial focus started with the possibility of utilizing singly-charged ion (SCI) beams for surface modification of polymers where inert gases, such as Ar, are used. SCI beam processing methods traditionally rely on an ion's kinetic energy to effect material modifications. In the last decade, there have been significant advances in the field of ion beam assisted etching, especially where the ion is missing multiple electrons and arrives at the surface as a multicharged ion (MCI). Multicharged ion beams are generated in tools such as electron beam ion trap (EBIT) systems, which operate at ultra-high vacuum (UHV).

In addition, biocompatible polymer materials such as highly oriented polygraphite (HOPG), diamond-like carbon (DLC), and polycarbonate (PC) surfaces were treated with multiply-charged ion beams of argon and oxygen with varying kinetic energies at Clemson University Electron Beam Ion Trap (CUEBIT) system [135, 136]. Researchers from CUEBIT facility have recently demonstrated multi-charged ion-induced surface modifications of PC surfaces which showed a significant enhancement in surface modifications well beyond those observed for similar, low flux singly-charged ions. The stored potential energy of MCIs, and the summed binding energies of the missing electrons (re-neutralization energy), can be utilized to initiate chemistry on the surface of the target that can otherwise happen only under extreme temperature conditions by dissipation of its energy within a small, atom-sized volume during interaction with the target. MCIs enhance the scission of C-C/C-H bonds on the surface and facilitate the formation of C-O and C=O bonds through the cross-linking mechanism on PC surface that exceeds those of SCI treatment methods on a per ion basis. MCIs effects are charge state-dependent and can be easily enhanced with the charge state [136–139]. The post-etch surfaces were found to have significantly less carbonaceous contamination. This could be a potential technology in the coming years for advanced semiconductor applications.

7.5.2 Aluminum Films on Bisphenol A Polycarbonate (PC)

Koh *et al.* [139], studied the possibility of modifying the PC surface with Ar^+ ion irradiation to improve the adhesion between the Al film and PC substrate. 1 keV Ar^+ ion beams with fluences of 10^{14} to 5×10^{16} ions/cm^2 in various oxygen flow rates (0 to 4 sccm) were used to irradiate the polymer surface to improve its wettability.

From Figure 7.12, it is evident that the water wetting angle of the PC before ion irradiation was 78° and with Ar^+ ions at 1 keV irradiation, the wetting angle was reduced to 50°, and then the wetting angle remained almost constant until the ion fluence of 5×10^{16} ions/cm^2. However, with the injection of oxygen at 4 sccm near the polymer surface during Ar^+ ion irradiation, the wetting angle reduced to 21° at 5×10^{14} ions/cm^2 at 1 keV. A minimum value of wetting angle ~ 12° was observed with 1 keV Ar^+ irradiation with ion fluence of 10^{16} ions/cm^2 along with 4 sccm of oxygen injection [139].

The improvements by energetic ion interaction are mainly due to bond scission, cross-linking, carbonization, and chemical reactions on the surface of polymer matrix.

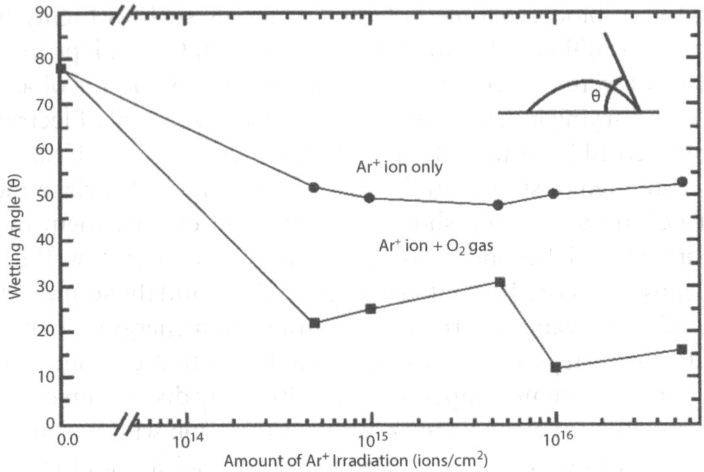

Figure 7.12 Contact (wetting) angle of PC with Ar^+ irradiation at 1 keV in oxygen flow. Adapted from [139].

The ion irradiation of polymer has mainly two effects:

1. Nucleation effect (physical interaction between energetic Ar^+ ions and atoms of polymer molecules).
- In this case, the incident energetic Ar^+ knocks out the nuclei atoms from the polymer matrix. The possibility of the nucleation effect is very low due to the high energy needed to knock out the nuclei atoms. In general, most commonly used ion beams in modification of polymer surfaces (such as Ar^+) do not possess such high energies needed to knock out the nuclei atoms.

2. Electronic effect (chemical interaction between incident particles and polymer).
- The dominant mechanism responsible for the modification of polymer surfaces is generally the electronic effect. Here, as the energetic Ar^+ ions pass through the polymer material that consists of C-C, C-H, and C-O bonds with bonding energies of about a few eV, and these could initiate bond scission of the polymer. This phenomenon would occur until the energetic Ar^+ ions drain their energy due to the interaction with the polymer material.

Figure 7.12 shows that the wettability could be enhanced by Ar^+ ion irradiation, which is attributed to the chemical interaction between incident particles and polymer surface. To further understand the modified polymer surface, aluminum was deposited by thermal evaporation on the ion beam treated surface and the untreated surface. The untreated samples failed the Scotch® tape test, which means that the adhesion between Al and the PC was not good, whereas the ion-treated samples did not display any failure by the Scotch® tape test [139].

In summary, higher wettability to water implies higher concentration of the hydrophilic groups on the polymer surface, and the ion beam irradiation before metal deposition can enhance adhesion between the deposited metal and the PC [139].

7.5.3 Indium Tin Oxide (ITO) Films on Bisphenol A Polycarbonate (PC)

PC is a polymeric material of interest for various applications such as biosensors, display panels, optical lenses, solar cells, etc. [139–141]. PC has high transmittance of about 90% in the visible region with high deformation temperature. However, pristine PC is considered to be unsuitable and also not widely used in many applications due to limited adhesion between coatings and PC material. The poor adhesion of PC is mainly attributed to the hydrophobic nature of the surface and large difference in the thermal expansion coefficients between the coating and the PC. Deposition of ITO coatings on the pristine PC results in non-adhering layers because of the PC surface microstructure. Methods such as hydrolysis, heat treatment, chemical cleaning techniques have been studied to improve adhesion [139–145]. However, ion beam treatment provides several advantages in achieving controlled chemical modification of surface to improve the poor adhesion property and facilitate thin film growth.

Highly conducting indium tin oxide (ITO) films possess low electrical conductivity with high transmissivity in the visible region which have been widely used in a variety of electrical and optoelectronic applications. The deposition of ITO films on polymer substrates such as PC is challenging because of its non-adhering nature. Cho and Koh [147] modified the PC surface to achieve good adhesion by irradiating the PC surface with an Ar^+ ion beam of varying fluences in the order of 10^{14} -10^{15} ions/cm^2 in the presence of oxygen flow. Pristine PC surface generally consists of C-C, C-O, C=O, C-N hydrophilic groups. From XPS results, the percentages of peak areas of C-C, C-O, and C=O bonds are found to be 84%, 13%,

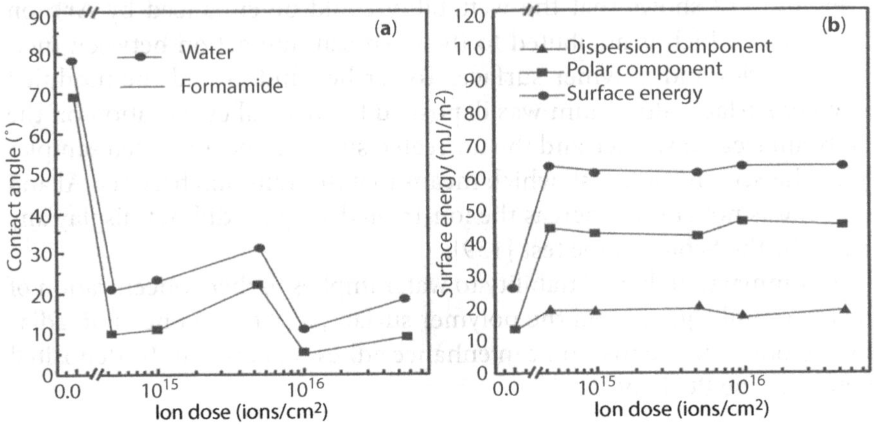

Figure 7.13 Contact angle (a) and surface energy (b) of PC as a function of Ar$^+$ irradiation dose. Adapted from [147].

and 3% respectively. However, after the Ar$^+$ irradiation, the C=O bond percentage increased from 3 to 17% which signified that Ar$^+$ ions were responsible for the chain scission of C-C bonds on the surface along with the formation of new C=O groups due to the presence of oxygen gas in the chamber.

From Figure 7.13, it is evident that the water and formamide contact angles have decreased from 78° to 12° and 68° to 5°, respectively which makes the surface more adhesionable. Also, the surface energy largely increases post Ar$^+$ irradiation. Peel test was performed to understand the improved adhesion quality of ITO films deposited on pristine and Ar$^+$ irradiated PC surfaces. Partial detachment and buckling of contact film (ITO) were observed on the pristine PC surface which is attributed to the weak adhesion at the interface. The ITO film remained intact for the case of Ar$^+$ irradiated PC surfaces that showed enhanced adhesion [142].

In summary, Ar$^+$ irradiation increases the interlayer adhesion and decreases the water contact angle. The newly formed C=O groups interact with the ITO film at the interface, resulting in enhanced interlayer adhesion [142].

7.5.4 Polyimide Films

With the increased usage of polyimides in microelectronics applications, the need to modify the polyimide surface for potential applications has intensified. Bachman and Vasile [34], studied the possibility of modifying polyimide surfaces with Ar$^+$ ion irradiation of varying ion fluence and energy. Two distinct polyimides: pyromellitic dianhydride-oxydianiline (PMDA-ODA) and biphenyl tetra-carboxylic dianhydride-phenylene diamine (BPDA-PDA) were grown using a wet chemical route on a Si wafer for this study.

PMDA-ODA and BPDA-PDA were reduced to nearly identical structures by an argon-ion bombardment of sufficient dose and energy. The carbonyl functional groups (C-O and C=O) of both polymers reduced in intensity with increasing ion dose. The modifications in the carbon bonds intensities depend on ion dose or ion energy. At high doses, the surface resembles graphite resulting from elimination of the carbon(yl)-oxygen (C-O and C=O) groups; whereas at small, low-energy ion doses, there is hardly any removal of carbonyl groups with almost no aromatic (C-C) structure. The high dose promotes surface roughening and increases the mechanical strength of the surface. Low ion dose helps to improve bonding when depositing metal in a non-UHV environment since adsorbed gases will be removed, and the carbonyl groups are still preserved [34].

7.5.5 Cu/Polyimide Films

Copper and copper silicides have been widely used in the semiconductor and electronics industries due to their low electrical resistance and excellent electromigration resistance. A recent study from Srinadhu *et al.* [148] revealed the complex growth mechanism and shapes of Cu_3Si (copper silicides) nanostructures on SiO_2/Si substrates for potential applications in the energy storage sector such as batteries, solar cells, etc. The resultant nanostructures depend on the density of voids on the SiO_2 surface, the interfacial adhesion strength between copper and Si, and the substrate type.

Chang *et al.* [35] explored Cu/polymer systems as promising candidates for potential applications in the microelectronics industry such as multilevel interconnects in very large scale integrated (VLSI) devices. However, the adhesion of Cu to polymers is still challenging and poses technical challenges. To overcome the poor adhesion, ion beam mixing methods were explored. In particular, their research revealed that increased adhesion strength of Cu/Al/polyimide system could be achieved through N_2^+ ion beam treatment. The adhesion strengths of Cu/polyimide and Cu/Al/polyimide systems are shown in Figure 7.14 as a function of ion dose. The untreated system at the beginning (where the dose is 0) possesses minimum adhesion strength and a significant increase in the adhesion strength has been observed for both systems for increasing the N_2^+ ion beam dose to 5×10^{15} cm^{-2}. After that, a decrease in the adhesion strength was observed

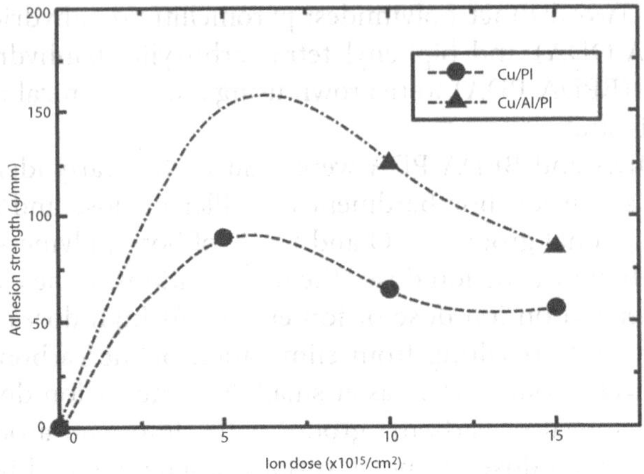

Figure 7.14 Adhesion strengths as a function of N_2^+ ion beam dose of the Cu/PI (polyimide) and Cu/Al/polyimide systems. Figure adapted from [35].

with a further increase of ion beam dose. The initial increase in adhesion strength at low ion dose (below 5×10^{15} cm^{-2}) is attributed to the formation of adhering $CuAl_2O_4$ interfacial layer, while the later decrease in the adhesion strength at high ion dose (above 5×10^{15} cm^{-2}) is due to modification of the chemical structure in the polyimide substrate itself, where amorphous carbon was formed.

7.5.6 PVA/PANI Polymer Composite Films

Polyaniline (PANI) & (PVA) poly(vinyl alcohol) polymers are made of chains of hydroxyl groups with carbon bonds and are good flexible polymers with excellent qualities such as low cost, high electrical conductivity, lightweight, etc. PVA and PANI have a vast range of applications in microelectronics, biomaterials, and semiconductors. However, there are drawbacks associated with these polymers such as low mechanical properties, low solubility, etc. [148–151]. Meanwhile, nanoparticles and their interactions with the polymers have attracted interest among researchers for their ability to change the structural, electrical, and optical properties, especially Ag nanoparticles into the polymers have shown promising applications for photonics, storage devices, and medical applications, etc. [152–155]. For this purpose, PVA/PANI/Ag nanocomposite thin films were fabricated as discussed in the reference [156]. Irradiation of the polymer composite containing nanoparticles can significantly improve the performance of the material especially for optoelectronic applications [157]. PVA/PANI/

Ag thin films are irradiated with oxygen ion beams of influences 8×10^{17}, 16×10^{17}, and 24×10^{17} ions/cm^2. It has been shown that Ag nanoparticles have improved the electrical and dielectric properties of the PVA and PANI polymers. The oxygen ion beam irradiation has increased the surface amorphicity due to a decrease in the crystalline size. Also, the irradiated films showed much better electrical and dielectric properties compared to the unirradiated films making them desirable for many potential applications in the microelectronic and semiconductor industries [148].

7.5.7 Multiple Ion Beam Treatment of Polymers

In contrast to singly-charged ions where the kinetic energy of the projectile constitutes the important parameter, potential energy plays a key role in the highly charged ions (HCIs) interaction. HCIs possess large potential energies, which are sum of the ionization energies of the removed electrons. This potential energy is liberated during the impact of the ions with the solid surfaces and is being deposited into nanometer sized region on the surface. The deposited potential energy induces radiation effects such as craters, hillocks, pits etc. and can facilitate chemical reactions that require high temperatures [39–49, 132]. This makes HCIs a promising tool for surface modifications. The effects due to HCIs on polymers have not been explored much. In this section, we discuss some of the recent work on the polymer surface modification using highly charged ion beams.

7.5.7.1 Kapton H, Teflon PFA, Tefzel and Mylar Polymers

Lee *et al.* [159] have investigated four different polymer materials: Kapton H (aromatic polyimide, polypyromellitimide), Teflon PFA (perfluoroalkyl copolymer resin), Tefzel (tetrafluoroethylene/ethylene copolymer), and Mylar (PET, poly(ethylene terephthalate)) for the improvements of surface-sensitive properties by ion beam methods such as ion implantation. Chemical structures of four polymer materials (Kapton H, Teflon PFA, Tefzel, and Mylar) are shown in Figure 7.15. Surface smoothness, hardness and wear resistance of the pristine polymer materials and those modified with ion beams of B, N, C, Si, and Fe with doses of ~3×10^{15} ions/cm^2 have been analyzed.

Polymer films subjected to ion beams showed a rapid color change, occurring at a dose of 1×10^{15} ions/cm^2. The color of implanted polymers depended on polymer type rather than implantation ion species. All polymers except Teflon PFA showed graphite-like blackish color after ion

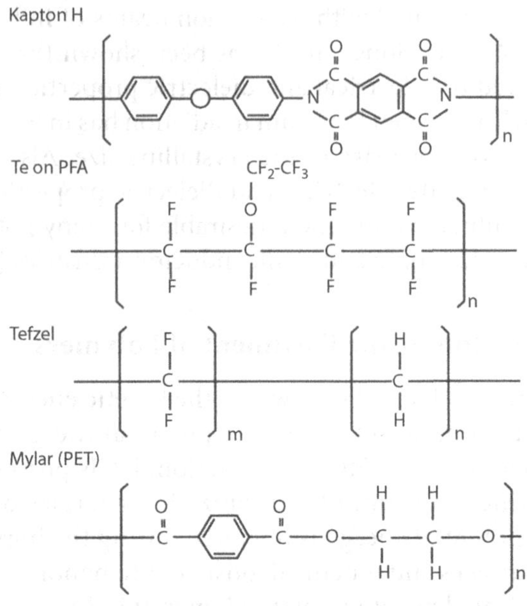

Figure 7.15 Chemical structures of polymers (Kapton H, Teflon PFA, Tefzel and Mylar). Adapted from [159].

bombardment; the other 3 polymers (Kapton H, Tefzel, and Mylar) also showed smoother surfaces after ion implantation, as revealed by optical microscopy.

Further test results showed remarkable improvement in surface smoothness, hardness, and wear resistance. The observed enhancements in the surface properties were attributed to the chain scission and crosslinking mechanisms due to ion beam implant and subsequent microalloying that led to the formation of new chemical bonds on the surface. The enhanced surface hardness is attributed to the randomly distributed diamond-like tetragonal bonding on the surface that led to three-dimensional rigid ladder structure formation. FTIR and Raman spectroscopy analyses confirmed the presence of new unsaturated or graphite-type bonding which is believed to be due to ion-beam alloying.

7.5.7.2 Polycarbonate (PC, Lexan)

Srinadhu *et al.* [137] studied the effects of multicharged ion irradiation on a polycarbonate surface using the HCI beams of Ar and O. The experiments were conducted at Clemson University Electron Beam Ion Trap Facility. PC samples were irradiated with O^{3+}, O^{5+} and Ar^{8+} ions beams at low kinetic energy (0.375 - 1.0 keV). The pressure in the chamber was maintained in the low 10^{-9} Torr range and the total ion doses used in these experiments were in the range of 10^{12} - 10^{13} ions per cm^2. Unlike the singly-charged ion beams where the presence of oxygen gas during the irradiation is not needed with HCIs technique. Post irradiation effects were tracked primarily using the SEM and XPS techniques.

Polycarbonate structure and bond assignments are notated by assigned XPS peak number(s). XPS peak assignment shown in Figure 7.16 corresponds to

1. Aromatic C-C/C-H
2. Aliphatic C-C/C-H
3. Aromatic C-O
4. Carbonate O-(C = O)-O
5. Aliphatic C = O
6. O-C = O
7. $\pi - \pi$ satellites

The formation of a shoulder at 286.2 eV (peak 3) accompanied by a decrease in the double peaks at 284.5 eV and 285 eV (peaks 1 and 2) was

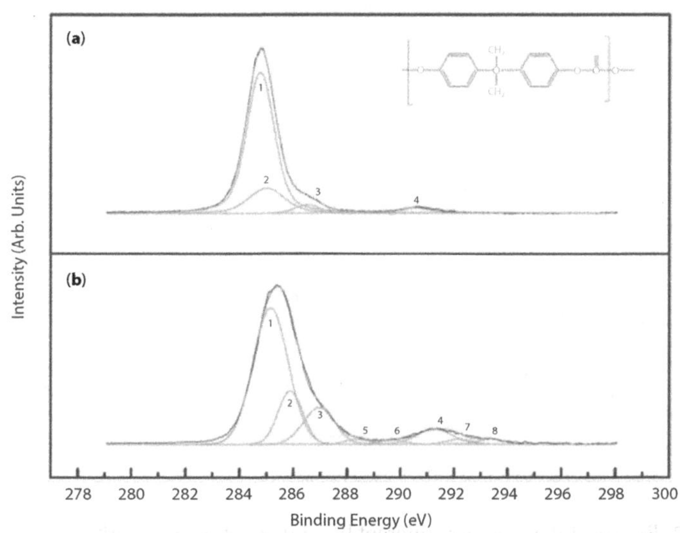

Figure 7.16. XPS spectra obtained on (a) pristine and (b) irradiated polycarbonate samples. The irradiated sample was exposed to 1000 eV Ar^{8+} ions with a dose of 1.41×10^{12} cm^{-2}. Adapted from [137].

primarily observed post irradiation which is also in agreement with the previous studies [45, 46, 124, 159].

All the singly-charged ion irradiations performed so far were accompanied by the simultaneous application of an oxygen gas flow across the sample. Hence, singly-charged ion beam impacts were studied using a two-step model where bond breaking due to the irradiation leads to chemically active dangling bonds that react with the flowing oxygen to form C-O which appears as a shoulder in the XPS spectra. It is important to note that without the presence of oxygen flow, there is no significant change in the polycarbonate samples.

Srinadhu *et al.* [137] performed experiments with HCIs and without oxygen gas flow unlike the singly charged studies. The relative contributions of the C-C, C-O and C=O peaks to the measured XPS spectra as a function of multicharged Ar and O ion fluence is shown in Figure 7.17.

HCIs interaction is widely explained in the literature using a well-known 'over-the-barrier' interpretation of their interaction with targets which defines a critical radius, $R_c = \sqrt{2Q/W}$, at which an impinging ion can capture a surface electron, where Q is the ion charge state and W is the surface work function (both in atomic units). They have also claimed that the secondary reionization and deexcitation steps as well as the collision of

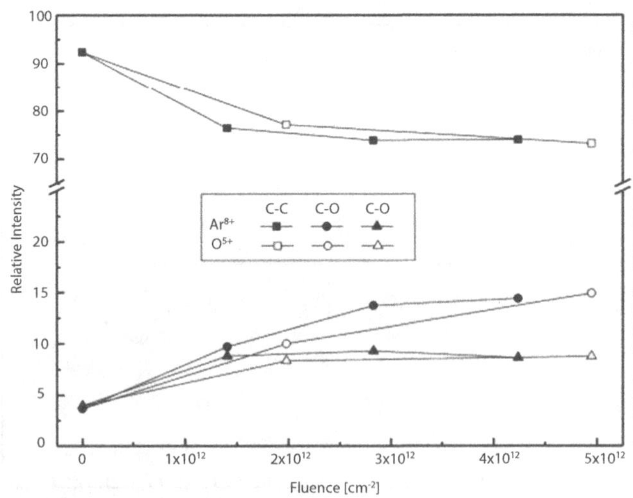

Figure 7.17 Relative intensities for different bond types at the polycarbonate surface as extracted from XPS spectra. The data are for O^{5+} and Ar^{8+} ion irradiations as a function of fluence. Adapted from [137].

the ion itself must be considered to capture a full picture of HCI-surface collision.

In summary, highly charged ion beams irradiation of polycarbonate substrates results in a significant enhancement in surface modification by initiating bond-breaking at the polymer surface without the presence of secondary oxygen flow and the results are for singly-charged ions. The results showed that the effects are initiated at a significantly lower (3 orders of magnitude) fluence when compared with singly charged ions. The enhanced surface effects observed indicate that the charge state plays a dominant role in the formation of pores that grow with the incident charge state. This method sheds light onto alternate polymer surface modification technique which is more efficient, and superior compared to the existing singly-charged ion beam techniques [137].

7.6 Summary

In this chapter, industrial applications of ion beam-assisted modification of polymer materials have been discussed extensively. Applications of ion beam technology spread across microelectronics, medical and optical fields, biomaterials, and thin film technologies to list a few. With state-of-the-art facilities and novel materials, ion beam technology has constantly evolved over several decades. In general, ion beam applications are typically aimed to promote adhesion by simple decontamination of the surfaces, generating a modified surface condition such as increasing the surface roughness, increasing the surface energy, rendering the surface hydrophilic, removal of metal oxides by physical and/or chemical interaction. The ion beam applications in microelectronics and medical sectors discussed in the above sections can be attributed to surface etching and modification of surface chemistry. The authors have also discussed the fundamentals of ion beams, mechanisms, software simulations, and advantages of ion beam surface modifications of selective polymers.

References

1. R.E. Lake, *Highly charged ion interactions with ultrathin dielectric films*, Ph.D. thesis, Clemson University, Clemson, SC, USA (2012).
2. D.P.R Thanu, E.S. Srinadhu, M. Zhao, N.V. Dole and M. Keswani, Fundamentals and applications of plasma cleaning, in: *Developments in*

Surface Contamination and Cleaning, Vol. 11, R. Kohli and K.L. Mittal (Eds.), pp.289-353, Elsevier, Oxford, UK (2019).

3. J.F. Ziegler, J.P. Biersack, and M.D. Ziegler, SRIM – *The Stopping and Range of Ions in Matter*, Lulu Press Co., Morrisville, NC (2008).
4. R. Shyam, *Energy loss of ions implanted in MOS dielectric films*, Ph.D. thesis, Clemson University, Clemson, SC, USA (2014).
5. D. Fink, P.S. Alegaonkar, A. Petrov, M. Wilhelm, P. Szimkowiak, M. Behar, D. Sinha, W.R. Fahrner, K. Hoppe and L.T. Chadderton, High energy ion beam irradiation of polymers for electronic applications, *Nucl. Instrum. Methods. Phys. Res. B*, 236, 11-20 (2005).
6. K.L. Mittal and W.-S. Lei (Eds.) *Laser Technology: Applications in Adhesion and Related Areas*, Wiley-Scrivener, Beverly, MA (2018)
7. A.Yen and H.I. Smith, An anti-reflection coating for use with PMMA at 193 nm, *J. Electrochem. Soc.*, 139, 616-619 (1992).
8. J.E. Griffith, Y. Qiu and T.A. Tombrello, Ion beam enhanced adhesion in the electronic stopping region, *Nucl. Instrum. Methods. Phys. Res. B*, 198, 607-609 (1982).
9. G.M. Wallner, W. Platzer and R.W. Lang, Structure-property correlations of polymeric films for transparent insulation wall applications. Part 2: Infrared optical properties, *Solar Energy*, 79, 593-602 (2005).
10. M.R. Cleland, L.A. Parks and S. Cheng, Applications for radiation processing of materials, *Nucl. Instrum. Methods. Phys. Res. B*, 208, 66-73 (2005).
11. J.M. Schneider, W.D Sproul and A. Matthews, Reactive ionized magnetron sputtering of crystalline alumina coatings, *Surf. Coat. Technol.*, 98, 1473-1476 (1998).
12. G.F. Smith, J. Mahood, M. Pixon, A.S. Zerda and M.A. Sanner, High heat polycarbonates, in: *Proc. ANTEC*, 1150-1155 (2007).
13. S. Vinoth, G. Kanimozhi, K. Hari Prasad, Kumar Harish, E.S. Srinadhu and N. Satyanarayana, Enhanced ionic conductivity of electrospun nanocomposite (PVD- HFP + TiO_2 nanofibers fillers) polymer fibrous membrane electrolyte for DSSC application, *Polym. Composites*, 40, 1585-1594(2018).
14. G. Kanimozhi, S. Vinoth, Kumar Harish, E.S. Srinadhu and N. Satyanarayana, Conductivity and dielectric permittivity studies of KI-based nanocomposite (PEO/ PMMA/KI/I_2/ZnO nanorods) polymer solid electrolytes, *Polym. Composites*,40, 2919-2918 (2019).
15. F. Dhaye, E. Balanzat, E. Ferain and R. Legras, Chemial modification induced in bisphenol A polycarbonate by swift heavy ions, *Nucl. Instrum. Methods. Phys. Res. B*, 209, 103-112 (2003).
16. B.W. Muir, S.L. McArthur, H. Thissen, G.P. Simon, H. Griesser and D.G. Castner, Effects of oxygen plasma treatment on the surface of bisphenol A polycarbonate: A study using SIMS, principal component analysis, ellipsometry, XPS and AFM nanoindentation, *Surf. Interface Anal.*, 38, 1186-1197 (2006).

17. M.A. Adams and A. Garton, Surface modification of bisphenol A polycarbonate by far UV-radiation, *Polym. Degrad. Stab.*, 41, 265-273 (1993).
18. B. Jaleh, P. Parvin, N. Sheikh, M. Hajivaliei and E. Hasani, Surface modification of Lexan treated by RF plasma, *Surf. Coat. Technol.*, 203, 2759-2762 (2009).
19. C. Kulshrestha, G. Agarwal, K. Awasthi, D. Vyas and Y.K. Vijay, Microsocopic studies of electron and ion irradiated polymeric films, in: *Microscopy: Science, Technology, Applications and Education*, Vol. 3, A. Mendez-Vilas and J. Diaz (Eds.), pp. 1696-1703, Formatex Research Center, Badajoz, Spain (2010).
20. S. Burkert, M. Kuntzsch, C. Bellmann, P. Uhlmann and M. Stamm, Tuning of surface properties of thin polymer films by electron beam treatment, *Appl. Surface Sci.*, 255, 6256-6261 (2009).
21. C. Darraud, B. Bennamane, C. Gagnadre, J.L. Decossas and J.C. Vareille, Optical modifications of polymers by ion beam irradiation, *Polymer*, 35, 2447-2451 (1994).
22. Y. Wang, Ion beam analysis of ion-implanted polymer thin films, *Nucl. Instrum. Methods. Phys. Res. B*, 161, 1027-1032 (2000).
23. R. Navarro-Gonzalez, P. Coll and R. Aliev, Pyrolysis of γ-irradiated bisphenol-A polycarbonate, *Polymer Bull.*, 48, 43-51 (2002).
24. F.D. Egitto and L.J. Matinezo, Plasma modification of polymer surfaces for adhesion improvement, *IBM J. Res. Develop.*, 38, 423-439 (1994).
25. J.C. Pivin, Hardening and embrittlement of polyimides by ion implantation, *Nucl. Instrum. Methods. Phys. Res. B*, 84, 484–490 (1994).
26. H. Dong and T. Bell, State of the art overview: Ion beam surface modification of polymers towards improving tribological properties, structural and chemical modification of polymer composite by proton irradiation, *Surf. Coat. Technol.*, 111, 29-40 (1999).
27. R. L. Clough, High-energy radiation and polymers: A review of commercial processes and emerging applications, *Nucl. Instrum. Methods. Phys. Res. B*, 184, 8-33 (2001).
28. S. Shah, A. Qureshi, N.L. Singh, P.K. Kulriya, K.P. Singh and D.K. Avasthi, Structural and chemical modification of polymer composites by proton irradiation, *Surf. Coat. Technol.*, 203, 2595-2599 (2009).
29. L. Calcagno, G. Compagnini and G. Foti, Structural modification of polymer films by ion irradiation, *Nucl. Instrum. Methods. Phys Res. B*, 65, 413-422 (1992).
30. H. Kupfer and G.K.Wolf, Plasma and ion beam assisted metallization of polymers and their application, *Nucl. Instrum. Methods. Phys. Res. B*, 166-167, 722-731 (2000).
31. T. Asmus and G.K.Wolf, Modification and structuring of conducting polymer films on insulating substrates by ion beam treatment, Plasma and ion beam assisted metallization of polymers and their application, *Nucl. Instrum. Methods. Phys. Res. B*, 166-167, 732-736 (2000).

32. J.H. Lee, J.-S. Cho, S.-K. Koh and D. Kim, Improvement of adhesion between plastic substrates and antireflection layers by ion-assisted reaction, *Thin Solid Films,* 449, 147-151 (2004).
33. B.S. Elman, D.J. Sandman and M.A. Newkirk, Transport properties of an ion implanted polydiacetylene, *Appl. Phys. Lett.*, 46, 100-102 (1985).
34. B.J. Bachman and M.J. Vasile, Ion bombardment of polyimide films, *J. Vac. Sci. Technol. A*, 7, 2709-2716 (1989).
35. G.S. Chang, K.H. Chae, C.N. Whang, E.Z. Kurmaev, D.A. Zatsepin, R.P. Winarski, D.L. Ederer, A. Moewes and Y.P. Lee, Mechanism for interfacial adhesion strength of an ion mixed Cu/polyimide with a thin buffer layer, *Appl. Phys. Lett.*, 74, 522-524 (1999).
36. J.K. Cho, Y.W. Beag, K.H. Kim, S. Han, J. Chao and S.K. Koh, High surface energy polymers obtained by bombardment with a keV ion beam in a reactive gas environment, in: *Polymer Surface Modification: Relevance to Adhesion*, Volume 2, K. L. Mittal(Ed.), pp.393-407, CRC Press, Boca Raton, FL (2000).
37. E.H. Lee, Ion beam modification of polyimides, in: *Polyimides: Fundamentals and Applications*, M.K. Ghosh and K.L. Mittal (Eds.) pp. 471-503, Marcel Dekker, New York (1996).
38. F. Yang, J.G.C. Wolke and J.A. Jansen, Biomimetic calcium phosphate coating on electrospun poly (ε-caprolactone) scaffolds for bone tissue engineering, *Chem. Eng. J.*, 137, 154-161 (2008).
39. X. Qian, O.G. Kravchenko,D. Pedrazzoli and I. Manas-Zloczower, Effect of polycarbonate film surface morphology and oxygen plasma treatment on mode 1 and II fracture toughness of interleaved composite laminates, *Composites Part A:* 105, 138-149(2018)
40. B. Jaleh, P. Parvin, N. Sheikh, M. Hajivaliei and E. Hasani, Surface modification of Lexan treated by rf plasma, *Surf. Coat. Technol.* 203, 759-762(2009).
41. S. Burkert, M. Kuntzsch, C. Bellmann, P. Uhlmann, and M. Stamm, Tuning of surface properties of thin polymer films by electron beam treatment, *Appl. Surface Sci.* 255, 6256-6261(2009).
42. C. Darraud, B. Bennamane, C. Gagnadre, J. L. Decossas, and J. C. Vareille, Optical modifications of polymers by ion-beam irradiation. *Polymer*, 35,2447-2451(1994).
43. R. Navarro-Gonzalez, P. Coll, and R. Aliev, Pyrolysis of gamma-irradiated bisphenol-A polycarbonate, *Polym. Bulletin*, 48, 43-51(2002).
44. K. Hareesh, C. Ranganathaiah, P. Ramya, R. Nair, G. G. Sangappa, and G. Sanjeev, Variation of Lexan polycarbonate properties by electron beam, *J. Appl. Polym. Sci.*127,2010-2018(2012)
45. A. Srivastava, T. V. Singh, S. Mule, C. R. Rajan, and S. Ponrathnam, Study of chemical, optical and thermal modifications induced by 100 meV silicon ions in a polycarbonate film, *Nuc. Instrum. Methods Phys. Res.* 192,402-406(2002).

46. A. Poirier, G. G. Ross, P. Bertrand, and V. Wiertz, Modification of wetting property of polycarbonate by means of ion beam implantation and storage in different gas environments, *MRS Proc.* 504, 425-430(1998)
47. B. S. Rathore, M. S. Gaur, and K.S. Singh. Dielectric properties and surface morphology of swift heavy ion beam irradiated polycarbonate films, *J. Thermal Anal. Calorimetry,* 111, 647-653(2013).
48. Z. Zhao, Y. Liu, J. Dai, and Y. Chen, The investigation of energy loss and damage in polycarbonate induced by meV carbon clusters, *Nucl. Instrum. Methods Phys. B* 262, 205-208(2007).
49. E. Hug, S. Thibault, D. Chateigner and L. Maunoury, Nitriding aluminum alloys by N-multicharged ions implantation: Correlation between surface strengthening and microstructure modifications, *Surf. Coat. Technol.* 206,5028-5035 (2012).
50. A. Srivastava, T.V. Singh, S. Mule, C.R Rajan and S. Ponrathnam, Study of chemical, optical and thermal modifications induced by 100 MeV silicon ions in a polycarbonate film, *Nucl. Instrum. Methods Phys. Res.* 192, 402–406(2002).
51. H.S. Sofi, R. Ashraf, A.H. Khan, M.A. Beigh, S. Majeed and F.A. Sheikh, Reconstructing nanofibers from nature polymers using surface functionalization approaches for applications in tissue engineering, drug delivery and biosensing devices, *Mater. Sci. Eng. C*, 94, 1102-1124 (2019).
52. F. Awaja, M. Gilbert, G. Kelly, B. Fox and P.J. Pigram, Adhesion of polymers, *Prog. Polym. Sci.*, 34, 948-968 (2009).
53. K.S. Siow, L. Britcher, S. Kumar and H.J. Griesser, Plasma methods for the generation of chemically reactive surfaces for biomolecule immobilization and cell colonization – A review, *Plasma Process. Polym.*, 3, 392-418 (2006).
54. N.Y. Cui and N.M.D. Brown, Modification of the surface properties of a polypropylene (PP) film using an air dielectric barrier discharge plasma, *Appl. Surf. Sci.*, 189, 31-38 (2002).
55. J. Lai, B. Sunderland, J. Xue, S. Yan, W. Zhao, M. Folkard, B.D. Michael and Y. Wang, Study on hydrophilicity of polymer surfaces improved by plasma treatment, *Appl. Surf. Sci.*, 252, 3375-3379 (2006).
56. D. Hegenmann, H. Brunner and C. Oehr, Plasma treatment of polymers for surface and adhesion improvement, *Nucl. Instrum. Meth. Phys. Res. B*, 208, 281-286 (2003).
57. J.M. Grace and L.J. Gerenser, Plasma treatment of polymers, *J. Dispersion Sci. Technol.*, 24, 305-341 (2003).
58. A.Vesel, I. Junkar, U. Cvelbar, J. Kovac and M. Mozetic, Surface modification of polyester by oxygen- and nitrogen-plasma treatment, *Surf. Interface Anal.*, 40, 1444-1453 (2008).
59. Forschungszentrum Juelich. Physicists develop new unique scanning tunneling microscope with magnetic cooling to study quantum effects. *ScienceDaily*, 31 (August 2021).

60. M. Nasrollahzadeh, M. Atarod, M. Sajjadi, S. M. Sajadi and Z. Issaabadi, Plant-mediated green synthesis of nanostructures: mechanisms, characterization, and applications, in: *An Introduction to Green Technology*, Vol.28, M. Nasrollahzadeh, S.M. Sajadi, M. Sajjadi, Z. Issaabadi and M. Atarod(Eds.), pp.199-322, Elsevier (2019).
61. A. Ranella, M. Barberoglou, S.Bakogianni, C. Fotakis and E. Stratakis, Tuning cell adhesion by controlling the roughness and wettability of 3D micro/nano silicon structures, *Acta Biomaterialia*, 6,2711-2720(2010)
62. X.M Yang, Z. W. Zhong, E.M. Diallo, Z.H. Wang and W.S. Yue, Silicon wafer wettability and aging behaviors: Impact on gold thin-film morphology, *Mater. Sci. Semiconductor Proc.* 26, 25–32(2014)
63. A.Behroudj, P.Salimitari, M.Nilsen and S.Strehle, Exploring nanowire regrowth for the integration of bottom-up grown silicon nanowires into AFM scanning probes, *J. Micromech. Microeng.* 31, 055010-055021(2021)
64. J.I.Paredes, A.Matrinez-Alonso and M.D.Tascon, Visualizing the porous structure of different carbon materials: a scanning tunneling microscopy study, *Studies in Surface Science and Catalysis,* 144,529-536(2002)
65. S.K. Rahi, E.S. Hassan and A.N. Abd, Spectroscopic methods studies in polymer, *World Scientific News*, 42, 41-53 (2016).
66. C.L. McCarthy, Y. Uchihara, M. Vlychou, G. Grammatopoulos and N.A. Athanasou, Development of malignant lymphoma after metal-on-metal hip replacement: a case report and review of the literature. *Skeletal Radiol,* 46, 831–836(2017).
67. L. Bacakova, E. Filova, M. Parizek, T. Ruml and V. Svorcik. Modulation of cell adhesion, proliferation and differentiation on materials designed for body implants. *Biotechnol. Advances* 29, 739–767(2011).
68. B. N. Brown and S.F. Badylak. Expanded applications, shifting paradigms and an improved understanding of host–biomaterial interactions. *Acta Biomater.* 9, 4948–4955(2013).
69. O. Kylian; A. Choukourov and H. Biderma. Nanostructured plasma polymers. *Thin Solid Films* 548, 1–17(2013).
70. L. Velard, A. Lorusso, F. Paladini, M.V. Siciliano, M. Giulio, A. Raino and V. Nassisi. Modification of polymer characteristics by laser and ion beam. *Radiat. Effects and Defects Solids*, 165, 637–642(2010).
71. L. Bacakova, V. Mares, M.G. Bottone, C. Pellicciari, V. Lisa and V Svorcik, Fluorine ion-implanted polystyrene improves growth and viability of vascular smooth muscle cells in culture. *J. Biomed. Mater. Res.* 49, 369–379(2000).
72. P. Slepička, I. Michaljaničová, N. Slepičková Kasálková, Z. Kolská, S. Rimpelová, T. Ruml and V. Švorčík. Poly-L-lactic acid modified by etching and grafting with gold nanoparticles. *J. Mater. Sci.* 48, 5871–5879(2013).
73. S. Bauer, P. Schmuki, V. Mark and K. Park, Engineering biocompatible implant surfaces Part I: Materials and surfaces. *Prog. Mater. Sci.* 58, 261–326 (2013).

74. E.R. Santos, E.C. Burini and S.H. Wang. UV-ozone generation from modified high intensity discharge mercury vapor lamps for treatment of indium tin oxide films. *Ozone Sci. Eng.* 34, 129–135(2012).
75. S. Gregoire, M. Boudinet, F. Pelascini, F. Surma, V. Detalle and Y. Holl. Laser-induced breakdown spectroscopy for polymer identification. *Anal. Bioanal. Chem.* 400, 3331–3340(2011).
76. N. Gour, K.X. Ngo and C. Vebert-Nardin. Anti-infectious surfaces achieved by polymer modification. *Macromol. Mater. Eng.* 299, 648–668(2014).
77. A. Reznickova, O. Kvitek, K. Kolarova, Z. Smejkalova and V. Svorcik. Cell adhesion and proliferation on polytetrafluoroethylene with plasma-metal and plasma-metal-carbon interfaces. *Jpn. J. Appl. Phys.* 56, 1–6(2017).
78. M. Parizek, N. Slepickova Kasalkova, L. Bacakova, Z. Svindrych, P. Slepicka, M. Bacakova, V. Lisa and V. Svorcik. Adhesion, growth, and maturation of vascular smooth muscle cells on low-density polyethylene grafted with bioactive substances. *BioMed Res. Int.* 371430-371448(2013).
79. L. Bacakova, E. Filova, F. Rypacek, V. Svorcik and V. Stary. Cell adhesion on artificial materials for tissue engineering. *Physiol. Res.* 53, 35–45(2004).
80. K.Anselme, A. Ponche and M. Bigerelle, Relative influence of surface topography and surface chemistry on cell response to bone implant materials-Part2: Biological aspects. *J. Engineering in Medicine* 224, 1487–1507(2010).
81. K. Walachova, L. Bacakova, B. Dvorankova and V. Svorcik. Biocompatibility of polymers modified by high-energy ions. *Chem. Listy* 96, 19–24(2002).
82. E. Carella, M. Leon, T. Sauvage, and M. Gonzalez. On ion implantation and damage effect in Li_2TiO_3 as a fusion breeder blanket: A technological approach for degradation testing. *Fusion Eng. Design* 89, 1529–1533 (2014).
83. Z. Chang and J.A. Laverne Hydrogen production in gamma-ray and helium-ion radiolysis of polyethylene, polypropylene, poly(methyl-methacrylate) and polystyrene. *J. Polym. Sci. Polym. Chem.* 38, 1656–1661(2000).
84. A.G. Nikolaev, G.Y. Yushkov, E.M. Oks, A. Oztarhan, A. Akpek, E. Hames-Kocabas, E.S. Urkas and I.G. Brown. Modification of anti-bacterial surface properties of textile polymers by vacuum arc ion source implantation. *Appl. Surf. Sci.* 310, 51–55(2014).
85. L. Bacakova, V. Svorcik, V. Rybka, I. Micek, V. Hnatowicz, V. Lisa and F.Kocourek, Adhesion and proliferation of cultured human aortic smooth muscle cells on polystyrene implanted with N^+, F^+ and Ar^+ ions: Correlation with polymer surface polarity and carbonization. *Biomaterials.* 17, 1121–1126(1996).
86. N. Shekhawat, S. Aggarwal, A. Sharma and K.G.M. Nair. Surface hardening in N+ implanted polycarbonate. *J. Mater. Sci.* 50, 3005–3013(2015).
87. A. Avalos, A.I. Haza and P. Morales. Cytotoxicity of silver nanoparticles on tumoral and normal human cells. *Toxicol. Lett.* 211, 202–203(2012).
88. J.S. Hayes, E.M. Czekanska and R.G. Richards, C. Kasper, F.Witte and R. Pörtner, *Tissue Engineering III: Cell-Surface Interactions For Tissue Culture,* Springer, 1st edition, Vol. 126, Heidelberg, Germany(2012).

89. M. Gosau, M. Haupt, S. Thude, M. Strowitzki, B. Schminke and R. Buergers Antimicrobial effect and biocompatibility of novel metallic nanocrystalline implant coatings. *J. Biomed. Mater. Res.* B 104, 1571–1579 (2016).
90. R.S. Kim, Surface modification of poly(tetrafluoroethylene) film by chemical etching, plasma, and ion beam treatments. *J. Appl. Polym. Sci.* 77, 1913–1920 (2000).
91. V. Svorcik, K. Proskova, V. Hnatowicz and V. Rybka. Iodine penetration and doping of ion-modified polyethylene. *Nucl. Instrum. Methods Phys. Res.* B, 149, 312–318(1999).
92. P. Predeep and A.M. Mathew. Intrinsically conducing rubbers: toward micro applications. *Rubber Chem. Technol.* 84, 366–401(2011).
93. V. Svorcik, K. Proskova, V. Rybka and V. Hnatowicz, Water diffusion in polyethylene modified by ion irradiation. *Polym. Degrad. Stab.* 60, 431–435(1998).
94. V. Svorcik, V. Rybka, V. Hnatowicz and B.L.Polarity, Resistivity and biocompatibility of polyethylene doped with carbon black. *J. Mater. Sci. Lett.* 14, 1723–1724(1995).
95. V. Svorcik, E. Arenholz, V. Rybka and V. Hnatowicz. AFM surface morphology investigation of ion beam modified polyimide. *Nucl. Instrum. Methods Phys. Res. B*, 122, 663–667(1997).
96. T. Paterbauer, S. Yakunin, Siegel and J. Heitz, Dynamics of the alignment of mammalian cells on a nanostructured polymer surface. *Macromol Symp.*, 296, 272–277(2010).
97. S. Mattioli, S. Martino, F. D'Angelo, C. Emiliani, J.M. Kenny and I. Armenato. Nanostructured polystyrene films engineered by plasma processes: Surface characterization and stem cell interaction. *J. Appl. Polym. Sci.* 131,40427 (2014).
98. V. Svorcik, R. Endrst, V. Rybka and V. Hnatowicz. Time dependence of the number of unpaired electrons and the sheet resistance in ion irradiated polymers. *Mater. Lett.* 28, 441–444(1996).
99. J.L. Alonso and W.H. Goldmann. Cellular mechanotransduction. *AIMS Biophys.* 3, 50–62(2016).
100. G.M. Gencer, S. Karadeniz, F.Y. Lambrecht, H. Havitcioglu, S. Ozkal and H. Baskin. The effects of plasma immersion ion implantation and deposition (PIII&D) process voltages on the morphology, phase formation and E. coli adhesion of Ag coatings obtained on the surface of Ti6Al4V orthopedic implant material in nitrogen plasma. *J. Fac. Eng. Archit.* Gazi 32, 231–241(2017).
101. L. Chvatalova, R. Cermak, A. Mracek, O. Grulich, A. Vesel, P. Ponozil, A. Minarik, U. Cvelbar, L. Benicek, and P. Sajdl. The effect of plasma treatment on structure and properties of poly(1-butene) surface. *Eur. Polym J.* 48, 866–874(2012).
102. I. Junkar, Plasma treatment of amorphous and semicrystalline polymers for improved biocompatibility. *Vacuum* 98, 111–115(2013).
103. F. Kessler, D. Steffens, G.A. Lando, P. Pranke and D.E. Weibel. Wettability and cell spreading enhancement in poly(sulfone) and polyurethane surfaces by

UV-assisted treatment for tissue engineering purposes. *J. Tissue Eng. Regen. Med.* 11, 23–31(2014).

104. D.V. Bax, D.R. McKenzie, M.M.M. Bilek and A.S. Weiss. Directed cell attachment by tropoelastin on masked plasma immersion ion implantation treated PTFE. *Biomaterials,* 32, 6710–6718(2011).
105. I.Takano, Y. Arai, M. Sasaki, Y. Sawada, K. Yamada, T. Yagasaki and Y. Kimura, Surface modification of biodegradable plastics by ion beams, *Vacuum*, 80, 788-792 (2006).
106. K.L. Mittal (Ed). *Polymer Surface Modifications: Relevance to Adhesion*, Vol.5, CRC Press, Boca Raton, FL(2009)
107. E. Sokullu, Ion beam modification of biodegradable polymeric biomaterials, *Current Trends Biomedical Eng. Biosci. (CTBEB)*, 1, 555567(1)-(3) (2017).
108. T. Yotoriyama, A. Nakao, Y. Suzuki, T. Tsukamoto and M. Iwaki, Analysis of cell adhesion surface induced by ion-beam irradiation into biodegradable polymer, *Nucl. Instrum. Methods. Phys. Res. B,* 242, 51-54 (2006).
109. E. Sokullu-Urkac, A. Oztarhan, F. Tihminlioglu, A. Nikolaev and I. Brown, Oxidation behavior of C-and Au-ion-implanted biodegradable polymers, *IEEE Trans. Plasma Sci.*, 40, 863-869 (2012).
110. V. Svorcik, K. Proskova, V. Hnatowicz and V. Rybka, Iodine penetration and doping of ion-modified polyethylene, *Nucl. Instrum. Methods. Phys. Res. B,* 149, 312–318 (1999).
111. J. Zhou, A.V. Ellis and N.H. Voelcker, Recent development in PDMS surface modification for microfluidic devices, *Electrophoresis*, 31, 2-16 (2010).
112. J. Yang, G. Shi, J. Bei, S. Wang, Y. Cao, Q. Shang, G. Yang and W. Wang, Fabrication and surface modification of microporous poly(L-lactic acid) and poly(L-lactic-co-glycolic acid) (70/30) cell scaffolds for human skin fibroblast cell culture, *J. Biomed. Mater. Res.*, 62, 438-446 (2002).
113. A. Chapiro, General consideration of the radiation chemistry of polymers, *Nucl. Instrum. Methods. Phys. Res. B,* 105, 5-7 (2005).
114. S.H. Tan, N.-T. Nguygen, Y.C. Chua and T.G. Kang, Oxygen plasma treatment for reducing hydrophobicity of a sealed polydimethylsiloxane microchannel, *Biomicrofluidics*, 4, 032204-1 – 8 (2010).
115. F.Z. Cui and Z.S. Luo, Biomaterials modification by ion-beam processing, *Surf. Coat. Technol.* 112, 278–285(1999).
116. P. Sioshansi and E.J. Tobin, Surface treatment of biomaterials by ion beam processes, *Surf. Coat. Technol.* 83,175–182 (1996).
117. T. Hasebe, S. Nagashima, Y. Yoshimoto, A. Hotta and T. Suzuki, Tailoring surface topographies of polymers by using ion beam: Recent advances and the potential applications in biomedical and tissue engineering, *Nucl. Instrum. Methods Phys. Res. B* 282,134-136 (2011).
118. M.W. Moon, S.H. Lee, J.Y. Sun, K.H. Oh, A. Vaziri and J.W. Hutchinson, Wrinkled hard skins on polymers created by focused ion beam, *Proc. Natl Acad. Sci. USA* 104,1130-1133(2007).

119. E.K. Her, H. S. Chung, M.W. Sk. F. Ahmed, G.-H. Rho, J.Y. Lee, S.J. Kim, H.-Y. Kim, Y.J. Jang, M.-W. Moon and K.-R. Lee, Nano-embossed structure on polypropylene induced by low energy Ar ion beam irradiation, *Surf. Coat. Technol.* 205, 104-108(2010).
120. A. Kosińska, J. Jagielski, D.M. Bielinski, O. Urbanek, M.Wilczopolska, M. Frelek-Kozak, A. Zaborowska, E. Wyszkowska, and I. Jozwik, Structural and chemical changes in He+ bombarded polymers and related performance properties, *J. Appl. Phys.* 132, 074701 (2022).
121. S. Yi and J. Kim, Bonding strengths at plastic encapsulant-gold-plated copper leadframe interface, *Microelectronics Reliability,* 40,1207-1214(2000).
122. M. White, The removal of die bond/epoxy bleed material by oxygen plasma, Proceedings 32nd IEEE Electronic Components Conference, 262,72-75(1982).
123. M. Strobel, C.S. Lyons and K.L. Mittal (Eds.) *Plasma Surface Modification of Polymers: Relevance to Adhesion*, CRC Press, Boca Raton, FL (1994).
124. P.K. Goyal, V. Kumar, R. Gupta, S. Mahendia, Anita and S. Kumar, Modification of polycarbonate surface by Ar^+ ion implantation for various opto-electronics applications, *Vacuum* 86, 1087-1091(2012).
125. H. M. Abourayana and D. P. Dowling, Plasma processing for tailoring the surface properties of polymers, in: *Surface Energy*, M. Aliofkhazraei (Ed.), pp. 123-152, IntechOpen Limited, London, UK (2015).
126. C.M. Chan, T.M. Ko and H.Hiroka, Polymer surface modification by plasmas and photons, *Surf. Sci. Reports* 24, 1-54(1996).
127. K.L. Mittal and W.-S. Lei (Eds.) *Laser Technology: Applications in Adhesion and Related Areas*, Wiley-Scrivener, Beverly, MA(2018).
128. E.S.Srinahdu, D.D. Kulkarni, D.A. Field, J.E. Harriss, and C.E.Sosolik, Multicharged ion processing for targeted nanostructure formation, *J. Appl. Phys.*,132, 155302-1,155302-6(2022).
129. F. F. Chen and J. P. Chang, *Lecture Notes on Principles of Plasma Processing*, Springer Science+Business Media, New York(2003).
130. R. Arumugam, E.S. Srinadhu, B. Subramanian and S. Nallani, β-PVDF based electrospun nanofibers–A promising material for developing cardiac patches, *Medical Hypotheses*, 122, 24-31 (2019).
131. R. Arumugam, V. Subramanyam, R.K. Chinnadurai, E.S. Srinadhu, B. Subramanian and S. Nallani, Development of novel mechanically stable porous nanocomposite (PVDF-PMMA/HAp/TiO2) film scaffold with nanowhiskers surface morphology for bone repair applications, *Mater. Letters*, 236, 694-696 (2019).
132. R. Arumugam, R.K. Chinnadurai, B.N. Subramaniam, B. Devaraj, V. Subramanium, E.S. Srinadhu and S. Nallani, Scalable novel PVDF based nanocomposite foam for direct blood contact and cardiac patch applications, *J. Mech. Behav. Biomed. Mater.*, 88, 270-280 (2018).
133. E.S. Srinadhu, Defect assisted growth of copper-silicide nanostructures on Si(100) and Si(111), Ph.D. thesis, Clemson University, Clemson, SC, USA (2016).

134. W.K. Choi, S.K. Koh and H.J. Jung, Surface chemical reaction between polycarbonate and kilo-electron-volt energy Ar^+ ion in oxygen environment, *J. Vac. Sci. Technol. A*, 14, 2366-2371 (1996).
135. R. Shyam, D.D. Kulkarni, D.A. Field, E.S. Srinadhu, D.B. Cutshall, W.R. Harrell, J.E. Harriss and C.E. Sosolik, First multicharged ion irradiation results from the CUEBIT Facility at Clemson University, *AIP Conf. Proc.*, 1640, 129-135 (2015).
136. R. Shyam, D.D. Kulkarni, D.A. Field, E.S. Srinadhu, D.B. Cutshall, J.E. Harriss, W.R. Harrell, and C.E. Sosolik, Encapsulating ion-solid interactions in metal-oxide- semiconductor (MOS) devices, *IEEE Trans. Nuclear Sci.*, 62, 3346-3352 (2015).
137. E.S. Srinadhu, D.D. Kulkarni, D.A. Field, J.E. Harriss and C.E. Sosolik, The effects of multicharged ion irradiation on a polycarbonate surface, *Radiat Effects Defects Solids*, 174, 205-213(2019).
138. F. Dehaye, E. Balanzat, E. Ferain and R. Legras, Chemical modifications induced in bisphenol A polycarbonate by swift heavy ions, *Nucl. Instrum. Methods Phys. Res.* 209, 103-112(2003).
139. S.K. Koh, S.K. Song, W.K. Choi, and H.J. Jung, Improving wettability of polycarbonate and adhesion with aluminum by Ar^+ ion irradiation, *J. Mater. Res.*, 10, 2390-2394 (1995).
140. K.L. Mittal and T. Bahners (Eds.), *Laser Surface Modification and Adhesion*, Wiley-Scrivener, Beverly, MA (2015).
141. K.L. Mittal and R. Jaiswal (Eds.), *Particle Adhesion and Removal*, Wiley-Scrivener, Beverly, MA (2015).
142. S. Lerouge, A.C. Fozza, M.R. Wertheimer, R. Marchand and L H. Yahia, Plasma sterilization: A review of parameters, mechanisms, and limitations, *Plasmas Polymers*, 6,175-188 (2001).
143. J.H. Young, New sterilization technologies, in: *Sterilization Technology for Health Care Facility*, 2nd edition, M. Reichert and J.H. Young (Eds.), pp. 228-235, Aspen Publishers, Gaithersburg, MD (1997).
144. W.A. Rutala, M.F. Gerden and D.J. Weber, Comparative evaluation of the sporicidal activity of new low-temperature sterilization technologies: Ethylene oxide, two plasma sterilization systems and liquid peracetic acid, *Amer. J. Infect. Control*, 26, 393-398 (1998).
145. M. Goldman, M. Lee, R. Gronsky and L. Pruitt, Oxidation of ultrahigh molecular weight polyethylene characterized by Fourier transform infrared spectrometry, *J. Biomed. Mater. Res.*, 37, 43-50 (1998).
146. M. Thomas and K.L. Mittal (Eds), *Atmospheric Pressure Plasma Treatment of Polymers*, Wiley-Scrivener, Beverly, MA (2013).
147. J.S. Cho and S.K. Koh, Enhanced adhesion between polycarbonate substrate and tin-doped indium oxide films by ion assisted reaction, *Nucl. Instrum. Methods. Phys. Res. B*, 175-177, 505-510 (2000).
148. E.S. Srinadhu, J.E. Harriss and C.E. Sosolik, Shape transitions of Cu_3Si islands grown on Si(1 1 1) and Si(1 0 0). *Appl. Surf. Sci.*, 465, 201–206 (2019).

149. J. Bhadra, N.J. Al-Thani and S Karmakar, Photo-reduced route of polyaniline nano fiber synthesis with embedded silver nanoparticles. *Arabian J Chem.* 12, 4848–4860(2019).
150. P Liu, J Yan and Z Guang, Recent advancements of polyaniline-based nanocomposites for supercapacitors, *J Power Sources.* 424, 108–130(2019).
151. J Bhadra, N.J. Al-Thani and N.K. Madi, Effects of aniline concentrations on the electrical and mechanical properties of polyaniline polyvinyl alcohol blends. *Arabian J Chem.*,10,664–672(2017).
152. S El-Sayed, T Abel-Baset and AA Elfadl, Effect of nanosilica on optical, electric modulus and AC conductivity of polyvinyl alcohol/polyaniline films. *Physica B,* 464,17–27(2015).
153. L Shubha and P Madhusudana Rao, Temperature characterization of dielectric permittivity and AC conductivity of nano copper oxide-doped polyaniline composite. *J Advanced Dielectrics,* 06,1650018(2016).
154. V Anju and SK Narayanankutty, Polyaniline coated cellulose fiber / polyvinyl alcohol composites with high dielectric permittivity and low percolation threshold. *AIP Advances* 6,015019(2016).
155. S Elashmawi and AA Menazea, Different time's Nd:YAG laser-irradiated PVA/Ag nanocomposites: structural, optical, and electrical characterization. *J Mater Res Technol.* 8,1944–1951(2019).
156. M Abdelhamied, A. Atta, A. Abdelreheem, A.Farag and M.El Okr, Synthesis and optical properties of PVA/PANI/Ag nanocomposite films. *J Mater Sci: Mater Electron.*, 31,22629-22641(2020).
157. Abdel Reheem, M.M. Ahmed, M.M. Abdelhamid and A.H.Ashour, Verification of high efficient broad beam cold cathode ion source. *Rev Sci Instrum.*, 87, 083302 (2016).
158. T. Hanawa, Metals and medicine, Materials Trans., 62, 139-148(2021).
159. E.H. Lee, M.B. Lewis, P.J. Blau and L.K. Mansur, Improved surface properties of polymer materials by multiple ion beam treatment. *J. Mater. Res.*, 6, 610-628 (1991).

8

Polymer Surface Modification by Charged Particles from Plasma Using Plasma-Based Ion Implantation Technique

Takeshi Tanaka[1*], Koji Kakugawa[1] and Katia Vutova[2]

[1]Hiroshima Institute of Technology, Miyake, Saiki-ku, Hiroshima, Japan
[2]Institute of Electronics, Bulgarian Academy of Sciences, Tzarigradsko, Chaussee, Sofia, Bulgaria

Abstract

Engineering polymers are characterized by favorable properties such as machinability, dimensional stability, nonflammability, and resistance to corrosion, most acids, solvents and heat. However, some of their surface properties including wear resistance, wettability, surface conductivity, etc., have to be improved in several applications. Possible ways of modifying advantageously their surface layers and improving their surface properties have been reported e.g., plasma surface modification, atmospheric pressure plasma modification, and plasma based ion implantation (PBII) and its deposition variant (PBII&D) have been used by many researchers.

Plasma-based ion implantation and deposition is known by a variety of names, acronyms, and trademarks, including, but not limited to, the following: plasma source ion implantation (PSII), plasma immersion ion implantation (PIII or PI^3), plasma ion implantation (PII or PI^2), plasma ion plating (PIP), plasma immersion ion implantation and deposition (PIIID), metal plasma immersion ion implantation and deposition (MePIIID), IonClad, plasma doping (PLAD), plasma ion immersion processing (PIIP).

Principle of the technique, generation of charged particles, equipment needed, and various factors affecting the outcome using this technique are reported.

Various effects produced by charged particles treatment of polymers regarding increase in wettability, increase in surface free energy, surface roughness, and generation of polar chemical groups are also presented.

**Corresponding author*: tanaka@cc.it-hiroshima.ac.jp

K.L. Mittal and A.N. Netravali (eds.) Polymer Surface Modification to Enhance Adhesion: Techniques and Applications, (329–364) © 2024 Scrivener Publishing LLC

Furthermore, various surface analysis techniques to analyze charged particles treated polymers are reported, and adhesion data showing enhancement of adhesion of treated polymers are presented.

Finally, the future prospects of this technology are reported.

Keywords: Plasma, polymer, surface modification, plasma based ion implantation, adhesion

8.1 Introduction

Engineering polymers are characterized by favorable properties such as machinability, dimensional stability, nonflammability, and resistance to corrosion, most acids, solvents and heat. However, some of their surface properties including wear resistance, wettability, surface conductivity, etc., have to be improved in several applications. Possible ways of modifying advantageously their surface layers and improving their surface properties are plasma surface modification [1, 2], atmospheric pressure plasma treatment [3], laser surface modification [4, 5] and plasma based ion implantation (PBII) and its deposition variant (PBII&D) [6].

Plasma-based ion implantation and deposition is known by a variety of names, acronyms, and trademarks, including, but not limited to, the following: plasma source ion implantation (PSII), plasma immersion ion implantation (PIII or PI^3), plasma ion implantation (PII or PI^2), plasma ion plating (PIP), plasma immersion ion implantation and deposition (PIIID), metal plasma immersion ion implantation and deposition (MePIIID), IonClad, plasma doping (PLAD), plasma ion immersion processing (PIIP) [7].

In the review [7], a brief introduction to the physical background and history is given as follows

Before the physics of PBII and PBIID is explained in some detail, a look at the historical development is warranted. The physical roots of plasma based ion implantation can be traced back several decades, and in particular to simulation work in the 1960s by Widner *et al.* [8] who investigated the generation of ion acoustic waves in plasma by applying negative voltage pulses to electrode plate. Another body of work was ion implantation research by Adler and Picraux [9] at Mission Research, in the early 1980s. They used metal ions from a pulsed metal plasma source and pulse-biased the substrate holder up to -80 kV, achieving ion implantation without the use of ion extraction in an ion source. The historically important impetus came with the work of Conrad *et al.* [10] at the University of Wisconsin, in 1987, when the PBII principle was clearly spelled out and applied to ion treatment of 3-D components. Conrad and his group made major contributions to the physics of the high-voltage sheath [11] and materials modification using the PSII principle [12, 13] (initially, the acronym PSII was preferred). This stimulated research at several laboratories throughout the world [14–21], which eventually led to the start of the PBII-Workshop Series in 1994 [22].

Plasma based ion implantation (PBII) is a rapidly advancing technique that is highly suitable for surface treatment of three-dimensional workpieces [10, 11]. In the PBII process, high-voltage negative pulses of 2~30 μs duration are applied to a target immersed in plasma. The high negative voltage attracts positive ions from the surrounding plasma to the target, and the ions strike all portions of the target simultaneously at normal incidence, making the technique ideal for treating three-dimensional shapes. Furthermore, the process is relatively simple; thin films of high quality can be produced at low temperature (room temperature), including an ion implantation layer that does not come off. A wide range of plasma sources have been employed by many authors, including filament discharge, radio-frequency (rf) discharge, microwave discharge, metal arc discharge, and more recently, a pulsed inductively coupled plasma (ICP) source [23, 24]. Pulsed plasmas are used for PBII processing, because high-density plasma created within the pulse duration can be obtained at low average power. A carbon thin layer of high quality is typically produced by ionizing acetylene (C_2H_2) in an rf field and attracting positive ions by high negative voltage.

Also, many authors have investigated thin diamond-like carbon (DLC) layers deposited on Si wafers or metals, such as Al, Cu, and Fe, which are conducting materials. These studies have been undertaken mainly for the purpose of attaining surface coating with DLC in order to improve friction, erosion, and mechanical properties.

8.2 Overview of Literature About Polymer Surface Modification by Charged Particles from Plasma Using Plasma-Based Ion Implantation [6]

It was reported that more than 150 publications on PBII and PBII&D of polymers [25–169] were found [6]. The majority of them deal with real engineering polymers. However, a number of articles dealing mainly with bioengineering/biomedical aspects are also considered here, since a part of information included can be of relevance also in the engineering field.

Figure 8.1 demonstrates a rapid increase in the cumulative number of publications appearing in the 13 years. As a subset, the evolution of the number of publications dealing with the deposition variant of the method is also shown in this figure. Starting from 2004, the share of publications describing deposition type treatments is about one-fourth of the total number of related papers.

Figure 8.2 shows the distribution of publications according to the polymers studied. Clearly, the most intensely studied polymers are poly (ethylene terephthalate) (PET) [34, 35, 73, 76–78, 86–124] and the various forms of polyethylene (PE) [53, 68–107]. These are followed by polyimide (PI) [35, 121–123, 125–135], polystyrene (PS) and its copolymers [34, 121, 123, 138–149], polytetrafluoroethylene (PTFE) and poly (vinylidene difluoride) (PVDF) [26, 74, 150–155], polycarbonate (PC), including poly(bisphenol A carbonate) and Columbia Resin No. 39 (CR-39) [29–36], poly(etheretherketone) (PEEK) [35, 79–85], poly(dimethyl-siloxane) (PDMS) and silicones [34,37–41], poly(methyl methacrylate) (PMMA) and other polyacrylates [33, 36, 136–137], polyurethane (PU) and its copolymers [124, 163–166], polyamide (PA) and its copolymers [25–28], and poly(vinyl chloride) (PVC) [34, 167–169]. Furthermore, a few other PBII-treated polymers can also be mentioned like amorphous

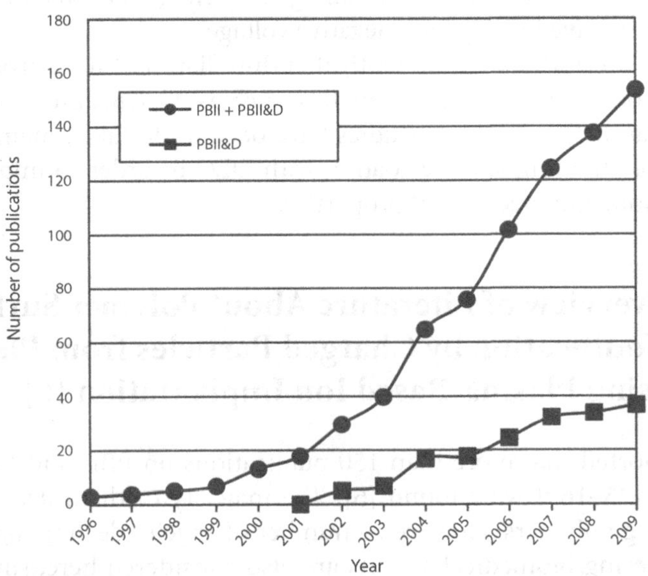

Figure 8.1 Publications on PBII and PBII&D of polymers vs. time.

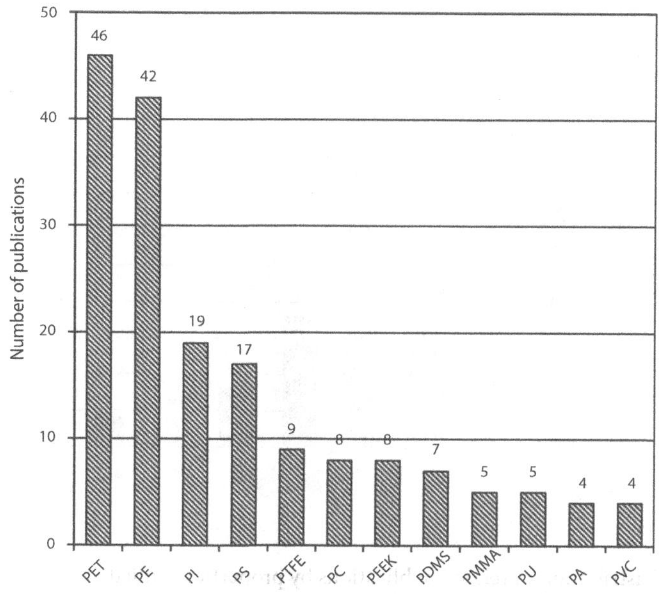

Figure 8.2 Distribution of related publications by polymers modified.

polyolefins (APOs) [156, 157], cellulose nitrate [158], ethylene–propylene–diene monomer (EPDM) rubber [40], ethylene–vinyl alcohol copolymer (EVOH) [159, 160], poly(butylene succinate) (PBSu) [161], poly(ethylene naphthalene) [34], polypropylene (PP) [78, 122, 162] and poly(phenylene oxide) (PPO) and its modified versions [28, 121, 123].

Figure 8.3 demonstrates the distribution of publications according to the modified properties studied. The overwhelming majority of the papers report results of surface chemical investigations. Further characteristics studied are (in decreasing order): topological and morphological ones [26–30, 35, 37–40, 44–46, 48, 49, 51, 57, 60, 63, 68, 72–74, 78, 80, 83, 92, 95–100, 108, 110, 119–121, 123, 124, 127, 128, 130, 132, 135, 141–144, 146, 151, 154–157, 159–161, 163, 164, 169], wettability and hydrophobic recovery [25, 32–34, 36, 38, 42, 45, 57, 60, 65, 68, 71, 72, 76, 81, 82, 84, 93, 95–99, 101, 109, 117, 118, 126, 134, 139, 141, 143, 144, 147, 148, 150, 151, 153–155, 159–161, 163, 167, 171], biomedical properties (like attachment of proteins, platelets or cells, cell growth, antibacterial property, controlled drug release, patterning, etc.) [38, 39, 55–57, 59, 60, 62–64, 65, 68–70, 72, 74, 81, 87, 90, 93, 96, 97, 99, 100, 108, 109, 124, 142, 145, 146, 153, 155, 161, 163, 165, 166, 169], mechanical and tribological properties (like hardness, elastic modulus, friction, wear, stress, etc.) [26, 27, 29, 30, 36, 42–44, 46, 48,

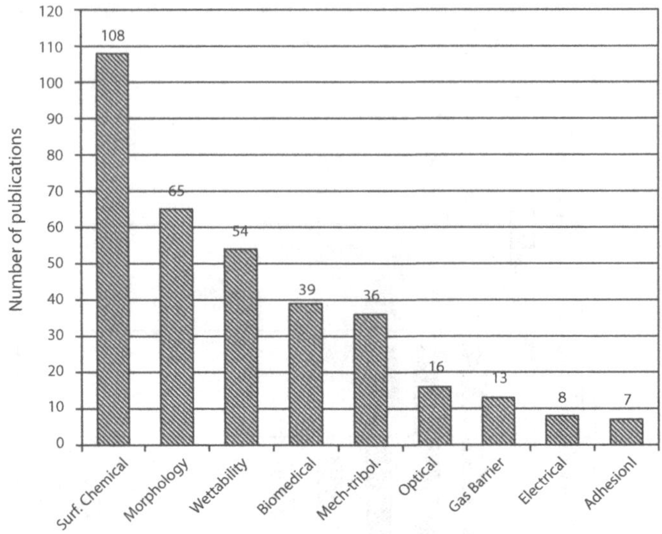

Figure 8.3 Distribution of related publications by properties studied.

50–53, 58, 61, 73, 75, 80, 83, 85, 98, 107, 111, 112, 115, 117, 125, 152, 156, 157], optical properties [36, 45, 82, 85, 112, 116, 120, 135, 140–142, 144, 146, 152, 162], gas barrier properties (to O_2, CO_2, H_2O, etc.) [75, 86, 88, 89, 91, 92, 94, 104, 105, 107, 112, 116, 122, 159, 160], electrical properties [110, 114, 121, 123, 126, 127], and adhesion to various materials [26, 73, 75, 128–130, 132, 166].

Finally, several theoretical and experimental studies deal with the processing conditions of polymers and insulating materials in general [31, 103, 106, 113, 139, 164, 170–178]. These include studies on the effects of applied voltage, plasma density and pulse duration [31, 139, 170, 176] on application of conducting mesh-assisted processes [171–175] and of various deposited and embedded conducting layers [177–178].

8.3 Principle of PBII: Advantages and Limitations [7]

In the PBII technique, substrates are immersed directly in the plasma and biased negatively with high-voltage pulses. The plasma conformally surrounds the substrate; its whole surface is implanted at the same time (nonuniformities at small features are discussed later). PBII eliminates the intermediate stages of beam extraction, focusing, scanning, as well as substrate or wafer manipulation.

The PBII technique has many advantages over the conventional implantation, such as high ion current density and relatively short processing time (e.g., minutes), high dose rates (e.g., 10^{14} cm^{-2} s^{-1}), wide ion energy range (up to about 100 keV), large implant areas (100s of cm^2), and treatment of 3-D workpieces with complex shapes.

However, a few limitations and disadvantages of PBII in comparison to beam-line ion implantation must be pointed out.

1. The PBII concept does not allow for ion-mass separation. All ions at the plasma sheath edge are accelerated and implanted, regardless of ion mass and charge state. In the case of molecular gases, a mixture of molecular and atomic ions is present, which necessarily implies at least two peaks in the energy (per mass) distribution.
2. Even for atomic gases, the ion energy distribution is not mono-energetic and it depends on the gas pressure and bias pulse shape.
3. The production of secondary electrons under ion impact and their acceleration in the sheath leads to large loads on the pulse modulator, making the process less economical and additionally causing undesirable X-rays when the sheath voltage exceeds about 20keV, and the issue becomes very important when the voltage approaches the 100 keV range.
4. Biasing of dielectric objects is not possible, and therefore, the treatment of insulators is limited to thin sheets or requires special setups with meshes.

Despite these disadvantages, PBII appears particularly attractive for certain processing steps in the automotive, semiconductor, display, and biomedical industries.

8.4 Equipment Needed

Figure 8.4 is a schematic diagram of the PSII apparatus. The chamber dimensions are 450 mm in height, 590 mm in width, and 470 mm in depth. The rf antenna for the ICP plasma source (North Star Res. Inc.) was wound on the inside of the upper lid, with one end grounded. The antenna itself is a 5-turn copper coil, approximately 250 mm in diameter and operating at

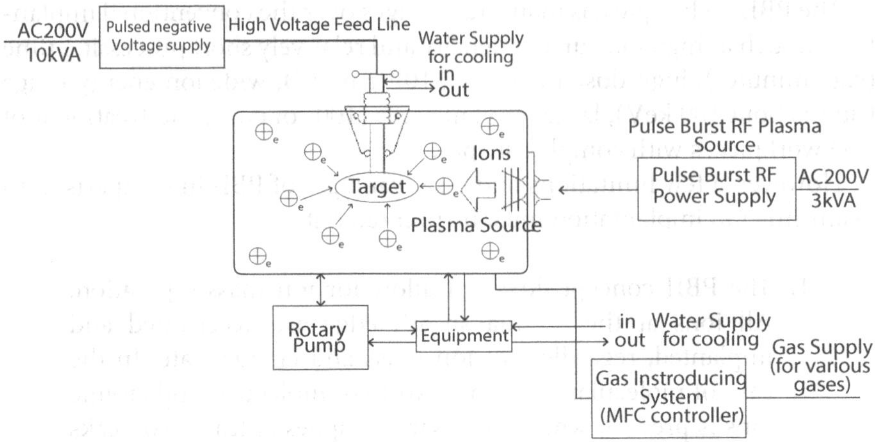

Figure 8.4 Experimental PSII apparatus.

222 kHz. The PET film targets were supported in the center of the chamber by a conducting copper rod that is isolated from the chamber. A pulsed negative voltage of up to -15 kV was applied to the target in pulse widths of 2~30 μs by a high-voltage pulse modulator (Kurita Inc.), with a maximum current of approximately 8 A. The timing sequence for applying the high-voltage pulse was controlled by the computer control system. The target chamber was evacuated by a diffusion pump to a base pressure of 4.0×10^{-4} Pa. The gas pressure was maintained at about 1.0-5.0 Pa during ion implantation.

The light absorption (200~900nm) of all samples was measured by a visible/ultraviolet spectrophotometer, and surface morphology of the film was observed by scanning electron microscopy (SEM). The chemical bonding features (characteristics) of the carbon film were examined by X-ray photoelectron spectroscopy (XPS) and laser Raman spectrometry. The Raman spectra were measured by use of 514.5nm wavelength light from an Ar^+ laser. The sample PET films were set on the quartz glass.

The Raman spectra were fitted with 2 Gaussian functions corresponding to the D and G bands, after the baseline was made flat in the range of 900~1.800 cm^{-1}. The integrated C-H stretch absorption of the DLC layers was measured by Fourier transform infrared spectroscopy (FT-IR). The peaks corresponding to the C-H bonding were carefully observed qualitatively so as to determine the differences between DLC layers deposited under different conditions.

• Conditions
Typical conditions employed for the PSII experiments were as follows:

- Pulse width : 10μs
- Pulse delay (delay time from start of rf pulse) : 50μs
- Pulse repetition : 300 pps
- Pulse voltage (peak) : -10kV
- Pulse current (peak) : 8A
- Gas : pretreatment N_2, Ar,
  DLC deposition C_2H_2, C_2H_2+Ar (30%)
- Gas pressure : 2Pa for pretreatment
  5Pa for DLC deposition
- Deposition time : 30 minutes
- PET film thickness : 38μm
- rf discharge : 5.5kV, 3.4A (strong discharge), 0.5kV, 0.2A (weak discharge)
- rf pulse : 20μs (~50μs after glow)

The PET film was set on the copper rod, and attached completely without any intervening space. The treatment area of the PET film was a circle of 50mm diameter. The copper rod was set on the stainless steel substrate, to which the negative high voltage was applied. In order to determine the effect of rf discharge, rf voltage was varied from 0.5kV to 5.5kV.

• Discharge properties
In this study, two drastically different discharge operations were performed in both the pretreatment and the DLC deposition experiments. One operation was strong ICP (rf) plasma discharge, with rf voltage set to the maximum value of 3.5kV, and the other was weak ICP plasma discharge, with rf voltage set to 0.5kV.

Figure 8.5 shows typical waveforms of C_2H_2 discharge for both strong and weak rf discharge. When rf discharge is strong, the negative voltage pulse rises rather slowly; the flat, top portion of the waveform cannot be reached within 10μs. The current pulse increases rapidly and decreases slowly as voltage increases. In contrast, when rf discharge is weak, the negative voltage pulse rises quite rapidly and shows a very narrow peak (~1μs) followed by the flat, top portion of the waveform. Current also rises rapidly and shows a flat, top region. These findings indicate that plasma density is quite high under strong rf discharge, lowering plasma impedance. When the negative voltage pulse is applied to this strong rf discharge plasma, pulse current becomes higher with relatively low voltages, increasing voltage

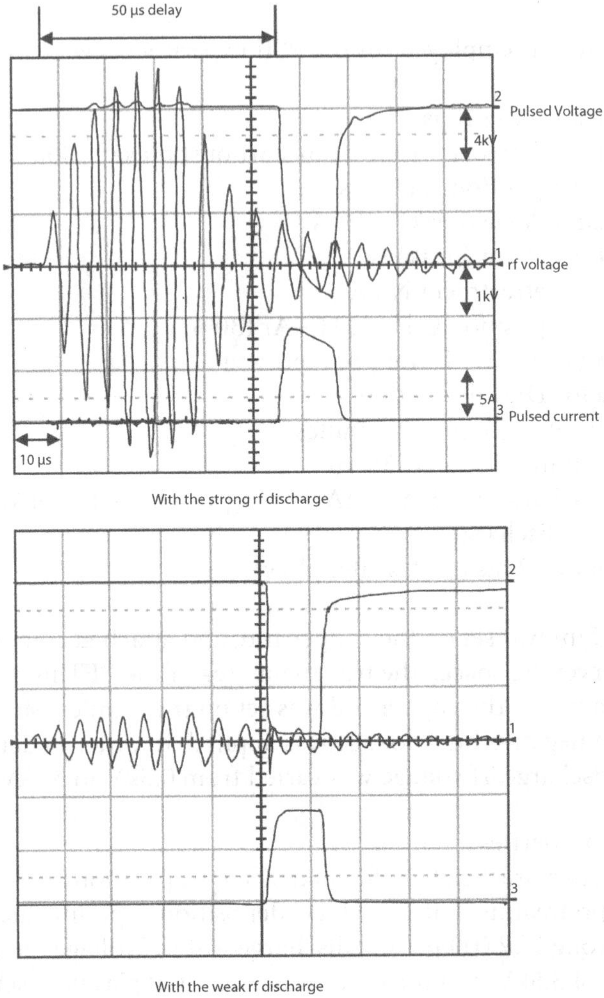

Figure 8.5 Waveforms of the discharges with C_2H_2.

rise-time. Plasma impedance is much lower under weak rf discharge, thus increasing pulse voltage.

Moreover, when rf discharge is weak, the pulse voltage applied to the substrate is stable, inducing rather high ion current to the substrate even when the amount of plasma in the vicinity of the substrate is less before the pulsed voltage is on. This indicates that the negative high voltage creates the plasma present in the vicinity of the substrate, which can be explained by the plot of voltage vs. current in the flat, top region. Figure 8.6 shows this relation for the case where Ar is used.

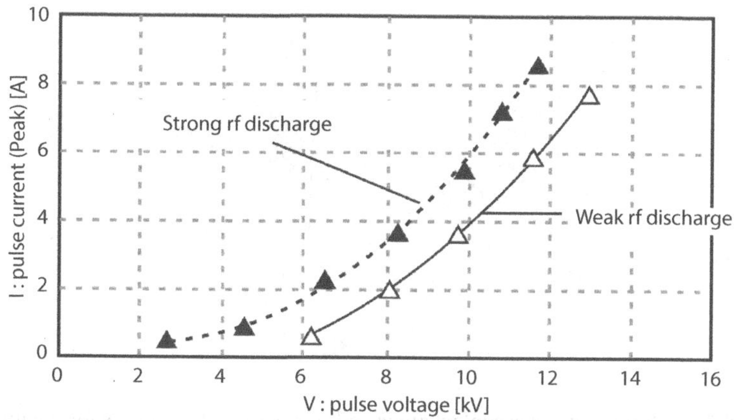

Figure 8.6 Characteristic I-V curves for the pulse. ▲: Strong rf discharge, △: Weak rf discharge.

Pulse current is higher for strong rf discharge than for weak rf discharge. In both cases, however, the following relation holds between pulse current I and pulse voltage V.

$$I \propto V^2 \tag{8.1}$$

If the plasma around the substrate is created mainly by the rf source, equation (8.1) should approximate I ∝ V (Ohm's law). Therefore, as pulse voltage V increases, plasma density increases, so that the ion (pulse) current obeys equation (8.1) for this experiment. For a given value of pulsed current, weak rf discharge yields higher pulsed voltage.

8.5 Factors Influencing the Outcome/Results

8.5.1 Wettability Studies [6]

For the untreated PC, the values of contact angles obtained with water and diiodomethane were $\Theta_w = 85.0°$ and $\Theta_{dim} = 28.9°$, respectively. The total surface free energy, calculated by the method of Owens and Wendt [179] was $\gamma_{tot} = 46.1$ mJm^{-2}, while its polar and dispersion components were $\gamma_{pol} = 1.4$ mJm^{-2} and $\gamma_{dis} = 44.7$ mJm^{-2}. The corresponding data for PA were as follows: $\Theta_w = 60.3°$, $\Theta_{dim} = 33.3°$, γ_{tot}=54.0, γ_{pol}=11.2 and $\gamma_{dis} = 42.8$ mJ m^{-2}. These data are in agreement with the available literature values [180, 181].

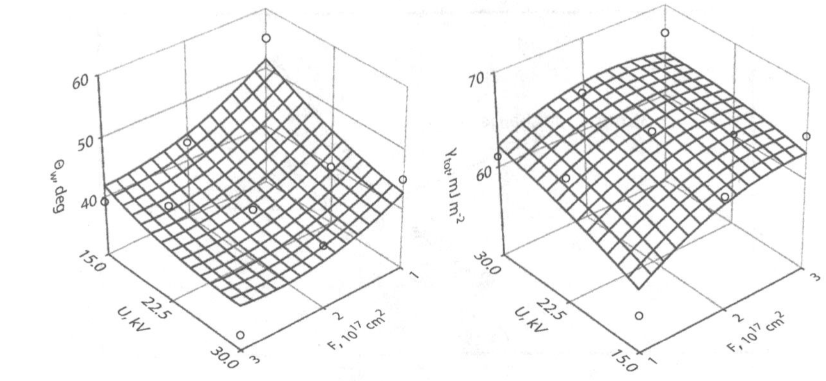

Figure 8.7 Evolutions of water contact angle and total surface free energy for the nitrogen PBII-treated PC.

As a result of PBII treatment of PC, Θ_w decreased to values ranging from 32.9° to 55.7° and Θ_{dim} increased to 37.2 – 47.1°. The corresponding values of surface free energy and its components varied in the following ranges: γ_{tot} between 54.7 and 67.4 mJ m^{-2}, γ_{pol} between 14.7 and 27.6 mJ m^{-2} and γ_{dis} between 39.2 and 41.0 mJ m^{-2}. For the PBII-treated PC, the wettability and surface free energy depended mainly on voltage and fluence. The evolutions of Θ_w and γ_{tot} are shown in Figure 8.7, which clearly testify to the positive correlation of hydrophilic property with voltage and fluence. In agreement with this, $\Theta_{w,min}$ and $\gamma_{tot,max}$ were obtained at U_{max} and F_{max} applied. U_{max} (30 kV in Figure 8.7) is the maximum value of acceleration voltage (U). F_{max} (= 3 × 10^{17} cm^{-2} in Figure 8.7) is the maximum value of particle fluence (F). The wettability increased also for the PBII-treated PA, but the data exhibited a higher scatter (probably because Θ_w values were relatively small), and therefore are not discussed here.

8.5.2 Surface Free Energy and Wettability [100]

The total solid surface free energy and its components are determined using double-distilled water and diiodomethane as test liquids. This determination is based on Young's equation [182] that relates the surface tension of a liquid in equilibrium with its vapor. Surface free energies together with the contact angles of water and diiodomethane are listed in Table 8.1. After C_2H_2 PIII-D, the polar component γ_p of surface free energy increases from 2.4 mJm^{-2} to 12.3 mJm^{-2} and γ_p/γ_d increases from 0.06 to 0.35. The wettability of the PET with the PLC coating is improved by C_2H_2 PIII-D.

Table 8.1 Contact angle and surface free energy of the untreated and the modified PET film.

	Contact angle (°)		Surface free energy (mJm^{-2})			
Substrate	**Water**	**Diiodomethane**	γ_s	γ_d	γ_p	γ_p/γ_d
PET	83.5	30.2	43.3	40.9	2.4	0.06
PIII-PET	64.8	32.4	47.1	34.8	12.3	0.35

8.5.3 Surface Morphology

The surface morphology of the films was observed by SEM. Figure 8.8(a) depicts an untreated PET film, and Figure 8.8(b) depicts a PET film pre-treated with N_2 ions under weak rf discharge. Further, Figure 8.8(c) depicts a PET film after the thin carbon layer was deposited with C_2H_2 under strong rf discharge, and Figure 8.8(d) depicts a PET film treated in the same manner as in Figure 8.8(c) except under weak rf discharge. Finally, Figure 8.8(e) depicts a PET film after the thin carbon layer was deposited with C_2H_2 + Ar (20%) under weak rf discharge.

The surface of the untreated PET film has many cracks and is found not to be smooth. After pretreatment by PSII with N_2, the surface of the PET film became much smoother and no cracks were found, except for dots scattered on the surface. Carbon layers covering the surface of the PET films were observed from Figures 8.8(c)~8.8(e), and the carbon layer surface produced from C_2H_2 under weak rf discharge was found to be smoother than that produced under strong rf discharge. Also, the carbon layer produced from C_2H_2 + Ar (20%) has a slightly rough surface. The reason for this is thought to be Ar^+ ions attacking the surface of the carbon layer during the process.

In all the cases shown in Figures 8.8(c)~(e), the surface of the carbon/PET film is smooth and exhibits a closely packed structure. These characteristics may reduce the oxygen permeability of the film.

Figures 8.9(a) and (b) contrast the 535 mm^2 surface topography as measured by AFM without any filtering [100]. The surface of the untreated PET sample exhibits a rectangular structure. Pinnacle-like structure can be observed in the image of PIII-treated PET. Each AFM image is analyzed in terms of surface average roughness. The average roughness (Ra) of the PET surface modified by acetylene PIII-D decreases from 58.9 nm to 11.4 nm. This result indicates that the surface morphology of the PET film is significantly affected by C_2H_2 PIII-D and a carbon layer is deposited on the PET film.

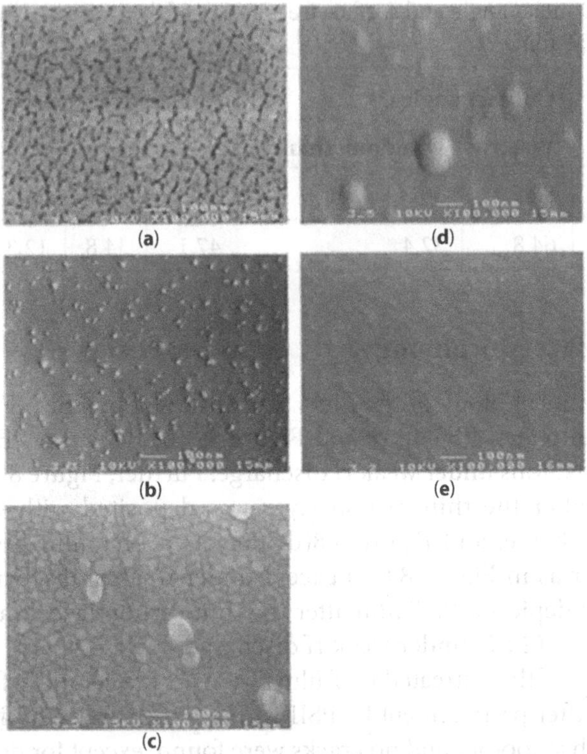

Figure 8.8 SEM pictures of various surfaces (X100,000). (a) Untreated PET film. (b) Treated with N_2 rf discharge. (c) C_2H_2 with strong rf discharge. (d) C_2H_2 with weak rf discharge. (e) C_2H_2 + Ar (20%) rf discharge.

8.5.4 XPS Spectra

The thin carbon layers deposited on the PET films were examined by XPS in order to identify the chemical bonding structure. As shown in Figure 8.10, the XPS spectra of Cls, representing the interatomic bonds of the carbon layer, could be resolved into three bonds: C-C (sp^3 bonds), C=C (sp^2), and C=O [183].

Figure 8.10 shows three XPS spectra for different conditions.

- Case 1 : Carbon films deposited from C_2H_2 under strong rf discharge.
- Case 2 : Carbon films deposited from C_2H_2 under weak rf discharge.
- Case 3 : Carbon films deposited from C_2H_2+Ar (20%) under weak rf discharge.

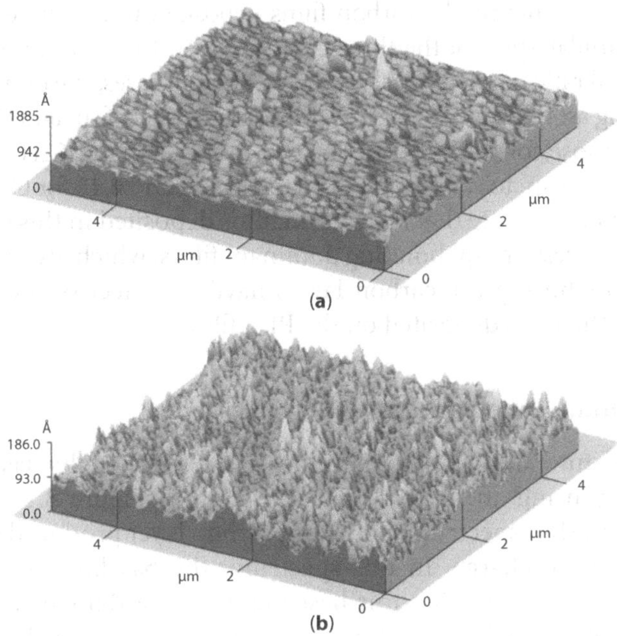

Figure 8.9 AFM photomicrographs of (a) PET control and (b) PIII-PET.

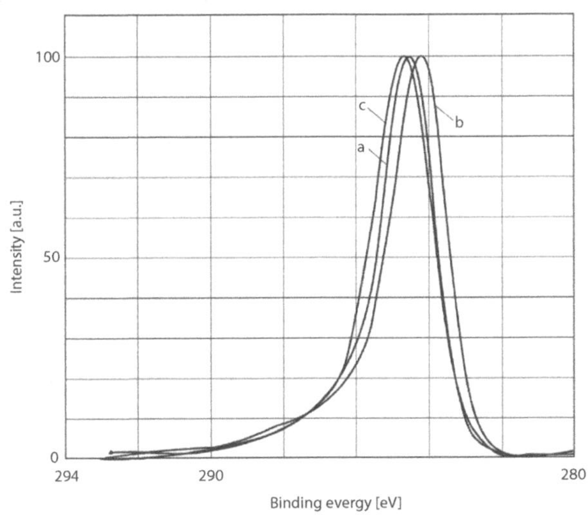

Figure 8.10 XPS spectra of carbon films. (a) C_2H_2 with strong rf discharge. (b) C_2H_2 with weak rf discharge. (c) C_2H_2 + Ar (20%) rf discharge.

As seen in the figure, the carbon films deposited under these conditions show very similar spectra; the three cases are seen to show almost identical compositional ratios of C-C, C=C, and C=O. For all cases, the dominant peak is sp^2 bonding around 284.3~284.7 eV with small peaks of sp^3 and C=O, which are close to those for ICP PBII performed using CH_4 and C_7H_8 gases [184, 185]. The ratio of sp^3 to sp^2 bonds is 0.13 for Case 2 and 0.45 for Case 3 [185].

These results indicate that the carbon films deposited in this experiment by the PSII process are sp^2 bonding dominant films, which are graphite-like films. Ar ions hitting the carbon layers have the effect of increasing sp^3 structure in the DLC deposited on the PET film.

8.5.5 Raman Spectrometry [186–188]

Crystal structure of the carbon films was examined by laser Raman Spectrometry using an Ar^+ ion laser (514.5mm). Figure 8.11 shows the typical Raman shift spectrum of the thin carbon layer produced from C_2H_2 under weak rf discharge. In this spectrum, the baseline increases with wavenumber. By subtracting this baseline from the Raman shift spectra, the spectra revealing the crystal structure were obtained for three cases, as described in the section titled XPS spectra (Case 1~Case 3).

Figure 8.12 shows the Raman shift spectra with flat baseline. The spectrum for the carbon layer produced from C_2H_2 under strong rf discharge exhibits small, sharp peaks at 1300, 1600, and 1700 cm^{-1}, which are the

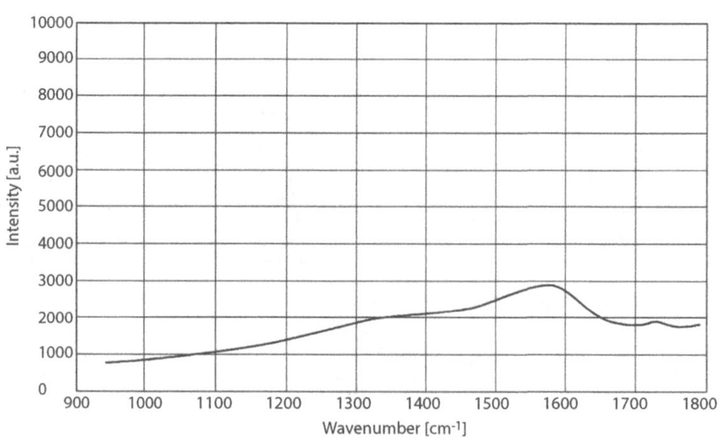

Figure 8.11 Raman shift spectrum of the thin carbon layer produced from C_2H_2 with weak rf discharge.

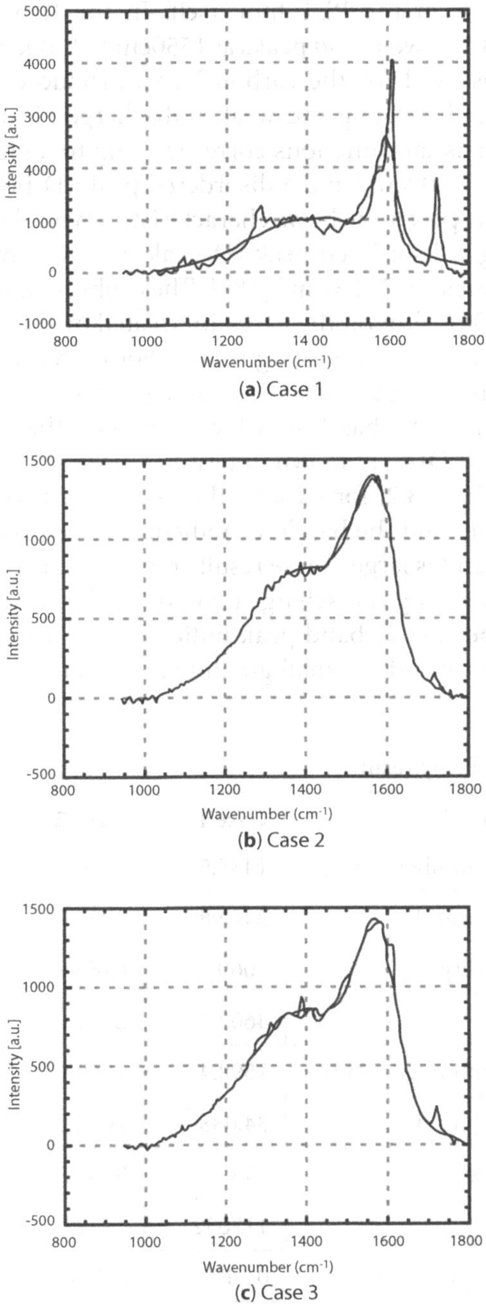

Figure 8.12 Raman shift spectra. (a) Case 1: C_2H_2 with strong rf discharge. (b) Case 2: C_2H_2 with weak rf discharge. (c) Case 3: C_2H_2 + Ar (20%) rf discharge.

characteristic peaks of the PET film itself. In the Raman shift spectra shown in Figure 8.12, two broad peaks at 1550cm^{-1}~1600cm-1 and at about 1400cm-1 are observed for the carbon layers produced from C_2H_2 and C_2H_2+Ar (20%) under strong and weak rf discharges. These spectra were fitted with two Gaussian functions corresponding to a graphite-like peak (G-bond) at 1550~1600cm^{-1} and a disordered peak (D-bond) at 1400cm^{-1} [186]. The Raman spectrum shows characteristics typical of diamond-like carbon, including a disordered peak (D-peak) at 1403 cm^{-1} and a graphite-like peak (G-peak) at 1553 cm^{-1} [183]. These observations confirm formation of a thin DLC film on the PET film treated by the PSII process.

Table 8.2 shows the spectra fitting results. For both the D-peak and the G-peak, peak wavenumber, deviation, and peak intensity are clarified. Also, the gradient of the baseline, which expresses the concentration of C-H bonding in the DLC, is shown in terms of arbitrary units.

As shown in Table 8.2, for Case 1, the D- and G-band peaks shift to higher wavenumber and the baseline gradient is large. Also, the area ratio of D-band to G-band is large. These results reveal that H concentration (as C-H) is higher for strong rf discharge than for weak rf discharge. Also, the ratio of D-band peak to G-band peak indicates that the DLC structure is more strongly dominated by small graphite crystals in the case of weak rf

Table 8.2 Spectra fitting results.

	Data name	**Case 1**	**Case 2**	**Case 3**
D-peak	Peak wavenumber [cm^{-1}]	1435.5	1400.6	1406.4
	Deviation [cm^{-1}]	244.98	194.87	207.51
	Peak Intensity	1060	816.99	870.34
	Integral	460269	282187	320113
G-peak	Peak wavenumber [cm^{-1}]	1598.4	1569.7	1574.5
	Deviation [cm^{-1}]	34.088	70.772	65.853
	Peak Intensity	1937	998.73	972.75
	Integral	117032	125281	113541
Ratio	I_D/I_G (peak point ratio)	0.547	0.818	0.895
	S_D/S_G (peak area ratio)	3.933	2.252	2.819
	Baseline Gradient [a. u.]	6.7989	1.2206	1.0571

discharge. Moreover, the baseline gradient and the area ratio of D,G bands decrease for the case of C_2H_2+Ar (20%) (Case 3).

8.5.6 FT-IR Measurements

Chemical bonding of C-H in the DLC was measured by FT-IR spectroscopy using the surface reflection method, since the PET film also has C-H bonding. Figure 8.13 shows the IR spectra from PET films with and without DLC films, as well as for DLC films produced solely from C_2H_2 and those produced from C_2H_2 and Ar, both under weak rf discharge.

In the figure, the peaks of sp^3-C-H and sp^2-C- H_2 are observed clearly for the case of C_2H_2+Ar (20%), and these peaks are smaller for the case of C_2H_2. These data reveal qualitatively that the C-H bonds were formed in greater number with Ar than with C_2H_2 only discharge. This could be due to Ar ions hitting the carbon layer to break the chemical bonding of C-O, C-C, and C=C to create C-H bonding.

However, we should note that the hydrogen atoms present in the DLC are not necessarily chemically bonded to carbon atoms; some H atoms or H_2 molecules could be present in the DLC without bonding to other atoms.

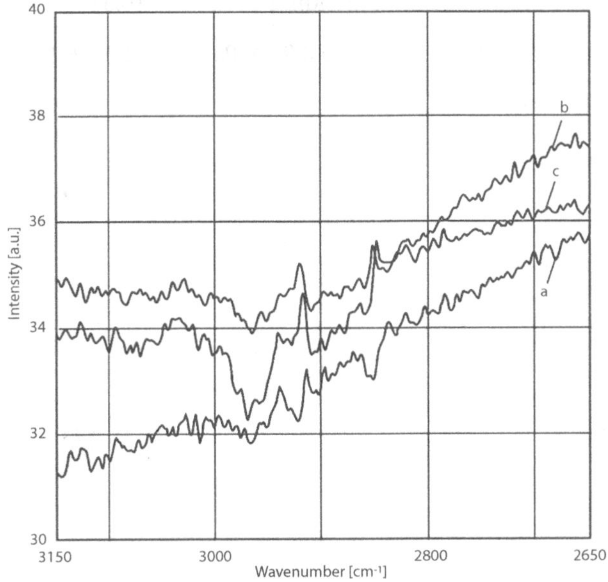

Figure 8.13 FT-IR spectra by reflection method. (a) Untreated PET film. (b) C_2H_2 with weak rf discharge. (c) C_2H_2 + Ar (20%) with weak rf discharge.

8.6 Results Showing Adhesion Improvement after PBII Treatment

In Table 8.3, we show the results of the tensile tests for the examined material pairs with treated and untreated base plates. PBII treatment had no effect in the case of PA6-t/ABS specimens and had a negative effect on bonding in ABS-t/ABS specimens. We found that PBII treatment significantly improved bonding in ABS-t/PA6 specimens (there was no bonding without PBII treatment and bonding strength was 12 MPa with PBII treatment).

Table 8.3 Summary of the effect of PBII treatment on bond strength for the different combinations of acrylonitrile-butadiene-styrene resin (ABS) and polyamide 6 (PA6).

The structural part of the specimen		**Overmolded rib**	
		PA6	**ABS**
Base plate	PA6-u	39.62 ± 0.56 MPa	No bonding
	PA6-t	37.26 ± 0.81 MPa	No bonding
	ABS-u	No bonding	16.31 ± 2.87 MPa
	ABS-t	12.76 ± 0.95 MPa	11.34 ± 2.19 MPa

u : untreated t : treated

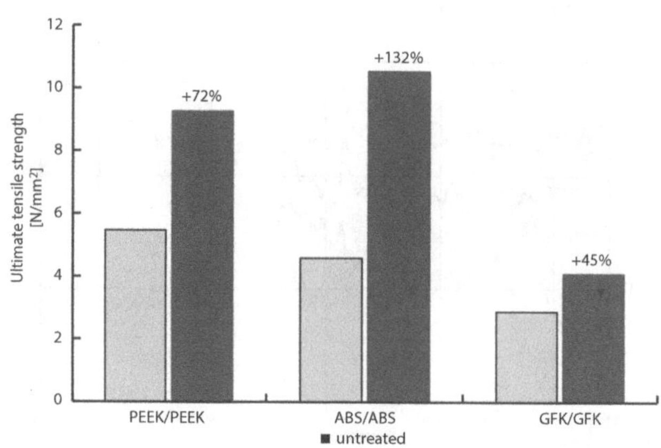

Figure 8.14 Ultimate tensile strength [N/mm²] for poly(etheretherketone) resin (PEEK) / PEEK, ABS / ABS, and GlasfaserverstärkterKunststoff in German (GFK) / GFK.

Figure 8.14 shows a comparison of the pull-off bond strength of bonding untreated plastics and the same plastics with PBII treatment with 2K epoxy adhesive. The comparison shows a significant improvement in bond strength [189].

8.7 Prospects

In recent years, the density and speed of integrated circuits have increased significantly. In order to support miniaturization and high-speed operation of individual devices in integrated circuits, device structures, semiconductor manufacturing equipment, semiconductor processes, and materials used are continuously being developed. In addition, semiconductor process modeling, high-precision equipment control, and high-purity materials have progressed, and once the process was determined, software that can simulate the device structure (such as Technology Computer Aided Design (TCAD)) has appeared. Furthermore, the use of material informatics knowledge is being investigated to derive the optimal parameters in a short time for optimizing the increasing number of parameters.

If the technique developed for the production of the above-mentioned integrated circuit and other materials and techniques developed are applied to the adhesiveness of polymers using plasma, it will be possible to clean the polymer surface, modify the surface, and alter the surface structure. Further improvement can be expected by using thin film deposition.

Also, in order to combine various materials and experimental parameters, find new functions, and reduce the resources required for experiments, the introduction of materials informatics and various simulations will be necessary.

In addition, we believe that the introduction of data science insights into the data obtained for complex polymer surface behavior will dramatically improve these techniques.

8.8 Summary

Engineering polymers are characterized by favorable properties such as machinability, dimensional stability, non-flammability, corrosion resistance, and resistance to most acids, solvents, and heat. However, some of the surface properties including abrasion resistance, wettability, surface conductivity, etc. need to be improved for some applications. Plasma surface modification, atmospheric pressure plasma modification, laser surface modification, and plasma-based ion implantation (PBII) and its deposition variant (PBII&D) are used to beneficially modify their surface layers and improve their surfaces. Many researchers have reported possible ways to improve the properties.

This chapter reports on the principles of plasma-based ion implantation and deposition technique, required equipment, discharge characteristics, etc. Various effects caused by charged particle treatment of polymers in terms of increased wettability, increased surface free energy, surface roughness, and generation of polar chemical groups are discussed.

In addition, various analytical techniques, X-ray photoelectron spectroscopy, Raman spectroscopy, and FT-IR spectroscopy have been reported to analyze charged particle-treated polymers. We also presented an example showing enhanced adhesion of the polymer.

Finally, we report on future prospects for this technique.

References

1. M. Strobel, C. S. Lyons and K. L. Mittal (Eds.) *Plasma Surface Modification of Polymers: Relevance to Adhesion,* CRC Press, Boca Raton, FL (1994).
2. K. L. Mittal (Ed.) *Polymer Surface Modification: Relevance to Adhesion, Vol. 5,* CRC Press, Boca Raton, FL (2009).
3. M. Thomas and K. L. Mittal (Eds.) *Atmospheric Pressure Plasma Treatment of Polymers: Relevance to Adhesion,* Wiley-Scrivener, Beverly, MA (2013).
4. K. L. Mittal and Thomas Bahners (Eds.) *Laser Surface Modification and Adhesion*, Wiley-Scrivener, Beverly, MA (2015).
5. K. L. Mittal and W. –S. Lei (Eds.) *Laser Technology: Applications in Adhesion and Related Areas*, Wiley-Scrivener, Beverly, MA (2018).
6. A. Tóth, K.Kereszturi, M. Mohai, and I. Bertóti, Plasma based ion implantation of engineering polymers, *Surf. Coat. Technol.* 204, 2898-2908 (2010).
7. J. Pelletier and A. Anders, Plasma-based ion implantation and deposition: A review of physics, technology, and application, *IEEE Trans. Plasma Sci.* 33, 1944-1959 (2005).
8. M. Widner, I. Alexeff, W. D. Jones, and K. E. Lonngren, Ion acoustic wave excitation and ion sheath evolution, *Phys. Fluids* 13, 2532-2540 (1970).
9. R. J. Adler and S. T. Picraux, Repetitively pulsed metal ion beams for ion implantation, *Nucl. Instrum. Meth. Phys. Res. B* 6, 123-128 (1985).
10. J. R. Conrad, J. L. Radtke, R. A. Dodd, F. J. Worzala and N. C. Tran, Plasma source ion-implantation technique for surface modification, *J. Appl. Phys.* 62, 4591-4596 (1987).
11. J. R. Conrad, Sheath thickness and potential profiles of ion-matrix sheaths for cylindrical and spherical electrodes, *J. Appl. Phys.* 62, 777-779 (1987).

12. J. R. Conrad, R. A. Dodd, S. Han, M. Madapura, J. Scheuer, K. Sridharan, and F. J. Worzala, Ion beam assisted coating and surface modification with plasma source ion implantation, *J. Vac. Sci. Technol. A* 8, 3146-3151 (1990).
13. J. R. Conrad, Plasma source ion implantation: a new approach to ion beam modification of materials, *Mater. Sci. Eng.* A116, 197-203 (1989).
14. B. Mizuno, I. Nakayama, N. Aoi, M. Kubota and T. Komeda, New doping method for subhalf micron trench sidewalls by using an electron cyclotron resonance plasma, *Appl. Phys. Lett.* 53, 2059-2061 (1988).
15. M. A. Lieberman, Model of plasma immersion ion implantation, *J. Appl. Phys.* 66, 2926-2929 (1989).
16. X. Y. Qian, N. W. Cheung, M. A. Lieberman, M. I. Current, P. K. Chu, W. L. Harrington, C. W. Magee, and E. M. Botnick, Sub-100 nm p+/n junction formation using plasma immersion ion implantation, *Nucl. Instrum. Meth. Phys. Res. B* 55, 821-825 (1991).
17. N. W. Cheung, Plasma immersion ion implantation for ULSI processing, *Nucl. Instrum. Methods Phys. Res. B* 55, 811-820 (1991).
18. I. G. Brown, X. Godechot and K. M. Yu, Novel metal ion surface modification technique, *Appl. Phys. Lett.* 58, 1392-1394 (1991).
19. S. Qin and C. Chan, Plasma immersion ion implantation doping experiments for microelectronics, *J. Vac. Sci. Technol. B* 12, 962-968 (1994).
20. A. Collins and J. Tendy, Measurements of the potentials and sheath formation in a plasma immersion ion implantation, *J. Vac. Sci. Technol. B* 12, 875-879 (1994).
21. A. Anders, S. Anders, I. G. Brown, M. R. Dickinson and R. A. MacGill, Metal plasma immersion ion implantation using vacuum arc plasma sources, *J. Vac. Sci. Technol. B* 12, 815-820 (1994).
22. Papers From the 1st International Workshop on Plasma-based Ion Implantation, *J. Vac. Sci. Technol. B* 12, 815-998 (1994).
23. S. Miyagawa, S. Nakao, M. Ikeyama, and Y. Miyagawa, Dynamic MC simulations of diamond-like carbon film synthesis by plasma-based ion implantation, *Surf. Coat. Technol.* 136, 122-126 (2001).
24. S. Miyagawa, S. Nakao, K. Saitoh, K. Baba, and Y. Miyagawa, Deposition of diamond-like carbon films using plasma source ion implantation with pulsed plasmas, *Surf. Coat. Technol.* 128-129, 260-264 (2000).
25. R.K.Y. Fu, I.T.L. Cheung, Y.F. Mei, C.H. Shek, G.G. Siu, P.K. Chu, W.M. Yang, Y.X. Leng, Y.X. Huang, X.B. Tian, and S.Q. Yang, Surface modification of polymeric materials by plasma immersion ion implantation, *Nucl. Instrum. Meth. Phys. Res. B* 237, 417-421 (2005).
26. K. Baba and R. Hatada, Deposition of diamond-like carbon films on polymers by plasma source ion implantation, *Thin Solid Films* 506–507, 55-58 (2006).
27. A. Kondyurin, P. Volodin, and J. Weber, Plasma immersion ion implantation of Pebax polymer, *Nucl. Instrum. Meth. Phys. Res. B* 251, 407-412 (2006).
28. A. Igarashi, H. Hayashi, T. Yamanobe, and T. Komoto, Structure and morphology of diamond-like carbon coated on nylon 66/poly(phenylene ether) alloy, *J. Molecular Structure* 788, 238-245 (2006).

29. C. Chaiwong, D.R. McKenzie, and M.M.M. Bilek, Cracking of titanium nitride films grown on polycarbonate, *Surf. Coat. Technol.* 201, 5596-5600 (2007).
30. C. Chaiwong, D.R. McKenzie, and M.M.M. Bilek, Study of adhesion of TiN grown on a polymer substrate, *Surf. Coat. Technol.* 201, 6742-6744 (2007).
31. M.M.M. Bilek, P. Evans, D.R. McKenzie, D.G. McCulloch, H. Zreiqat, and C.R. Howlett, Metal ion implantation using a filtered cathodic vacuum arc, *J. Appl. Phys.* 87, 4198-4204 (2000).
32. N. Mireault and G.G. Ross, Modification of wetting properties of CR39 by plasma source ion implantation, *Appl. Surf. Sci.* 254, 6908-6914 (2008).
33. N. Mireault, G. Abel, L. Andrzejewski, and G. Ross, Measurements of advancing and receding contact angles of water on PMMA and CR-39 at various g-levels, *Microgravity Sci. Technol.* 15, 219-223 (2005).
34. S. Han, Y. Lee, H. Kim, G.-H. Kim, J. Lee, J.-H. Yoon, and G. Kim, Polymer surface modification by plasma source ion implantation, *Surf. Coat. Technol.* 93, 261-264 (1997).
35. Z. Iskanderova, J. Kleiman, Y. Gudimenko, R.C. Tennyson, and W.D. Morison, Comparison of surface modification of polymeric materials for protection from severe oxidative environments using different ion sources, *Surf. Coat. Technol.* 127, 18-23 (2000).
36. X.-M. He, M. Hakovirta, and M. Nastasi, Hardness, hydrophobic and optical properties of fluorine and boron co-alloyed diamond-like carbon films, *Mater. Lett.* 59, 1417-1421 (2005).
37. I.F. Husein, C. Chan, S. Qin, and P.K. Chu, The effect of high-dose nitrogen plasma immersion ion implantation on silicone surfaces, *J. Phys. D: Appl. Phys.* 33, 2869-2874 (2000).
38. T. Kobayashi, T. Yokota, R. Kato, Y. Suzuki, M. Iwaki, T. Terai, N. Takahashi, T. Miyasato, and H. Ujiie, Surface modification of silicone medical materials by plasma-based ion implantation, *Nucl. Instrum. Meth. Phys. Res. B* 257, 128-131 (2007).
39. T. Kobayashi, R. Katou, T. Yokota, Y. Suzuki, M. Iwaki, and T. Terai, Surface modification of silicone sheets and tubes using plasma-based ion implantation, *Surf. Coat. Technol.* 201, 8039-8042 (2007).
40. I.F. Husein, C. Chan, and P.K. Chu, Surface energy and chemistry of ethylene-propylene-diene elastomer (EPDM) treated by plasma immersion ion implantation, *J. Mater. Sci. Lett.* 21, 1611-1614 (2002).
41. I.F. Husein, C. Chan, and P.K. Chu, Chemical structure modification of silicone surfaces by plasma immersion ion implantation, *J. Mater. Sci. Lett.* 19, 1883-1885 (2000).
42. H. Dong and T. Bell, State-of-the-art overview: ion beam surface modification of polymers towards improving tribological properties, *Surf. Coat. Technol.* 111, 29-40 (1999).
43. H. Dong, T. Bell, C. Blawert, and B.L. Mordike, Plasma immersion ion implantation of UHMWPE, *J. Mater. Sci. Lett.* 19, 1147-1149 (2000).

44. W. Shi, X.Y. Li, and H. Dong, Improved wear resistance of ultra-high molecular weight polyethylene by plasma immersion ion implantation, *Wear* 250, 544-552 (2001).
45. A. Kondyurin, V. Karmanov, and R. Guenzel, Plasma immersion ion implantation of polyethylene, *Vacuum* 64, 105-111 (2001).
46. X. Liu, T. Bell, D.G. Chetwynd, and X.Y. Li, Characterisation of engineered surfaces by a novel four-in-one tribological probe microscope, *Wear* 255, 385-394 (2003).
47. K.G. Kostov, M. Ueda, I.H. Tan, N.F. Leite, A.F. Beloto, and G.F. Gomes, Structural effect of nitrogen plasma-based ion implantation on ultra-high molecular weight polyethylene, *Surf. Coat. Technol.* 186, 287-290 (2004).
48. A.R. Marcondes, M. Ueda, K.G. Kostov, A.F. Beloto, N.F. Leite, G.F. Gomes, C.M. Lepienski, Improvements of ultra-high molecular weight polyethylene mechanical properties by nitrogen plasma immersion ion implantation, *Brazilian J. Phys.* 34, 1667-1672 (2004).
49. W. Shi, X.Y. Li, and H. Dong, Preliminary investigation into the load bearing capacity of ion beam surface modified UHMWPE, *J. Mater. Sci.* 39, 3183-3186 (2004).
50. M. Ueda, L.A. Berni, and R.M. Castro, Application of Plasma Immersion Ion Implantation for improved performance of tools and industrial components, *Surf. Coat. Technol.* 200, 517-520 (2005).
51. J.Y. Sze and B.K. Tay, Carbon ion implantation of ultra-high molecular weight polyethylene using filtered cathodic vacuum arc with substrate pulse biasing, *Surf. Coat. Technol.* 200, 4104-4110 (2006).
52. A. Tóth, M. Mohai, T. Ujvári, and I. Bertóti, Advanced surface modification of ultra-high molecular weight poly(ethylene) by helium plasma immersion ion implantation, *Polym. Adv. Technol.* 17, 898-901 (2006).
53. A. Tóth, M. Mohai, T. Ujvári, and I. Bertóti, Hydrogen plasma immersion ion implantation of ultra-high molecular weight polyethylene, *Surf. Interface Anal.* 38, 898-902 (2006).
54. M. Ueda, R.M. Oliveira, J.O. Rossi, C.M. Lepienski, and W.A. Vilela, Effects of ion energy on nitrogen plasma immersion ion implantation in UHMWPE polymer through a metal grid, *AIP Conf. Proc.* 866, 241-244 (2006).
55. W. Zhang, P.K. Chu, J. Ji, Y. Zhang, S.C. Ng, and Q. Yan, Surface antibacterial characteristics of plasma-modified polyethylene, *Biopolymers* 83, 62-68 (2006).
56. W. Zhang, P.K. Chu, J. Ji, Y. Zhang, R.K.Y. Fu, and Q. Yan, Antibacterial properties of plasma-modified and triclosan or bronopol coated polyethylene, *Polymer* 47, 931-936 (2006).
57. W. Zhang, Y.-H. Zhang, J.-H. Ji, J. Zhao, Q. Yan, and P.K. Chu, Antimicrobial properties of copper plasma-modified polyethylene, *Polymer* 47, 7441-7445 (2006).
58. I. Bertóti, M. Mohai, A. Tóth, and T. Ujvári, Nitrogen-PBII modification of ultra-high molecular weight polyethylene: Composition, structure and nanomechanical properties, *Surf. Coat. Technol.* 201, 6839-6842 (2007).

59. J.P.Y. Ho, N.J. Nosworthy, M.M.M. Bilek, B.K. Gan, D.R. McKenzie, P.K. Chu, and C.G. dos Remedios, Plasma-treated polyethylene surfaces for improved binding of active protein, *Plasma Proc. Polym.* 4, 583-590 (2007).
60. N.J. Nosworthy, J.P.Y. Ho, A. Kondyurin, D.R. McKenzie, and M.M.M. Bilek, The attachment of catalase and poly-l-lysine to plasma immersion ion implantation-treated polyethylene, *Acta Biomater.* 3, 695-704 (2007).
61. A. Tóth, I. Bertóti, M. Mohai, and T. Ujvári, Surface modification of polyethylene by nitrogen PIII: Surface chemical and nanomechanical properties, *Mater. Sci. Forum* 537–538, 255-262 (2007).
62. W. Zhang, J. Ji, Y. Zhang, Q. Yan, E.Z. Kurmaev, A. Moewes, J. Zhao, and P.K. Chu, Effects of NH_3, O_2, and N_2 co-implantation on Cu out-diffusion and antimicrobial properties of copper plasma-implanted polyethylene, *Appl. Surf. Sci.* 253, 8981-8985 (2007).
63. W. Zhang, Z. Wu, C. Liu, S. Pu, W. Zhang, and P.K. Chu, Experimental and theoretical investigation of the effects of sample size on copper plasma immersion ion implantation into polyethylene, *J. Appl. Phys.* 101, 113302 (2007).
64. W. Zhang, Y. Zhang, J. Ji, Q. Yan, A. Huang, and P.K. Chu, Antimicrobial polyethylene with controlled copper release, *J. Biomed. Mater. Res. A* 83, 838-844 (2007).
65. B.K. Gan, N.J. Nosworthy, D.R. McKenzie, C.G. dos Remedios, and M.M.M. Bilek, Plasma immersion ion implantation treatment of polyethylene for enhanced binding of active horseradish peroxidase, *J. Biomed. Mater. Res. A* 85, 605-610 (2008).
66. J.O. Rossi, M. Ueda, C.B. Mello, A.R. Marcondes, A. Tóth, and G. da Silva, Short repetitive pulses of 50–75 kV applied to plasma immersion implantation of aerospace materials, *IEEE Trans.Plasma Sci.* 37, 204-210 (2009).
67. M. Veres, M. Füle, S. Tóth, I. Pọcsik, M. Koós, A. Tóth, M. Mohai, and I. Bertóti, Raman scattering of ultra-high molecular weight polyethylene treated by plasma-based ion implantation, *Thin Solid Films* 482, 211-215 (2005).
68. W. Zhang, Y. Luo, H. Wang, J. Jiang, S. Pu, and P.K. Chu, Ag and Ag/N_2 plasma modification of polyethylene for the enhancement of antibacterial properties and cell growth/proliferation, *Acta Biomater.* 4, 2028-2036 (2008).
69. W. Zhang and P.K. Chu, Enhancement of antibacterial properties and biocompatibility of polyethylene by silver and copper plasma immersion ion implantation, *Surf. Coat. Technol.* 203, 909-912 (2008).
70. P.K. Chu, Applications of plasma-based technology to microelectronics and biomedical engineering, *Surf. Coat. Technol.* 203, 2793-2798 (2009).
71. A. Kondyurin, P. Naseri, K. Fisher, D.R. McKenzie, and M.M.M. Bilek, Mechanisms for surface energy changes observed in plasma immersion ion implanted polyethylene: The roles of free radicals and oxygen-containing groups, *Polym. Degr. Stab.* 94, 638-646 (2009).
72. W. Zhang, Y. Luo, H. Wang, S. Pu, and P.K. Chu, Biocompatibility of silver and copper plasma doped polyethylene, *Surf. Coat. Technol.* 203, 2550-2553 (2009).

73. A.L. Volynskii, S.V. Moiseeva, A.I. Dement'ev, D.A. Panchuk, O.V. Lebedeva, L.M. Yarysheva, and N.F. Bakeev, The structure and properties of polymer-metallic coating interphase, *Polym. Sci. A* 48, 731-737 (2006).
74. A. Kondyurin, E. Pecheva, and L. Pramatarova, Calcium phosphate formation on plasma immersion ion implanted low density polyethylene and polytetrafluorethylene surfaces, *J. Mater. Sci. Mater. Med.* 19, 1145-1153 (2008).
75. G.H. Shin, Y.H. Lee, J.S. Lee, Y.S. Kim, W.S. Choi, and H.J. Park, Preparation of plastic and biopolymer multilayer films by plasma source ion implantation, *J. Agricult. Food Chem.* 50, 4608-4614 (2002).
76. Y. Kim, Y. Lee, S. Han, and K.-J. Kim, Improvement of hydrophobic properties of polymer surfaces by plasma source ion implantation, *Surf. Coat. Technol.* 200, 4763-4769 (2006).
77. M. Ueda, W.K. Takahashi, A.R. Marcondes, I.H. Tan, and G. Silva, Surface protection and improved performance of satellite components as well as mitigation of space environmental pollution by plasma ion implantation, *AIP Conf. Proc.* 1087, 691-699 (2009).
78. M. Ueda, I.H. Tan, R.S. Dallaqua, J.O. Rossi, J.J. Barroso, and M.H. Tabacniks, Aluminum plasma immersion ion implantation in polymers, *Nucl. Instrum. Methods Phys. Res. B* 206, 760-766 (2003).
79. S. Mändl, B. Stritzker, and B. Rauschenbach, In situ diffusion measurements during RBS analysis of polymers coated with aluminium, *Nucl. Instrum. Meth. Phys. Res. B* 190, 728-731 (2002).
80. M.M.M. Bilek, D.R. McKenzie, R.N. Tarrant, T.W.H. Oates, P. Ruch, K. Newton-McGee, Y. Shi, D. Tompsett, H.C. Nguyen, B.K. Gan, and D.T. Kwok, Practical plasma immersion ion implantation for stress regulation and treatment of insulators, *Contr. Plasma Phys.* 44, 465-471 (2004).
81. M.M.M. Bilek, K. Newton-McGee, R.N. Tarrant, D.R. McKenzie, and C. dos Remedios, Surface modification of biomaterials using plasma immersion ion implantation (PIII), *Trans. 7th World Biomater. Congr.*, 639-639 (2004).
82. D.R. McKenzie, K. Newton-McGee, P. Ruch, M.M.M. Bilek, and B.K. Gan, Modification of polymers by plasma-based ion implantation for biomedical applications, *Surf. Coat.Technol.* 186, 239-244 (2004).
83. R.C. Powles, D.R. McKenzie, N. Fujisawa, and D.G. McCulloch, Production of amorphous carbon by plasma immersion ion implantation of polymers, *Diamond Related Mater.* 14, 1577-1582 (2005).
84. M.M.M. Bilek, K. Newton-McGee, D.R. McKenzie, and D.G. McCulloch, The role of energetic ions from plasma in the creation of nanostructured materials and stable polymer surface treatments, *Nucl. Instrum. Meth. Phys. Res. B* 242, 221-227 (2006).
85. R.C. Powles, D.R. McKenzie, S.J. Meure, M.V. Swain, and N.L. James, Nanoindentation response of PEEK modified by mesh-assisted plasma immersion ion implantation, *Surf. Coat. Technol.* 201, 7961-7969 (2007).

86. T. Tanaka,M. Yoshida,M. Shinohara, and T. Takagi, Diamondlike carbon deposition on plastic films by plasma source ion implantation, *J. Vac. Sci. Technol. A* 20, 625-633 (2002).
87. J. Wang, N. Huang, S. Hong, F. Chen, P. Yang, Y.X. Leng, J.Y. Chen, G.J. Wang, Z.Y. Liu, X. Wu, and P.K. Chu, Surface characterization and antibacterial adhesion of poly (ethylene terephthalate) modified by acetylene plasma immersion ion implantation, *Proc. IEEE Int. Conf. Plasma Sci.*, 311-311 (2002).
88. S. Watanabe, M. Shinohara, H. Kodama, T. Tanaka, M. Yoshida, and T. Takagi, Amorphous carbon layer deposition on plastic film by PSII, *Thin Solid Films* 420–421, 253-258 (2002).
89. M. Yoshida, T. Tanaka, M. Shinohara, S. Watanabe, J.W. Lee, and T. Takagi, Improvement of oxygen barrier of polyethylene terephthalate film by plasma-source ion implantation of carbon, *J. Vac. Sci. Technol. A* 20, 1802-1807 (2002).
90. N. Huang, Y.X. Leng, P. Yang, J.Y. Chen, H. Sun, G.J. Wan, J. Wang, P.K. Chu, and I.G. Brown, Plasma surface modification of biomaterials applied for cardiovascular devices, *Proc. IEEE Int. Conf. Plasma Sci.*, 439-439 (2003).
91. N. Sakudo, D. Mizutani, Y. Ohmura, H. Endo, R. Yoneda, N. Ikenaga, and H. Takikawa, Surface modification of PET film by plasma-based ion implantation, *Nucl. Instrum. Meth. Phys. Res. B* 206, 687-690 (2003).
92. M. Yoshida, S. Watanabe, T. Tanaka, T. Takagi, and M. Shinohara, and J.W. Lee, Investigation of diamond-like carbon formed on PET film by plasma-source ion implantation using C_2H_2 and CH_4, *Nucl. Instrum. Methods Phys. Res. B* 206, 712-716 (2003).
93. N. Huang, P. Yang, Y.X. Leng, J. Wang, J.Y. Chen, H. Sun, and G.J. Wan, Application of plasma immersion ion implantation to surface modification of biomaterials, *Trans. 7th World Biomater. Congr.*, 638-638 (2004).
94. M. Suehiro, S. Watanabe, T. Tanaka, and T. Takagi, Characteristics of carbon thin film formed on PET by plasma based ion implantation, *J. Vac. Soc. Jpn.* 47, 220-223 (2004).
95. M. Ueda, K.G. Kostov, A.F. Beloto, N.F. Leite, and K.G. Grigorov, Surface modification of polyethylene terephthalate by plasma immersion ion implantation, *Surf. Coat. Technol.* 186, 295-298 (2004).
96. J. Wang, N. Huang, P. Yang, Y.X. Leng, H. Sun, Z.Y. Liu, and P.K. Chu, The effects of amorphous carbon films deposited on polyethylene terephthalate on bacterial adhesion, *Biomaterials* 25, 3163-3170 (2004).
97. J. Wang, S.C.H. Kwok, N. Huang, P. Yang, Y.X. Leng, J.Y. Chen, H. Sun, and P.K. Chu, The effect of amorphous carbon films deposited on polyethylene terephthalate on blood compatibility, *Proc. IEEE Int. Conf. Plasma Sci.*, 299-299 (2004).
98. J. Wang, Y.X. Leng, J.Y. Chen, G.J. Wan, P. Yang, H. Sun, N. Huang, and P.K. Chu, Enhanced mechanical properties of biomedical poly(ethylene terephthalate) surface modified by acetylene plasma immersion ion implantation-deposition, *Proc. IEEE Int. Conf. Plasma Sci.*, 299 (2004).

99. J. Wang, C.J. Pan, P. Li, H. Sun, and N. Huang, Antibacterial properties of amorphous carbon films deposited on polyethylene terephthalate by C_2H_2 plasma immersion ion implantation-deposition, *J. Funct. Mater.* 35, 563-565 (2004).
100. J. Wang, C.J. Pan, S.C.H. Kwok, P. Yang, J.Y. Chen, G.J. Wan, N. Huang, and P.K. Chu, Characteristics and anticoagulation behavior of polyethylene terephthalate modified by C_2H_2 plasma immersion ion implantation-deposition, *J. Vac. Sci. Technol. A* 22, 170-175 (2004).
101. J. Wang, N. Huang, C.J. Pan, S.C.H. Kwok, P. Yang, Y.X. Leng, J.Y. Chen, H. Sun, G.J. Wan, Z.Y. Liu, and P.K. Chu, Bacterial repellence from polyethylene terephthalate surface modified by acetylene plasma immersion ion implantation–deposition, *Surf. Coat. Technol.* 186, 299-304 (2004).
102. J. Wang, S.C.H. Kwok, N. Huang, P. Yang, and P.K. Chu in vitro Platelet adhesion and activation of amorphous carbon films deposited on polyethylene terephthalate by plasma immersion ion implantation - deposition, *Proc. 7th World Biomater. Congr*, 1412-1412 (2004).
103. Y. Miyagawa, M. Tanaka, H. Nakadate, M. Ikeyama, and S. Miyagawa, Particle-in-cell/Monte-Carlo simulation of plasma and neutral gas flow for the inner coating of PET bottles by plasma-based ion implantation. N_2 gas plasma, *Plasma Proc. Polym.* 2, 480-484 (2005).
104. N. Sakudo, H. Endo, R. Yoneda, Y. Ohmura, and N. Ikenaga, Modifying inner surface of a PET bottle by plasma-based ion implantation, *J. Advanced Oxidation Technol.* 8, 65-70 (2005).
105. N. Sakudo, H. Endo, R. Yoneda, Y. Ohmura, and N. Ikenaga, Gas-barrier enhancement of polymer sheet by plasma-based ion implantation, *Surf. Coat. Technol.* 196, 394-397 (2005).
106. Y. Miyagawa, M. Tanaka, M. Ikeyama, S. Nakao, J. Choi, and S. Miyagawa, Analysis of plasma and neutral gas flow inside of a PET bottle under PIII condition by particle-in-cell/Monte Carlo simulation, *Nucl. Instrum. Methods Phys. Res. B* 242, 341-345 (2006).
107. S. Okuji, M. Sekiya, M. Nakabayashi, H. Endo, N. Sakudo, and K. Nagai, Surface modification of polymeric substrates by plasma-based ion implantation, *Nucl. Instrum. Methods Phys. Res. B* 242, 353-356 (2006).
108. J. Wang, J.Y. Chen, P. Yang, Y.X. Leng, G.J. Wan, H. Sun, A.S. Zhao, N. Huang, and P.K. Chu, In vitro platelet adhesion and activation of polyethylene terephthalate modified by acetylene plasma immersion ion implantation and deposition, *Nucl. Instrum. Meth. Phys. Res. B* 242, 12-14 (2006).
109. J. Wang, C. Pan, P. Li, Y. Leng, J. Chen, G. Wan, P. Yang, H. Sun, and N. Huang, The effect of surface free energy parameters of diamond-like carbon films deposited on medical polyethylene terephthalate on bacterial adhesion, *J. Biomed. Eng.* 23, 342-345 (2006).
110. Z.J. Han, B.K. Tay, P.C.T. Ha, M. Shakerzadeh, A.A. Cimmino, S. Prawer, and D. McKenzie, Electronic conductance of ion implanted and plasma modified polymers, *Appl. Phys. Lett.* 91, 052103 (2007).

111. M. Ikeyama, S. Miyagawa, S. Nakao, J. Choi, and T. Miyajima, DLC coatings on inner walls of PET bottles by a simplified PBII technique, *Nucl. Instrum. Methods Phys. Res. B* 257, 741-745 (2007).
112. M. Ikeyama, S. Miyagawa, Y. Miyagawa, Y. Hayakawa, and T. Miyajima, DLC coatings on inner walls of PET bottles by a simplified PBII technique, *Surf. Coat. Technol.* 201, 8112-8115 (2007).
113. Y. Miyagawa, M. Ikeyama, S. Miyagawa, M. Tanaka, and H. Nakadate, Plasma analysis for the plasma immersion ion implantation processing by a PIC-MCC simulation, *Computer Phys. Comm.* 177, 84-87 (2007).
114. Z.J. Han and B.K. Tay, Electrical conductivity of poly(ethylene terephthalate) modified by titanium plasma, *J. Appl. Polym. Sci.* 107, 3332-3336 (2008).
115. K. Kereszturi, A. Tóth, M. Mohai, and I. Bertóti, Surface chemical and nanomechanical alterations in plasma immersion ion implanted PET, *Surf. Interface Anal.* 40, 664-667 (2008).
116. K. Mizuno, B.K. Gan, A. Kondyurin, M.M.M. Bilek, and D.R. McKenzie, Reducing water permeability while maintaining transparency of PET: A plasma immersion ion implantation study. *Plasma Proc. Polym.* 5, 834-839 (2008).
117. G. Kalácska, L. Zsidai, K. Kereszturi, M. Mohai, and A. Tóth, Sliding tribological properties of untreated and PIII-treated PETP, *Appl. Surf. Sci.* 255, 5847-5850 (2009).
118. X. Tian, C. Gong, L. Liu, S. Yang, R.K.Y. Fu, and P.K. Chu, DLC coatings on inner walls of PET bottles by a simplified PBII technique, *Surf. Coat. Technol.* 203, 2751-2754 (2009).
119. P.H. Bae, Y.J. Hwang, H.J. Jo, H.J. Kim, Y. Lee, Y.K. Park, J.G. Kim, and J. Jung, Size removal on polyester fabrics by plasma source ion implantation device, *Chemosphere* 63, 1041-1047 (2006).
120. Z.A. Iskanderova, J.I. Kleiman, Y. Gudimenko, A. Tkachenko, R.C. Tennyson, I.G. Brown, and O.R. Monteiro, Metal ion implantation and dynamic ion mixing for the protection of high-performance polymers from severe oxidative environment. *Nucl. Instrum. Meth. Phys. Res. B* 148, 1090-1096 (1999).
121. H. Lim, Y. Lee, S. Han, Y. Kim, J. Cho, and K.-J. Kim, Reduction in surface resistivity of polymers by plasma source ion implantation, *Surf. Coat. Technol.* 160, 158-164 (2002).
122. I.H. Tan, M. Ueda, K. Kostov, P.A.P. Nascente, and N.R. Demarquette, Nuclear science, plasmas and electric discharges-polymer treatment by plasma immersion ion implantation of nitrogen for formation of diamond-like carbon film., *Jpn. J. Appl. Phys.* 43, 6399-6404 (2004).
123. Y. Lee, S. Han, H. Lim, Y. Kim, and H. Kim, Surface analysis of polymers electrically improved by plasma-source ion-implantation, *Anal. Bioanal. Chem.* 373, 595-600 (2002).
124. N. Huang, P. Yang, Y.X. Leng, J. Wang, H. Sun, J.Y. Chen, and G.J. Wan, Surface modification of biomaterials by plasma immersion ion implantation, *Surf. Coat.Technol.* 186, 218-226 (2004).

125. J.N. Matossian and R. Wei, Challenges and progress toward a 250 kV, 100 kW plasma ion implantation facility, *Surf. Coat. Technol.* 85, 111-119 (1996).
126. Y. Lee, S. Han, H. Lim, H. Jung, J. Cho, and Y. Kim, Surface characterization of polymers modified by keV and MeV ion beams, *J. Adhesion Sci. Technol.* 15, 1079-1089 (2001).
127. Y. Lee, S. Han, and M.-H. Kwon, TOF-SIMS study of modified polymer surfaces, *Appl. Surf. Sci.* 231–232, 278-282 (2004).
128. I.H. Tan, M. Ueda, R.S. Dallaqua, J.O. Rossi, A.F. Beloto, M.H. Tabacniks, N.R. Demarquette, and Y. Inoue, Treatment of polymers by plasma immersion ion implantation for space applications, *Surf. Coat. Technol.* 186, 234-238 (2004).
129. I.H. Tan, M. Ueda, R. Dallaqua, J.O. Rossi, A.F. Beloto, N.R. Demarquette, and L. Gengembre, Aluminum implantation in Kapton® for space applications: magnetic field effects on implantation in vacuum arcs, *Jpn. J. Appl. Phys.* 44, 5211-5215 (2005).
130. J.H. Hong, Y. Lee, S. Han, and K.-J. Kim, Improvement of adhesion properties for Cu films on the polyimide by plasma source ion implantation, *Surf. Coat. Technol.* 201, 197-202 (2006).
131. M. Kwon, Y. Lee, Y. Kim, S. Han, and H. Kim, Selective detection of organic compounds on modified polymer surfaces using TOF-SIMS in combination with derivatization, *Appl. Surf. Sci.* 252, 6689-6692 (2006).
132. Y. Lee, J.H. Hong, H. Chun, and S. Han, Investigation of improved adhesion between Cu film and polyimide by the PSII-EIAMAD technique, *Surf. Interface Anal.* 38, 343-347 (2006).
133. Z.J. Han, B.K. Tay, P.C.T. Ha, J.Y. Sze, and D.H.C. Chua, XPS studies on aluminum ions modified polyimide with the PIII technique, *J. Appl. Phys.* 101, 053301 (2007).
134. Z.J. Han, B.K. Tay, J.Y. Sze, and P.C.T. Ha, Passivation layer on polyimide deposited by combined plasma immersion ion implantation and deposition and cathodic vacuum arc technique, *J. Vac. Sci. Technol. A* 25, 411-414 (2007).
135. Y. Huang, X. Tian, S. Yang, R.K.Y. Fu, and P.K. Chu, Optical and mechanical properties of alumina films fabricated on Kapton polymer by plasma immersion ion implantation and deposition using different biases, *Appl. Surf. Sci.* 253, 9483-9488 (2007).
136. J.-M. Song, Y. Kim, S. Han, Y. Lee, and J.-S. Kim, Hydrophobic recovery behavior of PMMA and PEA ionomers treated with PSII, *J. Appl. Polym. Sci.* 89, 3100-3106 (2003).
137. H. Lim, Y. Lee, S. Han, J. Cho, and K.-J. Kim, Surface treatment and characterization of PMMA, PHEMA, and PHPMA, *J. Vac. Sci. Technol. A* 19, 1490-1496 (2001).
138. Y. Lee, S. Han, J.-H. Lee, J.-H. Yoon, H.E. Lim, and K.-J. Kim, Surface studies of plasma source ion implantation treated polystyrene, *J. Vac. Sci. Technol. A* 16, 1710-1715 (1998).

139. A. Lacoste, F. Le Coeur, Y. Arnal, J. Pelletier, and C. Grattepain, PBII processing of dielectric layers: physical aspects limitations and experimental results, *Surf. Coat. Technol.* 135, 268-273 (2001).
140. B.K. Gan, M.M.M. Bilek, A. Kondyurin, K. Mizuno, and D.R. McKenzie, Etching and structural changes in nitrogen plasma immersion ion implanted polystyrene films, *Nucl. Instrum. Methods Phys. Res. B* 247, 254-260 (2006).
141. A. Kondyurin, B.K. Gan, M.M.M. Bilek, K. Mizuno, and D.R. McKenzie, Etching and structural changes of polystyrene films during plasma immersion ion implantation from argon plasma, *Nucl. Instrum. Methods Phys. Res. B* 251, 413-418 (2006).
142. B.K. Gan, A. Kondyurin, and M.M.M. Bilek, Comparison of protein surface attachment on untreated and plasma immersion ion implantation treated polystyrene: protein islands and carpet, *Langmuir* 23, 2741-2746 (2007).
143. Z.J. Han, B.K. Tay, P. Ha, M. Shakerzadeh, and G.F. You, Structural and wetting properties of metal-polymer nanocomposites, *Proc. IEEE Int. Conf. Nanoelectronics*, 64-68 (2008).
144. A. Kondyurin, B.K. Gan, M.M.M. Bilek, D.R. McKenzie, K. Mizuno, and R. Wuhrer, Argon plasma immersion ion implantation of polystyrene films, *Nucl. Instrum. Methods Phys. Res. B* 266, 1074-1084 (2008).
145. D.V. Bax, D.R. McKenzie, A.S. Weiss, and M.M.M. Bilek, Linker-free covalent attachment of the extracellular matrix protein tropoelastin to a polymer surface for directed cell spreading, *Acta Biomater.* 5, 3371-3381 (2009).
146. Z.J. Han and B.K. Tay, Ti–PS nanocomposites by plasma immersion ion implantation and deposition, *Nucl. Instrum. Methods Phys. Res. B* 267, 496-501 (2009).
147. J.-S. Kim, M.-C. Hong, Y.H. Nah, Y. Lee, S. Han, and H.E. Lim, Wetting properties of polystyrene ionomers treated with plasma source ion implantation, *J. Appl. Polym. Sci.* 83, 2500-2504 (2002).
148. Y. Lee, S. Han, M.-H. Kwon, H. Lim, Y.-S. Kim, H. Chun, and J.-S. Kim, Structural characterization of various ionomers by time-of-flight secondary ion mass spectrometry, *Appl. Surf. Sci.* 203–204, 875-879 (2003).
149. A. Kondyurin, M. Bilek, A. Janke, M. Stamm, and V. Luchnikov, Nanostructured carbonized thin films produced by plasma immersion ion implantation of block-copolymer assemblies, *Plasma Proc. Polym.* 5, 155-160 (2008).
150. E.C. Rangel, W.C.A. Bento, M.E. Kayama, W.H. Schreiner, and N.C. Cruz, Enhancement of polymer hydrophobicity by SF_6 plasma treatment and argon plasma immersion ion implantation, *Surf. Interface Anal.* 35, 179-183 (2003).
151. R.K.Y. Fu, Y.F. Mei, G.J. Wan, G.G. Siu, P.K. Chu, Y.X. Huang, X.B. Tian, and S.Q. Yang, Surface composition and surface energy of Teflon treated by metal plasma immersion ion implantation, *Surf. Sci.* 573, 426-432 (2004).

152. T.L. Schiller, D. Sheeja, D.R. McKenzie, D.G. McCulloch, D.S.P. Lau, S. Burn, and B.K. Tay, Plasma immersion ion implantation of poly(tetrafluoroethylene), *Surf. Coat. Technol.* 177–178, 483-488 (2004).
153. A. Kondyurin, N.J. Nosworthy, and M.M.M. Bilek, Attachment of horseradish peroxidase to polytetrafluorethylene (teflon) after plasma immersion ion implantation, *Acta Biomater.* 4, 1218-1225 (2008).
154. D.T.K. Kwok, H. Wang, Y. Zhang, K.W.K. Yeung, and P.K. Chu, Effects of long pulse width and high pulsing frequency on surface superhydrophobicity of polytetrafluoroethylene in quasi-direct-current plasma immersion ion implantation, *J. Appl. Phys.* 105, 053302-1-4 (2009).
155. S. Okuji, H. Boldyryeva, Y. Takeda, and N. Kishimoto, Characteristics of poly(vinylidene difluoride) modified by plasma-based ion implantation, *Nucl. Instrum. Methods Phys. Res. B* 267, 1557-1560 (2009).
156. M. Tonosaki, H. Okita, Y. Takei, A. Chayahara, Y. Horino, and N. Tsubouchi, Nano-indentation testing for plasma-based ion-implanted surface of plastics, *Surf. Coat. Technol.* 136, 249-251 (2001).
157. M. Tonosaki, K. Kitagawa, and Y. Takei, Mechanical and chemical properties of PBIID-treated plastics, *Surf. Coat. Technol.* 156, 338-342 (2002).
158. K.F. Chan, J.P.Y. Ho, W.Y. Li, B.M.F. Lau, A.K.W. Tse, W.F. Fong, M.M.M. Bilek, D.R. McKenzie, P.K. Chu, and K.N. Yu, Investigation of cytocompatibility of surface-treated cellulose nitrate films by using plasma immersion ion implantation, *Surf. Coat. Technol.* 201, 6897-6900 (2007).
159. K. Kim, W. Sung, H. Park, Y. Lee, S. Han, and W. Kim, Hydrophobic properties of ethylene–vinyl alcohol copolymer treated with plasma source ion implantation, *J. Appl. Polym. Sci.* 92, 2069-2075 (2004).
160. S.I. Hong, K.B. Kim, Y. Lee, S.Y. Cho, J.A. Ko, S.K. Hong, and H.J. Park, Surface modification of ethylene-vinyl alcohol copolymer treated with plasma source ion implantation, *J. Appl. Polym. Sci.* 113, 2988-2996 (2009).
161. H. Wang, J. Ji, W. Zhang, Y. Zhang, J. Jiang, Z. Wu, S. Pu, and P.K. Chu, Biocompatibility and bioactivity of plasma-treated biodegradable poly(butylene succinate), *Acta Biomater.* 5 279-287 (2009).
162. A.P. Kauling, G.V. Soares, C.A. Figueroa, R.V.B. de Oliveira, I.J.R. Baumvol, C. Giacomelli, and L. Miotti, Polypropylene surface modification by active screen plasma nitriding. *Mater. Sci. Eng. C* 29, 363-366 (2009).
163. J. Wang, P. Yang, H. Sun, J.Y. Chen, Y.X. Leng, G.J. Wan, and N. Huang, Surface modification of medical polyurethane by acetylene plasma immersion ion implantation, *Proc. IEEE Int. Conf. Plasma Sci.*, 282-282 (2003).
164. Y.-S. Cho, H.J. Lee, C.-H. Park, H.-J. Lee, and S.-K. Lee, Ion beam generation from sheath field of grid electrode and its application to surface treatment, *J. Appl. Phys.* 100, 123303 (2006).
165. A.V. Kondyurin, M.F. Maitz, V.A. Romanova, V.P. Begishev, I.V. Kondyurina, and R. Guenzel, Drug release from polyureaurethane coating modified by plasma immersion ion implantation, *J. Biomater. Sci. Polym. Ed.* 15, 145-159 (2004).

166. A. Kondyurin, V. Romanova, V. Begishev, I. Kondyurina, R. Guenzel, and M.F. Maitz, Crosslinked polyurethane coating on vascular stents for enhanced x-ray contrast, *J. Bioact. Compat. Polym.* 20, 77-93 (2005).
167. W.C.A. Bento, R.Y. Honda, M.E. Kayama, W.H. Schreiner, N.C. Cruz, and E.C. Rangel, Hydrophilization of PVC surfaces by argon plasma immersion ion implantation, *Plasmas Polym.* 8, 1-11 (2003).
168. W. Zhang, P.K. Chu, J. Ji, Y. Zhang, and Z. Jiang, Effects of O_2 and H_2O plasma immersion ion implantation on surface chemical composition and surface energy of poly vinyl chloride, *Appl. Surf. Sci.* 252, 7884-7889 (2006).
169. W. Zhang, P.K. Chu, J. Ji, Y. Zhang, X. Liu, R.K.Y. Fu, P.C.T. Ha, and Q. Yan, Plasma surface modification of poly vinyl chloride for improvement of antibacterial properties, *Biomaterials* 27, 44-51 (2006).
170. X.B. Tian, R.K.Y. Fu, J. Chen, P.K. Chu, and I.G. Brown, Charging of dielectric substrate materials during plasma immersion ion implantation, *Nucl. Instrum. Methods Phys. Res. B* 187, 485-491 (2002).
171. X.B. Tian, K.Y. Fu, P.K. Chu, and S.Q. Yang, Plasma immersion ion implantation of insulating materials, *Surf. Coat. Technol.* 196, 162-166 (2005).
172. N.W. Cheung, Processing considerations with plasma immersion ion implantation, *Surf. Coat. Technol.* 156, 24-30 (2002).
173. A. Lacoste, and J. Pelletier, Processing considerations with plasma-based ion implantation of polymers: theoretical aspects, limitations, and experimental results, *Nucl. Instrum. Methods Phys. Res. B* 208, 260-266 (2003).
174. R.K.Y. Fu, X. Tian, and P.K. Chu, Enhancement of implantation energy using a conducting grid in plasma immersion ion implantation of dielectric/polymeric materials, *Rev. Sci. Instrum.* 74, 3697-3700 (2003).
175. R.K.Y. Fu, P.K. Chu, and X. Tian, Influence of thickness and dielectric properties on implantation efficacy in plasma immersion ion implantation of insulators, *J. Appl. Phys.* 95, 3319-3323 (2004).
176. J.O. Rossi, I.H. Tan, and M. Ueda, Plasma implantation using high-energy ions and short high voltage pulses, *Nucl. Instrum. Methods Phys. Res.B* 242, 328-331 (2006).
177. T.W.H. Oates, D.R. McKenzie, and M.M.M. Bilek, Plasma immersion ion implantation using polymeric substrates with a sacrificial conductive surface layer, *Surf. Coat. Technol.* 156, 332-337 (2002).
178. T.W.H. Oates and M.M.M. Bilek, Insulator surface charging and dissipation during plasma immersion ion implantation using a thin conductive surface film, *J. Appl. Phys.* 92, 2980-2983 (2002).
179. D.K. Owens and R.C. Wendt, The two-liquid method, *J. Appl. Polym. Sci.* 13, 1741-1747 (1969).
180. S. Kitova, M. Minchev, and G. Danev, RF plasma treatment of polycarbonate substrates, *J. Optoelectr. Adv. Mater.* 7, 2607-2612 (2005).
181. C.W. Extrand, Water contact angles and hysteresis of polyamide surfaces, *J. Colloid Interface Sci.* 248, 136-142 (2002).

182. M. P. Ferraz, F. J. Monteiro, A. P. Serro, B. Saramago, I. R. Gibson, and J. D. Santos, Effect of chemical composition on hydrophobicity and zeta potential of plasma sprayed HA/CaO–P_2O_5 glass coatings, *Biomaterials* 22, 3105-3112 (2001).
183. K. Baba and R. Hatada, Formation of amorphous carbon thin films by plasma source ion implantation, *Surf. Coat. Technol.* 103-104, 235-239(1998).
184. S. Miyagawa, S. Nakao, M. Ikeyama, and Y. Miyagawa, Dynamic MC simulations of diamond-like carbon film synthesis by plasma-based ion implantation, *Surf. Coat. Technol.* 136, 122-126 (2001).
185. S. Miyagawa, S. Nakao, K. Saitoh, K. Baba, and Y. Miyagawa, Deposition of diamond-like carbon films using plasma source ion implantation with pulsed plasmas, *Surf. Coat. Technol.* 128-129, 260-264 (2000).
186. M. Yoshikawa, Y. Mori, H. Obata, M. Maegawa, G. Katagiri, H. Ishida, and A. Ishitani, Raman scattering from nanometer-sized diamond, *Appl. Phys. Lett.*, 67, 694-696 (1995).
187. N. Wada, P. J. Gaczi and S. A. Solin, "Diamond-like" 3-fold coordinated amorphous carbon, *J. Non-crystalline Solids*, 35-36, 543-548 (1980).
188. M. Yoshikawa, G. Katagiri, H. Ishida, and T. Akamatsu, Raman spectra of diamondlike amorphous carbon films, *J. Appl. Phys.*, 64, 6464-6468 (1988).
189. https://www.relyon-plasma.com/applications/adhesion-improvement/?lang=en

9

Laser Surface Engineering of Polymeric Materials for the Modification of Wettability and Adhesion Characteristics

D.G. Waugh[1]* and J. Lawrence[2]

[1]*Manufacturing Technology Centre, Coventry, UK*
[2]*School of Computing, Engineering & Physical Sciences, University of the West of Scotland (UWS), Paisley, UK*

Abstract
The popularity of surface engineering has increased significantly over the past 10 years. This is usually owed to the fact that, in many instances, no more can be done to improve a material's adhesion performance through manipulation of the bulk material. Due to this, many industrialists and academics are turning their attentions to controlling material surfaces to optimize their properties, including wettability and adhesion. With its numerous benefits, when it comes to materials processing, laser surface engineering is now becoming more widely used to bring about the necessary required surface modifications for a wide range of applications such as biomedical, microbiological, tooling, battery development and for applications in the electronics industry. This Chapter discusses the implementation of lasers for the surface engineering of polymeric materials along with contact angle goniometry and tensiometry, two common methods for analysis of the wettability and adhesion of materials that will lead to the unlocking of new knowledge about wetting and adhesion regimes.

Keywords: Laser, surface engineering, wettability, contact angle, topography, wetting regime, adhesion

**Corresponding author*: David.Waugh@the-mtc.org

K.L. Mittal and A.N. Netravali (eds.) Polymer Surface Modification to Enhance Adhesion: Techniques and Applications, (365–388) © 2024 Scrivener Publishing LLC

9.1 Introduction

Polymer materials possess poor adhesion characteristics and are difficult to wet, having inferior bond quality. As a direct result of this, several researchers have utilised various surface engineering techniques to manipulate the surface of polymers, improving wettability characteristics and adhesion property. This has been carried out for various applications such as bioengineering [1–3], solar cells [4], electronics [5] and various applications within the automotive industry [6]. The need for surface engineering has been specifically identified in the bioengineering industry, for example, where polymeric biomaterials have sufficient bulk properties but lack adequate surface properties with regards to biological cell response. As these polymeric biomaterials are not optimised for their intended use, it is crucial for these polymeric biomaterials to undergo surface engineering to optimise their surface properties for their intended use.

Surface coatings [7, 8], plasma [9–13] and lithography [14, 15] are all examples of techniques which have been developed to manipulate the surface properties of polymeric materials. Laser surface engineering [16–27] is an alternative method which offers accuracy and processing flexibility, all whilst not affecting the bulk of the target material. Laser surface engineering is also a relatively clean surface processing technique when compared to alternative, competing techniques, minimizing the time required for post-processing.

A large volume of work has been carried out which has determined that both the wettability characteristics and adhesion property can be modulated by means of laser surface engineering [28–30]. The way a material responds to an incident laser beam is highly dependent on a number of parameters. For polymeric materials which interact with infra-red (IR) laser sources, the IR lasers produce a temperature increase in the material resulting from bond and lattice vibrations which manifest itself as resonant coupling. This is due to weak absorption of the laser photons by the polymer, creating vibrational modes from the absorbed energy [31]. With most lasers it is seen that the smallest achievable features are on the micrometre scale; however, nano-structures have been achieved using a laser emitting at a wavelength of 157 nm [32] and with those lasers which can be used to create laser-induced periodic surface structures (LIPSS) [33, 34].

Given the increased interest and importance to laser surface engineering of polymeric materials, this Chapter will introduce two main methods for analysis of the wettability characteristics and adhesion property of solid materials, namely, contact angle goniometry and tensiometry. This Chapter will also discuss recent advances in laser surface engineering of polymeric materials and the latest understanding of using laser surface engineering to modify the wettability characteristics and adhesion property of polymeric materials.

9.2 Methods for Measuring Wettability and Adhesion Characteristics

9.2.1 Contact Angle Goniometry

Contact angle goniometry is an optical technique which uses shadowography to image liquid droplets on solid surfaces. Such imaging of liquid droplets on a solid surface enables the various surface and interfacial parameters to be derived experimentally. The contact angle, θ, is determined by the triple line of the three phases (liquid, solid and vapour/gas) and characterises the wetting behaviour of the solid with the liquid used to analyse it. Figure 9.1 shows typical images of Figure 9.1(a), a hydrophobic

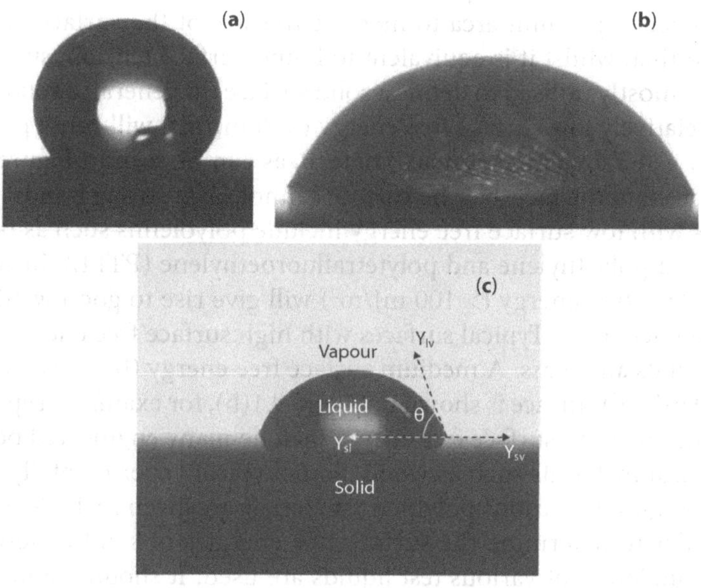

Figure 9.1 Typical example goniometer images of an incident droplet on a polymer sample. (a) Polytetrafluoroethylene (PTFE) displaying a hydrophobic surface, (b) Polyamide 6,6 showing a hydrophilic surface, and (c) A diagram showing the triple line of the three phases (liquid, solid and vapour/gas).

surface where θ > 90°, Figure 9.1(b), a hydrophilic surfaces where θ < 90° and Figure 9.1(c), a diagram showing the triple line and the experimental definition of the contact angle, θ.

As can be seen from Figure 9.1(c), the contact angle, θ, is determined by the equilibrium of force vectors. These force vectors, as seen in Figure 9.1(c) are defined as γ_{sl}, γ_{sv} and γ_{lv}. The surface free energy of the solid, γ_{sv}, acts along the solid surface. The solid-liquid interfacial free energy, γ_{sl}, acts in the opposite direction to γ_{sv} whereas the liquid surface tension γ_{lv} acts tangential to the liquid droplet on the surface. These force vectors can also be defined by Young's Equation [35] :

$$\gamma_{sv} = \gamma_{sl} + \gamma_{lv} \cos\theta \tag{9.1}$$

Within the bulk form of many materials, the atoms within the bulk of the materials tend to be stable with balanced bonds and interactions. Surface atoms, however, tend to be incomplete and have an unbalanced set of bonds/interactions. With surfaces having dormant, incomplete bonds, the surface free energy can be classed as a relative measurement of the energy at a surface which has resulted from the dormant, incomplete bonding. Surface free energy, with units of mJ/m^2, is often defined as the energy needed per unit area to increase the size of the surface. It should be noted that, whilst it is equivalent to liquid surface tension, surface free energy is mostly utilised to define a solid surface. In general, a solid surface with a relatively low surface free energy (< 36mJ/m^2) will cause poor wetting and generate a high contact angle, θ, as can be seen in Figure 9.1(a). This is owed to the fact that the surface cannot form strong bonds. Typical surfaces with low surface free energy include polyolefins such as polypropylene and polyethylene and polytetrafluoroethylene (PTFE). In contrast, high surface free energy (> 100 mJ/m^2) will give rise to good wetting and a low contact angle. Typical surfaces with high surface free energy include many metals and glass. A medium surface free energy (between 36 mJ/m^2 and 100 mJ/m^2) surface is shown in Figure 9.1(b), for example. Typical surfaces with medium surface free energy include many engineered polymers and natural materials such as wood. Some typical experimentally derived contact angles for various polymeric materials are given in Table 9.1.

In order to determine the surface free energies of solid materials, the contact angles, θ, of various test liquids are used. It should be noted that the surface tensions, along with the dispersion and polar parts for each of the test liquids should be known. This is owed to the fact that the dispersion and polar parts of the test liquids are implemented to calculate the solid-liquid interfacial free energy, γ_{sl}.

Table 9.1 Typical experimentally determined water contact angles of different selected as-received polymers using a DataPhysics OCA100 Contact Angle Goniometer.

Polymer	Contact angle (°)
Polyamide (PA) 6,6	56
Poly(methyl methacrylate) (PMMA)	90
Polytetrafluoroethylene (PTFE)	115
Ultra-high-molecular-weight polyethylene (UHMWPE)	103

9.2.2 Tensiometry

The basis of tensiometry is high precision weight measurement for the analysis of wettability characteristics and adhesion property of a solid sample. As result, tensiometry can be seen as an alternative to contact angle goniometry for the analysis of solid surfaces. A tensiometer can be used to analyse a solid surface by measuring the tensile force created on a test liquid droplet incident on a solid surface, creating a hysteresis curve as the tensiometer device moves the liquid droplet towards the solid surface, enables contact between the liquid droplet and solid surface and then pulls the droplet away from the surface. A typical tensiometer adhesion force hysteresis curve for water on Polytetrafluoroethylene (PTFE), measured by a DataPhysics Instruments DCAT25F tensiometer device is shown in Figure 9.2. As the droplet is brought into contact with the solid surface, the surface charge acting from the solid surface on the liquid droplet initially increases the measured force. Once the liquid droplet comes into contact with the solid surface, there is a sudden increase in force known as the snap-in force. As the liquid droplet is then pulled away from the surface, the measured force begins to gradually increase until the liquid droplet is pulled away from the solid surface. The maximum force that is achieved during this experimentation, after the initially measured snap-in force, is known as the maximum pull-off force. As a result, tensiometer devices have the capability to directly measure the adhesion force for solid surfaces with various test liquids. Some typical experimentally determined adhesion force measurements for various polymers are given in Table 9.2.

Dipping of solid samples into a bath of test liquid can also be carried out with tensiometry devices to determine the dynamic contact angles. That is, the solid sample is dipped into the test liquid, with known surface tension, and pulled back out again. During this process the tensiometer weighing system identifies the buoyancy of the solid material along with the lamella

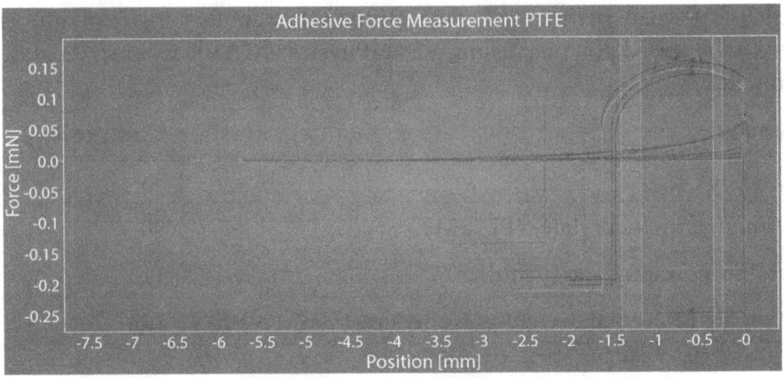

Figure 9.2 Typical adhesion force hysteresis curve for water on Polytetrafluoroethylene (PTFE), measured by a DataPhysics Instruments DCAT25F tensiometer.

Table 9.2 Typical experimentally determined adhesion force measurements for selected polymers.

Polymer	Snap-in force (mN)	Max pull-off force (mN)
Poly(methyl methacrylate) (PMMA)	0.135	0.160
Polytetrafluoroethylene (PTFE)	0.092	0.149
Ultra-high-molecular-weight polyethylene (UHMWPE)	0.172	0.182

weight. Whilst the device removes any buoyancy contribution, the tensiometer device makes use of the Wilhelmy equation to calculate both the advancing contact angle and the receding contact angle.

9.3 Laser Surface Engineering of Polymeric Materials

9.3.1 Common Surface Engineering Techniques

Owed to the requirement of being able to manipulate surface properties, for numerous applications, there are now numerous well-known surface engineering techniques. Some of the most common surface engineering techniques, along with their advantages and disadvantages are given in Table 9.3. Table 9.3 also provides an overview of the types of materials

Table 9.3 A summary of common surface engineering techniques.

Process	Materials	Typical smallest feature size	Advantages	Disadvantages
Contact Printing	Polymers.	Mainly sub-μm but nm possible.	Low capital cost; high repeatability at both μm and sub-μm scale.	Limited precision/ repeatability with sub-μm and nm techniques.
Ion Beam/ Electron Beam Processing	Ceramics; metals; polymers	Mainly sub-μm. nm achievable.	Highly selective and accurate processing; repeatability; potential for scaling-up.	Processing depths are small; capital costs and running costs are high.
Laser Surface Engineering	Ceramics; metals; polymers.	Mainly μm. Sub-μm and nm possible.	Flexibility; ease of scale up; non-contact; accuracy and repeatability; can modify the surface topography and chemistry, simultaneously.	Non-homogeneous energy distribution; wavelength/ material absorption dependency - reflective materials difficult to process.
Lithography	Polymers.	Mainly sub-μm and nm.	Fast processing leading to cost-effectiveness; well-proven technology.	Equipment set-up can be complex; requires a completely flat substrate initially; extremely clean operating conditions needed.
Plasma Treatment	Ceramics, metals, polymers.	Mainly μm. Sub-μm possible.	Easily combines chemical and physical modification,	Treats only small surface areas which can lead to long processing times; complex equipment set-up; high initial capital costs.
Radiation Grafting	Polymers	Induces chemical modification on the μm scale.	Well established; can induce surface chemistry changes with minimal effect to surface topography.	Lack of accuracy leading to a risk of unwanted areas of modification; running costs can be high.

Note: μm = micrometer and nm = nanometer.

which can be surface engineered with the competing surface engineering techniques, in addition to highlighting the typical smallest feature size which can be achieved with the various surface engineering techniques.

Through laser surface engineering, the chemical properties of a sample surface can be altered by laser-induced melting. The surface topography can also be varied on either a micro- or nano-scale (dependant on laser wavelength used and the target materials optical properties). It should be noted that both the modification of the surface chemistry and the modification of surface topography can lead to a modification of the wetting contact angle and material surface adhesion property.

9.3.2 CO_2 Laser and Ultraviolet (UV) Excimer Laser Surface Engineering of Polymeric Materials

Prior to the introduction of ultrafast lasers (ps pulsed and fs pulsed lasers), considerable research was carried out via ultraviolet (UV) excimer lasers to surface engineer polymeric materials [19, 31]. This can mainly be attributed to the highly efficient absorption characteristics of the UV wavelengths for polymeric materials. In terms of laser-material interaction, the UV surface engineering process is an etching process arising from ablative photodecomposition. It has also been demonstrated that UV laser systems can be utilised for micro-metre scale direct-write into polymeric materials [19]. Leading on from this, it has been demonstrated that both 2-D and 3-D micro-scale surface features can be achieved through the use of UV laser systems [22, 26]. To reduce production time and to compensate for any non-linearity in the polymer response, during the UV laser surface engineering, diffractive phase masks can be used [36, 37].

Both CO_2 lasers and UV lasers are highly effective in modifying the surfaces of polymeric materials. For example, three typical 3-D profiles are given in Figure 9.3 for CO_2 laser surface engineered polyamide 6,6 (Figure 9.3(a)), KrF UV excimer laser surface engineered polyamide 6,6 (Figure 9.3(b)) and F_2 excimer laser surface engineered polyamide 6,6 (Figure 9.3(c)). As it can be seen from Figure 9.3(a), the CO_2 laser surface engineered polyamide 6,6 gives rise to melting, giving rise to protrusions on the polyamide 6,6 surface promoting a considerable increase in surface roughness. Whereas, it can be seen in Figure 9.3(b) and Figure 9.3(c) that a UV excimer laser gives rise to a much more structured surface with material being removed layer-by-layer. IR laser sources, such as the CO_2 lasers, are highly effective at giving rise to melting and solidification of polymeric materials and can subsequently give rise to the modification of surface chemistry [31, 38–40]. IR laser sources for

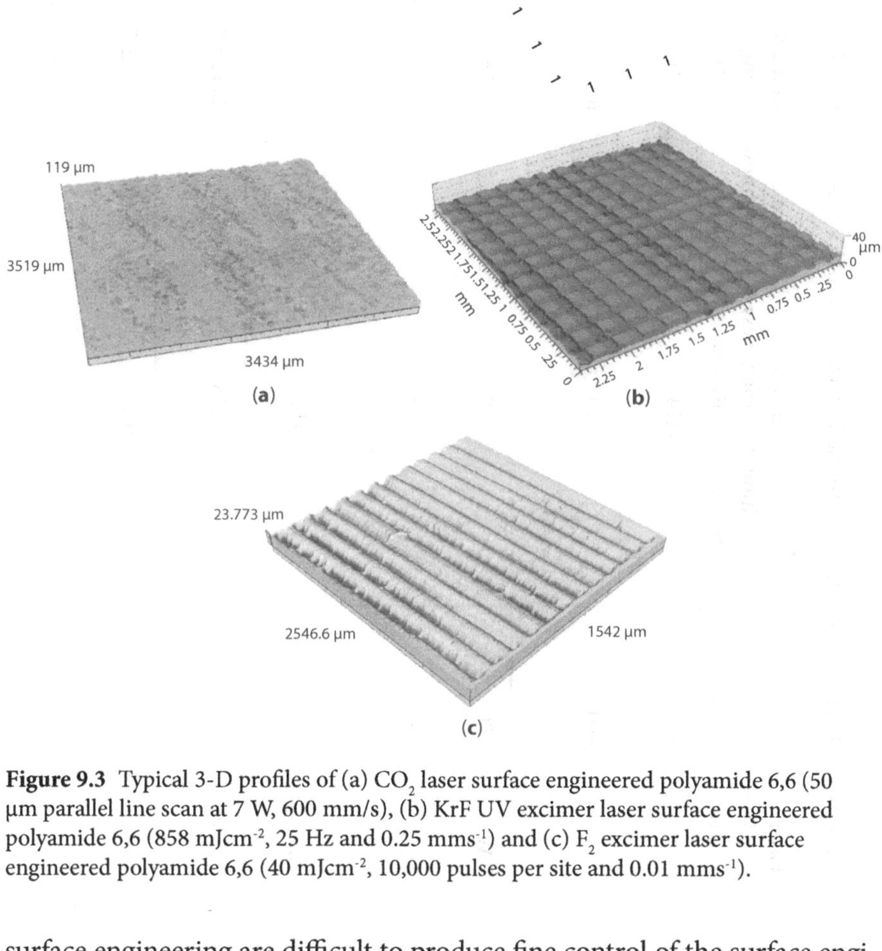

Figure 9.3 Typical 3-D profiles of (a) CO_2 laser surface engineered polyamide 6,6 (50 µm parallel line scan at 7 W, 600 mm/s), (b) KrF UV excimer laser surface engineered polyamide 6,6 (858 $mJcm^{-2}$, 25 Hz and 0.25 mms^{-1}) and (c) F_2 excimer laser surface engineered polyamide 6,6 (40 $mJcm^{-2}$, 10,000 pulses per site and 0.01 mms^{-1}).

surface engineering are difficult to produce fine control of the surface engineering. UV lasers, on the other hand, offer much more control due to the nature of the laser-material interaction compared to that of IR laser sources.

Table 9.4 gives typical laser surface engineering parameters for a CO_2 laser, a KrF UV excimer laser and an F_2 UV excimer laser and the subsequent effects on the surface roughness and the contact angle of polyamide 6,6. For all laser systems detailed in Table 9.4, it can be seen that they can be employed to significantly modulate the contact angle and wettability characteristics of polyamide 6,6. Whilst the CO_2 laser system can be used to vary the contact angle between -13° and +10°, the largest modulation in contact angle arose with the UV excimer lasers with changes in contact angle between -19° and +24°, when compared to the respective as-received samples. The smallest achievable contact angle was achieved with

Table 9.4 Typical laser surface engineering parameters with subsequent variations in wettability characteristics of polyamide 6,6.

Laser type	Scan/processing strategy	Spot size	Laser power/ energy	Scan speed (mms^{-1})	Pulse repetition rate (Hz)	Surface roughness, Ra (μm)	Surface roughness, Sa (μm)	Contact angle (°)
CO_2 Laser	Parallel line scan, 50 μm spacing	95 $μm^2$	7 W	600	-	0.148	0.636	66.0±4.0
CO_2 Laser	Parallel hatch scan, 50 μm spacing	95 $μm^2$	7 W	600	-	0.103	0.423	65.8±2.9
CO_2 Laser	Parallel line scan, 100 μm spacing	95 $μm^2$	7 W	600	-	0.185	0.297	57.5±2.4
CO_2 Laser	Parallel hatch scan, 100 μm spacing	95 $μm^2$	7 W	600	-	0.155	0.326	62.2±2.3
CO_2 Laser	Large area processing	5 mm^2	125 W	150	-	0.060	0.111	55.8±2.7
CO_2 Laser	Large area processing	5 mm^2	125 W	100	-	0.158	0.100	54.6±3.5
CO_2 Laser	Large area processing	5 mm^2	125 W	75	-	0.092	0.101	56.4±3.5

(Continued)

Table 9.4 Typical laser surface engineering parameters with subsequent variations in wettability characteristics of polyamide 6,6. (*Continued*)

Laser type	Scan/processing strategy	Spot size	Laser power / energy	Scan speed (mms^{-1})	Pulse repetition rate (Hz)	Surface roughness, Ra (μm)	Surface roughness, Sa (μm)	Contact angle (°)
CO_2 Laser	Large area processing	5 mm^2	125 W	50	-	0.139	0.341	64.1±2.2
CO_2 Laser	Large area processing	5 mm^2	125 W	25	-	1.236	4.356	46.6±4.7
CO_2 Laser	Large area processing	5 mm^2	125 W	20	-	1.335	3.201	43.0±3.7
KrF Excimer Laser	Non-contact mask projection etching. Parallel line pattern	50 μm wide trench spaced by 50μm	80 mJ	0.125 (constant 10 pulses per area)	25	0.312	0.281	60.6±3.0
KrF Excimer Laser	Non-contact mask projection etching. Hatch pattern	50 μm wide trench spaced by 50μm	80 mJ	0.125 (10 pulses per area)	25	0.548	0.739	75.1±3.9
KrF Excimer Laser	Non-contact mask projection etching. Parallel line pattern	100 μm wide trench spaced by 100μm	80 mJ	0.250 (constant 10 pulses per area)	25	0.679	1.320	80.8±5.2

(*Continued*)

Table 9.4 Typical laser surface engineering parameters with subsequent variations in wettability characteristics of polyamide 6,6. *(Continued)*

Laser type	Scan/processing strategy	Spot size	Laser power / energy	Scan speed (mms^{-1})	Pulse repetition rate (Hz)	Surface roughness, Ra (μm)	Surface roughness, Sa (μm)	Contact angle (°)
KrF Excimer Laser	Non-contact mask projection etching. Hatch pattern	100 μm wide trench spaced by 100 μm	80 mJ	0.250 (constant 10 pulses per area)	25	1.032	1.530	73±3.7
KrF Excimer Laser	Large area processing	23 mm x 12 mm	100 mJ	0 mms^{-1}, 100 pulses per site.	25	0.029	0.119	52.8±1.1
KrF Excimer Laser	Large area processing	23 mm x 12 mm	150 mJ	0 mms^{-1}, 100 pulses per site.	25	0.039	0.092	49.1±4.7
KrF Excimer Laser	Large area processing	23 mm x 12 mm	200 mJ	0 mms^{-1}, 100 pulses per site.	25	0.030	0.108	48.9±2.6
KrF Excimer Laser	Large area processing	23 mm x 12 mm	250 mJ	0 mms^{-1}, 100 pulses per site.	25	0.031	0.107	50.3±1.5

(Continued)

Table 9.4 Typical laser surface engineering parameters with subsequent variations in wettability characteristics of polyamide 6,6. (*Continued*)

Laser type	Scan/processing strategy	Spot size	Laser power / energy	Scan speed (mms^{-1})	Pulse repetition rate (Hz)	Surface roughness, Ra (μm)	Surface roughness, Sa (μm)	Contact angle (°)
KrF Excimer Laser	Large area processing	23 mm x 12 mm	250 mJ	0 mms^{-1}, 500 pulses per site.	25	0.039	0.164	42.9±1.6
KrF Excimer Laser	Large area processing	23 mm x 12 mm	250 mJ	0 mms^{-1}, 1000 pulses per site.	25	0.036	0.198	37.6±3.2
F_2 Excimer Laser	Non-contact mask projection etching. Parallel line pattern	50 μm wide spaced by 50μm	35 mJ	0.01 mms^{-1}, 1000 pulses per site	20	0.253	0.248	66.7±2.3
F_2 Excimer Laser	Non-contact mask projection etching. Parallel line pattern	50 μm wide spaced by 50μm	35 mJ	0.01 mms^{-1}, 10000 pulses per site	20	2.947	2.647	72.6±1.7

NOTE: For as-received polyamide 6,6 surface roughness (Sa) = 0.126 μm, surface roughness (Ra) = 0.029 μm and contact angle is 56.4°±1.2°. Mean contact angles calculated by data taken from 10 contact angles for each sample.

the large area KrF UV excimer laser processing with a mean contact angle of 37.6°±3.2°. The largest contact angle was achieved by the KrF UV excimer laser parallel line pattern with a mean contact angle of 80.8° ±5.2. It should be noted that the largest contact angle achieved is likely due to the structured surface created by the UV laser-material interaction. As the surface roughness did not vary significantly for the large area KrF UV excimer laser processing, it is highly likely that variations in surface chemistry, from the laser-material interaction, gave rise to the significant reduction in the contact angle.

Current theory states that for a hydrophilic material, such as polyamide 6,6, an increase in surface roughness should bring about a reduction in contact angle, θ [3]. However, Table 9.4 shows that this is not the case for the CO_2 laser-patterned samples (parallel line scan and hatch) as the contact angle had increased for each sample. As hypothesized previously [41–43], this phenomenon can be explained by the existence of a mixed-state wetting regime or variation in wetting transition [44–46]. This mixed-state wetting regime can also account for any observations in the reduction of apparent polar component, γ^P, and apparent total surface free energy, γ^T.

9.3.3 Ultrafast Lasers for Surface Engineering of Polymeric Materials

The advancement of laser technology over the last 10 years has meant that ultrafast lasers (both picosecond and femtosecond lasers) have become more widely available, both in academia and industry [47]. Such new technologies can provide two or even multi-photon absorption during the laser processing in a timeframe which is faster than the energy exchange between the lasers photoelectrons and the lattice ions of the materials surface. This enables ultrafast lasers to be more competitive compared to the older generation of laser systems, operating at longer pulse durations, as it enables significantly more control over the thermal nature of the laser-material interaction and more control over the photoionization taking place. Owed to this level of control, ultrafast lasers can give rise to spatial resolutions between 0.1 and 1.0 μm [47]. It should be noted, however, that whilst there are numerous benefits of using ultrafast lasers for surface engineering of polymeric materials, significantly and relatively high capital costs, compared to other laser surface engineering techniques, and other competing techniques, are a common barrier for industry to take up this technology. Having said that, as this relatively new technology matures, it is highly likely that the capital costs for ultrafast laser implementation will be reduced over time.

9.3.4 Currently Available Literature Related to Laser Surface Engineering of Polymeric Materials

Table 9.5 collates the most recent experimental data for implementing laser surface engineering of polymeric materials for the modification of wettability parameters and adhesion characteristics. Considering Table 9.5 and the available literature, it can be seen that both micro-structuring and nano-structuring of polymeric materials, by implementing various laser sources, can give rise to contact angles ranging between 14° to 160°. This demonstrates that laser surface engineering of polymeric materials can alter significantly the adhesion characteristics of polymeric materials. In recent literature there is very much emphasis on the development and application of laser-induced periodic surface structures (LIPSS) for the manipulation of wettability characteristics [33]. It should be noted, though, that a number of the techniques provided in Table 9.5 are not LIPSS-based but still give rise to a large variation in the contact angle, the subsequent wetting regimes, wetting transitions and variations in adhesion characteristics. LIPSS techniques though do offer considerable control over the surface engineering technique to produce sub-micrometre and nanometre scale variations in surface topography.

Whilst collating the information to produce Table 9.5, reviewing the available literature, it was apparent that there was a lack of discussion and subsequent conclusions regarding the wetting regimes taking place. This is of extreme importance when considering both wetting theory and adhesion characteristics. Whilst some authors did consider the wetting regime [2, 48–50], others did not discuss or refer to it [25, 29, 51, 52]. It is imperative that wetting regimes, and any wetting transitions occurring, be accounted for and discussed fully to ensure that the correct wetting theories are used for the analysis of laser surface engineered polymeric materials. It has also been noted that there does not seem to be an overall conclusive correlation between the surface roughness derived from the laser surface engineering techniques and the resulting wettability characteristics. However, there has been a significant study which demonstrates that the microtexture height of the surface can be correlated with the contact angle value of the surface, in addition to the arising wetting regimes and wetting transitions [48]. Whilst further study is needed to confirm this hypothesis, it could provide a future means for being able to accurately estimate the contact angle, wetting regime and adhesion characteristics based solely on the surface roughness.

There has been significant developments in the understanding of the wetting behaviour of rough surfaces [55, 56]. On some occasions, this

Table 9.5 Literature data detailing various laser surface engineered materials, the main surface engineering results and associated contact angles.

Laser type (wavelength)	Polymeric material	Surface features (micro/nano-structuring)	Smallest surface feature	Water contact angle range	Wetting regime	Reference
ArF Excimer Laser (193 nm)	Polylactide (PLA)	Micro- and nano-structuring	11 nm up to 27 nm	65° to 100°	Not stated or discussed	[52]
CO_2 Laser (10.6μm)	Silicone	Micro-structuring	Not stated	65° to 140°	Wenzel regime to Cassie-Baxter regime	[53]
Femtosecond Laser (800 nm)	Poly(dimethylsiloxane) (PDMS)	Micro-structuring	Microtexture heights of up to 40 μm.	90° to 154°	Wenzel regime to Cassie-Baxter regime	[48]
Femtosecond Laser (775 nm)	Poly(methyl methacrylate) (PMMA)	Micro-structuring	~2.00 μm	76.8° to 124.6°	Brief discussion on wetting regimes	[29]
Femtosecond Laser (800 nm)	Polytetrafluoroethylene (PTFE)	Micro-structuring	Not stated	14° to 170°	Not stated or discussed	[54]

(Continued)

Table 9.5 Literature data detailing various laser surface engineered materials, the main surface engineering results and associated contact angles. (*Continued*)

Laser type (wavelength)	Polymeric material	Surface features (micro/nano-structuring)	Smallest surface feature	Water contact angle range	Wetting regime	Reference
KrF Excimer Laser (248 nm)	Poly(dimethylsiloxane) (PDMS)	Micro-structuring	Not stated	50° to 155°	Wenzel regime to Cassie-Baxter regime	[49]
KrF Excimer Laser (248 nm)	Polyimide (PI)	Micro-structuring	Up to 6.5 μm	75° to 160°	Wenzel regime to Cassie-Baxter regime	[50]
Nd:YAG Laser (1064 nm)	Poly(etheretherketone) PEEK	Micro-structuring	Not stated	25° to 85°	Not stated or discussed	[25]
Nd:YAG Laser (1064 nm)	Polypropylene (PP)	Micro-structuring	4.04 μm	86° to 94°	Not stated or discussed	[24]
Nd:YAG Laser (1064 nm)	Ultra-high-molecular-weight polyethylene (UHMWPE)	Micro-structuring	2.30 μm	56° to 85°	Not stated or discussed	[24]

(*Continued*)

Table 9.5 Literature data detailing various laser surface engineered materials, the main surface engineering results and associated contact angles. (*Continued*)

Laser type (wavelength)	Polymeric material	Surface features (micro/nano-structuring)	Smallest surface feature	Water contact angle range	Wetting regime	Reference
Nd:YAG Laser (532 nm)	Poly(1-methoxy-4-(O-disperse Red 1)-2,5-bis(2-methoxyethyl) benzene) (PODR1)	Micro-structuring	~450 nm	115° to 157°	Cassie-Baxter regime	[2]
Nd:YAG Laser (532 nm)	Polypropylene (PP)	Micro-structuring	4.91 μm	86° to 94°	Not stated or discussed	[24]
Nd:YAG Laser (532 nm)	Ultra-high-molecular-weight polyethylene (UHMWPE)	Micro-structuring	1.70 μm	56° to 85°	Not stated or discussed	[24]
Nd:YAG Laser (355 nm)	Polypropylene (PP)	Micro-structuring	1.88 μm	86° to 94°	Not stated or discussed	[2]
Nd:YAG Laser (355 nm)	Ultra-high-molecular-weight polyethylene (UHMWPE)	Micro-structuring	1.90 μm	56° to 85°	Not stated or discussed	[24]

seems to have been overlooked by some researchers implementing lasers to conduct surface engineering applications. From the literature pertaining to the wetting behaviour of rough surfaces, those applying lasers for the modification of wettability and adhesion characteristics should also consider accounting for time to contact [57], energy paths [46], wetting transitions [46, 55, 56, 58], external factors such as pressure and vibrations [58] and the definition of various wetting states [56, 58]. Another aspect for consideration is that surfaces, following surface engineering of any kind, need to produce stable wetting transitions and stable wetting regimes, especially when considering applications requiring omniphobic materials [55].

Through the analysis of the droplet impact and the contact line velocity for the droplet on a rough surface, it has been shown that these parameters, in addition to pressure and wetting area, have a significant impact on the wetting regime and resulting contact angle [57]. This is a highly significant finding as those working in the field of laser surface engineering of polymeric materials, for the manipulation of the wettability characteristics, should consider this when analysing and estimating wetting regimes taking place on the polymeric surfaces. With this in mind, it has also been demonstrated that grooved surfaces, which can arise from laser surface engineering, can elicit a localized sagging of the incident liquid droplet [46] and have a significant impact on the wetting regimes and wetting transitions taking place. Rough surfaces, especially those with a hierarchical structure, can also elicit a reduction in slippage, making the surfaces less susceptible to wetting transitions [55].

There have been a number of studies which have shown that the modified contact angles, on the laser surface engineered polymeric materials, have not followed any of the current theories of wetting, with the authors hypothesizing a wetting transition from a Wenzel regime to a mixed-state wetting regime [41] Also there have been significant studies on Cassie impregnating wetting regimes [56, 58]. It is highly advisable that those working in the field of laser surface engineering for the modification of wettability characteristics be mindful of these hypotheses and, to do so, these researchers would need to account for the dimensions of wetting transitions, energy barrier and critical pressure.

9.4 Summary

From the work detailed in this Chapter it is clear that laser surface engineering as a means to modify the wettability and adhesion characterisitics of polymeric materials has tremendous potential for both direct application into industry, and as a route for generating new science and technology within the field. The latter area of potential is extremely important as we know already that a number of different parameters come into play when polymer surfaces are functionalized with laser beams to achieve bespoke wettability and adhesion performance. Many of these are still to be scientifically verified to confirm the wettability regimes taking place across the solid-liquid interface, especially identifying the presence and nature of mixed-state wetting regimes in which both Wenzel and Cassie-Baxter regimes form. The authors have spent the last dozen years in this endeavor and, as in many fields, the scientific instrumentation and computational methods are now arriving that will facilitate this major breakthrough in the scientific field of the wettability and adhesion characteristics of laser surface engineered polymeric materials.

Acknowledgements

The authors would like to kindly thank DataPhysics Instruments GmbH. for their support, expertise and access to their DCAT tensiometer devices within their Applications Laboratory.

References

1. A. Abdal-hay, H.R. Pant, and J.K. Lim, J. Super-hydrophilic electrospun nylon-6/hydroxyapatite membrane for bone tissue engineering. *Europ. Polym. J.* 49, 1314-1321 (2013).
2. M.R. Cardoso, V. Tribuzi, D.T. Balogh, L. Misoguti and C.R. Mendonça, Laser microstructuring for fabricating superhydrophobic polymeric surfaces. *Appl. Surf. Sci.* 257, 3281-3284 (2011).
3. L. Hao and J. Lawrence, *Laser Surface Treatment of Bio-Implant Materials*, John Wiley & Sons, Hoboken, NJ, USA (2005).
4. P. Parvin, A. Reyhani, M. Mehrabi, M. Refahizadeh, S.Z. Mortazavi and A. Ranjbar, Efficiency enhancement using ArF laser induced micro/nanostructures on the polymeric layer of solar cell. *Optics Laser Technol.* 88, 242-249 (2017).
5. H. Niino and A. Yabe, Chemical surface modification of fluorocarbon polymers by excimer laser processing. *Appl. Surf. Sci.* 96-98, 550-557 (1996).
6. A. Patil, A. Patel and R. Purohit, An overview of polymeric materials for automotive applications. *Mater Today: Proc.* 4, 3807-3815 (2017).
7. C. Chang and J. Duh, Duplex coating technique to improve the adhesion and tribological properties of CrAlSiN nanocomposite coating. *Surf. Coat. Technol.* 326, 375-381 (2016).

8. A.M. Telford, S.C. Thickett and C. Neto, Functional patterned coatings by thin polymer film dewetting. *J. Colloid Interface Sci.* 507, 453-469 (2017).
9. M. Thomas and K.L. Mittal (Eds.), *Atmospheric Pressure Plasma Treatment of Polymers: Relevance to Adhesion*, Wiley-Scrivener, Beverly, MA, USA (2013).
10. T.V. Ivanova, G. Baier, K. Landfester, E. Musin, S.A. Al-Bataineh, D.C. Cameron, T. Homola. J.D. Whittle and M. Sillanpaa, Attachment of poly(l-lactide) nanoparticles to plasma-treated non-woven polymer fabrics using inkjet printing. *Macromol. Biosci.* 15, 1274-1282 (2015).
11. C. Chan, T. Ko and H. Hiraoka, Polymer surface modification by plasmas and photons. *Surf. Sci.Reports* 24, 1-54 (1996).
12. J. Lai, B. Sunderland, J. Xue, S. Yan, W. Zhao, M. Folkard, B.D. Michael and Y. Wang, Study on hydrophilicity of polymer surfaces improved by plasma treatment. *Appl. Surf. Sci.* 252, 3375-3379 (2006).
13. E. Biazar, M. Heidari, A. Asefnezhad and N. Montazeri, The relationship between cellular adhesion and surface roughness in polystyrene modified by microwave plasma radiation. *Int. J. Medicine.* 6, 631-639 (2011).
14. Y.H. Sung, Y.D. Kim, H. Choi, R. Shin, S. Kang and H. Lee, Fabrication of superhydrophobic surfaces with nano-in-micro structures using UV-nanoimprint lithography and thermal shrinkage films. *Appl. Surf. Sci.* 349, 169-173 (2015).
15. F. Yu, P. Li, H. Shen, S. Mathur, C. Lehr, U. Bakowsky and F. Mucklich, Laser interference lithography as a new and efficient technique for micropatterning of biopolymer surface. *Biomaterials* 26, 2307-2312 (2015).
16. D.G. Waugh and J. Lawrence, Laser surface processing of polymers for biomedical applications. In: *Laser-Assisted Fabrication of Materials*, J.D. Majumdar and I. Manna (Eds.), pp. 275-318 Springer, Berlin (2012).
17. D.G. Waugh and J. Lawrence, Chapter 3: Wettability characteristics of laser surface engineered polymers. In: *Laser Technology: Applications in Adhesion and Related Areas*, K.L. Mittal, W.S. Lei (Eds.), pp. 99-122. Wiley-Scrivener Publishing LLC, Beverly Mass. USA (2018).
18. D.G. Waugh and J. Lawrence, CO2 laser surface patterning of nylon 6,6 and subsequent effects on wettability characteristics and apatite response. *Surface Engineering* 27, 724-728 (2011).
19. K. Callewaert, Y. Martelé, L. Breban, K. Naessens, P. Vandaele, R. Baets, G. Geuskens and E. Schacht, Excimer laser induced patterning of polymeric surfaces. *Appl. Surf. Sci.* 208, 218-225 (2003).
20. A.G. Demir, P. Maressa and B. Previtali, Fibre laser texturing for surface functionalization. *Phys.Procedia* 41, 759-768 (2013).
21. H. Mirzadeh and M. Dadsetan, Influence of laser surface modifying of polyethylene terephthalate on fibroblast cell adhesion. *Radiation Phys. Chem.* 67, 381-385 (2003).
22. A.C. Duncan, F. Weisbuch, F. Rouais, S. Lazare and C. Baquey, Laser microfabricated model surfaces for controlled cell growth. *Biosensors Bioelectronics* 17, 413-426 (2002).

23. K.L. Mittal and T. Bahners (Eds.), *Laser Surface Modification and Adhesion*, Wiley-Scrivener, Beverly, MA, USA (2015).
24. A. Riveiro, R. Soto, R. Comesaña, M. Boutinguiza, J. del Val J, F. Quintero, F. Lusquinos and J. Pou, Laser surface modification of PEEK. *Appl. Surf. Sci.* 258, 9437-9442 (2012).
25. A. Wilson, I. Jones, F. Salamat-Zadeh and J.F. Watts, Laser surface modification of poly(etheretherketone) to enhance surface free energy, wettability and adhesion. *Int. J. Adhesion Adhesives* 62, 69-77 (2015).
26. W. Pfleging, M. Bruns, A. Welle and S. Wilson, Laser-assisted modification of polystyrene surfaces for cell culture applications. *Appl. Surf. Sci.* 253, 9177-9184 (2007).
27. K.S. Tiaw, M.H. Hong and S.H. Teoh, Precision laser micro-processing of polymers. *J. Alloys Compounds* 449, 228-231 (2008).
28. D.G. Waugh, J. Lawrence and P. Shukla, Modulating the wettability characteristics and bioactivity of polymeric materials using laser surface treatment. *J. Laser Applications* 28, 022502 (2016).
29. Z.K. Wang, H.Y. Zheng and Y.C. Lam, Investigation on femtosecond laser irradiation energy in inducing hydrophobic polymer surfaces. *Appl. Surf. Sci.* 257, 10427-10433 (2011).
30. D.G. Waugh and J. Lawrence, On the use of CO2 laser induced surface patterns to modify the wettability of poly(methyl methacrylate) (PMMA). *Optics Lasers Eng.* 48, 707-715 (2010).
31. C.D. Skordoulis, M. Makropoulou and A.A. Serafetinides, Ablation of nylon-6,6 with UV and IR lasers. *Appl. Surf. Sci.* 86, 239-244 (1995).
32. E. Sarantopoulou, Z. Kollia, A.C. Cefalas, A.M. Douvas, M. Chatzichristidi, P. Argitis and S. Kobe, Polymer self-assembled nano-structures and surface relief gratings induced with laser at 157nm. *Appl. Surf. Sci.* 253, 7884-7889 (2007).
33. J. Cui, A. Nogales, T.A. Ezquerra and E. Rebollar, Influence of substrate and film thickness on polymer LIPSS formation. *Appl. Surf. Sci.* 394, 125-131 (2017).
34. A. Cunha, A. Elie, L. Plawinski, A.P. Serro, A.M. Botelho do Rego, A. Almeida, M.C. Urdaci, M.-C. Durrieu and R. Vilar, Femtosecond laser surface texturing of titanium as a method to reduce the adhesion of Staphylococcus aureus and biofilm formation. *Appl. Surf. Sci.* 360, 485-493 (2016).
35. M.E. Schrader, Young-Dupre revisited. *Langmuir* 11, 3585-3589 (1995).
36. C. David, J. Wei, T. Lippert and A. Wokaun, Diffractive grey-tone phase masks for laser ablation lithography. *Microelectronic Eng.* 57, 453-460 (2001).
37. I. Zergioti, S. Mailis, N.A. Vainos, A. Ikiades, C.P. Grigoropoulos and C. Fotakis, Microprinting and microetching of diffractive structures using ultrashort laser pulses. *Appl. Surf. Sci.* 138–139, 82-86 (1999).
38. D.G. Waugh and J. Lawrence, Wettability characteristics variation of nylon 6,6 by means of CO2 laser generated surface patterns, in: Proc. International Congress on the Application of Lasers and Electro-Optics (2008).

39. D.G. Waugh, J. Lawrence, C.D. Walton and R.B. Zakaria, On the effects of using CO2 and F2 lasers to modify the wettability of a polymeric biomaterial. *Optics Laser Technol.* 42, 347-356 (2010).
40. S. Dadbin, Surface modification of LDPE film by CO2 pulsed laser irradiation. *European Polym. J.* 38, 2489-2495 (2002).
41. D.G. Waugh, I. Hussain, J. Lawrence, G.C. Smith, D. Cosgrove and C. Toccaceli, *In vitro* mesenchymal stem cell response to a CO2 laser modified polymeric material. *Mater. Sci. Eng. C.* 67 727-736 (2016).
42. D.G. Waugh, J. Lawrence and E.M. Brown. Osteoblast cell response to a CO2 laser modified polymeric material. *Optics Lasers Eng.* 50, 236-247 (2012).
43. D.G. Waugh and J. Lawrence. *Laser Surface Treatment of a Polymeric Biomaterial: Wettability Characteristics and Osteoblast Cell Response Modulation.* Old City Publishing, Philadelphia, PA, USA (2014).
44. G. Pashos, G. Kokkoris and A.G. Boudouvis. A modified phase-field method for the investigation of wetting transitions of droplets on patterned surfaces. *J. Computational Phys.* 283, 258-270 (2015).
45. R. Pant, S. Singha, A. Bandyopadhyay and K. Khare, Investigation of static and dynamic wetting transitions of UV responsive tunable wetting surfaces. *Appl. Surf. Sci.* 292, 777-781 (2014).
46. G. Pashos, G. Kokkoris and A.G. Boudouvis, Minimum energy paths of wetting transitions on grooved surfaces. *Langmuir* 31, 3059-3068 (2015)
47. M. Malinauskas, A. Zukauskas, S. Hasegawa, Y. Hayasaki, V. Mizeikis, R. Buividas and S. Juodkazis,Ultrafast laser processing of materials: From science to industry. *Light* 5, e16133 (2016).
48. B.K. Nayak, P.O. Caffrey, C.R. Speck and M.C. Gupta, Superhydrophobic surfaces by replication of micro/nano-structures fabricated by ultrafast-laser-microtexturing. *Appl. Surf. Sci.* 266, 27-32 (2013).
49. B. Farshchian, J.R. Gatabi, S.M. Bernick, S. Park, G. Lee, R. Droopad and N. Kim, Laser-induced superhydrophobic grid patterns on PDMS for droplet arrays formation. *Appl. Surf. Sci.* 396, 359-365 (2017).
50. V. Oliveira, S.P. Sharma, M.F.S.F. de Moura, R.D.F. Moreira and R. Vilar, Surface treatment of CFRP composites using femtosecond laser radiation. *Optics Lasers Eng.* 94, 37-43 (2017).
51. A. Riveiro, R. Soto, J. del Val, R. Comesaña, M. Boutinguiza , F. Quintero, F. Lusquinos and J. Pou, Texturing of polypropylene (PP) with nanosecond lasers. *Appl. Surf. Sci.* 374, 379-386 (2016).
52. P. Rytlewski, W. Mróz, M. Żenkiewicz, J. Czwartos and B. Budner, Laser induced surface modification of polylactide. *J. Mater. Process. Technol.* 212, 1700-1704 (2012).
53. B.S. Yilbas, O. Keles and A.Y. Toprakli, Surface engineering towards self-cleaning applications: Laser textured silicon surface. *Procedia Eng.* 184, 716-724 (2017).
54. S. Falah Toosi, S. Moradi, S. Kamal and S.G. Hatzikiriakos, Superhydrophobic laser ablated PTFE substrates. *Appl. Surf. Sci.* 349, 715-723 (2015).

55. E. Bormashenko, Progress in understanding wetting transitions on rough surfaces. *Adv. Colloid Interface Sci.* 222, 92-103 (2015).
56. C. Ishino, K. Okumura and D. Quéré, Wetting transitions on rough surfaces. *Europhys. Lett.* 68, 419-425 (2004).
57. C. Lee, Y. Nam, H. Lastakowski, J. Hur, S. Shin and A.L. Biance, Two types of Cassie-to-Wenzel wetting transitions on superhydrophobic surfaces during drop impact. *Soft Matter* 11, 4592-4599 (2015).
58. E. Bormashenko, Wetting transitions on biomimetic surfaces. *Philos. Trans. Royal Soc. London A* 368, 4695-4711 (2010).

10

Competition in Adhesion between Polysort and Monosort Functionalized Polyolefin Surfaces Coated with Vacuum-Evaporated Aluminium

Jörg Florian Friedrich

Technical University Berlin, Polymer Technique and Polymer Physics, Berlin, Germany

Abstract

Polyolefins do not possess functional groups and, therefore, they are not able to interact with metals, fibers, inorganics, or polymers. Key issue of polyolefin composites and laminates with improved properties and high mechanical strength is the adhesion at the interface between polymer and composite partner. Therefore, the most urgent task is the introduction of reactive or at least interactive functional groups onto the polyolefin surface to generate points of interaction.

Introduction of such groups onto the polyolefin surface increases its surface free energy and makes it able to interact physically and chemically with metal coatings. Redox reactions between introduced functional groups and aluminium coating change the surface free energy of polyolefin and that of metal along the interface. Introduced oxygen containing functional groups at the polyolefin surface oxidize the deposited aluminium along the interface. Thus, the high surface free energy of aluminum is strongly decreased and matches more with that of the oxidized polyolefin. In this way, a better thermodynamic compatibility is established.

In this sense, thermal, chemical, plasma-chemical oxidations as well as autooxidation are suited to generate polysort functional groups at the polyolefin surface, i.e. a modification with oxygen-containing functional groups of different types without dominance of one sort.

**Corresponding author*: joerg.florian.friedrich@gmx.de

Dedicated to PhD Dr.h.c. Kash Mittal's 75th birthday

K.L. Mittal and A.N. Netravali (eds.) Polymer Surface Modification to Enhance Adhesion: Techniques and Applications, (389–458) © 2024 Scrivener Publishing LLC

Monosort functionalization instead of polysort, however, allows generation of very strong metal-polymer covalent bonds in comparably high concentration as with polysort ones, however 100 times stronger than with physical forces. For this purpose, the type of synthesized functional groups must allow the formation of stable covalent metal-polymer bonds across the interface.

Appropriate monosort functional groups are COOH and OH groups. They can then react chemically with evaporated aluminium as monodentate, bidentate, or chelating bidentate bonds.

Flexible aliphatic bifunctional spacer molecules can be grafted onto the monosort functionalized polyolefin surface forming a "new" flexible monosort functionalized surface over the original polyolefin surface. Such flexibilization of attached functional groups along the interface is needed to compensate the mechanical stress caused, for example, by different thermal expansions of the laminate partners.

Hydrophobic aliphatic spacers can also repel moisture from the interface. Using silane as terminal group of spacers, a Si-O diffusion barrier can be formed to prevent ions and water from attacking the polar metal-polymer covalent bonds. Thus, monosort silanol groups bond to the aluminium.

Keywords: Adhesion, polymer composites, polymer laminates, covalent bonding, monosort functional groups, plasma

10.1 Introduction

The key issue in composites and laminates is the adhesion at the interface between polymer and the other component. Sufficient mechanical strength and improved properties are desired in composite or laminate formation. Polyolefins generally have the tendency to have no adhesion with metals, resins, or carbon or glass fibers. The absence of adhesion-promoting functional groups at the polyolefin surface is the reason for this behavior. Other problems are the very different properties of polyolefins and metal deposits. Therefore, they are completely incompatible and the adhesion of metal coatings to polyolefins is negligible or totally absent.

Polyolefins consist of very long linear or branched alkane chains. These aliphatic macromolecules stick together by very weak Heitler-London forces, which are of a short range. Physical coiling of macromolecules contributes also to the polymer strength.

Interactions between polyolefins and metals and inorganics are not possible. The exception is formation of a few fragile Al-C bonds [1].

The different chemical nature of polyolefins and metals (aluminium) is also reflected in difference in surface free energies. Polyolefins have very low surface free energies in contrast to clean and freshly evaporated metals (aluminium) having very high surface free energies.

From a thermodynamic point of view, the situation can be improved by lowering the high surface free energy of metals and the oxidation of polyolefin surface. Such surface oxidations increase the surface free energy of polyolefin considerably.

Oxidations produce polysort O-functional groups at the polyolefin surface. Polysort means that different O-groups exist side by side, whereby no one type of O-groups dominate. The mishmash of O-functional groups forms weak physical interactions with the evaporated metal [2–7]. Some reactive groups among the polycot groups, for example COOH groups, may additionally form single chemical bonds [8].

Now, it is obvious to impart monosort functional groups at the surface of polyolefin. Then, exclusively strong covalent metal-polymer bonds can be formed across the interface. It can be expected that now strong adhesion will be generated.

Some problems exist:

- Finding a well-suited pretreatment for generation of monosort functional groups at the surface of the polyolefin.
- Which type of functional group should be preferred?
- Are different types of functional groups (polycot) also suited?
- Is a single type of group (monosort) the best solution?
- How can the functional group be bonded onto the aliphatic chains?
- How can the molecular weight degradation of polyolefin can be avoided and thus formation of a "Weak Boundary Layer" (WBL)?
- How can the formed metal-polymer bond be protected against dissociation (hydrolysis) and redox reactions?

To begin with, the variants should be summarized which exist to equip polyolefin surfaces with functional groups (Figure 10.1).

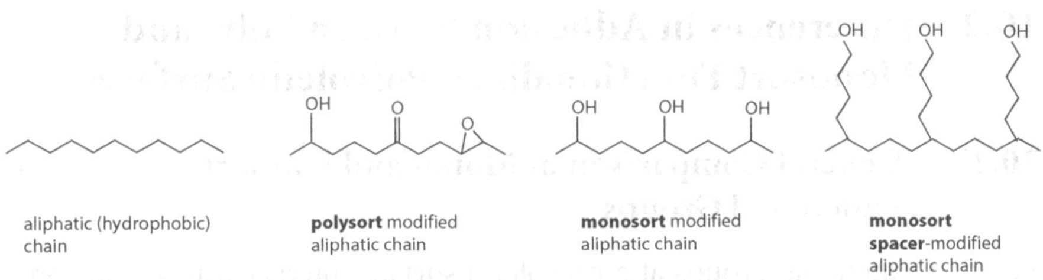

Figure 10.1 Polysort, monosort and spacer bonded monosort functionalization of polyolefin surfaces.

Oxidation is the simplest and easiest way for introduction of polar O-functional groups resulting in a polycot functionalization of the polyolefin surface [10–17].

Monosort functionalization is not possible with any oxidation process. The enthalpy gain in any oxidation, such as burning, flaming, oxygen plasma, etc. is too high for a selective reaction to monosort O-functional groups, e.g. COOH groups [18]. "Selective end-products" of such oxidation are gaseous CO_2 and H_2O. The slower process of auto-oxidation via radicals, peroxy and hydroperoxide groups and oxygen from air is also not selective. A broad variety of O-functional groups are formed.

To achieve monosort functionalization, a two-step combination of plasma or flame oxidation with wet-chemical reduction to OH-groups was utilized [19, 20]. Other ways to obtain monosort functional groups is the use of classic polymers with functional groups applied by coating, dipping or using copolymers, etc. [1]. Selective plasma processes are bromination [21, 22] and use of plasma polymerization of appropriate monomers for its deposition as adhesion promoting thin films [23–25]. Electrospray ionization allows deposition of classic polymers without plasma and side-reactions [25, 26].

Finally, the monosort functionalization can be complemented by grafting of spacer molecules onto the monosort functional groups and thus obtaining monosort functionalization at the polyolefin surface but at a distance from the polyolefin surface corresponding to the length of spacer molecules (see Figure 10.1) [27–29]. These "new" monosort functional groups are flexibly bonded to the polyolefin surface via aliphatic spacer molecules in contrast to monosort functional groups which are directly bonded to the surface [25–27]. Moreover, aliphatic spacer molecules are hydrophobic and water/moisture repelling. Thus, they can hinder the diffusion of water molecules to the interface. Additionally, they protect metal-polymer bonds from hydrolysis, too.

10.2 Differences in Adhesion between Poly- and Monosort Functionalized Polyolefin Surfaces

10.2.1 General Comparison of Mono- and Polysort Functional Groups

Polysort functional groups at a polyolefin surface interact with metals predominantly by different physical interactions such as dipole-dipole, acid-base, or hydrogen bonds. All these bonds are relatively weak but a high

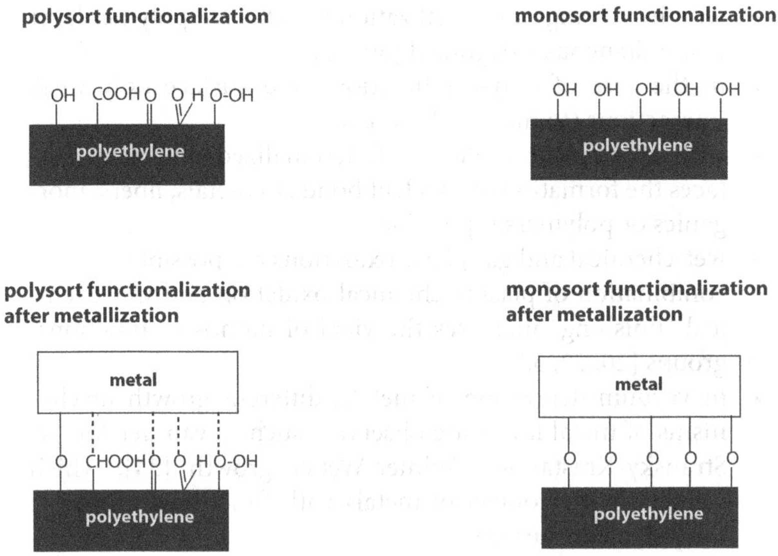

Figure 10.2 Poly- and monosort functionalization of polyolefin surfaces before and after metallization.

density of such weak interactions can balance these low physical interaction forces and, nevertheless, appropriate adhesion can be produced. Using polysort functionalized surfaces only a small portion of groups are able to react with the functional groups of the other component (Figure 10.2). However, all physical (and chemical) interactions are very sensitive towards traces of moisture. Thus, the adhesion is eliminated and the coating lifts off.

Monosort functional groups, such as OH and COOH groups, at polymer surfaces can also form chemical bonds with aluminum. Such bonds are 100 times stronger than physical interactions, such as van der Waals forces [2–7].

Monosort functionalization is difficult to realize at polyolefin surfaces. Hydrocarbons are less reactive. Moreover, as described in the following section, a balance exists between the number of introduced monosort functional groups and the degradation of the polyolefin layer which carries these monosort functional groups. It should be added that the generation of polysort functional groups has the same problem of polymer degradation.

Thus, a number of problems exist:

- for introduction of functional groups onto inert polyolefin surfaces much energy is needed such as by plasma or flame oxidation

- with increasing functionalization the affected polymer layer is simultaneously degraded [30, 31]
- in the case of polysort functionalized surface only weak interactions (to metals) dominate
- with appropriate monosort functionalized polyolefin surfaces the formation of covalent bonds to metals, fibers, inorganics or polymers is possible
- wet-chemical and gas phase oxidations are possible
- combination of plasma-chemical oxidation and wet-chemical "finishing" increases the yield of monosort functional groups [20, 32, 33]
- in vacuum-deposition of metals, different growth mechanisms of metal layers are observed, such as van der Merve, Stransky-Krastanoff, Vollmer-Weber growth [34], which influences the contact of metals with functional groups at the polyolefin surface.

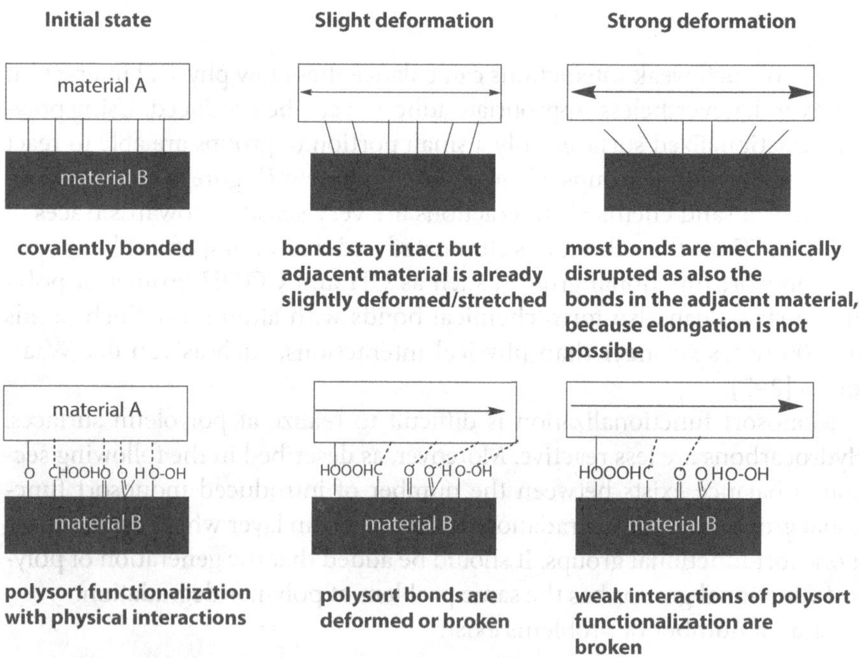

Figure 10.3 Behavior of physical interactions and covalent bonds at increasing temperature and existence of strongly different thermal expansion coefficients of materials A and B, such as metals and polyolefins or carbon fibers and polymers.

Another problem is the direct coupling of metal and polymer via short-range covalent bonds or physical interactions. Covalent bonds across the interface link the materials very strongly. In case of temperature fluctuations the strongly different thermal expansion coefficients of polymers/polyolefins, for instance polyethylene (measured at room temperature) $170*10^{-6}$ K^{-1}, polypropylene (measured at 80°C) $610*10^{-6}$ K^{-1}, metals, for instance aluminum (measured at room temperature) $23.3*10^{-6}$ K^{-1} and carbon fibers (measured at room temperature) $-0.5*10^{-6}$ K^{-1} lead to the deformation of the near-interface polymer phase and bond rupture near the interface occurs (near-interface failure) (Figure 10.3). The same is true for mechanical loading.

10.2.2 Methods for Modifying Polyolefin Surfaces with Polysort Functional Groups

As already stated in the foregoing section the oxidation of polyolefin surfaces is an appropriate method to insert polysort functional groups into the polyolefin surface.

Oxidation reactions to generate polysort O-functional groups at the polyolefin surface are:

- flaming
- plasma oxidation using pure oxygen gas or air
- chemical oxidation, for example using chromo-sulphuric acid, hydrogen peroxide, etc. [57]
- radical generation via ultraviolet irradiation or electron beam irradiation followed by oxidation with diffusing oxygen from air to trapped radicals in near and moderate distance to the polyolefin surface (auto-oxidation) on exposure to air.

The surface oxidation of ethane as an example for alkanes and polyolefins by flame, atmospheric air plasma, or low-pressure oxygen plasma proceeds rapidly to gaseous CO_2 and H_2O (Figure 10.4). The intermediate oxidation state remains in a steady-state oxygen functionalization and is continuously renewed following gasification and newly restored oxidized polyolefin surface.

All types of oxidations are accompanied by chain scissions as appearing by the formation of carboxyl, aldehyde and primary hydroxyl groups, and are followed by further scissions of the macromolecular chain to fragments

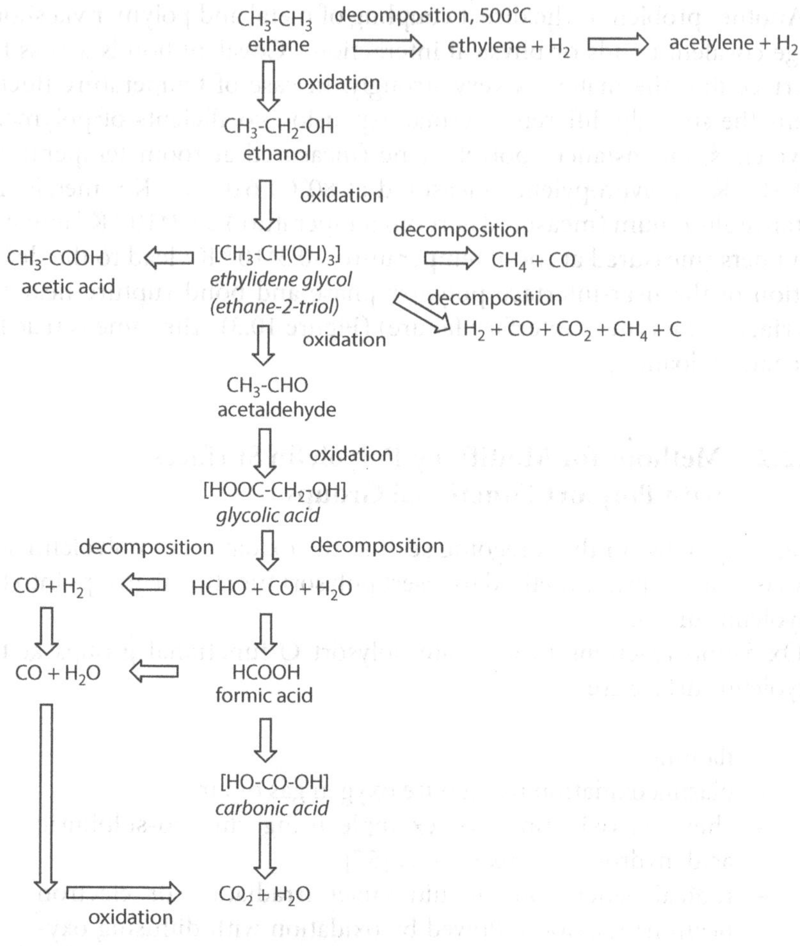

Figure 10.4 Gas phase oxidation of ethane.

and finally the polyolefin is etched and gaseous products are formed. Polymer degradation is connected with shortening of chain length, which is associated again with strong reduction of the mechanical strength of polymers, thus counteracting the improved adhesion by introduction of functional groups (Figure 10.5) [31, 56].

Epoxy, keto and peroxy groups can be generated without chain scissions. These groups are also exclusively formed in pure oxygen atmosphere, for example, in oxygen plasma. Here, hydrogen is initially absent in the oxygen plasma atmosphere. OH or COOH functional groups can only be formed in an indirect way after release of hydrogen from polyolefin. The most probable pathway, however, is post-plasma oxidation on exposure

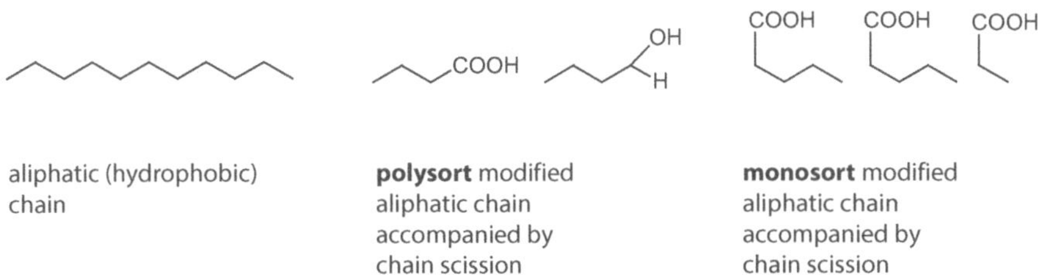

Figure 10.5 Poly- and monosort functionalizations are not advantageous because of their association with chain scission.

to air on the auto-oxidation via peroxide-hydroperoxide groups becomes operative.

This second post-plasma oxidation is often observed especially for plasma processes. The auto-oxidation is based on reactions of trapped radicals in the polymer with diffusing oxygen into the polymer bulk on exposing the sample for long time to air.

Such oxidation via attachment of molecular oxygen onto radicals is a milder and slower oxidation of polyolefins than flaming. It is known as "auto-oxidation" initiated by photons in the plasma, generation of (trapped) radicals and on post-plasma exposure to oxygen from air. This process occurs also on ageing and weathering of polymers and paraffins.

Fatty acids can also be synthesized in an industrial process by oxidation of paraffins using heat and oxygen (from air). The oxidation mechanism was firstly formulated by Langenbeck and Pritzkow (Figure 10.6) [12, 15]. In principle, it is the same/similar mechanism as occurring in long-term ageing of polymers.

Surface oxidation of polyolefin foils in industry is accomplished by oxidative flaming or by plasma oxidation using a dielectric barrier discharge (DBD) often incorrectly termed "corona discharge" (Figure 10.7) [24].

A few examples present typical O/C values of plasma oxidized polyolefin surfaces published in a variety of papers (Table 10.1), confirming that the polysort surface functionalization occurs without a dominating species.

The oxidation level of C-H and C-C bonds is characterized in the C1s peak of X-ray photoelectron spectra (XPS) with a binding energy (BE) of 285.0 eV and comprises C, CH, CH_2, CH_3, C-C, C=C (C≡C) bonds (group 1a). To be more exact, at a slightly higher binding energy of 0.1 to 0.8 eV, the ß-shifted carbon (for example $\underline{C}$-COOH) appears as group 1b between group 1 and the next group 2. The second group or category contains singly bonded carbon atoms to oxygen at a BE≈286.3 eV, such as C-OH, C-O-OH, C-O-C, and epoxy. The next group (3a) has two bonds to oxygen

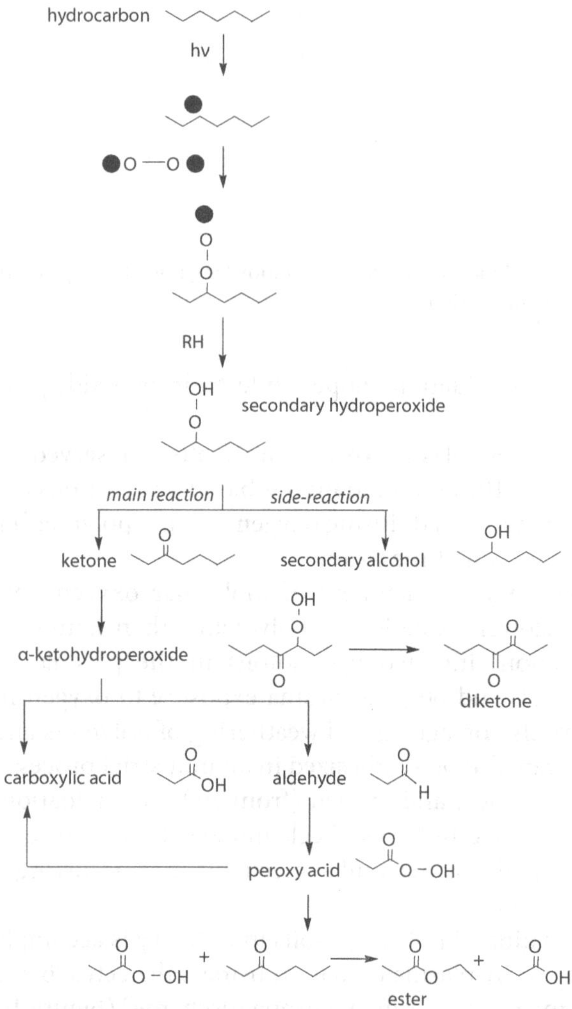

Figure 10.6 Auto-oxidation initiated by chain scission in presence of oxygen (air).

(2 single or 1 double) and appears at about 287.5 eV, such as CHO, >C=O, -O-C-O-, -CO-Cl and so on. It should be mentioned that phenols with a single C-O bond also appear at this binding energy, as an exception, classified as group 3b. Fourth group of C1s peaks contains 3 single or 1 single + 1 double carbon-oxygen bonds, such as O-C(O)-O, COOH, COOR, which appear in the range of BE around 289.0 eV (group 4a). Fifth group contains 4 single or 2 single + 1 double O-C bonds, occurring in organic carbonates

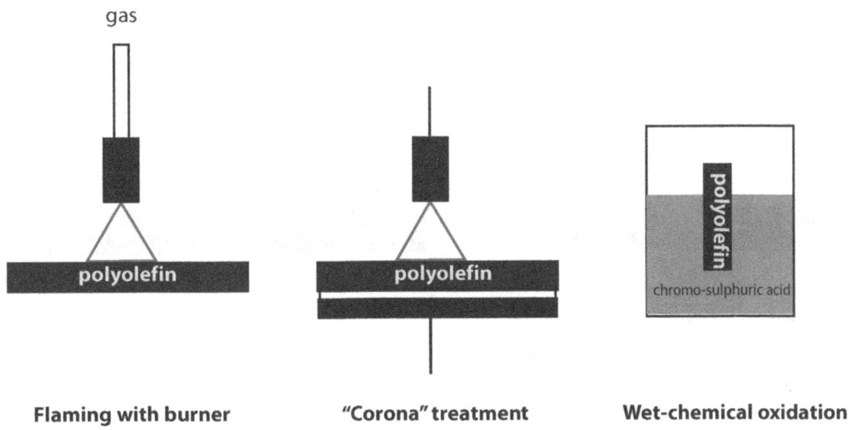

Figure 10.7 Methods to oxidize polyolefin surfaces.

Table 10.1 A few representative examples for O/C and C1s fitted results of plasma oxidation of certain polymers (PE=polyethylene, PS=polystyrene).

(O/C) [%]	C-H, C-C	C-O	C=O	COO	O-CO-O	Reference
25	100	10.4	4.0	10.0		35
36 (PS)	100	3.6	3.2	2.6	4.5	36
25.1 (PE)	100	12.8	5.2	3.7	5.0	1
28 (PE)	100	30	20	19	3	36

(O-CO-O) at BE≈290.4 eV. Exceptions are peroxyacids, peroxyesters and benzoic acid appearing slightly below BE=290 eV (group 4b).

Another example of polysort functionalization is the exposure of polyolefins to ammonia plasma [37]. In this case, a rapid primary functionalization in the plasma with various N species is observed followed by a slow but intense auto-oxidation on post-plasma exposure of the sample to air. Thus, an undesired mixture of different N and O functional groups is present instead of 100% desired primary amino groups ($-NH_2$) as expected (Figure 10.8).

It should be added that each CH_2 group in polyethylene produces about $\Delta H_{CH2}\approx$-650 kJ/mol from its complete oxidation, thus the oxidation progresses further (and is self-sustaining in case of flame).

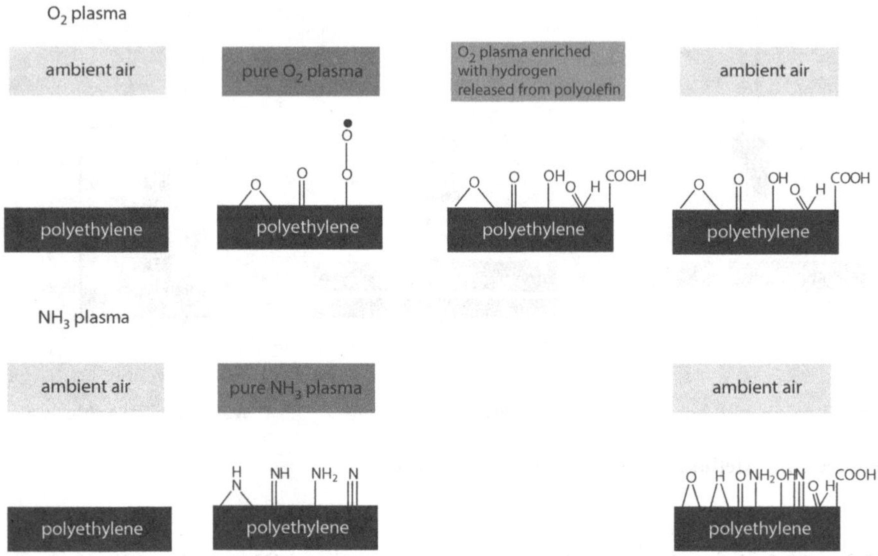

Figure 10.8 Plasma- and post-plasma formed polysort functional groups on exposure of polyolefins to oxygen or ammonia low-pressure plasma.

10.2.3 Polysort Functionalization of Polyolefins and the General Problem with Polymer Degradation

Flame, air, and oxygen plasma treatments, chemical oxidation, and polymer ageing produce polysort O-functional groups at the polyolefin surface. However, there are general problems with the attachment of functional groups X by substitution of H atoms (⋽C-H→⋽C-X) from the aliphatic backbone of polyolefins. The oxidation proceeds for example in the oxygen plasma starting from:

- fixation of O atomic and O_2-molecular species as polysort functionalized polyolefin surface
- release of hydrogen from polyolefin to oxygen plasma by CH_2→CO substitution, CH_2→CH•, crosslinking CH• + CH•→ CH-CH, formation of polyenes $-(CH_2\text{-}CH_2)_n-$→ $-(CH{=}CH)_n-$
- chemical modification of a near-surface polymer layer by plasma-vacuum ultra-violet radiation (formation of trapped radicals, crosslinking, and formation of multiple C-C bonds=polyenes)

- saturation with O functional groups, also in a few molecular layers beneath the topmost surface, by oxygen diffusion (or with dissolved oxygen in the polymer) and its attachment as peroxy group to radicals generated by plasma UV radiation and starting the auto-oxidation
- etching of polymer to CO_2 and H_2O
- establishing a steady-state between O-functionalization and polymer etching whereby the oxidized layer remains unchanged during its progression into the polymer bulk (Figure 10.9).

Principally, the C-C bond of the polyethylene backbone has a lower binding energy than the C-H bond in alkanes, here expressed as standard dissociation energy $\Delta_D H$ ($\Delta_D H_{C\text{-}C} \approx 348$ kJ/mol) in comparison to the C-H bond in CH_2-groups ($\Delta_D H_{C\text{-}H} \approx 396$ kJ/mol) [38]. Shielding of backbone by hydrogen atoms of the CH_2 group may slightly reduce the polymer degradation but not eliminate. Thus, the oxidation may easily take place and also the complementary chain scission. Therefore, highly densely functionalized low-molecular weight oxidized material (LMWOM) is generated as the first step of gasification to CO_2.

In Figure 10.4, the oxidation behavior was schematically depicted which leads to the steady-state between functionalization and etching as shown in Figure 10.9.

Figure 10.10 presents the time-dependence of oxygen introduction into the polypropylene surface as measured with X-ray photoelectron spectroscopy (XPS) and the loss in molecular order of polypropylene on exposure to low-pressure oxygen plasma [39–42]. The significant difference in oxidation rate between atmospheric flaming and exposure to atmospheric

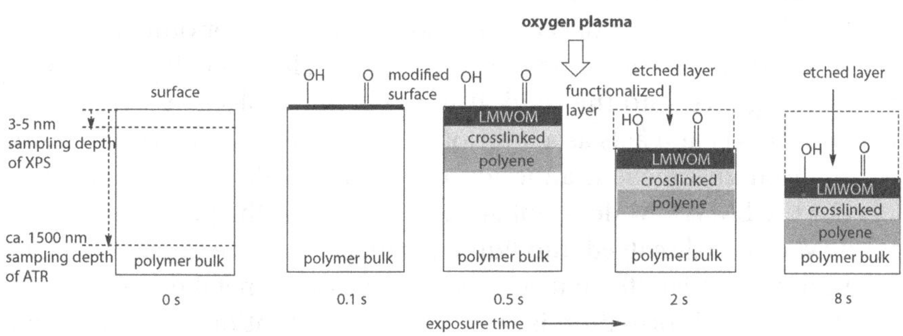

Figure 10.9 Development of a steady-state between introduction of oxygen functional groups into the polyolefin surface and etching of this O-modified layer by advance of the oxidized layer into the polymer matrix.

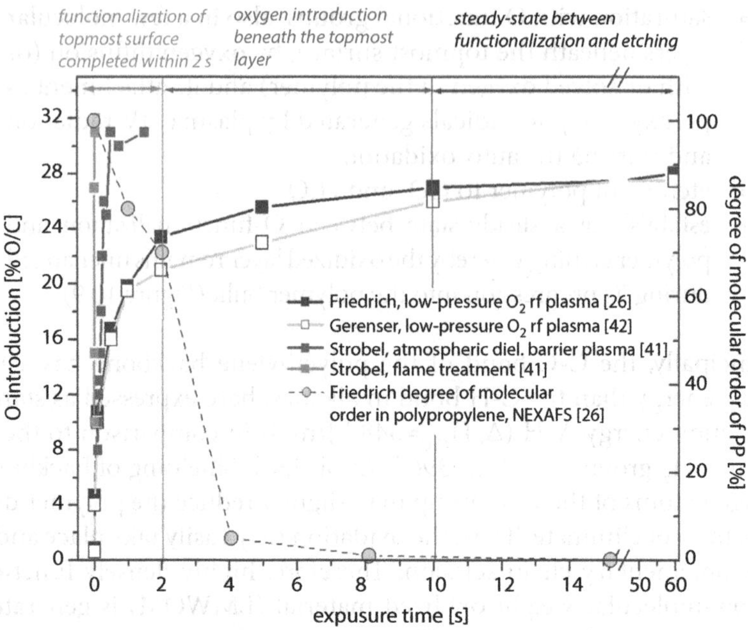

Figure 10.10 Time dependence of O-insertion into polypropylene and simultaneous loss in macromolecular orientation (estimated data adapted from M. Strobel-unpublished [41] and L.J. Gerenser [42]).

dielectric barrier plasma and low-pressure oxygen plasma is evident. The loss in molecular orientation in polypropylene is surely synonymous with the formation of loosely or not bonded low-molecular weight oxidized material (LMWOM) at the polymer surface. This LMWOM hinders an effective adhesion of any coating to the polymer matrix because of its weak bonding to the matrix itself.

LMWOM can interact or react very well with metals or other functional groups of other polymer composite partners. However, it is no longer strongly integrated into the solid. It has very low molecular weight (500-2000 g/mol) [41] and acts as a layer of dust or molecular debris. Thus, one has produced an excellent adhesion of the LMWOM to metals or polymers but the LMWOM does not adhere strongly to the polyolefin matrix. LMWOM can be dissolved into liquid coatings before being hardened and thus it cannot hinder efficient adhesion as in case of metal deposition.

Such LMWOM formation is relevant for modification with strong oxidizing reagents such as oxygen or fluorine plasma (Figure 10.11); however, it appears also on using the only selective bromine plasma as discussed later in detail. LMWOM formation is observed both at flaming of polyolefin foils

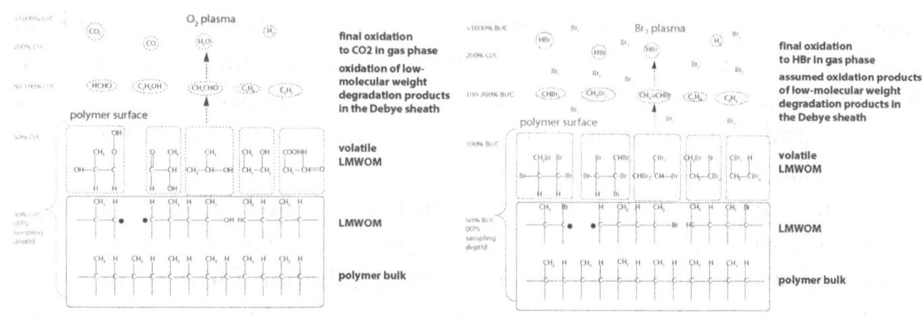

Figure 10.11 Schematic models for the formation of low-molecular weight oxidized material (LMWOM) on exposure to oxygen and bromine plasmas.

as well as on their exposure to the dielectric barrier discharge ("corona") working at atmospheric air pressure.

The following cases must be considered:

- Very short exposure times to flame or atmospheric plasma are needed to avoid polymer degradation. However, thus introduced concentration of O functional groups is low using such treatment conditions. The mechanical adhesion is also not so high.
- Exhaustive treatment promotes extensive polymer degradation, extra-large LMWOM formation and etching. Such LMWOM adheres well to aluminium but not to the unmodified polymer beneath. However, LMWOM formation relocates the mechanical failure at peeling to the LMWOM-unmodified polyolefin interface. Now, a near-interface failure is observed.
- Wiping, extraction or ultrasonic treatment with solvents can remove the majority of LMWOM. Thus, after removal of LMWOM, the density of functionalization at the polymer surface is lower. Thus, the adhesion is also low.

Such undesired LMWOM formation appears in both cases of surface modification i.e., the formation of mono- or polysort functional groups.

10.2.4 Polysort Functionalization and Surface Free Energy

Until now, adhesion was maximized/optimized most often using a simple polysort modification of the polyolefin surface. A rough indicator of surface adhesivity is the wetting with water and the resulting contact angle between water and substrate. Using further test liquids with lower surface tensions than water, the surface free energy can be calculated [43–45]. The determination of surface free energy is an often used indicator of expected adhesion and an integral measure of the oxidation (modification) of polyolefin surface.

This test result can often be related to the expected adhesion. There is the hypothesis that the thermodynamics prefers similar surface free energies of substrate and coating for maximum adhesion [46–53]. In such a case the compatibility should be optimum. The polarity of both materials should be equal. In such a case, it can be assumed that polar (physical) interactions are balanced and the chemical potentials, too.

Other factors are also important, the surface groups themselves may not be strongly bonded and the bulk of material is bonded only to low-molecular weight oxidized material (LMWOM), as described above as the phenomenon of "weak boundary layer" [9]. A further problem is that the surface free energy of polyolefin and perfluorinated polymers is very low ($\approx$20 mJ/m^2) and those of metals are very high (>1000 mJ/m^2). Chemical or plasma-chemical oxidation can increase the surface free energy of polyolefins considerably, however up to max. 45 mJ/m^2, otherwise etching occurs. Nevertheless, the problem of large difference in surface free energies is only marginally reduced by polar polyolefin surfaces (Figure 10.12).

Redox reactions at the interface reduce the difference in surface free energies as long as a difference in electrochemical potentials of aluminium and the functional groups at the polyolefin surface is present [54–56].

Figure 10.13 shows the relation between the polar contribution to surface free energy of polyolefins and the resulting adhesion using the examples of surface-oxidized polyethylene and polypropylene (O_2 plasma) coated with a thin thermally deposited (150 nm) aluminium film. The maximum peel strength and the polar contribution to the surface free energy are for 20% O/C.

Low O/C ratio associated with low polar forces produces nevertheless considerable adhesion (O/C= 2-18%), i.e. a few functional groups at the polyolefin surface are sufficient to produce relatively high peel strength [24, 56, 57]. Higher O/C values (>20% O/C) cause decrease in peel strength and also in polar contribution. It is assumed that this decrease is connected with the beginning of polymer decomposition (LMWOM formation).

Long exposure (t>10 s) to the oxygen plasma produces O functional groups at every fourth carbon atom at the surface ($\approx$25% OH/C). These groups are able to form physical interactions or covalent bonds to the coating.

Considering the surface of a commercial poly(vinyl alcohol) (PVA), it carries 1 OH-group at every second carbon atom (50% OH/C), in case

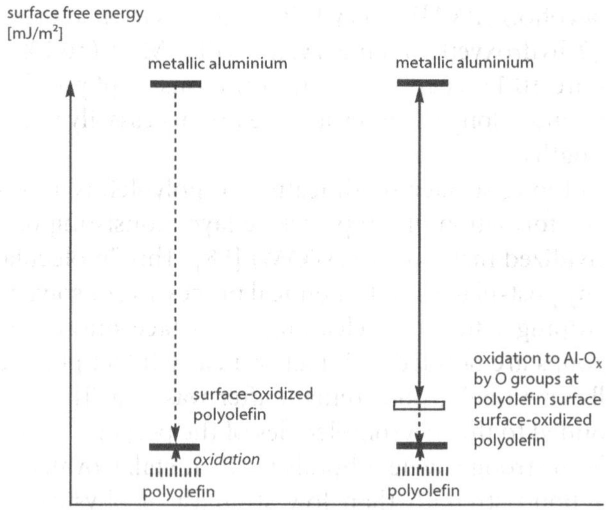

Figure 10.12 Difference in surface free energy between aluminium and polyolefin before and after surface oxidation.

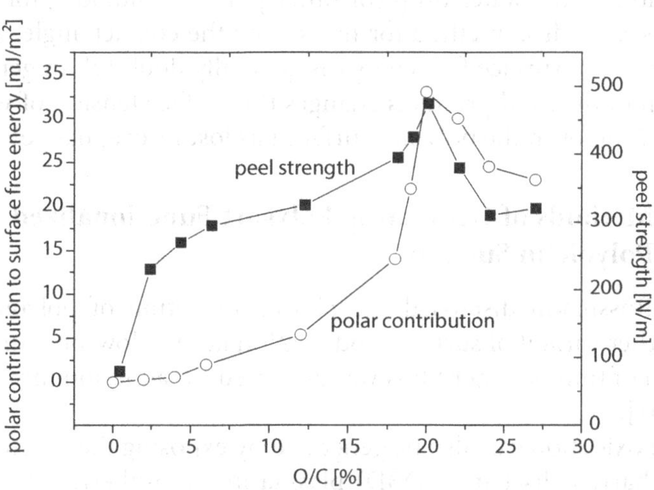

Figure 10.13 Peel strength of 100 nm Al on PP and polar contribution to surface free energy in dependence on XPS-measured O/C ratio.

of poly(allyl alcohol) (PAAl), every third (33% OH/C), and every sixth in case of poly(2-hydroxyethylmethacrylate) (PHEMA) (16.7% OH/C). As shown in Figure 10.13, such high concentrations of physical interactions and covalent bonds along the interface are not necessarily needed to reach high peel strengths.

The result of long surface modification of polyolefins and many other polymers is the formation of a top surface layer consisting of low-molecular-weight oxidized material (LMWOM) [58]. This "molecular debris" is removed by any post-plasma wet-chemical processing or solvent extraction or simply by wiping. After such "cleaning" of surface, only a few remaining functional groups are bonded to intact or nearly intact polyolefin chains. 20-30% of all functional groups remain after washing. They are obviously covalently bonded to the macromolecules of the polymer [59]. Thus, these groups can form strong covalent bonds to the metal. Covalent bonds have much higher bond strength than low strength of physical interactions. Therefore, covalently bonded Al-polyolefin laminates should present stronger adhesion. However, the concentration of remaining groups after exposure to solvents is often strongly lowered but most often are sufficient to allow enough covalent bonding.

It must be added that contact angle measurements at strongly oxidized polyolefin surfaces with large amount of loosely bonded LMWOM cannot be accurately made. Such strongly polar low-molecular-weight debris is easily soluble in the water drop (or other polar test liquids), for example using the sessile-drop method for measuring the contact angle. Thus, the determination of surface free energy is generally doubtful. In general, the dissolution of oxidized products changes the surface tension of water significantly. Moreover, the polymer surface has lost its original polarity.

10.2.5 Methods of Generating Polysort Functionalized Polyolefin Surfaces

In 1956, Rossmann discussed the plasma oxidation of polyolefin foils [60]. Another variant of surface modification i.e., the low-pressure plasma deposition of thin polymer films was also used for adhesion improvement in 1960 [61].

Surface oxidation can also be generated by exposing the polymer to the dielectric barrier discharge (DBD) plasma in atmospheric air, often also erroneously called "corona" treatment [62, 63]. This is a simple polymer surface treatment often used in industry. This result of this process is comparable to that of flaming.

Flaming is also characterized like DBD treatment by unspecific polysort oxidation at the polyolefin surface. Like DBD its advantages are the high rate of modification, and working at ambient pressure with simple gases such as methane, ethane, propane or butane, and air [39].

Oxyfluorination via gas phase or liquid phase is also used for adhesion improvement [64–66]. Either post-plasma auto-oxidation or use of fluorine-oxygen gas produces strong polar oxy-fluoro groups.

Laser treatment was extensively tested in the past [67]. It is based on roughening and polysort oxidation and planar laser-like light is produced by excimer lamps, working in UV-C or vacuum-UV region which has been widely used in the past [68].

There is also the flame-assisted deposition of nanometer thick SiO_2 layers equipped with Si-OH groups working as adhesion promoter, termed as Silicoater-process [69, 70]. Here, some silanol groups at the surface of SiO_2 ($\geqslant$Si-OH) generate adhesion, thus these can be classified as monosort functional groups.

In carbon fiber treatment, fibers were initially polysort oxidized (thermally, anodic, plasma-chemically...) or/and a size (adhesion promoter) was deposited, most often epoxy resin based [71]. The same is valid for treatment of carbon nanotubes [1].

Photo-chemical functionalization using sensitizers such as benzophenone has too low energy to attack aliphatic structures efficiently to generate monosort functionalization [72, 73].

Chemical etching and roughening of polymers associated with polysort O functionalization is an old but often used method to form strong adhesion via mechanical interlocking. In particular, metallization of acrylonitrile-butadiene-styrene copolymer in printed circuit boards by pickling with the toxic chromo-sulfuric acid has been often used [74–76].

Moreover, different spray treatments of plastics exist for improvement of adhesion, such as thermal, plasma, flame, laser, thermal and plasma spraying of different materials onto plastics or other substrates [77].

Last, but not least, underwater plasma produces predominantly polysort O-functionalization [78]. Here, the electrical plasma is initiated in water or in gas phase but with contact to the surface of liquid.

10.2.6 Monosort OH Groups

OH groups can form polar interactions as well as hydrogen bonds. Covalent bonds can also be formed in fact with alkylsilyl groups of silanes, isocyanates, carboxylic groups, etc. Hydroxyl groups often occur at the surfaces of inorganics and metals. As mentioned before, a few polymers also carry OH groups in their structure, such as poly(vinyl alcohol) (50% OH/C), poly(allyl alcohol) (33% OH/C) or poly(hydroxyethyl methacrylate) (17% OH/C).

Plasma polymerization of allyl alcohol allows thin-film deposition of such OH-carrying polymer layers with nearly expected stoichiometric OH concentration (80-90% of stoichiometric structure) in the deposited plasma polymer, i.e. the measured concentration amounts to 28-30% OH/C [79] in comparison to ideally 33% OH/C; but other investigation found lower percentage [80]. It is assumed that the loss of OH groups during plasma exposure is partially compensated by OH formation via post-plasma auto-oxidation as described before. Hydroxyethylmethacrylate (HEMA) is another monomer suited for plasma polymerization [81].

Precursor molecules not suitable for classic polymerization can also be plasma polymerized, for example ethanol. However, then the loss in OH groups during the polymerization process in comparison to the ideal stoichiometry is significant. Auto-oxidation of plasma polymerized layers subsequently adds some newly formed OH groups among other (polysort) O functional groups [82].

OH groups can interact with other materials via hydrogen bonds or simply via dipole-dipole interactions. OH groups can form covalent bonds with aluminium.

10.2.7 OH Groups by Wet-Chemical Post-Plasma Transformation of Polysort O Functional Groups into OH Groups

Reduction of polysort O functional groups to OH groups, however, is a further appropriate (chemical) way to generate OH groups at polyolefin surfaces [20]. The main problem is the selection of a moderate reducing agent weak enough to avoid the reduction of carbonyl, ether and hydroxyl groups to alkanes but strong enough to reduce carbonyls, epoxies, etc.

If a simple polysort oxidation of polyolefins was carried out by exposure to oxygen plasma, flaming, etc., a second treatment with hydrogen plasma to transform carbonyl species to OH groups however failed. Also the use of water or hydrogen peroxide as plasma gases (vapors) for direct generation of OH groups did not work satisfactorily as well as underwater plasma [24].

Only a separate and extra mild wet-chemical reduction generates predominantly the desired OH groups. Here, especially the use of diborane as a reducing agent was successful [32, 33, 83]:

▒-COOR + B_2H_6 →	▒-$CHOBH_2$ (H_2SO_4) →	▒-CH_2-OH	*slow*
▒-COOH + B_2H_6 →	▒-COBO + H_2 (H_2SO_4) →	▒-CH_2-OH	*fast*
▒-C=O + B_2H_6 →	▒-$CHOBH_2$ (H_2SO_4) →	▒-CH_2-OH	*fast*
▒>C=C< + B_2H_6 →	▒-adduct + H_2O_2 →	▒>CH-C(OH)<	*very fast*
▒$>C^{\underline{O}}C<$ + B_2H_6 →		▒>CH-C(OH)<	*very fast*

Carbonyl species, olefinic double bonds and epoxy ($>C^{\underline{O}}C<$) groups can be reduced to OH groups. Use of stronger reducing agents, such as $NaBH_4$ or $LiAlH_4$, leads to additional reduction of ether groups, thus degrading poly(ethylene glycol)s strongly [84]. Moreover, the loss of bonded oxygen (H_2O formation) is observed and alkanes are generated (Figure 10.14).

Exposure of polymer surfaces to oxygen plasma produces a low concentration of OH groups (≈2% OH/C) among the majority of other O functional groups (about 26-28% O/C) [34–36]. Transfer of H atoms released from polymer to O species is needed for formation of OH or COOH species in the pure oxygen plasma. Additionally, auto-oxidation of sample on exposure to air occurs.

As mentioned before, post-plasma wet-chemical reduction of oxygen plasma modified polyolefins with diborane enhances the OH concentration from 2% to 11% OH/C and with hydroxylation of C=C bonds to 14 OH/C, thus more than the half of all residual O functional groups are transferred to OH groups (>50% OH/O_{total}). Therefore, diborane post-plasma

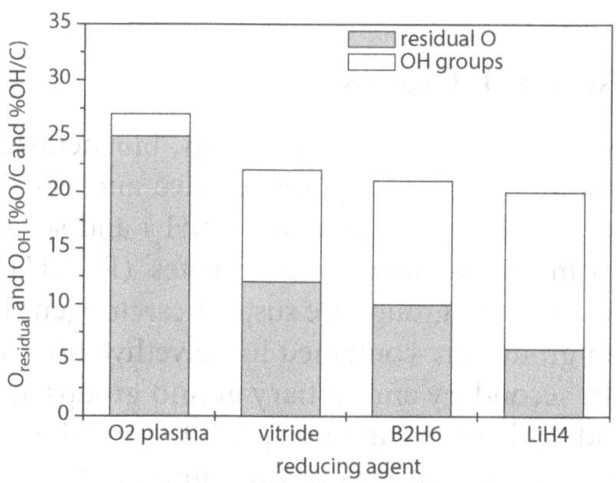

Figure 10.14 Wet-chemical post-plasma reduction of oxidized polyethylene [vitride= sodium bis(2-methoxyethoxy)aluminum hydride].

reduction of all carbonyl groups and hydroxylation of double bonds is a suitable way for going from polysort O groups to monosort OH-groups [18, 20, 36, 84].

About 25% of O species were lost by chemical reduction because of dissolution of co-produced LMWOM by wet-chemical processing with diborane and/or by total reduction to aliphatic C-H bonds (Figure 10.14).

10.2.8 Monosort COOH Groups

A selective post-plasma oxidation of polysort O functionalized polyolefin surfaces to COOH groups does not occur. Its association with a seamless transition to etching and release of CO_2 and H_2O is the limiting factor, i.e. the oxidation cannot be stopped at the level of COOH groups.

In industry, apart from the use of plasma processes, the copolymerization of olefins with maleic anhydride/maleic acid is often utilized for promotion of adhesion property by means of COOH insertion.

Thus, alternatively, the deposition of thin films of poly(acrylic acid) can equip the surface with monosort COOH groups. The deposition of acrylic acid is often carried out using plasma polymerization [85–87]. However, the number of original COOH groups is lowered by 25-50% through decarbonylation/decarboxylation of the COOH group [24, 88].

Another unusual route is the plasma bromination [24]. The introduced C-Br group at the surface of the polyolefin is then post-plasma wet-chemically substituted by amino acids leading to COOH groups [21].

COOH groups can strongly interact with metals and also covalently bond to metals and functional groups of polymers.

10.2.9 Monosort NH_2 Groups

NH_2 groups play an important role in biology, biochemistry and genetics, especially in amino acids. They are reactive and widely involved in biochemical processes. However, primary (-NH_2) and secondary (-NH-) amino groups form nitrosamines with nitrites (R_2-NH+HNO_2→R_2N-NO+H_2O). These >N-N=O groups are suspect carcinogens [89].

Primary amino groups are contained in polyethylenimine (PEI), however together with secondary and tertiary amino groups as also found in melamine-formaldehyde resins and polypeptides/proteins. Monosort primary amino groups are present in poly(vinyl amine) (PVAm), and poly(allylamine) (PAAm). Secondary and tertiary amino groups are also found in urea- or melamine-formaldehyde resins (UF, MF).

Amino groups cannot be simply generated at the surface of polyolefins on their exposure to ammonia plasma. The yield is only in the range of 2% NH_2/C [18].

It is a pity that ammonia plasma does not produce monosort but only polysort N functional groups. Additionally, diverse O functional groups are formed by auto-oxidation on exposure to air [90]. Comparable to about 2% OH/C groups formed at any oxidation of polyolefins, just the same, i.e., 2% NH_2/C groups are produced using the ammonia plasma.

The thermodynamics is the reason why the formation of secondary and tertiary amino groups is preferred [37]:

$$NH_3 \text{ plasma} \rightarrow NH_2\bullet + \bullet H, \Delta H=+440 \text{ kJ/mol}$$
$$NH_3 \text{ plasma} \rightarrow \bullet NH\bullet + H_2, \Delta H=+209 \text{ kJ/mol}$$
$$NH_3 \text{ plasma} \rightarrow N + 3/2\ H_2, \Delta H=+191 \text{ kJ/mol}$$

It is evident that the formation of primary amino groups is generally disadvantageous.

Wet-chemical post-plasma reduction of polyolefins pretreated in ammonia plasma with diborane is also a promising approach. It also effects rapid and complete reduction of amides and nitriles.

Indirect NH_2 generation via plasma is possible by bromination first and subsequent wet-chemical reaction with NH_3 or preferably with diaminoethane (ethylenediamine) (20% NH_2/C) [21].

Plasma polymerization of allylamine (AAm) [91] or diaminocyclohexane (DACH) [92] produces only a limited yield of remaining NH_2 groups, i.e, 25-50% of expected stoichiometry of monomers.

Primary amino groups are very reactive, thus they are used for adhesion promotion as anchoring points for grafting of spacer molecules.

The amino group itself cannot interact strongly with metal atoms. Exception is copper; where diamino or tetraamino copper (II) complexes promote adhesion (see Figure 10.15).

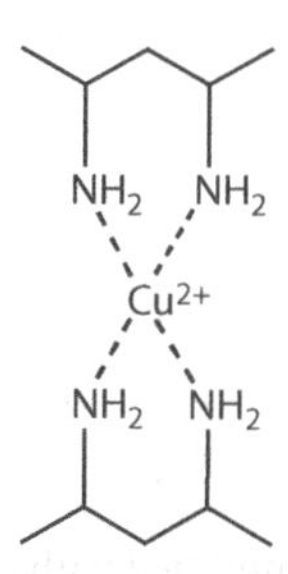

Figure 10.15 Complexation of Cu^{2+} with ethylenediamine units of poly(allylamine).

10.2.10 Other Methods for Selectively Modifying Polyolefin Surfaces with Monosort Functional Groups

Monosort functional groups are generally difficult to produce at polyolefin surfaces. There is almost no chemical, flame or plasma process which can produce monosort functional groups as mentioned before [93]. Monosort means that only one sort of functional groups is present at the surface or at least is strongly dominating (>50%). Such possibilities are shown schematically in Figure 10.16.

Such functional groups can be produced by:

- choice of compact classic polymers with (monosort) functional groups, however limited in mechanical strength, ageing behavior, water resistance, etc.
- copolymers with functional groups, such as copolymers with maleic anhydride
- calendering of classic polymer layers or glues onto polyolefin surfaces
- choice of classic polymers with functional groups as coating, deposited by dipping, spraying, spin-coating, etc.
- depositing ultra-thin classic polymers by atmospheric electrospray-ionization (ESI) [94, 95]

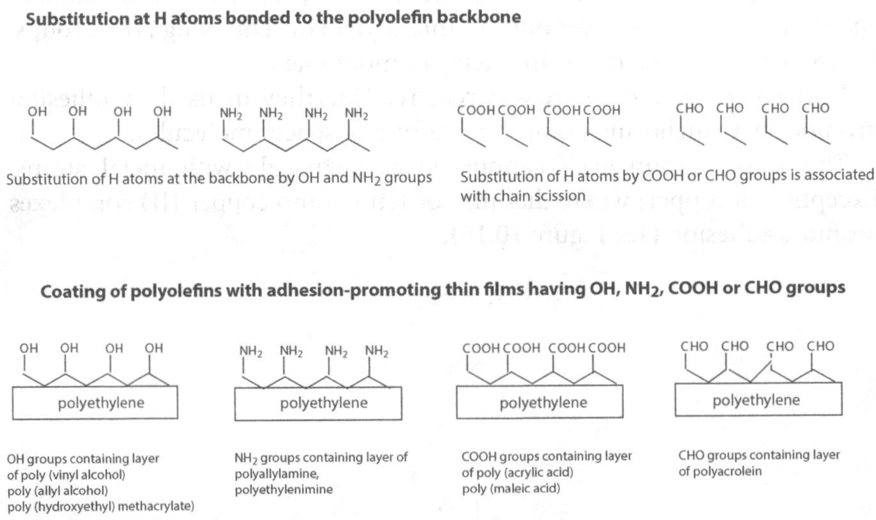

Figure 10.16 Examples of polyolefin modified with functional groups by H substitution or by coating with thin polymer layers.

- depositing ultra-thin classic polymers by atmospheric-pressure chemical-ionization (APCI), and deposition is comparable to ESI but requires assistance of plasma
- depositing classic polymers as thin films which are introduced into an atmospheric dielectric barrier discharge (DBD) as polymer solution
- depositing ultra-thin plasma polymers from low-molecular weight precursors (non-chemically polymerizable) using the atmospheric dielectric barrier discharge (DBD) or a low-pressure radio-frequency glow discharge
- wet- or plasma-chemical substitution reactions at polyolefin backbone
- gas or liquid phase halogenation and subsequent re-functionalization

Not well-suited are processes such as:

- gas-phase introduction of reagents able to form functional groups with assistance of photons, electrons, ultrasonic energy
- exposure to noble gas plasmas and addition of oxygen (peroxy formation)
- underwater plasma

Figure 10.17 represents some of the above-listed processes with their proposed schematic structures:

Using chemically manufactured classic polymers, such as poly(allyl alcohol) (PAAl), poly(vinyl alcohol) (PVA) or hydroxyethylmethacrylate (HEMA), one sort of functional group can be obtained. Copolymerization using such a monomer and an olefinic or diene comonomer allows varying the density of monosort functional groups in the resulting copolymer. The problem is that the whole polymer now has new (special) properties and composition. In contrast, modification of the polyolefins exclusively at their topmost surface has no consequence on the matrix which retains its original properties of polyolefins.

Calendering of polyethylene substrates with thin foils of poly(vinyl acetate) or poly(allylamine) may optimize the adhesion property, however, the adhesion between polyethylene and the calendered polymer is low (Figure 10.17). The same is true for dipping, spraying, spin-coating, etc.

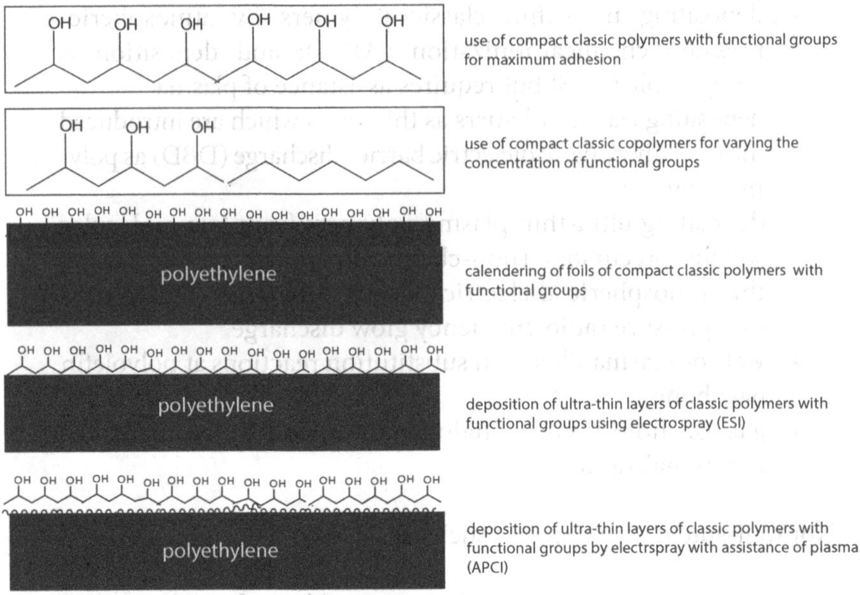

Figure 10.17 Poly(vinyl alcohol), poly(vinyl alcohol)-copolymer, polyethylene calendered with foils of poly(vinyl alcohol), electrospray-deposited ultra-thin layer of poly(vinyl alcohol) and electrospray assisted plasma deposition of poly(vinyl alcohol).

10.2.11 Monosort Bromination

However, there is one exception in plasma processing, i.e., the plasma treatment with bromine [24]. Chlorine reacts slightly exothermally with polyethylene and substitutes hydrogen, most often accelerated with the help of photons, electrons and/or heat as known from basic chemistry. Bromine reacts endothermally; it needs energy, for example plasma energy, to substitute hydrogen at aliphatic backbone. Fluorine reacts too violently. Its reaction is strongly exothermic and, therefore, undesired chain scissions are associated with post-plasma oxidation on exposure to air. This "oxyfluorination" is also achieved by addition of oxygen to fluorine and it generates very reactive and very polar O-F groups best suited for adhesion, however associated with strong polyolefin degradation and hydrolysis [62–65]. In the liquid phase, may be use of liquid Freons, the fluorination reaction can be cooled and thus better controlled [96].

Bromine (Br_2) or bromine releasing precursors, such as bromoform ($CHBr_3$), react in the plasma phase selectively at the polyolefin surface as already expected from its chemical nucleophilic substitution reactions. Bromine plasma can substitute up to 30% Br/H of all hydrogens, i.e. ca.

60% Br/C [21]. The chemical bromination of polyethylene (polyolefin) surfaces is thermodynamically not possible without additional enthalpy (endothermic). Looking to the example of polyethylene, the molar standard reaction enthalpy $\Delta_R H^\theta$ can be calculated approximately using the dissociation energies of bond scissions and forming new bonds following Hess' law:

$$Br_2 \rightarrow 2\ Br\bullet\ \ \Delta_R H^\theta = +190\ kJ/mol$$

$$Br\bullet + -CH_2\text{-}CH_2\text{-} \rightarrow HBr + -CH\bullet\text{-}CH_2\text{-}\ \ \Delta_R H^\theta = \Delta_D H^\theta_{CH2\text{-}CH\text{-}H} - \Delta_D H^\theta_{H\text{-}Br} = (395\text{-}363)\ kJ/mol = +32\ kJ/mol.$$

This reaction is endothermic and does not proceed spontaneously. The entropy does not play an important role because number of reactants = number of products.

C-C bond scission in polyethylene also does not proceed spontaneously because it needs also addition of energy:

$$Br\bullet + -CH_2\text{-}CH_2\text{-} \rightarrow -CH\bullet\text{-} + Br\text{-}CH_2\text{-}.$$

$$\Delta_R H^\theta = \Delta_D H^\theta_{CH2\text{-}CH2} - \Delta_D H^\theta_{CH2\text{-}Br} = (348\text{-}281)\ kJ/mol = +67\ kJ/mol.$$

In case of plasma exposure or UV irradiation, Br_2 can dissociate. Now, C-H bonds are scissioned and the bromination proceeds spontaneously:

$$Br_2\ (\textit{plasma}) \rightarrow 2\ Br\bullet$$

Chemical substitution proceeds as:

$$Br\bullet + H\text{-}C \rightarrow HBr + \bullet C$$

$$C\bullet + \bullet Br \rightarrow C\text{-}Br$$

Plasma initiates the following chemical reactions:

$$-CH_2\text{-}CH_2\text{-}\ (\textit{plasma}) \rightarrow -CH_2\text{-}CH\bullet\text{-} + H\bullet$$

$$-CH_2\text{-}CH_2\text{-}\ (\textit{plasma}) \rightarrow -CH_2\bullet + \bullet CH_2\text{-}$$

$$\Delta_R H^\theta = \Delta_D H^\theta_{CH2\text{-}CH\text{-}H} - \Delta_D H^\theta_{H\text{-}Br} = (0\text{-}363)\ kJ/mol = -363\ kJ/mol \text{ and}$$

$$\Delta_R H^\theta = \Delta_D H^\theta_{CH2\text{-}CH2} - \Delta_D H^\theta_{CH2\text{-}Br} = (0\text{-}281)\ kJ/mol = -281\ kJ/mol.$$

The high yield in monosort C-Br groups is caused by the preferred formation of C-Br, where Br has achieved a complete outer electron shell like krypton. The other reaction variant to fill the electron shell of Br is the addition of an electron and formation of Br^- such as in gaseous HBr, which is pumped-off from the plasma together with other gaseous reaction products.

Another important factor is the rapid quenching of any radicals generated by C-C scission with bromine leading also to C-Br groups. Thus, all radicals are saturated by Br attachment; therefore oxygen from air does not find post-plasma survived radicals for attachment as peroxy radicals.

Bromine attacks metal pipes only moderately, differently than commonly suggested. This is not a significant drawback. Purging with ammonia gas after plasma use is helpful to neutralize rest of HBr, as well as use of a cooling trap or zeolites.

The bromination proceeds rapidly. Within 10 s exposure to bromine plasma the steady-state maximum in C-Br concentration (60% Br/C) is reached but 35-40 %Br/C already after only 1-2 s exposure.

10.2.12 Advantages of Plasma Monosort Bromination

Chemical bromination of alkanes needs photons and electrons. It has lower yield compared to plasma. Chlorine is more reactive and high chlorination degree is achievable. However, the produced C-Cl group does not react wet-chemically as easily as the C-Br group. This high reactivity of C-Br bonds allows performing easily many chemical (post-plasma) reactions, such as transfer into monosort OH- or NH_2-groups. Moreover, bromine groups are well suited for grafting of spacer molecules with different terminal groups.

Bromine atoms react very rapidly with all plasma-generated radicals to form C-Br groups, even after switching-off the discharge. Thus, no free radical sites survive the plasma bromination process and cannot react with oxygen on exposure to air. It should be mentioned that using allyl bromide to deposit a monosort adhesion-promoting plasma polymer thin film onto the polyolefin with C-Br bonds, many radicals survive the deposition process and become trapped in the layer, cannot sufficiently be saturated with bromine and can subsequently react with oxygen on exposure to air. However, mixing of CH_2=CH-CH_2Br with Br_2 the post-plasma autooxidation is stopped because the radicals have reacted with the additional bromine. Moreover, this bromine addition allows producing of very high bromination degree. Using bromoform (120% Br-C/100C) instead of allyl bromide with addition of bromine, the highest bromination degree was reached (160-180 Br/C) (Figure 10.18).

However, the only drawback of all plasma reactions with polymers is also found with plasma bromination, i.e., polymer degradation by chain scissions as already mentioned. The reason why C-C chain scissions occur frequently on exposure of polyolefins to plasma the dissociation energies of C-H bonds are 385-411 kJ/mol and that of the C-C_{PE} bond is 348 kJ/mol in polyethylene (polyolefins). Therefore, C-C bond scission is preferred in the plasma compared to hydrogen abstraction. As already mentioned before shielding of C-C bonds by hydrogen atoms of the -[CH_2-CH_2]- backbone partially protects the C-C bonds but it does not prevent polymer degradation generally. As discussed in connection with the formation of LMWOM, this drawback of polymer degradation is true for plasma processes and also for all chemical substitution reactions. Therefore, the loss of brominated LMWOM by exhaustive washing or extracting of polyolefin surfaces with solvents is enormous, about 70-75% of C-Br is wahed off (Figure 10.19). However, the remaining about 25-30% Br/C are sufficient for transformation into other types of monosort functional groups in sufficient concentration or for serving as anchoring points for "grafting to" [97–99].

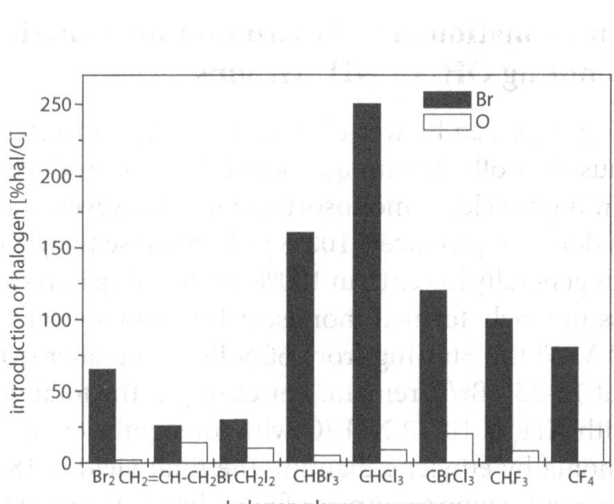

Figure 10.18 Yields of bromination and post-plasma oxidation in polyethylene.

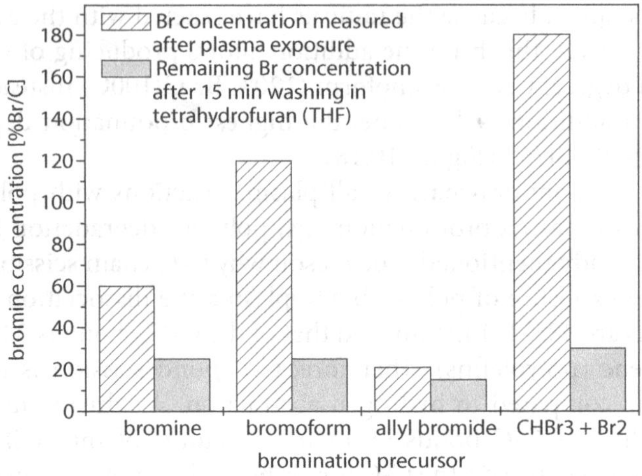

Figure 10.19 Loss in bromination at polyethylene surfaces by 15 min washing of brominated PE samples with THF. Note: $CHBr_3$ + Br_2 means 1:1 molar mixture of both components.

It must be added that brominated polyolefin surfaces bring no advantage for adhesion promotion of metal or polymer films to C-Br modified polyolefin surfaces. The C-Br group must be chemically converted into adhesion-promoting groups such as OH or COOH [21].

10.2.13 Transformation of C-Br Groups into Adhesion-Promoting OH- or NH_2-Groups

Monosort C-Br groups can be wet-chemically easily converted into other groups [21]. Thus, they offer the unique possibility to introduce other functional groups in high yield as monosort groups. However, such chemical transformation does not produce a 100% yield because the yield of chemical conversion is generally lower than 100%. Performing it in solvents there is a further loss in newly formed monosort functional groups by solvent dissolution of LMWOM. Starting from 60% Br/C and after extracting the LMWOM about 15-25%Br/C remain. For example, the total yield in reaction of C-Br with NH_3 is 10-12 NH_2/C without significant side-reactions. Replacing ammonia by ethylene diamine the total yield is 18-20 NH_2/C. The list below presents some examples of wet-chemical conversion of C-Br into other monosort functional groups, thus the range of possible monosort functional groups is much broadened:

$$\blacksquare\text{-Br} + NH_3 \rightarrow \blacksquare\text{-}NH_2 + HBr$$

$$\blacksquare\text{-Br} + NH_2\text{-}CH_2\text{-}CH_2\text{-}NH_2 \rightarrow \blacksquare\text{-NH-}CH_2\text{-}CH_2\text{-}NH_2 + HBr$$

$$\blacksquare\text{-Br} + NH_2\text{-}(CH_2)_6\text{-}NH_2 \rightarrow \blacksquare\text{-NH-}(CH_2)_6\text{-}NH_2 + HBr$$

$$\blacksquare\text{-Br} + NaNO_2 \rightarrow \blacksquare\text{-}NO_2$$

$$\blacksquare\text{-}NO_2 + NaBH_4 \rightarrow \blacksquare\text{-}NH_2$$

$$\blacksquare\text{-Br} + NaN_3 \rightarrow \blacksquare\text{-}N_3$$

$$\blacksquare\text{-}N_3\ (+\ H^+) \rightarrow \blacksquare\text{-}NH_2$$

$$\blacksquare\text{-Br} + KCN \rightarrow \blacksquare\text{-}C{\equiv}N$$

$$\blacksquare\text{-}C{\equiv}N \rightarrow \blacksquare\text{-}CH_2\text{-}NH_2$$

$$\blacksquare\text{-Br} + (Na^+)O^{(-)}\text{-}[CH_2]_n\text{-OH} \rightarrow \blacksquare\text{-O-}[CH_2]_n\text{-OH} + NaBr$$

$$\blacksquare\text{-Br} + (Na^+)O^{(-)}\text{-}CH_2\text{-}CH_2\text{-}[O\text{-}CH_2\text{-}CH_2]_n\text{-OH} \rightarrow \blacksquare\text{-O-}CH_2\text{-}CH_2\text{-}[O\text{-}CH_2\text{-}CH_2]_n\text{-OH} + NaBr$$

$$\blacksquare\text{-Br} + (Na^+)NH^{(-)}\text{-}[CH_2]_n\text{-}NH_2 \rightarrow \blacksquare\text{-NH-}[CH_2]_n\text{-}NH_2 + NaBr$$

$$\blacksquare\text{C-Br} + (H_2N\text{-phenylene})_8\text{-POSS} \rightarrow \blacksquare\text{C-NH-phenylene-POSS-}(\text{phenylene-}NH_2)_7 + HBr$$

$$\blacksquare\text{C-Br} + H_2N\text{-silane-Si}(OH)_3 \rightarrow \blacksquare\text{C-NH-silane-Si}(OH)_3 + HBr$$

The newly introduced functional groups are also monosort groups. Moreover, in the same way spacer molecules can be grafted onto the monosort C-Br groups, as well as also polymers (poly(ethylene glycol)-PEG).

10.2.14 Monosort Functionalization via Grafting of Flexible Spacer Molecules

After grafting onto monosort C-Br or other monosort groups, the newly generated flexible and water-repelling spacer surface shows again a terminal monosort functionalized surface [21, 99, 101]. Many types of monosort terminal groups of spacers were applied, such as OH, COOH, SH, NH_2 as shown below.

▒-Br + H_2N-PAMAM-$(OH)_{253}$ → ▒-NH-PAMAM-$(OH)_{253}$ + HBr (PAMAM= poly(amidoamine) with 253 OH groups

▒-Br + H_2N-POSS → ▒-NH-POSS + HBr (POSS=polyhedral oligomers of silsequioxanes with 8 NH_2 groups)

▒-Br + KN_3 → ▒-N_3 + KBr (-N_3=azide)

▒-N_3 + R_1-C≡C-R_2 → ▒-*cyclo*-$N_3C_2(R_1R_2)$ (formation of triazine rings by click chemistry)

Even poly(ethylene glycol) with a molecular weight 2000 g/mol could be grafted.

▒-Br + HO-$(CH_2\text{-}CH_2\text{-}O)_{\approx 50}$-OH → ▒-O-$(CH_2\text{-}CH_2\text{-}O)_{\approx 50}$-OH + HBr

This grafting can be continued by chain-extension with silanes, here with the head-group, such as NH_2, epoxy or vinyl. After anchoring to the polyolefin, hydrolysis of Si-O-R groups of silanes and subsequent condensation of silanol groups generates a hydrophobic, molecularly dense and unreactive intermediate 2D Si-O layer. Remaining Si-OH groups are able to form covalent bonds with vacuum-evaporated aluminium atoms of the deposit. An example shall explain this grafting: polyolefin→polyolefin-Br→polyolefin-spacer-NH_2→polyolefin-spacer-silane→polyolefin-spacer-siloxane/silane→polyolefin-spacer-siloxane/silane-aluminium.

10.2.15 Advantages of Monosort Functionalized Surfaces Formed by Spacer Molecules

Mechanical stress along the interface may be generated by different thermal expansions of composite or laminate partners or by simple mechanical loading. Forces appearing along the interface can be balanced by introduction of flexible spacer molecules as depicted in Figure 10.20, which can be compared with Figure 10.3 showing the conditions for direct coupling via (monosort) covalent bonds.

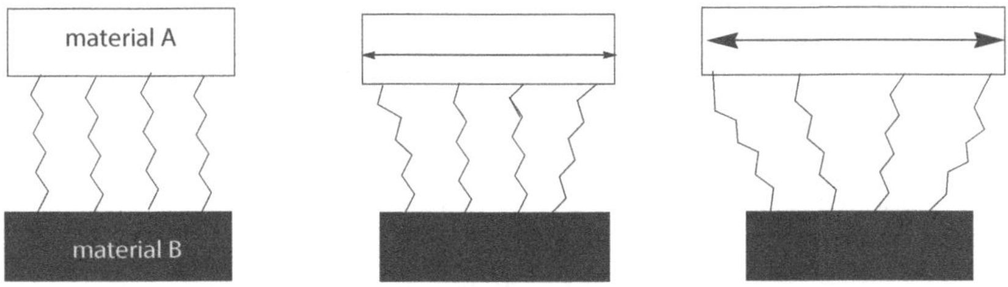

Figure 10.20 Bond breaking along the interface because of different thermal expansions of materials involved.

Moreover, additional helpful functions of spacers are presented for the final laminate between polyethylene and vacuum-evaporated aluminium film [19].

These functions of spacers are (Figure 10.21):

- flexibilization across the interface by aliphatic spacers
- water-repelling by hydrophobic aliphatic spacers
- water-repelling and hindrance to ion migration and initiation of polymer degrading redox reactions as well as protecting the polar covalent metal-oxgen-silicon bond from hydrolysis
- remaining silanol groups react chemically with aluminium (and other metal atoms).

10.2.16 Monosort Functional Groups Deposited via Plasma Polymers as Ultra-Thin Film Adhesion Promoters

Another variant for generating monosort functional groups is the deposition of a thin almost always strongly adhering plasma polymer film [59, 102, 103]. For preparing such plasma polymer films, it must be differentiated between classic monomers able to polymerize in a chemical way (chain-growth polymerization) and precursors which are not able to react chemically to form polymers, however, as exception being under plasma conditions. The precursor molecules become fragmented in the plasma and form a polymer-like deposit in a random fragmentation-recombination

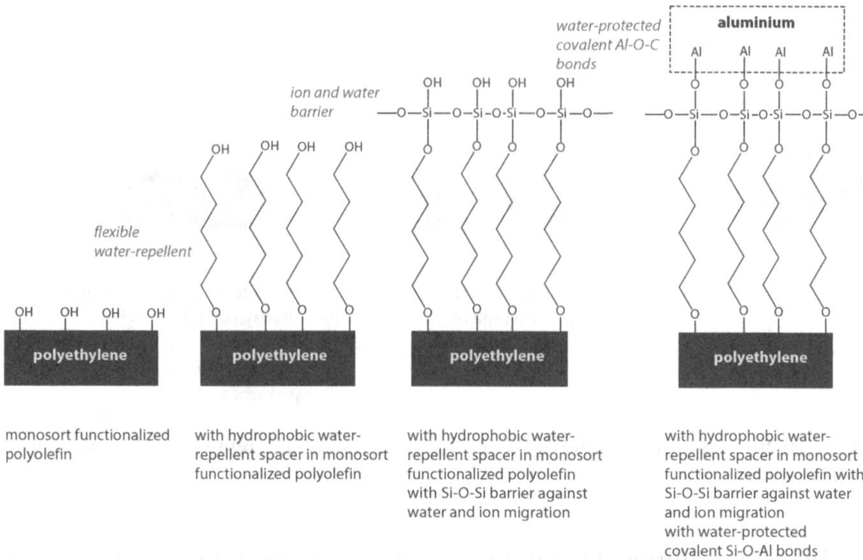

Figure 10.21 Schematic presentation of monosort functionalized polyethylene surface, with monosort spacers, with Si-O-O ion and water barrier as well as with water-protected Si-O-Al covalent bonds to aluminium metal coatings.

process. These plasma polymers have structures completely different from those of classic polymers. They are crosslinked but also trap low-molecular weight fragments and oligomers in their structure. Using a classic polymerizable monomer the resulting polymer structure also shows many irregularities in comparison to classic polymers.

Moreover, the polyrecombination of radicals is not complete, thus the remaining radicals are trapped in the growing layer during plasma polymerization. These remaining radicals can then react post-plasma with oxygen on contact with air. A slow auto-oxidation process starts. Linear sequences as product of regular chain-growth polymerization occurs using classic monomers; however, further plasma particle bombardment and irradiation with energy-rich vacuum UV radiation rearrange such regular chemical structures.

This random and stochastic formation of plasma polymers leads to a considerable loss of monosort functional groups by conversion into other groups or by elimination. However, this irregular polymer structure with trapped radicals may be responsible for good adhesion property. The introduction of considerable concentration of oxygen post-plasma by auto-oxidation changes the plasma polymers often from monosort films to partially or predominantly polysort functionalized films. Only in case

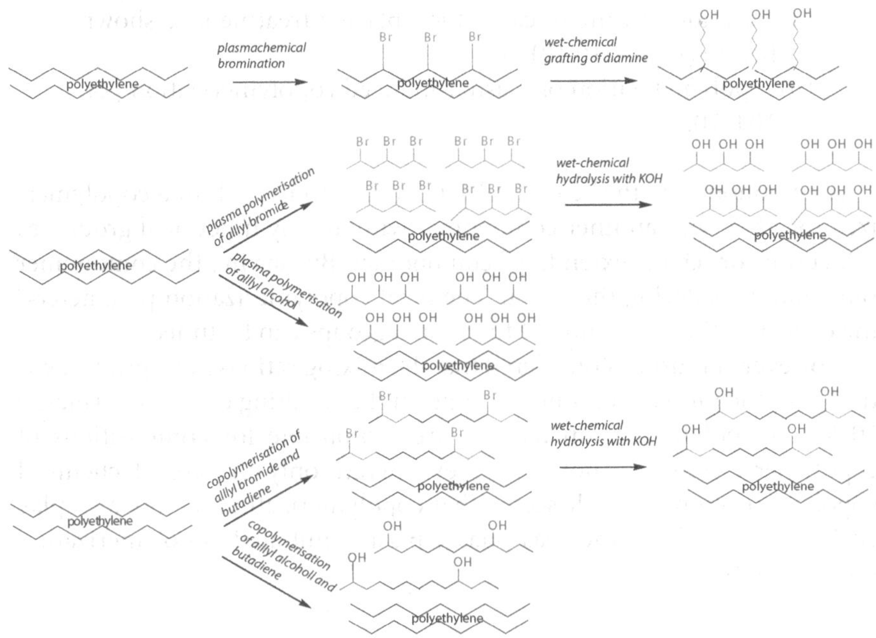

Figure 10.22 Bromination of polyethylene on exposure to a plasma of elemental bromine, post-plasma hydrolysis to OH groups using KOH, deposition of allyl bromide plasma polymer and its hydrolysis post-plasma, deposition of allyl alcohol plasma polymer, deposition of a copolymer of allyl bromide and butadiene and its post-plasma hydrolysis, as well as the deposition of a copolymer of allyl alcohol and butadiene.

of the plasma polymerization of allyl alcohol, the auto-oxidation produces also the desired OH groups. Thus, up to 80-90% OH groups are obtained compared to the OH content of classic poly(allyl alcohol). This is the sum of retained OH groups from the polymerized allyl alcohol polymer and newly formed OH groups by auto-oxidation.

The monomer used for such plasma polymer can be selected from a number of monomers with the desired functional groups (Figure 10.22).

It has been found that ultra-thin polymer layers with functional groups deposited at thicknesses of about 10 to 20 nm adhere well to polyethylene. It may be that at the beginning of deposition the plasma "activates" the polyolefin target, thus, a higher degree of adhesion results.

10.2.17 Variation of the Density of Monosort Functional Groups by Copolymerization

Variation of the density of monosort functional groups can be realized by

(a) variation of time in case of gas plasma treatment as shown before (Figure 10.10) or
(b) copolymerization or plasma-induced copolymerization [24, 104, 105].

A comonomer with the desired functional group is plasma copolymerized together with another comonomer without any functional group i.e., a "neutral" or "chain-extending" comonomer. By varying the comonomer ratio and considering the actual "chemical copolymerization parameters" the concentration of monosort functional groups can be tuned.

However, it must be considered that the mixing ratio is not a priori identical with the ratio of comonomer units in the resulting copolymer (Figure 10.23). "Copolymerization factors" are responsible for compositions of copolymers. However, normally they operate only in case of chemical copolymerization [105]. It seems that copolymerization factors also play an important role in the (gas phase) plasma-initiated copolymerization process [103].

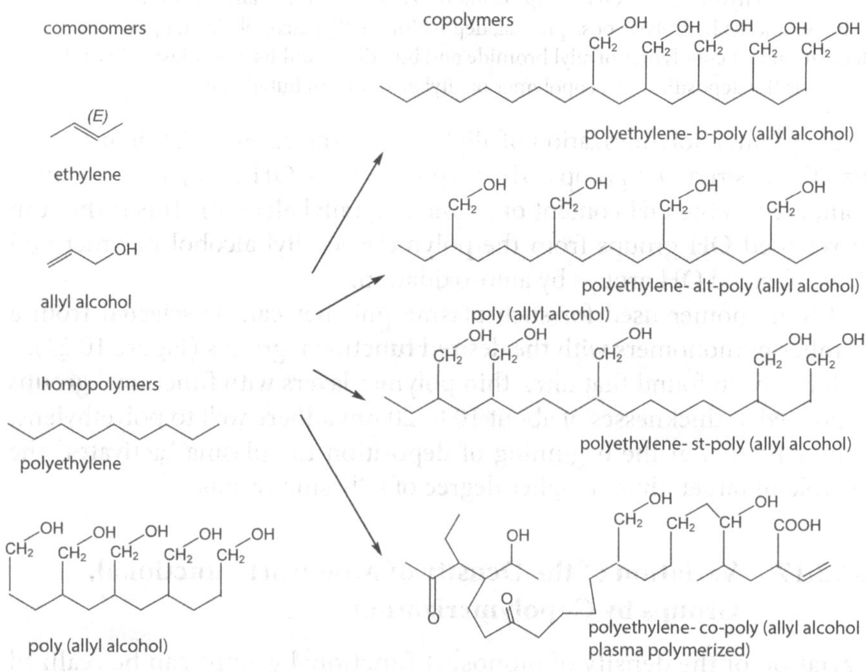

Figure 10.23 Models of copolymers synthesized chemically or plasma chemically.

Chemical copolymerization is not a simple mixing of two comonomers with a resulting copolymer showing the same ratio of comonomer units as the original mixing ratio. Sometimes the ratio of different comonomer units in the resulting copolymer is very different from the original mixing ratio of comonomers or even each comonomer forms only a homopolymer without copolymer sequences. Here, the "copolymerization rules" determine the process of copolymerization i.e., not every comonomer copolymerizes with another one to a form copolymer. Moreover, plasma conditions do not allow chain growth polymerization or copolymerization. Such classic and well-defined reactions are associated with and most often dominated by (co)monomer fragmentation and their random recombination to form irregular "polymers" and "copolymers". Nevertheless, it is possible to vary the monosort functional groups by plasma-induced copolymerization, also with high retention of monosort functional groups, especially in case of OH groups.

There is surely a high degree of homopolymerization instead of expected real chemical copolymerization (and plasma-caused simple crosslinking) but coiling of linear chains and only a few links (weak crosslinking) leads to the same behavior and group density as with copolymers. However, the desired density of functional groups must be determined beforehand by preparing calibration curves. Then, it is easy to prepare the desired density of monosort functional groups. The density or concentration of resulting functional groups is determined by derivatization of functional groups with fluorine-containing markers and measuring these with X-ray photoelectron spectroscopy (XPS) [106].

10.2.18 Plasmaless Coating with Classic Polymers Carrying Monosort Functional Groups

Dipping, spraying, spin-coating, lamination, evaporation and many other techniques are available [43–50]. On a molecular level of the topmost surface, the flame and plasma treatments are widely in use.

Instead of plasma polymerization with the assembling of polymers from monomer units, it is also possible to deposit intact polymers with high molecular weight in monomolecular dimensions with and without assistance of plasmas. With assistance of plasma the deposited polymer is degraded but adheres well to the substrate surface [107]. Some original sequences of the polymer survive in the deposited polymer layer. Deposition without polymer degradation (electrospray-ionization, ESI) with monomolecular layers of intact polymers leads also to high adhesion [94, 95]. The process with injection of a polymer solution into the plasma using the atmospheric dielectric barrier discharge (DBD) plasma results in deposition of strongly irregular polymer layers. Deposition of macromolecules after their singularization (separation of individual macromolecules) in the atmosphere by means of a high voltage field (ESI) and the additional activation of macromolecules by (corona) plasma is another possibility. It is applied to polymers soluble in water, known as atmospheric-pressure chemical ionization (APCI). The same process can also be applied without any plasma action, thus any polymer degradation is avoided. This process is known as electrospray ionization (ESI). It should be emphasized that this process produces pinhole-free layers on conducting surfaces.

10.3 Bonding of Metal Coatings to Polysort and Monosort Functionalized Polyolefins

10.3.1 Physical Bonding

The question concerning the most important interactions for maximum promotion of adhesion is: Are more but much weaker physical bonds most important for adhesion or a few but extraordinary strong chemical bonds?

Indeed, physical interactions are much weaker than chemical bonds (Table 10.2) [1–8].

Use of physical interactions at the interface of metal coating to polyolefins is the preferred way in industrial applications. However, chemical bonds dominate for adhesives used for bonding metal substrates.

All types of polar functional groups are suited to undergo physical interactions. It is possible and relatively simple to generate polar groups at polymer surfaces, most often by oxidation. Therefore, the common effort is to produce many polar groups to establish physical interactions in very high concentrations along the interface, after the motto "the more, the better". However, such physical interactions are eliminated in presence of humidity.

Notwithstanding the above, the hypothesis is to produce an equivalence of polar groups on both sides to obtain maximum adhesion. It is also expressed in terms of similar or equal surface free energies [49]. Generated high concentrations of polar groups are often associated with polymer degradation as discussed before in detail. Indeed, the thus generated LMWOM at polyolefin surfaces adheres very well to coatings and metals but not to the intact polyolefin matrix. On peeling the coating, the peel front is relocated from interface to the near-interface polymer layer resulting in only moderate adhesion. Generation of a low concentration of polar groups at

Table 10.2 Physical and chemical interactions and bonds.

Type of force	Maximal operating distance [nm]	Bond energy [kJ/mol]
physical forces		
van der Waals		
dispersion (Heitler-London)	0.3-0.5	0.5-2
interdiffusion [9]	several nanometers	
dipole-dipole (Keesom)	0.3-0.5	20
acid-base [10]	0.3-0.5	20
dipole-induced dipole (Debye)	0.3-0.5	2
hydrogen and electrostatic bonds		
hydrogen bond	0.3-0.5	50
electrostatic forces [11]	50	high
chemical forces/chemical bonds		
covalent (single C-bonds)	0.1-0.2	300-490
covalent (double C-bonds)	0.1	530-745
covalent (triple C-Bonds)	0.1	839-891
metallic	0.1-0.2	110-260
ionic	0.1-0.2	560-1000

the polyolefin surface may avoid too strong polymer degradation associated with loss of mechanical strength as well as LMWOM formation but the sum of interactions is too low for efficient adhesion.

10.3.2 Covalent Bonding along the Metal-Polymer Interface

Extensive covalent interlinking across the interface of a composite needs monosort reactive functional groups on both sides of interface. Most often, only one side or one component of the composite, here polyolefins, must be equipped with monosort functional groups in high density to have enough strong bonding sites.

Using the synthesis of monosort groups as described before only one type of covalent bonds can be established along the interface. The methods described for generation of monosort functional groups allow producing a high density of monosort groups at the surface in such a high concentration as with polysort oxidations. The concentration of polysort functional groups for maximum adhesion ranges from a few % to 20 % O/C (see Figure 10.13). By generating reactive monosort groups, such as COOH groups, which can react with the metal to form covalent bonds, the maximum concentration of COOH groups needed for adhesion is much lower than 20% O/C with polysort groups. Especially, such maximum adhesion, obtained with chemical bonds, shows a different behavior of failure on mechanical loading. Now, the weak point on mechanical loading is not the insufficient bonding force along the interface (interface failure) but the new weak point location is within the topmost polyolefin layer or even in the polymer matrix itself. With increasing bonding forces the locus of failure moves firstly into the polymer adjacent to the interface (near-interface failure). As shown later, after introduction of flexible spacer molecules into the interface the mechanical failure shifts further into the polymer bulk (cohesive failure).

The problem of bonding of metals (or other coatings) to polymers is that the covalent bonds formed have a strong partial ionic character. Such deformed bond orbitals can be attacked easily by water molecules because of their highly polar character. Therefore, the tendency of such covalent bond to hydrolyze is very high. Thus, a durable adhesion is only possible by protection of bonds from moisture.

Evaporated metal (Me) atoms impinging on the polyolefin surfaces are able to react with the topmost carbon atoms forming Me-C [1] or with oxidized polymer surfaces forming Me-O-C bonds [108–111]. Both types of metal-polymer bonds hydrolyze easily. After metal deposition the polymers show a lower binding energy component at 283.4 eV in the C1s signal of XPS spectra, which was attributed to the Al-C bond (carbide) [1]. In the presence of oxygen-containing groups on both sides i.e., the polymer as well as the metal side, for example poly(vinyl alcohol) (PVA) and oxidized aluminium, an $AlCO_{1.5}$ complex is formed at the aluminium-PVA interface. This stoichiometry corresponds to that of aluminium oxide (Al_2O_3) [53].

Using poly(ethylene terephthalate) as a substrate instead polyolefins, Al-O-C bonds can be formed by reaction of the ester (or carboxylic) group with metallic aluminium [109, 110]. This reaction involves one (monodentate) or two (bidentate) bonded Al atoms.

Using an aluminium substrate (pseudoboehmite surface) and depositing poly(acrylic acid) on it at ambient air conditions monodentate bonding was found (-CO-O-Al<) [108] using FTIR spectroscopy as well as bidentate ($>Al-O)_2C<$ formation too, using inelastic electron tunneling spectroscopy (IETS) [111] (Figure 10.24).

Aluminium has a strong electronegative redox potential (E=-1.66 V). Thus, it binds oxygen as oxide or hydroxide or fluorine as fluoride (AlF_3). At the interface of metallic aluminium e.g., vacuum-deposited, and polymers with oxygen- or fluorine-containing units or functional groups, the following tendency of redox reaction exists:

1 Physical interactions (Al^0) → 2 chemical bonds (Al^{-1}) → 3 full redox reaction to Al_2O_3 ($Al^{1.5}O$).

The driving force is the negative redox potential as also found with magnesium or in the extreme case with alkali metals. Thus, for adhesion purposes, redox reactions with extremely electronegative alkali metals are applied to make polytetrafluoroethylene bondable, for example: $-(CF_2\text{-}CF_2)_n\text{-} + 4\ Na \rightarrow (>C{=}C<)_n + 4\ NaF$.

It should be emphasized that physical interactions despite their high concentration after oxidation produce only moderate adhesion. The next step in redox reaction was establishing dense covalent Al-O-C bonds giving maximum adhesion. Completion of this redox reaction after physical interaction (1) and covalent bonding (2) leads to much weaker adhesion. The redox reaction is complete after formation of (presumably amorphous) Al_2O_3 (3). Thus, it can be concluded that the Al-O-C bonds are an unstable intermediate in the thermodynamic sense. They can only be preserved and stabilized along the interface by additional protection against further reduction (and hydrolysis).

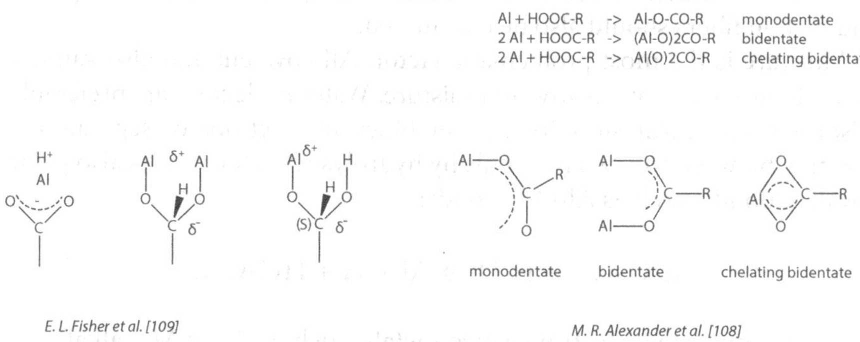

Figure 10.24 Proposed bonding of carboxylic groups of poly(acrylic acid) to metal substrates [108, 109].

Another but opposite example is Cu as copper (II) ions in CuO. Cu^2, on the contrary, has an electropositive redox potential. It tends to release oxygen from its most often oxidized metal surface (CuO). Copper ions can diffuse and transfer across the interface into polyolefins or other polymers and oxidize them along the interface also in a redox process [55]. Thus, an oxidized LMWOM is generated, reducing the adhesion.

Thus, there are many requirements along the interface between metal and polymer which must be fulfilled to reach a permanent and strong adhesion, for example to aluminium:

- formation of covalent Al-polyolefin bonds (Figure 10.20)
- prevention of complete redox reaction to Al_2O_3
- protecting against ion migration and starting of redox reaction
- protection against moisture by insertion of hydrophobic spacer chains
- flexibilization along the interface by insertion of freely rotatable alkane spacer molecules for damping of mechanical stress.

Although not often considered, differences in electrochemical potentials of metals and not defined (hypothetical) of functionalized polymer surfaces act as driving force for redox reactions also on metal-polymer interface. A defined redox potential cannot officially be assigned to polymers but functional groups or unsaturations or ionic sites in the polymer structure can undergo such redox reactions as known in chemistry from the quinone-hydroquinone redox system [37]. Thus, these constituents have a "quasi-electrochemical potential". Ion migration/diffusion occurs across the interface. Restoration of aromatic rings in substituted graphene and carbon fibers should also be considered.

Moisture is the most problematic factor. All covalent and also van der Waals bonds are very sensitive to moisture. Water molecules are preferably adsorbed onto polar sites. Then, they block interactions or separate the bonding between the two materials by hydrolysis. Water attacks also polar covalent bonds, such as Al-O-C bonds:

$$\text{Al-O-C}\lessdot + H_2O \rightarrow \text{Al-OH} + \text{HO-C}\lessdot.$$

As already stated, electronegative metals, such as Al or Mg, alkali elements, tend to pick up oxygen from the polymer. Noble and semi-noble metal ions, such as Cu^{2+}, also Cu bond in an oxide layer, oxidize polymers along the interface. Associated with this oxidation, ions begin to diffuse into the polymer, oxidize it partially. The polymer is now degraded and oxidized at the interface. Therefore, ionic fragments of the degraded polymer are formed. Such ions, carrying carboxylic groups COO^-, were found to be responsible for the electrical breakdown in cable insulations made from polymers and for provoking electrical breakdown [112]:

$$>CH_2 + x\ CuO \rightarrow Cu^0 + \text{-}COO^-Cu^+ \text{ or } (\text{-}COO^-)_2Cu^{2+} \text{ and}$$

$$>CH\text{-}O\text{-}CH< + x\ Al^0 \rightarrow >CH + Al_2O_3$$

As for inorganic materials, the most important examples are glass and carbon fibers in polymer or resin composites. In case of glass fibers the silanol groups (⋧Si-OH) are most important to form covalent bonds. Aminosilane is often used as primer for glass fibers used in epoxy resins:

glass fiber + $(HO)_3\text{-}Si\text{-}C_3H_6\text{-}NH_2$ + *epoxy-resin* → *glass fiber-O-*$Si(OH)_2\text{-}C_3H_6\text{-}NH\text{-}CH_2\text{-}CH(OH)$*-epoxy resin* [21–23].

Polymer-polymer chemical reactions along the interface via monosort functional groups are known as "grafting to" and shall not be further discussed in this context.

The adhesion phenomenon takes place at the interface of two materials in one atomic layer on each side of interface, thus emphasizing the role of functional groups at the topmost surface. Different forces responsible for adhesion have been discussed [2–7].

For a long time, chemical bonding at the interface was termed as the key factor for improved adhesion [47, 48]. Watts proposes the term "The Holy Grail of Adhesion" [52]. Chemical bonds as an essential part of adhesion have been assumed since 1960/1970s, in particular with the introduction of silane coupling agents [113–115], for example, glass fibers coated with aminosilane and embedded in epoxy resin showed covalent bonding between fiber and resin.

The proof for the real existence of chemical bonds is difficult and vague. It was presented by means of X-ray photoelectron spectroscopy (XPS) [116]. The role of chemical bonds in polymer-metal and polymer-silicate adhesion was emphasized by Plueddemann [113–115], Gähde [117] and Friedrich *et al.* [82].

10.4 Adhesion Results for Evaporated Aluminium Coating on Poly- and Monosort Functionalized Polyolefin Surfaces

10.4.1 Competition of Adhesion Efficiency between Strong Covalent Bonds and Polar Groups with Weak Interactions

Polysort functional groups at polyolefin surfaces were introduced by flame or atmospheric plasma exposure to achieve an improved adhesion; however it was realized by simple and inexpensive methods. Monosort functional groups were synthesized with the ambitious goal to form exclusively covalent bonds across the interface of metals and modified polyolefins. Single covalent carbon-carbon and carbon-hydrogen bonds have bond energies of about 400 kJ/mol, whereas physical interactions have bond energies of about 4 kJ/mol i.e., they are about 100 times weaker than covalent bonds. To have the same interaction along the interface i.e., the same adhesion, the polar groups must have a 100 times higher concentration than that of chemical bonds. Using classic copolymers or plasma modification of polyolefins, most often more than 10% groups/C are introduced and are available for bond formation. This is almost the same range of concentration as with polar groups. Limiting to 10% after plasma modification would prevent an excessive formation of LMWOM associated with the formation of a weak boundary layer (WBL) and reducing the adhesion.

Chemical bonding is associated with the displacement of failure at mechanical loading from the interface to near-interface zone in the polymer as mentioned before. The limiting factor of the polyolefin oxidation is the residual bonding of the highly functionalized low molecular weight oxidized material (LMWOM) to the underlying intact polymer bulk (Figure 10.22), which then determines the peel strength [41]. The highly functionalized LMWOM may form many covalent bonds with the coating; however because the LMWOM is only loosely bonded to the polymer matrix, therefore the covalent bonds at the interface are without effect (Figure 10.22).

It is evident that 20% O/C is the maximum for both the surface free energy (Figure 10.13) as well as the peel strength of thin evaporated aluminum films in case of polysort functionalization. It is remarkable that about 4% O/C generate only a low polar contribution to surface free energy but the peel strength is already strongly increased and levels off near the maximum (Figure 10.25).

In some cases of using classic primer systems, excessive amounts of functional groups decrease the adhesion strength, as shown for poly (vinyl chloride)-maleic acid/aluminium joints and for poly(methyl methacrylate-acrylamide/epoxy ester primed steel joints (Figure 10.26) [50, 118–120].

The decrease in peel strength at higher concentrations of functional groups at the surface of non-plasma treated classic copolymers must have another origin than plasma-formed LMWOM on exposure to oxygen plasma. One can argue that too many interactions and bonds make the interface region rigid and therefore brittle and inflexible. Another explanation would be that an exhaustive reaction of COOH groups with the metallic aluminium (PVC-COOH/Al) cannot be stopped at the level of mono-/bidentate bonds and thus a complete redox reaction could occur.

To sum up, the redox reaction along the interface could oxidize aluminum to its oxide/hydroxide and decompose the COOH groups and thus the molecular structure of the plasma-deposited poly(acrylic acid). Thus, a redox-formed LMWOM is produced:

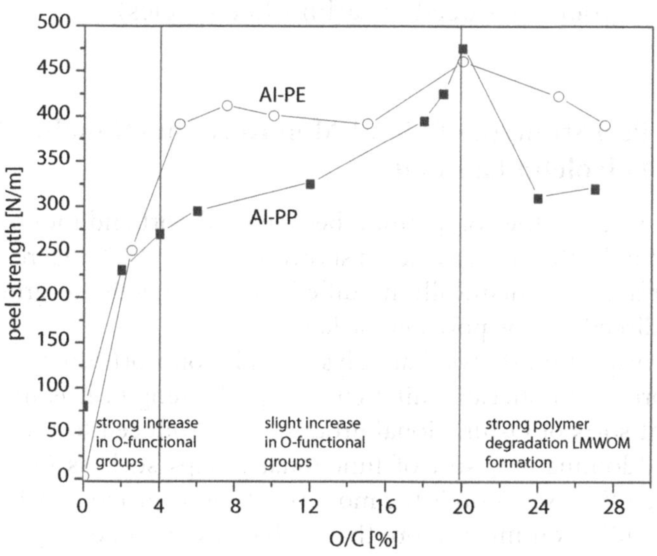

Figure 10.25 Peel strength of 150 nm thermally vacuum-evaporated aluminium deposited onto oxygen low-pressure plasma modified polyolefins.

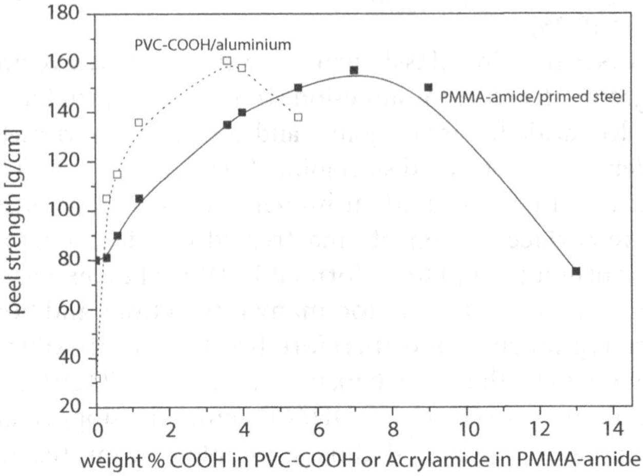

Figure 10.26 Peel strength in dependence on COOH concentration of the primer.

2 Al + HOOC → bidentate (>Al-O)$_2$C< → bonded carboxylic groups

Al + HOOC → monodentate Al-O-CO- → Al_2O_3 → reduced carboxylic groups (reduced to carbon-like species).

10.4.2 Peel Strength of Metal-Monosort Functionalized Polyolefin Laminates

A further result of the competition between polysort and monosort functionalized polyethylene surfaces is shown in Figure 10.27. Here, the polyolefin surface was chemically modified, i.e. the functional group X was attached directly to the polymer backbone.

A first inspection shows that polysort and monosort functional groups did not produce noticeable difference in peel strength. It is insignificant if different sorts of O-functional groups are introduced to the surface or only one (dominating) sort of functional groups such as OH, CHO or phenylene-OH. Exception is the monosort COOH group which promotes the Al-PE adhesion much more than polysort O functional groups or all other monosort O functional groups. Due to the significantly increased peel strength per single COOH group it can be assumed that in this case chemical bonds played an important role.

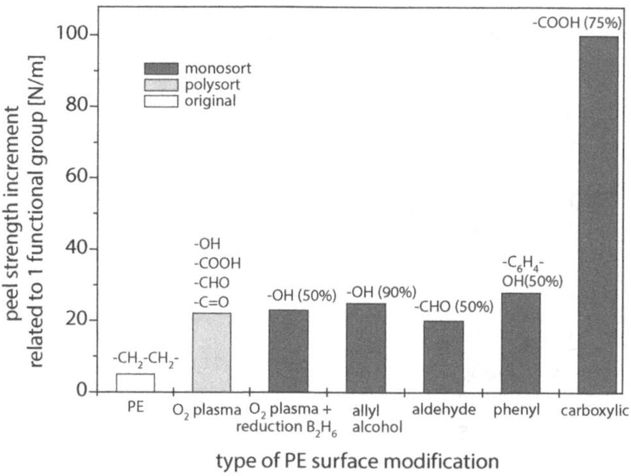

Figure 10.27 Competition for influence of mono- or polysort surface functionalization of polyethylene using different plasma techniques and the influence of the nature of each monosort functional groups referenced to 1 functional group with percents from total number of (all) O-functional groups.

Conversely, the nature of groups other than COOH has no significant influence on peel strength. It can be speculated that the other groups interact more physically.

Another variant was to coat the polyolefin surface with a thin plasma polymer layer (ca. 50 nm) using allyl alcohol for OH group functionalization and acryl acid for COOH functionalization as monomers. Using the plasma-initiated copolymerization with ethylene the concentration of OH or COOH groups in the resulting copolymer deposit could be varied for OH and COOH groups from 0 to 29 %OH/C and from 0 to 27% COOH/C.

Nearly 30% OH/C are needed to result in high peel strength and peeling occurs far from aluminium-polyethylene interface. Only 10% COOH were needed to measure (the same) high peel strength causing cohesive failure (near-interface). Higher COOH concentration did not further increase the peel strength because the cohesive strength of polyethylene near the interface was the limiting factor (Figure 10.28).

The failure was never observed at the interface between polyethylene and plasma polymer.

The maximum adhesion of COOH groups already at low concentration (<10% COOH/C) (Figure 10.28) confirms again the assumption of formation of COOH-Al covalent bonds. Highest OH concentration was reached

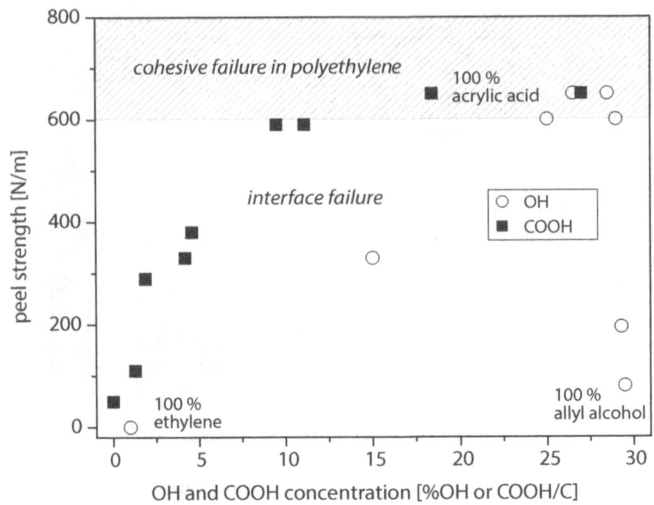

Figure 10.28 Ethylene-allyl alcohol and ethylene-acrylic acid copolymers with different concentrations of OH and COOH groups and their influence on the measured peel strength.

by deposition of pure (100%) ally alcohol plasma polymer. However, such layer was too weak and had no cohesive strength. Therefore, the peel strength decreased drastically (>28% OH/C). On the other hand, plasma polymers possess some flexibility, thus they can balance internal mechanical stress. However, mechanical stress in the near-interface polymer cannot be balanced because the functionalized polymer molecules are densely tacked to the metal coating via chemical bonds. Therefore, polymer molecules chemically bonded to the metal coating are stiff and inflexible.

10.4.3 Peel Strength of Monosort Terminal Groups of Flexible Spacers Grafted to the Polyolefin Surface

Figure 10.29 compares polysort, monosort and monosort-spacer equipped polyolefin surfaces.

As shown in Figure 10.28 for direct coupling of Al to different monosort O functional groups introduced to polyethylene surface, here, starting from brominated polyolefin surfaces, grafting of C_{10} spacers and the results in peel strengths are documented in Figure 10.30. 1 COOH spacer group produces the same peel strength as 10 COOH groups directly bonded to the polyolefin surface. Using OH terminated spacers a similar result is found.

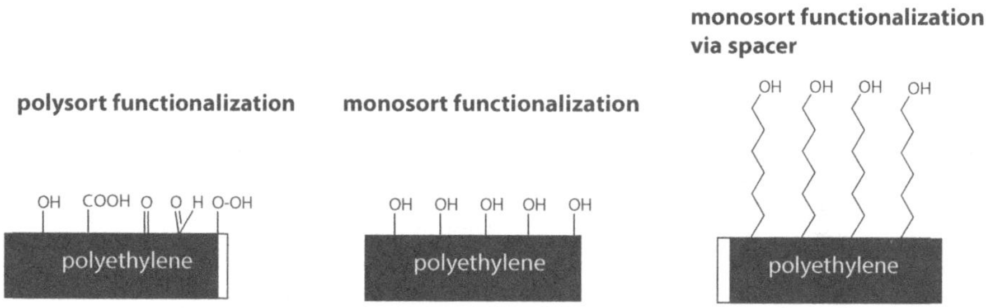

Figure 10.29 Schematics of polysort, monosort and monosort-spacer functionalizations.

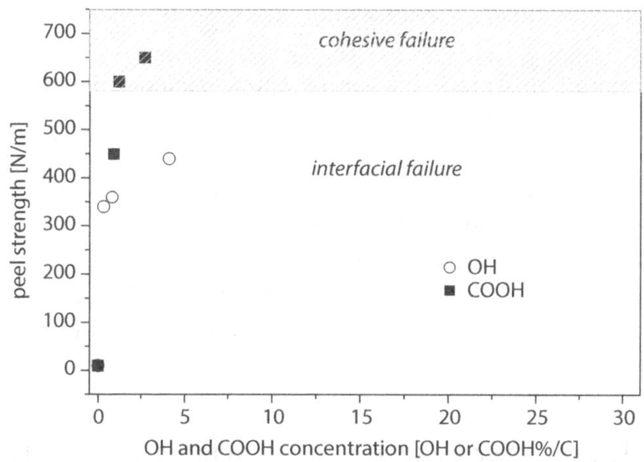

Figure 10.30 Ethylene-allyl alcohol and ethylene-acrylic acid copolymers grafted with spacers in different concentrations of OH and COOH terminal groups of spacers and their influence on measured peel strength

It is obvious that insertion of flexible aliphatic spacer molecules between polyolefin surface and monosort functional groups increases the resulting adhesion to aluminium dramatically. At first, the interface is transformed into a flexible and elastic interphase. Thus, mechanical stress is not focused on interface or near-interface polymer molecules but the elastic spacer molecules can better distribute and balance the stress. The spacer molecules are covalently grafted to the polyolefin and adhere also covalently to aluminium, for this reason the same bonding forces are present as in case of direct bonding without spacer (cf. Figure 10.28). Secondly, it can be further speculated that the monosort functional group at the end of the

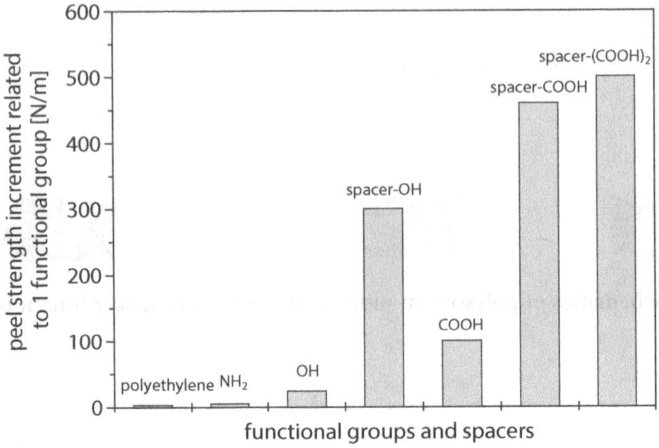

Figure 10.31 Comparison of peel strength increments for monosort directly bonded or by spacer to polyolefin bonded functional groups.

freely rotatable aliphatic spacer chain can better find and can get closer to anchoring points of the aluminium layer.

Figure 10.31 underlines the surprising increase in peel strength with spacers, here as peel strength increment related to 1 bonding terminal group. The hydroxyl groups bonded to spacer molecules also promote high adhesion.

10.4.4 Role of Spacer Length and Associated Flexibility

As discussed before, the abrupt transition of properties from inorganic or metallic solids at the interface to those of polymers must be (mechanically) distributed and balanced by flexible elements. Otherwise, the covalent bonds between the two different materials are disrupted mechanically on thermal exposure or mechanical loading. Aliphatic chains are suited as flexible and elastic molecular units. Free rotation around single bonds and cis-trans isomerism make aliphatic CH_2 sequences flexible. Figure 10.32 shows the maximum density (concentration) of a series of grafted spacer molecules in dependence on their chain length. The larger the coil volume is the smaller is the number of grafted molecules. Therefore, a compromise between needed flexibility and maximum graft density must be found.

To find out the influence of spacer chain length on adhesion, a series of ethylene glycols with different chain lengths were grafted onto the brominated polyolefin surface (Williamson's ether synthesis). After vacuum

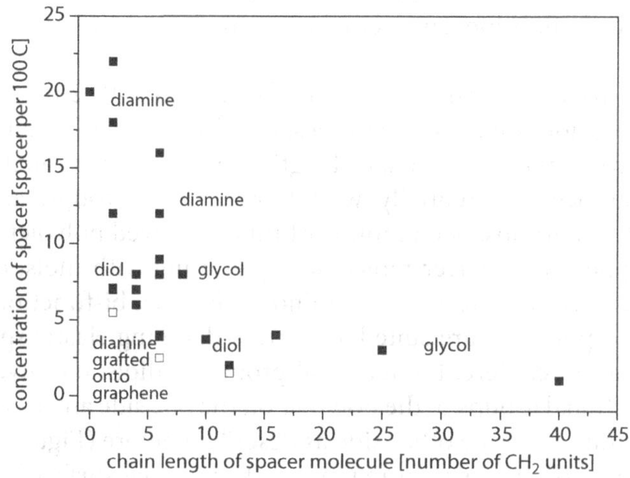

Figure 10.32 Maximum spacer concentration at polyolefin surface in dependence on the chain length of the spacer molecule.

evaporation with 150 nm aluminum, the peel strength of the laminates was measured. The results are depicted in Figure 10.33. The peel strength is significantly increased with increasing chain length. Longer chains could not be grafted because the same concentration of spacers could not be realized as for smaller ones. The space requirement would be too large.

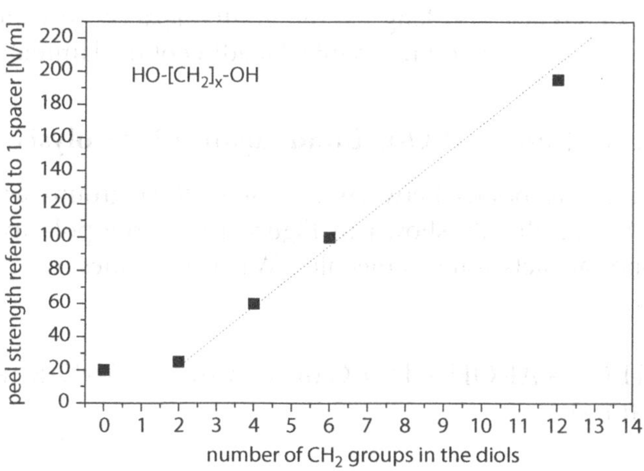

Figure 10.33 A series of ethylene glycols (diols) with different chain lengths as spacers and the influence of chain length on peel strength of aluminium coating.

Perfluorinated aliphatics are inappropriate because aluminium reacts completely with fluorine and removes it from the molecule ($Al + 3/2\ F_2 \rightarrow AlF_3$).

Aliphatic spacer molecules can also be equipped with 2 monosort terminal groups (diols, diglycols, diamines, dicarboxylic acids, dialdehydes). They can have variable CH_2 chain length. The monosort functionalized polyolefin surface preferentially with C-Br-, $-NH_2$ groups, also CHO, COOH, OH, SH groups. Such monosort functionalized polyolefin surface can be equipped with spacer molecules by grafting with diols, diamines, di-aldehydes, dicarboxylic acids or amino acids. Such bi-functional oligomers and also polymers are suited as a stress-damping elastic spacer unit across the interface. Here, the terminal groups of diols and glycols react with the functional groups of the polymer on the one side and also with the metal (aluminium) on the other side as described before (Figure 10.33).

Long-chain diols may also form loops at the polymer surface by reaction with an adjacent C-Br group:

$$▒\text{-Br} + HO\text{-}CH_2\text{-}CH_2\text{-}CH_2\text{-}CH_2\text{-}CH_2\text{-}CH_2\text{-}CH_2\text{-}OH \rightarrow$$

$$▒\underline{\text{-O-}CH_2\text{-}CH_2\text{-}CH_2\text{-}CH_2\text{-}CH_2\text{-}CH_2\text{-}CH_2\text{-O-}}▒$$

However, the reaction with the aluminium is preferred.

Glycols of different chain lengths are also used. The C-O-C unit is a rigid bond element, thus the two OH-end groups of the glycol cannot be bonded simultaneously to the surface as loop. Additionally, grafting a mixture of glycols with different chain lengths, the grafted spacers disrupt sequentially, not at the same time on mechanical loading of the laminate.

10.4.5 Protection of Al-O-C Bonds against Hydrolysis

Mono or bidentate bonds between OH or COOH groups and metals (aluminium) were already shown in Figure 10.24. The polarity of these covalent bonds attracts water molecules. Water molecules hydrolyze such bonds easily:

$$Al\text{-}O\text{-}C + H_2O \rightarrow Al\text{-}OH + HO\text{-}C \text{ or with silanes } Al\text{-}O\text{-}Si + H_2O \rightarrow Al\text{-}OH + HO\text{-}Si$$

Therefore, these bonds must be protected against water. The metal bulk is impermeable to water if the thickness is sufficient and if the layer is pinhole-free; however, the interface itself is a channel for water molecules. Si-O- structures, i.e. Si-O-Si layers can efficiently hinder diffusion of water molecules and that of ions from the metal side to the polymer side. Silanes are very suited to form such a barrier monolayer by condensation of silanol groups (Figure 10.34).

$$\text{R}'\text{-Si(OR)}_3 + \text{H}_2\text{O} \rightarrow \text{R}'\text{-Si(OH)}_3$$

$$\text{R}'\text{-Si(OH)}_3 + \text{(HO)}_3\text{Si-R}' \rightarrow \text{R}'\text{-Si(OH)}_2\text{-O-Si-(OH)}_2\text{-R}' + 2\ \text{ROH}$$

$$\text{condensation in 2D plane (R-Si (O-Si}'\ldots)_2\text{-OH and}$$

$$\text{(R-Si (O-Si}'\ldots)_2\text{-OH} + \text{Al} \rightarrow \text{(R-Si (O-Si}'\ldots)_2\text{-O-Al} <$$

The Si-O monolayer and also the aliphatic spacer molecules are hydrophobic and therefore water-repellent. The Si-O-C or the Si-NH-C bonds of the spacer to the polyolefin substrate are resistant against hydrolysis. Only the metal-spacer (Al-O-Si) bond is fragile and must be protected by Si-O layer.

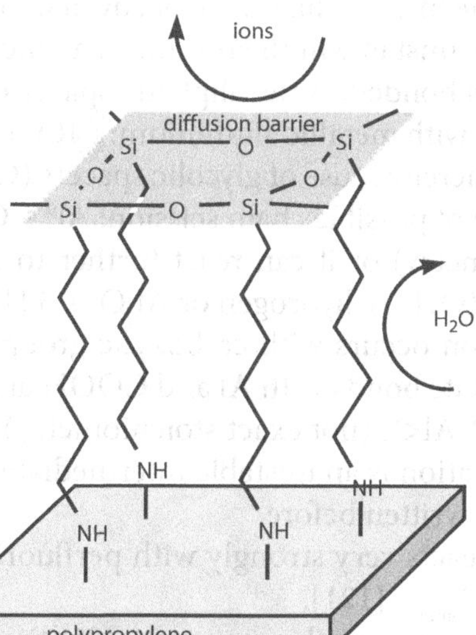

Figure 10.34 Water impermeable protection layer.

Use of silanes with one or two unhydrolysable Si-CH_3 groups has confirmed the significance of the formed (condensed)-Si-O layer for adhesion and durability of metal-polyolefin laminates.

10.4.6 Redox Reactions and Ion Migration

It was described before that Al-O-C or Al-O-Si bond must be protected from hydrolysis. It was also mentioned that besides redox reactions, ion migration from metal-side of the interface to the polymer bulk is also possible and causes problems, for example in the cable industry by ion migration through the polymer insulation causing electrical breakdown as already mentioned. Therefore, this migration must also be stopped.

The same Si-O-Si layer, which hinders water diffusion to the metal-polyolefin mono or bi or chelating bidentate bonds can hinder ion migration too (Figure 10.34).

As already mentioned, metallic aluminium (vacuum-evaporated) reacts with the oxygen of the silane spacer (or polymer [51]) because of its strong negative redox potential (E^0_{red}=-1.66 V) first to the desired covalent bond: ■Si-OH + Al^0 → ■Si-O-Al + hydrogen (monodentate bonds) or the complete redox reaction starts: ■Si-OH + Al^0 → ■Si + $Al(OH)_3$. In case of a complete redox reaction, chemical bonding is absent, only physical interactions are possible. Thus, the Si-O layer should stop the progress of the redox reaction at the intermediate state of covalent bond. Thus, mono- and bidentate bonds are unstable in the thermodynamic sense.

Further O atoms bonded in the aliphatic spacer or in the polymer substrate can also react with metallic aluminium: ▒R-OH + Al^0 → ▒(R-O)-Al + hydrogen (*slow*). Therefore, use of glycolic spacers (C-O-C) seems not to be well-suited because of possible chain scission: Al^0 + C-O-C → Al_2O_3 (without exact stoichiometry) or it can react further to aluminium hydroxide: Al^0 + 3 H_2O- → $Al(OH)_3$ + hydrogen or Al_2O_3 + 3 H_2O → 2 $Al(OH)_3$ [54].

A similar reaction occurs with carboxylic groups of poly(acrylic acid) layers. Here, bidentate bonds with Al and COOH are formed: ▒$(\text{-COOH})_2$ + 2 Al^0 → ▒-C<$(\text{O-Al}<)_2$ (not exact stoichiometry). It should be repeated that the bond formation is an unstable intermediate of the complete redox reaction as already written before.

Evaporated Al reacts very strongly with perfluorinated spacers: x Al^0 + $(CF_2)_n$ →x AlF_3 + C_{amorph} [121].

Nevertheless, there is a further connection to the thermodynamics. The complete redox reaction lowers the high surface free energy of Al by formation of Al_2O_3 (Figure 10.12) and the oxidation of polymer surfaces increases the surface free energy. This increased surface free energy is slightly decreased by the redox reaction and the removal of certain oxygen used for the oxidation of Al to Al_2O_3 resp. $Al(OH)_3$ across the interface (3 R-OH + $Al^0 \rightarrow 3$ R + $Al(OH)_3$). In general, the surface free energies of the polyolefin and the metal coating along the interface converge. Such reaction is complete and ends if surface free energies of laminate materials are equal. Thus, the surface energies of polyethylene ($\sigma \approx 30$ mJ/m^2) and aluminium ($\sigma \approx 1200$ mJ/m^2) change after oxidation and interface redox reaction to 45 mJ/m^2 for the oxidized polyethylene and to 70 mJ/m^2 for the oxidized aluminium, with a much smaller difference in surface free energies.

10.5 Realization of Ideal Covalently Bonded Interface

10.5.1 Schematic Structure of "Ideal" Interface

Considering all new insights, requirements, limitations and ideas described before, it was possible to construct an aluminium-polyolefin laminate which could not be peeled (or sheared). It was not possible to separate the metal without mechanical separation of the laminate deep in the polyolefin matrix also after long-term aging under high humidity and elevated temperature as well as cooling-heating cycles. Figure 10.35 presents the principle structure of such metal-polyolefin laminate.

To realize such ideal composite structure, first step was the introduction of monosort functional groups, preferably Br and OH, NH_2 or COOH groups. This monosort functionalization was an indispensable precondition for formation of a high density of covalent bonds between polymer and metals or inorganics, polymers.

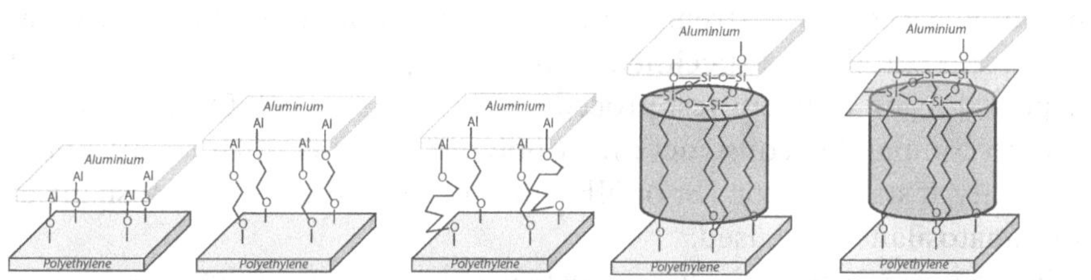

Figure 10.35 Principle structure of the interface between aluminium and polyethylene for fabricating mechanically inseparable laminates with long-term durability.

The second step was the consumption, i.e. covalent bonding of these monosort functional groups with aliphatic spacers of at least 3-6 CH_2 and with maximum 10 units for flexibilization of the interface by well-known wet-chemical or gas phase reactions (C-Br and ethylene diamine). The spacer molecules consist of a hydrophobic bifunctional alkane.

Third reaction step was the chain-extension by coupling the terminal group of the spacer with a silane. Here, the headgroup of silane reacted with the terminal group of the spacer molecule also covalently.

Fourth step was the hydrolysis of the silyl ether groups of the silane to silanol groups. Adjacent silanol groups condensed to a two-dimensional Si-O layer with covalent Si-O bonds, forming an ion- and moisture barrier layer.

Fifth step was the reaction of the remaining (third) silanol group with evaporated aluminium to form Al-monodentate/bidentate bonds (Si-O-Al). These bonds are covalent ones but with polar (ionic) share. These bonds are very sensitive to hydrolysis and, therefore, they were protected by a two-dimensional Si-O layer and hydrophobic (water-repellent) aliphatic spacer molecules against water diffusing along the interface. Thus, the diffusion of water along the interface and of ions across the interface was stopped in all directions.

Other realized variants of graft synthesis starting from $C-NH_2$ and C-Br groups at polyolefin surfaces were also successfully tested.

10.5.2 Results

The above explained "ideal" laminate showed very high peel strength, more exactly, the laminate could not be peeled or separated by available means. The laminate also withstood humid, warm climate, temperature change, rain in regular test scenarios, etc. at least for one year without measurable loss in peel strength, more exactly without mechanical separation [54]. Thus, the test specimen of the metal-polymer laminate could not be peeled using different test procedures (adhesive tape). Other chemical ways to comparably construct metal-polymer systems show equal/similar durability, if starting from Br or NH_2 groups. In such cases, aminosilane or isocyanatosilane were used.

Figures 10.35 shows the generally realized structures of interface design with different functions of molecular units:

- monosort functional groups for covalent bonding (for example Si-OH)
- flexible aliphatic chains for buffering mechanical stress $[-(CH_2)_{2-10}-]$

- monosort functional groups of aliphatic spacer (NH_2, OH…) chains for covalent bonding to the monosort functional groups at the surface of the polymer substrate (Br, NH_2) $(Si\text{-}O)_x$
- inorganic barrier layer against diffusion of ions and water molecules (Si-O-Si)
- covalent bonding of silanol groups (hydrolyzed silane) to aluminium (Si-O-Al)

It was found that all 3 silyl ether groups of the silane were needed [19]. The formation of all possible silanol groups of the alkoxy silane and their partial condensation to a two-dimensional $(Si\text{-}O)_x$ layer was significant for adhesion.

Similar results could be obtained by using plasma polymers with monosort functional groups as anchoring points for spacer and silane grafting. A problem was the lower durability on exposure to ageing conditions (humid hot climate) because of the long-term ageing of plasma polymer.

10.5.3 Comparison of Results

Figure 10.36 presents a survey on the results of differently designed interfaces in Al-polyolefin laminates.

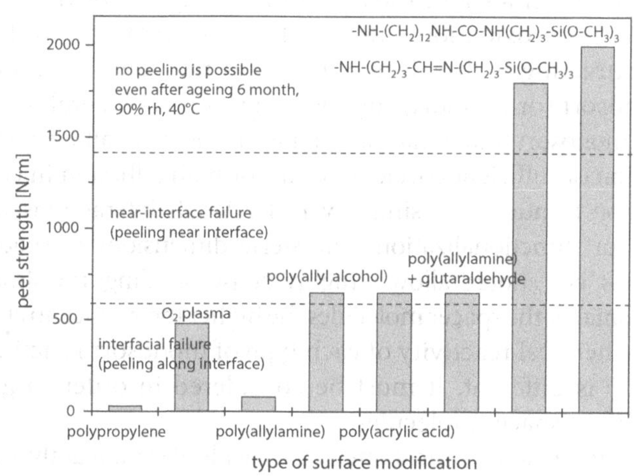

Figure 10.36 Survey on maximum peel strength results for differently modified aluminium-polypropylene laminates.

10.5.4 Limitations of Ideally Designed Interface

The drawbacks of the formation of the "ideal" metal-polyolefin laminate are:

1. The considerable experimental effort in such multi-level approach may be a significant drawback. First, the realization of mono-functionalization is difficult using plasma or chemical processes and secondly, the designing of the durable and mechanically inseparable "ideal" laminate is also cumbersome, expensive and costly.
2. The monosort functionalization is often not complete, with the exception of bromination. Mixture of dominating monosort and partial polysort functionalization occurs in some cases. Monosort was defined as when one sort of functional group dominates and accounts for more than 50% of all functional groups.
3. Plasma and chemical treatments are generally associated with chain scission, i.e. polymer degradation. This unintendedly generated low-molecular weight oxidized material (LMWOM) presents a potential weak boundary layer according to Bikerman [9]. Moreover, the dominant radical mechanism using plasma exposure leads to the generation of trapped radicals in the polymer (or in plasma polymers). Such trapped radicals react over long periods with oxygen from air and thus they initiate ageing of the laminates (auto-oxidation). Removal of LMWOM by washing or extracting with solvents decreases the concentration of monosort (or polysort) functional groups considerably.
4. The monosort functionalization must be dense to allow the formation of sufficient covalent bonds for high adhesion in comparison to numerous simple van der Waals interactions with polysort functionalization. The steric dimensions of spacers, silanes, etc. should allow a highly dense bonding; the shorter and smaller the spacer molecules the higher their concentration.
5. The chemical reactivity of each type of monosort functional group is different. It must be considered in order to graft sufficient spacer molecules.
6. Chemical reactions most often have yields of significantly lower than 100%. Moreover, steric effects can hinder a reaction.

10.6 Summary

Monosort versus polysort functionalization of polyolefin surfaces should be compared. Which type of surface modification of polyolefins produces a stronger and more durable adhesion to aluminium?

The answer is open in many cases. Exception is the monosort functionalization with carboxylic groups which leads to the formation of chemical bonds between COOH groups and Al (mono and bidentate). OH groups also form such covalent (polar) bonds (monodentate) with slightly lower efficiency in adhesion or higher concentration is needed.

It was assumed that many weak physical interactions can balance the much higher bond energy of a few chemical bonds. Here, it could be demonstrated that it is possible to generate a comparble high concentration of chemical bonds via monosort functionalization as in the case of polysort modification. Therefore, the covalently bonded metal-polyolefin laminates showed much higher adhesion and durability.

The limiting factor for the generation of polysort as well as monosort functionalization is the formation of a weak boundary layer, the formation of LMWOM, by preferred scission of C-C bonds because of the lower binding energy of C-C in polyolefins in comparison to the C-H bonds.

And last, but not least, the energy transfer from plasma to the polyolefin surface is very large and causes frequently C-C bond scissions (LMWOM formation) and also crosslinking and other side-reactions.

Another factor is the decomposition (and consumption) of functional groups by redox reactions which cannot be completely stopped at the level of unstable intermediate formation of covalent bonds. The redox reaction tends to proceed and to form aluminium oxide as end product.

Ways to produce polysort functional groups are flaming, chemical oxidation and exposure to oxygen plasma or underwater plasma. All sorts of oxygen-containing functional groups are formed without dominance of one sort. This oxidation uses the well-known pathways known from chemical oxidation of paraffins and also from auto-oxidation, occurring also post-plasma and started by reaction of trapped radicals with diffusing oxygen from air.

Monosort functionalization is absent on polyolefin oxidation generally. One exception is the exposure to elemental bromine plasma, which produces highly selective (up to 97%) high concentrations (60% Br/C) of C-Br bonds.

C-Br groups are chemically reactive and can be easily converted to other monosort functional groups by well-known wet-chemical reactions, for further modification or for grafting of amines and alcohols.

As known, the oxygen plasma generates a wide variety of O functional groups. Most of these groups can be post-plasma chemically reduced with diborane to OH-groups. The same is true for many N-functional groups which can be converted into primary amino groups. These wet-chemical processes increase the yield of OH or NH_2 groups considerably.

Another possibility to modify polyolefins with adhered monosort functional groups is the deposition of ultra-thin polymer layers with monosort functional groups. Dipping, spraying, etc. as well as plasma techniques are used.

Plasma polymers as thin coating deposited from functional group carrying monomers are a monosort functionalized layer on polyolefin surfaces. Drawbacks of such technique are the irregular structure of the deposited polymer layer, considerable loss in original monosort functional group,s and long-lasting post-plasma oxidation.

Injecting solutions of classic polymers into the atmospheric plasma increases only marginally the chance that longer sequences of the regular polymer survive the plasma deposition process and therefore more functional groups are obtained.

Another thin-film process also starts from polymer solutions. The deposition takes place in presence of high-voltage without plasma assistance (ESI). The sprayed polymer film retains its original structure and adheres well to polyolefins as ultra-thin film and serves as adhesion promoter. This process deposits singly or multiply ionized macromolecules step by step. The deposited molecules do not lose any functional groups.

The "royal pathway" for adhesion is the transfer of monosort functionalization from the polyolefin surface to monosort functionalization to one level higher, i.e., above the polyolefin surface by covalent bonding of flexible spacer molecules. These spacer molecules can balance all mechanical stress by their flexible aliphatic chains.

Covalent bonds between polyolefin surfaces and metal coatings, introduction of flexible long-chain spacers with water-repelling (hydrophobic) property, or segments and barrier-forming units produce non-peelable laminates with very high resistance towards moisture and long-range ageing (Figure 10.37).

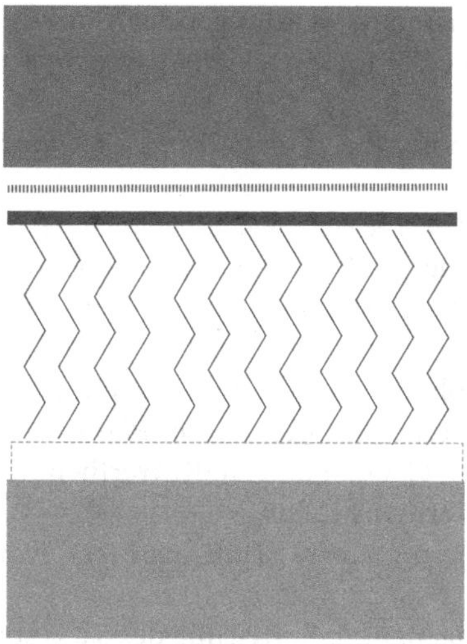

Figure 10.37 Scheme of the ideally designed interface for a metal-polymer laminate.

Acknowledgements

I am most thanksful to Gundula Hidde for competent, exact and extensive experimental work. Sascha Wettmarshausen contributed many interesting ideas, as well as Renate Mix, Asmus Meyer-Plath, Gerhard Kühn, and Renate Decker. Wolfgang Unger made valuable contributions to analytical investigations.

References

1. B. M de Koven and P. L. Hagans, XPS studies of metal/polymer interfaces—thin films of Al on polyacrylic acid and polyethylene, *Appl. Surf. Sci.* 27, 199-233 (1986).
2. W. H. Keesom, Die van der Waalsschen Kohäsionskräfte (van der Waals cohesion forces), *Physikalische Zeit* 22, 129-154 (1921).
3. P. J. W. Debye, Interatomic and intermolecular forces, in: *Adhesion and Cohesion*, P. Weiss (Ed.), pp. 1-17, Elsevier, New York, USA (1962).

4. F. London, Über einige Eigenschaften und Anwendungen der Molekularkräfte (Properties and applictaions of molecular forces), *Z. Physik. Chem.* B11, 222-251 (1930).
5. F. London, The general theory of molecular forces, *Trans. Faraday Soc.* 33, 8-26 (1937).
6. W. Heitler, *Quantentheorie und homöopolare chemische Bindung (Quantum theory and homopolar bonds)*, Akad. Verlagsgesellschaft, Leipzig, Germany (1934).
7. V. A. Parsegian, *Van der Waals Forces: A Handbook for Biologists, Chemists, Engineers, and Physicists*, Cambridge University Press, UK (2006).
8. J. Friedrich, R. Mix and G. Kühn, Polymer surface modification with monotype functional groups of different type and density, in: *Plasma Processing and Polymers*, R. d'Agostino, P. Favia, C. Oehr and M. R. Wertheimer (Eds.), pp. 3-21, Wiley-VCH, Weinheim, Germany (2005).
9. J. J. Bikermann, *The Science of Adhesive Joints*, Academic Press, New York, USA (1968).
10. C. Engler and W. Wild, Ueber die sogenannte Activirung des Sauerstoffs und über Superoxydbildung (Excitation of oxygen and superoxide formation). *Berichte der deutschen chemischen Gesellschaft* 30, 1669-1681 (1897).
11. A. N. Bach, On the role of peroxides in the processes of slow oxidation. *J. Russ. Phys.-Chem. Soc.* 29, 373–395 (1897).
12. W. Langenbeck and W. Pritzkow, Untersuchungen über den Mechanismus der Paraffinoxydation (Investigtaions on mechanism of paraffin oxidation). Teil 1, *Fette u. Seifen* 55, 435-440 (1953).
13. A. D. Walsh, Processes in the oxidation of hydrocarbon fuels-1. *Trans. Faraday Soc.* 42, 269-279 (1946).
14. D. G. Knorre, Some elementary stages in the liquid-phase oxidation of hydrocarbons, in: *The Oxidation of Hydrocarbons in the Liquid Phase*, N. N. Emanuel (Ed.), pp. 45-53, Pergamon Press, Oxford, UK (1965).
15. G. Wietzel, Herstellung synthetischer Fettsäuren durch Oxydation von paraffinischen Kohlenwasserstoffen mit molekularem Sauerstoff (Formation of synthetic fatty acids through oxidation of paraffinic hydrocarbons using molecular oxygen). *Chem. Eng. Sci.* 3, Supplement 1, 17–30, IN1-IN4 (1954).
16. F. Asinger, *Chemie und Technologie der Paraffinkohlenwasserstoffe (Chemistry and technology of paraffinic hydrocarbons)*, Akademie Verlag, Berlin, Germany (1956).
17. A. N. Bashkirov and V. Kamzolkin, Synthesis of higher aliphatic alcohols by direct oxidation of paraffinic hydrocarbons, *Proceed. World Petroleum Congress* 4, 175–183 (1959).
18. J. Friedrich, R. Mix and G. Hidde, Ways to generate monosort functionalized polyolefin surfaces. *Rev. Adhesion Adhesives.* 2, 311-370 (2014).
19. J. Friedrich, Tailoring of interface/interphase to promote metal-polymer adhesion, in: *Adhesion in Microelectronics*, K. L. Mittal and T. Ahsan (Eds.), pp. 67-135, Wiley-Scrivener, Beverly, MA, USA (2014).

20. R. G. Nuzzo and G. Smolinsky, Preparation and characterization of functionalized polyethylene surfaces, *Macromolecules* 17, 1013-1019 (1984).
21. S. Wettmarshausen, R. Mix, A. Meyer-Plath, H.-U. Mittmann and J. Friedrich, Plasmabromination - the selective way to monotype functionalized polymer surfaces, in: *Polymer Surface Modification: Relevance to Adhesion, Vol. V.*, K. L. Mittal (Ed.), pp. 3-18, CRC Press, Boca Raton, FL, USA (2009).
22. R. T. Chen, B. W. Muir, G. K. Such, A. Postma, R. A. Evans, S. M. Pereira, K. M. McLean and F. Caruso, Surface "click" chemistry on brominated plasma polymer thin films. *Langmuir* 26, 3388–3393 (2010).
23. J. Friedrich, Study of metal-polymer interfaces: Relevance to Adhesion, in: *Recent Advances in Adhesion Science and Technology in Honor of Dr. Kash Mittal*, V. Gutowski and H. Dodiuk (Eds.), pp. 3-33, CRC press, Boca Raton, FL, USA (2014).
24. J. F. Friedrich, R. Mix, R.-D. Schulze, A. Meyer-Plath, R. Joshi and S. Wettmarshausen, New plasma techniques for polymer surface modification with monotype functional groups, *Plasma Processes Polymers* 5, 407-423 (2008).
25. J. Friedrich and G. Hidde, Ultra-thin polymer films as novel class of adhesion promoters. *Rev. Adhesion Adhesives* 3, 1-52 (2015).
26. J. Friedrich and J. Meichsner, *Nonthermal Plasmas for Materials Processing*, Wiley-Scrivener, Beverly, MA (2022).
27. J. Friedrich, S. Wettmarshausen and M. Hennecke. Haloform plasmas applied to polymers, *Surface Coatings Technol.* 203, 3647-3655 (2009).
28. U. Resch-Genger, K. Hoffmann, R. Mix and J. F. Friedrich, Monitoring of amino-functionalities on plasma-chemically modified polypropylene supports with a chromogenic and fluorogenic pyrylium reporter. *Langmuir* 23, 8411-8418 (2007).
29. R. Mix, K. Hoffmann, U. Resch-Genger, R. Decker and J. F. Friedrich, Covalent coupling of fluorophores to surface-bonded functional groups, in: *Polymer Surface Modification: Relevance to Adhesion, Vol. 4,* K. L. Mittal (Ed.), pp. 171-192, CRC Press, Boca Raton, FL, USA (2007).
30. R. H. Hansen, J. V. Pascale. T. de Benedicts and P. M. Rentzepis, Effect of atomic oxygen on polymers. *J. Polym. Sci.: Part A* 3, 2205-2214 (1965).
31. J. Friedrich, G. Kühn and J. Gähde, Untersuchungen zur Plasmaätzung von Polymeren. I. Strukturveränderungen von Polymeren nach Plasmaätzung (Investigations on plasma etching of polymers. I. Structural changes of polymers after plasma etching). *Acta Polymerica* 30, 470-477 (1979).
32. H. C. Brown, *Hydroboration*, Benjamin, New York, USA (1962).
33. H. C. Brown, From Little Acorns to Tall Oaks - From Boranes Through Organoboranes. Nobel Lecture (December 8, 1979).
34. J. A. Venables, G. D. T. Spiller and M. Hanbucken, Nucleation and growth of thin films. *Reports Progr. Phys.* 47, 399-410 (1984).

35. R. Cueff, G. Baud, M. Benmalek, J. P. Besse, J. R. Butruille and M. Jacquet, X-ray photoelectron spectroscopy studies of plasma-modified PET surface and alumina/PET interface. *Appl. Surf. Sci.* 115, 292-298 (1997).
36. J. T. Hall and P. K. Hansma, Chemisorption of monocarboxylic acids on alumina: A tunneling spectroscopy study. *Surf. Sci.* 77, 61-76 (1978).
37. J. Friedrich, The formation processes of functional groups at polyolefin surfaces on exposure to oxygen or ammonia plasma: A critical review. *Rev. Adhesion Adhesives* 6, 253-328 (2018).
38. K. Schwetlick, *Organikum, 24th ed.*, Wiley-VCH, Weinheim, Germany (2015).
39. M. Strobel, C. Dunatov, J. M. Strobel, S. Lyons, S. J. Perron and M. C. Morgen, Low-molecular-weight materials on corona-treated polypropylene. *J. Adhesion Sci. Technol.* 3, 321-335 (1989).
40. M. Strobel, V. Jones, C. S. Lyons, M. Ulsh, M. J. Kushner, R. Dorai and M. C. Branch, A comparison of corona-treated and flame-treated polypropylene films. *Plasmas Polym.* 8, 61-95 (2003).
41. M. Strobel, S. M. Kirk and C. S. Lyons, Analysis of water-soluble, surface-oxidized polypropylene films. Unpublished results (2019).
42. L. J. Gerenser, X-ray photoemission study of plasma modified polyethylene surfaces. *J. Adhesion Sci. Technol.* 1, 303-318 (1987).
43. J. Friedrich, *Metal-Polymer Systems: Interface Design and Chemical Bonding*, Wiley-VCH, Weinheim, Germany (2017).
44. F. M. Etzler, Determination of the surface free energy of solids. *Rev. Adhesion Adhesives* 1, 3-45 (2013).
45. F. M. Etzler, Chracterization of surface free energies and surface chemistry, in: *Contact Angle, Wettability and Adhesion, Vol. 3,* K. L. Mittal (Ed.), pp. 219-264, CRC Press, Boca Raton, FL, USA (2003).
46. C. Bischof and W. Possart, *Adhäsion: Theoretische und experimentelle Grundlagen (Adhesion: Theoretical and Experimental Basics),* Akademie-Verlag, Berlin (1983).
47. G. Habenicht, *Kleben: Grundlagen, Technologien, Anwendungen (Gluing: Basics, Technologies, Applications)*, Springer, Heidelberg, Germany (2008).
48. D. E. Packham (Ed.), *Handbook of Adhesion*, Wiley, Chichester, UK (2003).
49. G. A. Dykerhoff and P.-J. Sell, Über den Einfluß der Grenzflächenspannung auf die Haftfestigkeit (Correlation between interface energy and measured adhesion). *Angew. Makromol. Chem.* 21, 169-185 (1972).
50. S. Wu, *Polymer Interface and Adhesion*, Marcel Dekker, New York (1982).
51. W. Possart (Ed.), *Adhesion: Current Research and Applications*, Wiley-VCH, Weinheim, Germany (2005).
52. J. F.. Watts, The interfacial chemistry in adhesion: Novel routes to the Holy Grail?, in: *Adhesion: Current Research and Applications,* W. Possart (Ed.), pp. 1-16, Wiley-VCH, Weinheim, Germany (2005).
53. K. L. Mittal, The role of the interface in adhesion phenomena. *Polym. Eng. Sci.* 17, 467-473 (1977).

54. J. Friedrich, I. Loeschcke and J. Gähde, Zur Adhäsion von Aluminium auf Polypropylen (Adhesion of aluminium to polypropylene). *Acta Polymerica* 37, 687-695 (1986).
55. J. Friedrich, B. Falk, I. Loeschcke, B. Rutsch, Kh. Richter, H.-D. Reiner, U. Throl and H. Raubach, Untersuchungen zur Haftung zwischen plasmachemisch oxidiertem Kupfer und Polyphenylchi-noxalin (Investigations on adhesion of plasma-chemically oxidized copper to poly(phenylquinoxaline). *Acta Polymerica* 36, 310-320 (1985).
56. J. F. Friedrich, W. E. S. Unger, A. Lippitz, R. Giebler, I. Koprinarov, St. Weidner and G. Kühn, Significance of interfacial redox reactions and formation of metal-organic complexes for the adhesion of metals on pristine and plasma-treated polymers, in: *Polymer Surface Modification: Relevance to Adhesion, Vol.* 2, K. L. Mittal (Ed.), pp. 137-172, CRC Press, Boca Raton, FL, USA (2000).
57. J. Friedrich, *The Plasma Chemistry of Polymer Surfaces*, Wiley-VCH, Weinheim, Germany (2012).
58. M. Strobel and C. S. Lyons, The role of low-molecular-weight oxidized materials in the adhesion properties of corona-treated polypropylene film. *J. Adhesion Sci. Technol.* 17, 15-23 (2003).
59. J. Friedrich, Beschichtung und Modifizierung von Kunststoffoberflächen (Coating and modification of plastics), in: *Vakuum-Plasma-Technologien Beschichtung und Modifizierung von Oberflächen (Vacuum and Plasma Technologies, Coating and Modification of Surfaces), Teil I und II*, G. Blasek and G. Bräuer (Eds.), pp. 537-600, Eugen-G-Leuze-Verlag, Saulgau, Germany (2010).
60. K. Rossmann, Improvement of bonding properties of polyethylene. *J. Polym. Sci.* 19, 141-144 (1956).
61. J. Goodman, The formation of thin polymer films in the gas discharge. *J. Polym. Sci.* 44, 551-552 (1960).
62. C. Y. Kim and D. A. I. Goring, Surface morphology of polyethylene after treatment in a corona discharge. *J. Appl. Polym. Sci.* 15, 1357-1364 (1971).
63. M. Strobel, M. C. Branch, M. Ulsh, R. S. Kapaun, S. Kirk, and C. S. Lyons, Flame surface modification of polypropylene film. *J. Adhesion Sci. Technol.* 10, 515-539 (1996).
64. T. Volkmann and H. Widdecke, Fluorination of polyethylene films, *Makromolekulare Chemie. Macromolecular Symposia* 25, 243-248 (1989).
65. J. F. Friedrich, B. Möller, G. Kühn, W. Unger and P. Krüger, Improvement of adhesion properties of polymers by producing oxyfluorinated surfaces, *Galvanotechnik* 92, 1334-1344 (2001).
66. S. Kirk, M. Strobel, C. Y. Lee, S. J. Pachuta, M. Prokosch, H. Lechuga, M. E. Jones, C. S. Lyons, S. Degner, Y. Yang and M. J. Kushner, Fluorine plasma treatments of polypropylene films, 1-surface characterization. *Plasma Processes Polymers* 7, 107-122 (2010).

67. R. Srinivasan and S. Lazare, Modification of polymer surfaces by far-ultraviolet radiation of low and high (laser) intensities. *Polymer* 26, 1297-1300 (1985).
68. H. Esrom and U. Kogelschatz, Modification of surfaces with new excimer UV sources, *Thin Solid Films*. 218, 231-246 (1992).
69. R. Musil and H.-J. Tiller, Die molekulare Kopplung der Kunststoff-Verblendung an die Legierungsoberfläche (Molecular coupling of plastic facing onto surfaces of alloys). *dental-Labor* 32, 1155-1161 (1984).
70. H.-J. Tiller, R. Göbel, B. Magnus, A. Garschke and R. Musil, A new concept of metal—resin adhesion using an intermediate layer of SiOx-C. *Thin Solid Films* 169, 159-168 (1989).
71. J. Gähde, H. Wittrich, E. Schlosser, J. Friedrich and G. Kaiser, Verfahren zur Herstellung von verstärkten Polymerformstoffen (Process of manufacturing of enforced polymer materials). GDR-patent: DD-WP 106 052 (1972).
72. O. N. Tretinnikov, V. V. Pilipenko and L. K. Prikhodchenko, Benzophenone-initiated grafting photopolymerization of acrylic acid on the surface of polyethylene from the monomer aqueous solution without its deaeration. *Polym. Sci. Ser. B* 54, 427–433 (2012).
73. S. A. Piletsky, H. Matuschewski, U. Schedler, A. Wilpert, E. V. Piletska, T. A. Thiele and M. Ulbricht, Surface functionalization of porous polypropylene membranes with molecularly imprinted polymers by photograft copolymerization in water. *Macromolecules* 33, 3092–3098 (2000).
74. V. G. Kämpf and H. Schuster, Struktur und Eigenschaften von ABS-Polymeren. IX. Quantitative Untersuchungen zur Morphologie von ABS-Polymerisaten (Structure and properties of ABS polymers. IX. Quantitative investigations on morphology of ABS-polymers). *Angew. Makromol. Chem.* 27, 81-98 (1972).
75. V. G. Kämpf and H. Schuster, Struktur und Eigenschaften von ABS-Polymeren. IV. Teilchengrößenbestimmungen an Butadien-Homo-und Pfropfpolymerisat-Latices mit dem Elektronenmikroskop (Structure and properties of ABS polymers. IV. Measuring of particle sizes of butadiene-homo- and graft polymer lattices using the electron microscopy). *Angew. Makromol. Chem.* 14, 111-129 (1970).
76. D. Rathmann, Kunststoffgalvanisierung (Plastics galvanization). *Chemie in unserer Zeit* 15, 201–207 (1981).
77. K. Biederbick, *Kunststoffe (Plastics), 4. ed.*, Vogel-Verlag, Würzburg, Germany (1977).
78. R. Joshi, R.-D. Schulze, G. Hidde and J. F. Friedrich, Selective surface modification of polypropylene with OH groups using the underwater capillary discharge. *Plasma Processes Polymers* 5, 695-707 (2008).
79. S. Swaraj, U. Oran, J. F. Friedrich, A. Lippitz and W. E. S. Unger, Surface analysis of plasma deposited polymer films, 6: Analysis of plasma deposited allyl alcohol films before and after aging in air. *Plasma Processes Polymers* 2, 572-580 (2005).

80. A. P. Ameen, R. D. Short and R. J. Ward, The formation of high surface concentrations of hydroxyl groups in the plasma polymerization of allyl alcohol. *Polymer* 35, 4382-4391 (1994).
81. H. Biederman and D. Slavınská, Plasma polymer films and their future prospects. *Surf. Coatings Technol.* 125, 371-376 (2000).
82. J. Friedrich, J. Gähde, H. Frommelt and H. Wittrich, Modifizierung von Feststoffoberflächen in einer HF-Entladung. II. Schichtuntersuchungen, Reaktionsmechanismus und Verbundergebnisse (Modification of solid surfaces on exposure to the RF plasma. II. Analysis of deposited layers, reaction mechanism and results of composite manufacturing). *Faserforsch. Textiltechn. / Z. Polymerenforsch.* 27, 599-603(1976).
83. J. Friedrich, Sh. Geng, W. Unger and A. Lippitz, Modelling the plasma-induced reactions on polymer surfaces by using aliphatic self-assembling and LB layers as substrates. *Surf. Coatings Technol.* 98, 1132-1141 (1998).
84. G. Kühn, St. Weidner, R. Decker, A. Ghode and J. Friedrich, Selective surface functionalization of polyolefins by plasma treatment followed by chemical reduction. *Surf. Coatings Technol.* 116-119, 796-801 (1999).
85. D. L. Cho, P. M. Claesson and C. G. Gölander, Structure and surface properties of plasma polymerized acrylic acid layers. *J. Appl. Polym. Sci.* 41, 1373-1390 (1990).
86. L. O'Toole, A. J. Beck and R. D. Short, Characterization of plasma polymers of acrylic acid and propanoic acid. *Macromolecules* 29, 5172–5177 (1996).
87. S. Swaraj, U. Oran, A. Lippitz, J. F. Friedrich and W. E. S. Unger, Study of influence of external plasma parameters on plasma polymerized films prepared from organic molecules (acrylic acid, allyl alcohol, allylamine) using XPS and NEXAFS. *Surf. Coatings Technol.* 200, 494-497 (2005).
88. A. Fahmy, R. Mix, A. Schönhals and J. F. Friedrich, Structure of plasma polymerized acrylic acid films. *Plasma Processes Polymers* 8, 147-159 (2011).
89. E Cameron, L Pauling and B Leibovitz, Ascorbic acid and cancer: a review. *Cancer Research*, 39, 663-681 (1979).
90. J. Friedrich, J. Gähde, H. Frommelt and H. Wittrich, Modifizierung von Feststoffoberflächen in einer HF-Entladung. III. Plasmachemisches Aufbringen funktioneller Gruppen und selektiver Abbau teilkristalliner Polymere (Modification of solid surfaces on exposure to the RF plasma. III.Plasmachemical grafting of functional groups and selective etching of semi-crystalline polymers) Analysis of deposited layers, reaction mechanism and results of composite manufacturing. *Faserforsch. Textiltechn. / Z. Polymerenforsch.* 27, 604-608 (1976).
91. W. R. Gombotz and A. S. Hoffman, Functionalization of polymeric films by plasma polymerization of allyl alcohol and allylamine. *J. Appl. Polym. Sci. Appl. Polym. Symp.* 42, 285-304 (1988).
92. M. Müller and C. Oehr, Plasma aminofunctionalisation of PVDF microfiltration membranes: comparison of the in plasma modifications with a grafting

method using ESCA and an amino-selective fluorescent probe. *Surf. Coatings Technol.* 116-119, 802-807 (1999).

93. J. Friedrich, M. Jablonska and G. Hidde, Plasma oxidation of polyolefins - Course of O/C ratio from unmodified bulk to surface and finally to CO_2 in the gas phase: A critical review. *Rev. Adhesion Adhesives* 7, 233-257 (2019).
94. J. F. Friedrich, S. Wettmarshausen, K. Altmann and G. Hidde, Adhesion promotion by ESI deposi-tion of ultra-thin polymer layers onto fibers. *Surface Innovations* 3, 215-227 (2015).
95. J. Friedrich, K. Altmann, S. Wettmarshausen and G. Hidde, Coating of carbon fibers with adhesion-promoting thin polymer layers using plasma polymerization or electrospray ionization technique-a comparison. *Plasma Processes Polymers* 16, 1600074 (2017).
96. K. Okazoe, Development of the "PERFECT" direct fluorination method and its industrial applications, *J. Fluorine Chem.* 174, 120-131 (2015).
97. S. Wettmarshausen, H.-U. Mittmann, G. Kühn, G. Hidde and J. F. Friedrich, Plasmabromination – the selective way to monotype functionalized polymer surfaces. *Plasma Processes Polymers* 4, 832-839 (2007).
98. J. F. Friedrich, S. Wettmarshausen, S. Hanelt, R. Mach, R. Mix, E. Zeynalov and A. Meyer-Plath, Plasma-chemical bromination of graphitic materials and its use for subsequent functionalization and grafting of organic molecules. *Carbon* 48, 3884-3894 (2010).
99. C. Heise, U. Schedler, S. Wettmarshausen and J. Friedrich, Plasma-brominated cyclo olefin polymer slides - Suitable macroinitiators for ARGET-ATRP. *J. Appl. Polym Sci.* 40662, 131 (2014).
100. J. F. Friedrich, G. Hidde, A. Lippitz and W. E. S. Unger, Plasma bromination of graphene for covalent bonding of organic molecules. *Plasma Chem. Plasma Process.* 34, 621-645 (2013).
101. M. C. Lukowiak, S. Wettmarshausen, G. Hidde, P. Landsberger, V. Boenke, U. Braun, J. F. Friedrich, A. Gorbushina and R. Haag, Polyglycerol coated polypropylene surfaces for protein and bacteria resistance. *Polymer Chemistry* 6, 1350-1359 (2015).
102. H. K. Yasuda, *Plasma Polymerization*, Academic Press, Orlando, FL (1989).
103. J. Friedrich, Mechanisms of plasma polymerization –a chemist's point of view, *Plasma Processes Polymers* 8, 783–802 (2011).
104. R. Mix, J. Falkenhagen, R.-D. Schulze, V. Gerstung and J. F. Friedrich, Plasma homo- and copolymerization of allyl alcohol and styrene, in: *Polymer Surface Modification, Relevance to Adhesion, Vol. V.*, K. L. Mittal (Ed.), pp. 317-340, CRC Press, Boca Raton, FL, USA (2009).
105. H.-G. Elias, *Makromoleküle (Macromolecules)*, 5th ed., Hüthig&Wepf, Basle, CH (1990).
106. D. Briggs, Applications of XPS in polymer technology, in: *Practical Surface Analysis*, 2nd ed., D. Briggs and M. P. Seah (Eds.), Chapter 9, pp. 437-484, Wiley, Chichester, UK (1990).

107. R. Mix, J. F. Friedrich and N. Inagaki Modification of branched polyethylene by aerosol-assisted dielectric barrier discharge. *Plasma Processes Polymers* 9, 406-416 (2012).
108. M. R. Alexander, G. Beamson, C. J. Blomfield, G. Leggett and T. M. Duc, Interaction of carboxylic acids with the oxyhydroxide surface of aluminium: Poly(acrylic acid), acetic acid and propionic acid on pseudoboehmite. *J. Electr. Spectr. Rel. Phenom.* 121, 19–32 (2001).
109. G. L. Fisher, A. E. Hooper, R. L. Opila, D. L. Allara and N. Winograd, The interaction of vapor-deposited Al atoms with CO2H groups at the surface of a self-assembled alkanethiolate monolayer on gold. *J. Phys. Chem. B*104, 3267-3273 (2000).
110. J. F. Silvain and J. J. Ehrhardt, An overview on metal/PET adhesion. *Thin Solid Films* 236, 230-235 (1993).
111. M. R. Alexander, S. Payan and T. M. Duc, Interfacial interactions of plasma-polymerized acrylic acid and an oxidized aluminium surface investigated using XPS, FTIR and poly (acrylic acid) as a model compound. *Surf. Interf. Anal.* 26, 961-973 (1998).
112. M. G. Chan and D. L. Allara, Infrared reflection studies of metal-polymer interfaces. *Polym. Eng. Sci.* 14, 12-15 (1974).
113. E. P. Plueddemann, H. A. Clark, L. E. Nelson and K. R. Hofmann, New silane coupling agents for reinforced plastics. *Modern Plastics* 39, 135-187 (1962).
114. E. P. Plueddemann, *Silane Coupling Agents*, Plenum Press, New York, USA (1982).
115. E. P. Plueddemann, Adhesion through silane coupling agents. *J. Adhesion* 2, 184-201 (1970).
116. G. A. Somorjai, Chemische Bindungen an Oberflächen (Chemical bonds atsurfaces). *Angew. Chemie* 77, 94-102 (1989).
117. J. Gähde, Chemische Bindungen bei Adhäsionsvorgängen (Role of chemical bonds in adhesion), in: *Adhäsion-Theoretische und experimentelle Grundlagen (Adhesion: Theoretical and Experimental Basics)*, C. Bischof and W. Possart (Eds.), Akademie-Verlag, Berlin, Germany (1982).
118. A. D. McLaren and C. J. Seiler, Adhesion. III. Adhesion of polymers to cellulose and alumina. *J. Polym. Sci.* 4, 63-74 (1949).
119. P. Weiss (Ed.), *Adhesion and Cohesion*, Elsevier, Amsterdam (1962).
120. I. Skeist (Ed.), *Handbook of Adhesives*, Reinhold, New York (1962).
121. J. F. Silvain, A. Arzur, M. Alnot, J. J. Ehrhardt and P. Lutgen, XPS study of Al/polyethylene terephtalate interface. *Surf. Sci.* 251-252, 787-783 (1991).

Part II

CHEMICAL TREATMENTS

11

Amine-Terminated Dendritic Materials for Polymer Surface Modification to Enhance Adhesion

Zaynab Daneshzand[1], Kiana Karimi[2], Somaye Akbari[1]* and Atefeh Solouk[2]

[1]Textile Engineering Department, Amirkabir University of Technology (Tehran Polytechnic), Tehran, Iran
[2]Biomedical Engineering Department, Amirkabir University of Technology (Tehran Polytechnic), Tehran, Iran

Abstract

Adhesion is a universal concept in science and industry that when controlled leads to the development of more efficient and durable products. Dendritic structures have attracted attention of many researchers due to their specific properties such as abundant active functional groups, lots of cavity between branches, high specific active area, and desired 3D architecture. Specifically, amine-terminated structures have beneficial properties for adhesion as they contain numerous surface amino functional groups, possess positive charge, and enhance hydrophilicity that can be exploited to alter surface chemical properties, as well as topography. Here, we review the structures of dendritic materials and investigate the effects of amine-terminated dendritic structures in five different fields; cell adhesion, bacterial adhesion, adhesion in composites (i.e. adhesion between matrix and reinforcing phase), adhesion in water treatment, and coating adhesion and how they can help to overcome adhesion related issues.

Keywords: Amine-terminated, dendritic structures, dendrimers, adhesion

**Corresponding author*: akbari_s@aut.ac.ir.

K.L. Mittal and A.N. Netravali (eds.) Polymer Surface Modification to Enhance Adhesion: Techniques and Applications, (461–486)

11.1 Introduction

The importance of adhesion is universal. This phenomenon appears in various fields; thus more understanding of adhesion leads to better knowledge for its use in many products. Centuries ago, human recognized adhesion and numerous theories have been developed to explain it ever since. Mechanical interlocking, electrostatic, theory of boundary layers and interphases, adsorption (van der Waals forces) theory, diffusion theory, and chemical bonding or acid-base theories are the six main theories to explain adhesion [1–4]. Various techniques are utilized for modification of surface chemistry and surface topography of polymers which can be beneficial to enhance adhesion property. These techniques vary from chemical treatment such as functionalization by grafting macromolecules and nanoparticles, photochemical, UV or laser irradiation, blending of surface active agent, and chemical vapor deposition as well as physical treatments. Plasma [5] and laser [6] ablations help to alter surface topography whereas chemical modification changes surface chemistry. Figure 11.1 illustrates the various techniques for surface modification.

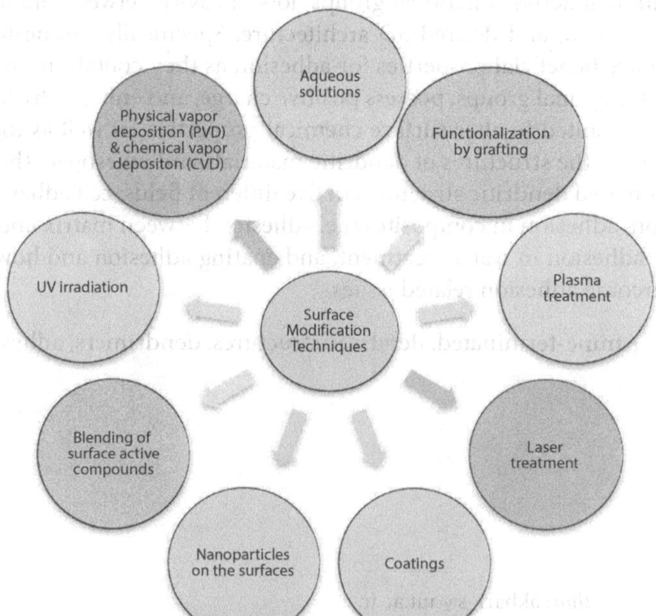

Figure 11.1 Various techniques for surface modification.

In the field of adhesion, the properties of adhesives, adherends, and interfaces have essential roles. Changes in chemical, physical, and micro-mechanical interactions can lead to changes in adhesion. Chemical interactions alter adhesion by forming new covalent bonds, electrostatic interactions, hydrogen bonding, and hydrophilic/hydrophobic attractions [7]. Physical interactions enhance adhesion via interfacial interlocking [8]. In this chapter, we investigate surface modification via amine-terminated dendritic groups to provide the possibility of establishing desired interactions to promote adhesion.

Dendritic structures are a class of nano-polymers with a specific architecture, introducing new possibilities for surface and bulk modification of materials. Dendrimers differ from classical random coil macromolecules by possessing three distinct architectural components: (a) an initiating core, (b) interior layers (generations), composed of repeat units radially attached to the initiating core, and (c) exterior layer containing terminal functional groups [9]. They are synthesized thorough two methods: convergent and divergent synthesis, depending on whether the molecule starts to grow from the core or the core is the last formed component [10].

The terminal functional groups determine the unique features of these compositions [11]. One of the widely used dendritic structures is amine-terminated such as polyamidoamine (PAMAM), polypropylenimine (PPI), and polyethylenimine (PEI) [12]. The terminal primary amine groups provide positive surface charges due to protonation, hydrophilic interactions due to hydrogen bonding, and new sites for covalent bonding as amine groups widely react with different functional groups. Various hyper-branches of dendrimers also can create physical entanglements depending on their environments [13].

11.2 Dendritic Materials

The architecture of dendritic molecules can be an excellent guide for finding the origin of naming these structures. It combines two Greek words, "Dendron" and "Meros", which mean tree-like. In that case, dendritic structures can be seen as one of the essential topographies observed in nature in different dimensions, from lung veins to roots of trees [14]. Plants provide the most surface area with a dendritic structure to absorb sunlight and water from the roots below the soil, and the lung veins create a branched design to improve oxygen transport [11].

Dendritic structures are characterized by a new set of unique properties. Based on their architecture, dendritic polymers are divided into four subclasses: random hyper-branched polymers, dendrigrafts, dendrons, and dendrimers [15]. Although, the first conceptualized hyper-branched (HB) polymer was discussed by Flory in 1952 [16], the first inspiration should have come from Tomalia leading to the creation of a specific core-shell macromolecular architecture, known as dendrimer [7, 17].

A highly branched structure with great regularity, compact shape, vacancies between the branches for taking up guest molecules are the main characteristics of dendritic structures. Dendrimer's reactive end groups increase exponentially in higher generations. This can be a very effective parameter in adhesion, and by varying the end group, it is possible to alter the force of adhesion [15].

Recently, the use of amine-terminated dendritic materials for polymer surface modification has become a hot research topic, especially in biomedical applications [18]. Surface alternations after using these materials can be divided into short-term and long-term effects which are explained further in the next section. This means that the changes directly from the presence of new groups on the surface, or created indirect effects on the final characteristics such as changes in crystallinity which we will explain further in the next section. Changing the physical and chemical surface properties influences adhesion forces. Controlling the interactions at interfaces to optimize adhesion is often challenging. Adhesion is crucial in different applications such as cell adhesion, bacterial adhesion [19], dentistry [20], and coatings [21], etc. For an application, the purpose of surface modification is to optimize adhesion. Among all possible ways that can help to achieve this goal, dendritic materials are a good candidate to increase or decrease adhesion by altering the various abundant functional groups which can be used to modify both the bulk and the surface of materials. A method of classifying dendritic materials is based on the end groups. Among the various possible functional groups, the terminal amino groups can be effective in many adhesion applications.

11.3 Amine-Terminated Dendritic Materials as Adhesion Modifiers

Amine-terminated dendritic materials can provide adhesion related surface properties in two ways:

1. Short-term effect: Modification based on the direct presence of amine-terminated dendritic materials on the surface,

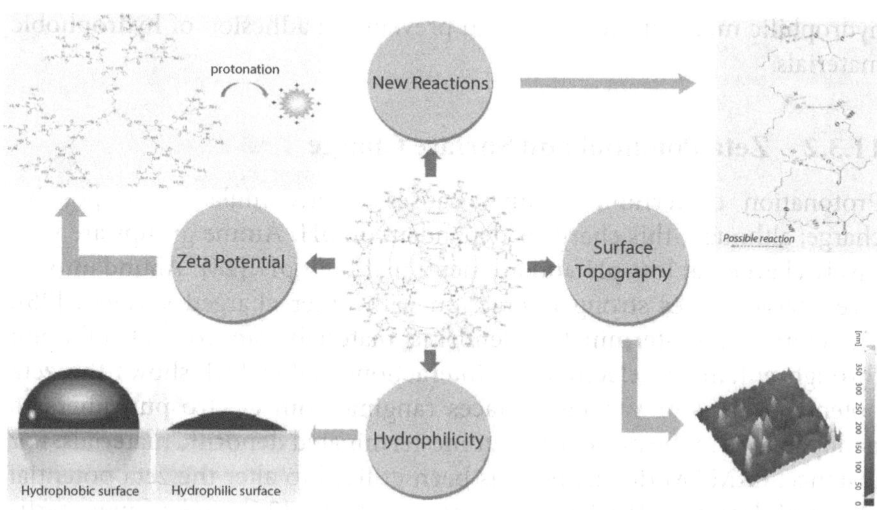

Figure 11.2 Dendritic characteristics affecting surface properties.

which play a role in determining properties such as hydrophilicity, zeta potential, etc.

2. Long-term effect: Use of dendrimer in the modification process so that it is effective in creating surface features such as roughness.

Incorporating these structures on surfaces have certain effects on adhesion as illustrated in Figure 11.2.

11.3.1 Hydrophilicity

Due to the presence of abundant polar terminal amino groups with the ability to establish hydrogen bonds, amine-terminated dendritic materials reduce the water contact angle and thus increase hydrophilicity [22]. This leads to interaction with hydrophilic groups and, on the other hand, repels hydrophobic groups from the surface. It seems that this effect on the surface is due to the direct presence of the dendrimer groups. For instance, growing PAMAM dendrimers on the carbon fiber surface by *in situ* polymerization enhances the adhesion between carbon fiber and matrix due to enhanced hydrophilicity. The evidence for enhanced hydrophilicity is revealed by reduction of water contact angle from 69.4° to 55.8° [23].

Looking at the mentioned interactions from the point of view of adhesion, one can say that increasing the surface hydrophilicity will increase hydrophilic materials adhesion and prevent the adhesion of hydrophobic materials.

11.3.2 Zeta Potential and Surface Charge

Protonation of terminal amines causes an accumulation of positive charge; although, this charge is dependent on pH. Amine groups are positively charged at low pH and are neutral at high pH [24]. Abundant surface charge causes strong interaction with other charged surfaces [25]. Therefore, amine-terminated dendritic materials can control adhesion through enhancing electrostatic interactions. Table 11.1 shows the zeta potential values of various surfaces ranging from electrospun fibers to nanoparticles in the presence of amine-terminated dendritic materials. For instance, PAMAM dendrimer has been utilized to alter the zeta potential of poly(L-lactic acid) (PLLA) electrospun fibers [26], and of nanoparticles such as halloysite nanotubes (HNTs) [27]. In the other work, modified halloysite nanotubes (HNTs) have been used in nanocomposite electrospun scaffolds [28]. All these samples reveal increasing zeta potential in the presence of amine-terminated materials as shown in Table 11.1.

11.3.3 Surface Topography

Amine-terminated dendritic materials can affect different surface characteristics such as roughness, steric hindrance, and entanglement between chains. Roughness is one of these features that can increase adhesion between materials due to the increased surface area. Akbari *et al.* [30] found that roughness of acrylic films increased with increasing generation number from zero to fourth generation. In other work, dendritic materials in the remineralization process of tooth enamel smoothen the surface and prevent bacteria adhesion [20]. It should be mentioned that the long chains of dendrimers also cause physical entanglement [31] between surfaces and helps to strengthen the adhesion between them [13]. Moreover, the specific surface area is increased due to the presence of the hollow interior spaces between dendritic polymer branches [27]. Table 11.2 shows the effect of surface topography and roughness on the adhesion force before and after modification by amine-terminated dendritic materials. These changes are related to interfacial interaction due to the available surface area. In the case of reduction of adhesion, it is obvious that by reducing roughness, the surface area decreases. As a result, adhesion force reduces due to the surface interaction.

Table 11.1 Zeta potential of surfaces by amine-terminated functionalization.

Sample	Zeta potential (mV)	Modified sample	Zeta potential (mV)	Reference
Poly(L-lactic acid) (PLLA)	-11.4	Poly(L-lactic acid) (PLLA)+ PAMAM	-1.7	[26]
HNTs-OH	-8	HNTs-NH_2	-4	[27]
Lignin	-23.3	Lignin with 1% PAMAM based on the lignin weight	-20.7	[29]

Table 11.2 The effect of surface topography modification by amine-terminated groups on the adhesion force.

Characteristic	Before modification	After modification	Reference
Surface topography (roughness)	289.48 ± 26.21 nm	203.10 ± 31.26 nm	[20]
Adhesion force	5.52 ± 1.62 nN	3.64 ± 1.52 nN	[20]
Surface topography (roughness)	99 nm	159 nm	[23]
Interlaminar shear strength (ILSS)	62.46 MPa	86.23 MPa	[23]
Surface free energy	42.01 mJ m^{-2}	43.4 mJ m^{-2}	[23]

11.3.4 New Interaction

The functional groups at the interface between two materials could be used to enhance adhesion. Considering the reactivity of amine groups, it is possible to establish new interaction with amine groups on the surface. For example, carbon fiber as a filler component in composites does not disperse well and interacts poorly with the matrix. Surface modification using amine-terminated materials improves filler adhesion to the matrix as well as dispersion quality due to the better interaction between filler and matrix [23].

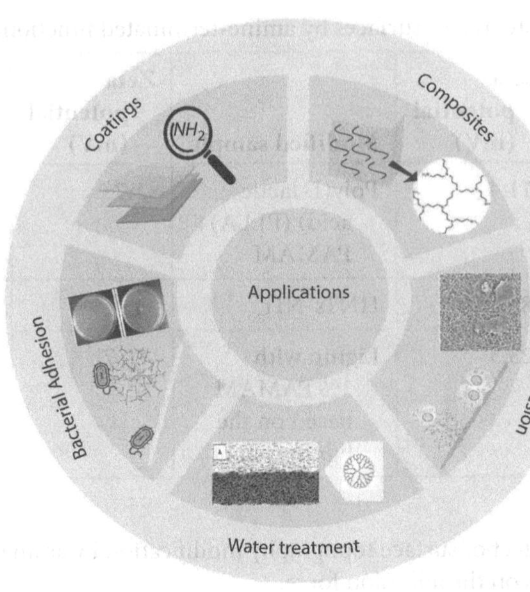

Figure 11.3 Applications of amine-terminated dendritic materials in adhesion.

It is assumed that all the aforementioned features can individually or in cooperation with one another affect adhesion. Therefore, in the next section, a literature review of recent studies is presented and is divided into five different areas illustrated in Figure 11.3.

11.4 Applications of Amine-Terminated Dendritic Materials in Adhesion

11.4.1 Cell Adhesion

Tissues are composed of two main parts: cells and extracellular matrix (ECM). The ultimate goal of tissue engineering is to prepare a suitable ECM mimicking substrate for cells to adhere and continue living to tackle the global problem of organ donor shortage and immunological obstacles [32]. Cell adhesion is a prerequisite for cell viability in anchorage-dependent cells. It is also crucial for cell-environment and cell-cell communications. Furthermore, cell adhesion can cause cell migration, proliferation, functionality, and differentiation (in the case of stem cells) [28, 29].

During the past decades, various polymers have been explored for tissue engineering applications, and researchers have tried to overcome their shortcomings via different approches. As was previously mentioned, surface charge, functional groups, and hydrophilicity are a few surface parameters that can directly affect cell adhesion and can be altered via amine-terminated dendritic structures. It all comes down to facilitating cell adhesion via the number of available amine groups. As mentioned, there are more amine terminal groups in higher generations. Thus, when using dendrimers to modify materials, cell adhesion will be increased with increasing dendrimer generation number [35]. Providing a more positive surface charge facilitates cell adhesion as cells have a negative membrane potential [36]. Increased hydrophilicity can also affect this phenomenon to some extent as it increases protein adsorption from culture medium [37]. Hence, cells can bind to the surface [38]. It must be kept in mind that there is an optimal surface density and generation for dendrimers as cell adhesion enhancers. Several studies have reported cell toxicity effect of these compositions, usually for generations higher than four [39].

Since the birth of tissue engineering, researchers have been studying various natural and synthetic polymers, trying to overcome their shortcomings for cell culture via different additives. Other than the batch-to-batch variation of natural polymers, they need cross-linking for compensating physical shortcomings. Linear polymer cross-linkers carry two functional end groups and crosslinking gelatin with these compounds can result in functional group deficiency that can hinder cell adhesion. In a study, collagen was cross-linked with polyethylenimine (PEI). Although the adhesion of human corneal epithelial cells was not statistically different in PEI and glutaraldehyde cross-linked samples, the latter showed consistently decreased cell viability after four days due to the release of unreacted glutaraldehyde from hydrogel [40]. Other than the bioadhesive property of these dendrimers that can interact with mucin covering corneal epithelium, they can solubilize poorly water-soluble drugs as a suitable solution for drug delivery purposes [5, 29]. Razmshoar *et al.* [41] aminolysed polycaprolactone (PCL) with G2-PAMAM through PCL ester bonds and prepared PCL-PAMAM nanofibers and investigated the effect of PCL/PAMAM ratio and mixing time on the properties of obtained fibers. Aminolysis made PCL more hydrophilic and decreased the water contact angle from 98° for pure PCL to 9° for PCL-PAMAM (65:35). The host-guest property of scaffold was also estimated with curcumin as an anticancer drug and the experiment revealed that with an increase in PAMAM content from 15% to 35%, the solubility and loading capacity of curcumin increased.

In another study, gelatin was cross-linked with PAMAM-G4, which resulted in higher keratinocyte adhesion. For fibroblasts, collagen type I gene expression also increased which can help cells to recreate ECM around them. PAMAM also affected water absorption and biodegradation of hydrogel, which is essential for mass transfer of cell wastes and nutrients within the scaffold [42]. Incorporating PAMAM-NH_2 in these structures not only provides enough amine groups to react with –COOH functional groups in gelatin as a cross-linker but also is a means for maintaining extra free amine groups for enhancement of cell attachment [43]. Agarwal *et al.* [44] functionalized cellulose filter paper via PAMAM-G4 to increase cell adhesion and used it for the culture of human hepatocyte cells to develop a 3D model of the liver for drug toxicity screening.

On the other hand, despite better processablity of synthetic polymers such as PLLA compared to natural polymers, they are often biophysically inert and cannot support proper cell adhesion. Surface aminolysis of PLLA via amine-terminated dendrimers will replace side chains of PLLA with relatively hydrophilic amide and hydroxyl groups and therefore will increase cell adhesion. Ganjalinia *et al.* [38] used G2-polypropylenimine as an aminolysing agent and modified PLLA film and electrospun scaffolds. This method not only increased human fibroblast adhesion, but also enhanced hydrophilicity and prevented massive immune cell infusion response to acidic degradation of PLA. By aminolysis reduction and degradation of the PLA chain, carboxylic acid end group converted to amine group and increased the pH. In a similar study, a gelatinated hybrid electrospun poly(lacticacid)/G2-PAMAM/gemini surfactant was prepared. The cell biocompatibility was investigated with PC-12 cells to reveal an increase in cell attachment in response to both the addition of dendrimer and gelatination of the scaffold [26]. This study was continued in another publication [35] with second, third and fifth generation of PAMAM dendrimers and PEI. Fibroblasts showed a greater affinity to scaffolds containing the fifth generation of PAMAM and PEI with higher cell adhesion.

In another study on PLLA, Delyanee *et al.* [45] prepared an electrospun scaffold containing different ratios of PEI to dendrimer. The presence of PEI resulted in an increase in mechanical properties and as well in surface capacity for gelatin grafting which led to 5 times higher cell adhesion. In a similar study, PEI modified halloysite nanotubes were added to PLA to prepare a hemostatic wound dressing while supporting cell adhesion. The blood clot formation time reduced from 9 to 4 minutes which is desirable for wound dressings [28]. In a recent study, the nanotubes were modified with different generations of PAMAM and were incorporated in the PLLA/halloysite nanocomposite. Hemostatic experiments showed that the sample containing G3-PAMAM modified composite had the shortest blood clotting time. Furthermore, the composite containing 10% of these nanotubes significantly improved cell adhesion. The *in vivo* studies revealed the promotion of angiogenesis, collagen expression, and cell proliferation as well as inhibition of inflammatory response in a rat model [46]. Another PLLA based study showed that the presence of PAMAM-G2 in an electrospun PLLA scaffold not only increased OE-mesenchymal stem cell viability but also enhanced fiber's ionic conductivity by nine times, making a proper platform for cell-cell signaling [47].

11.4.2 Bacterial Adhesion

Development of bacteria repellent and antibacterial surfaces has been widely investigated in various fields such as medicine [48], hospitals [49], food packaging [50], dairy, food and water system industries [51].

Bacteria tend to bind together and form a more complex structure to survive under harsh conditions. This structure, known as biofilm, develops through three stages: bacteria adhesion, maturation, and spreading. When a biofilm is formed, it is harder to remove bacteria as they become more resistant to antibacterial agents [52]. The first stage of biofilm formation is highly affected by surface properties such as topography, free energy, roughness, and hydrophilicity. Rough surfaces are more prone to biofilm formation irrespective of other surface properties. Therefore, it is suggested to use surfaces with roughness lower than 0.2 μm in the oral cavity [20]. As a case in point, the oral environment is a hydrated environment with different proteins due to the presence of saliva. Oral surfaces are covered with pellicles, a plasticizer film of proteins and glycoproteins that might contain bacteria as well and can be the starting point for bacteria colonization [53]. Studies show that some oral bacteria even have saliva-regulated genes, meaning saliva induces transcription of specific proteins known as adhesins, which affect surface colonization. These biofilms make a dam of ECM between the oral cavity and tooth surface, where their acidic metabolic byproducts are accumulated. Later, these will demineralize the tooth structure and result in dental caries [52], the most widespread chronic disease worldwide [54]. Thus, preventing bacteria adhesion can be a solution to reduce dental caries. As a short-term effect, PAMAM-grafted surfaces inhibit protein adhesion, similarly to the behavior of poly(ethylene glycol) coatings in water-based buffers. Grafting PAMAM on the surface provides a positively charged surface with a 3-dimensional flexible structure that can attract water molecules and ions present to create a hydrogen bonded layer which has to be overcome by proteins and bacteria prior to adhesion [55].

Dendritic structures have been used as antibacterial agents in biomedical applications. Post-surgery infections cause many problems, such as delayed or incomplete healing and secondary surgeries in more severe cases. These compounds disrupt the membrane integrity of bacteria and cause cell death [56]. Dai *et al.* [57] developed an oxidized polysaccharide-based hydrogel containing G5-PAMAM-NH_2 and silver nanoparticles with pH-responsive ability. In this study, the antibacterial activity of composite was investigated against *S. epidermidis* and *S. aureus*. Despite the usual belief that the more the Ag particles, the more the antibacterial activity, it was revealed that an optimum concentration of Ag particles was needed, which was enough for release over time without covering all terminal amine groups of PAMAM to hamper antibacterial activity. In another investigation, the adhesion of *Streptococcus gordonii*, an earlier colonizer responsible for infections after dental implantation, was evaluated for two linear and PAMAM grafted surface groups. PAMAM-NH_2 surfaces maintained their bacteria-repellent property even after saliva pretreatment despite surfaces were modified with linear structures. This paper stated that different compositions of salivary proteins were detected in pellicles, which might be the main reason why bacterial adhesion differed among the examined samples [55]. As surface roughness was previously discussed, these amine-terminated species can have a long-term effect on bacteria adhesion in the oral cavity by decreasing surface roughness over time. Researches show that the positive surface charge of PAMAM-NH_2 on the surface can attract Ca^{2+} and $(PO_4)^{3-}$ ions and remineralize enamel with calcium phosphate deposition mainly composed of hydroxyapatite crystals. AFM analysis reveals that as the surface roughness decreases, the adhesion of *Streptococcus mutans*, the bacteria responsible for dental carries, also decreases [20].

11.4.3 Adhesion in Composites

Composites are composed of two components: matrix and filler as reinforcing agent. One of the determining parameters for mechanical properties of composites are the interfacial interactions between these two components. Sometimes the lack of sufficient adhesion creates a challenge for various reasons such as low surface activity, high chemical inertness, high crystallinity, and lack of entanglements due to the great regularity of polymer chains. Even in some cases, the aggregation of fillers can prevent effective interfacial interaction. Surface modification can be an efficient solution to strengthen the interaction between two surfaces and also improve dispersion quality by avoiding agglomeration [58].

Polypropylene (PP) is a common thermoplastic resin used as a matrix in composites. This polymer has a low polarity and an inert nature, and this is the reason for the weak bond between the filler and PP. Investigations show the effectiveness of dendritic materials in improving the adhesion between the matrix and the filler [59] due to the terminal functional groups of the dendritic material, which enable establishment of new interfacial bonds [60].

In addition to modifying the matrix component to enhance interfacial interactions, the filler component can also be modified by various methods. For instance, carbon fiber (CF) is an ideal filler for both thermoset and thermoplastic composites [61]. The attachment of the dendrimer provides more active groups on the CF surface and increases the surface free energy [23]. CFs do not disperse well due to the van der Waals forces between them, stable π-π bonds and entanglement between fibers caused by high aspect ratio [62]. For instance, Liu *et al.* [61] showed that the presence of long chain dendrimers on the surface of CFs not only enhances the affinity but also creates a physical interaction between the CFs and the polyurethane matrix, which results in a homogeneous dispersion of the filler. It is worth mentioning that the functionalization with dendritic materials can be done in various ways, such as *in situ* polymerization with Michael addition [23], physical adsorption, chemical adsorption [63], chemical grafting [64], and layer-by-layer coating [59].

Other common fillers in polymer-based composites are inorganic oxides (e.g. silica) and HNTs, which are desirable for modifying thermal and mechanical properties. OH groups on the surface of silica particles lead to agglomeration [13]. Similarly to silica, the polar surface of the HNTs causes a strong interaction between them [65]. As mentioned earlier, the amine functional group helps to increase the surface free energy, and, consequently, the adhesion between the matrix and fillers increases [6, 57, 66].

In other work using nonpolar matrix, enhancing hydrophilicity of filler reduced adhesion due to lack of interaction between matrix and filler. For instance, by adding modified hydrophilic HNTs by growing PAMAM on the surface of the filler in a nonpolar matrix such as epoxy resin, varying the functional groups on the surface of the filler weakens the interaction between the fillers and matrix because polyamidoamine grafting on HNTs decreases surface polarity [65].

Interfacial adhesion between fibers and matrix has a strong influence on composite mechanical performance. The mechanical properties of the composite were characterized via the interlaminar shear strength (ILSS) and interfacial shear strength (IFSS) [22, 68]. The ILSS and IFSS data after modification by amine-terminated materials are summarized in Table 11.3.

The increase of the fiber surface free energy could lead to better wettability between CF and matrix, and improvement in the interfacial adhesion [64]. Ma *et al.* [69] found that the surface free energy of unmodified carbon fibers enhanced from 30.1 mJ.m^{-2} to 64.6, 74.6, and 70.5 mJ.m^{-2} by

Table 11.3 Interlaminar and interfacial shear strength data after modification by an amine-terminated material.

Untreated carbon fiber	Modified carbon fiber with amine-terminated dendrimer (MPa)	Reference
(ILSS) 65 MPa	85	[23]
(ILSS) 50 MPa	70	[64]
(IFSS) 64.9 MPa	99.7	[64]

modifying with the first, second, and third generation of PAMAM dendrimer, respectively.

11.4.4 Adhesion in Water Treatment

The availability of clean water is a global issue. Water purification requires several interconnected processes [70], where the concept of adhesion is discussed in two ways, firstly, by preventing adhesion of particles in the pores to prevent clogging or affecting the flux in the filtration process; secondly, by an efficient adhesion of pollutants to the adsorbents.

One of the troublesome challenges in membrane design is fouling, a classic interfacial phenomenon related to membrane separation. Interactions between fouling materials (i.e., foulants) and membrane; interactions among the foulants; and interactions among membranes, foulants, and water eventually lead to fouling [71].

Fouling is an inherent challenge that affects membrane permeability and interferes with the separation process and flux flow. By preventing fouling, one will see a significant impact on the lifetime of the membrane as well as energy storage. Evidence shows that fouling ultimately occurs due to adhesion between the foulants and the surface. Understanding this phenomenon provides a means for controlling adhesion force and even increase repulsion. Precipitation of sparingly soluble salts, adsorption, cake or gel formation, and pore blockage lead to fouling [2].

It should be noted that fouling may be the result of several simultaneous events and there is no simple way to determine the primary cause [71]. Generally, fouling sources are organic, colloidal, and mineral scales [72]. Reversible fouling is weak in bonding and can be cleaned by physical methods easily but irreversible fouling is tightly bonded and can only be removed by chemical methods [73]. The adhesion force between the foulant and the surface is influenced by various factors. Among these, one can mention particle dimension, membrane microstructure, interaction between membrane and solvent, roughness, porosity, and other surface-related features [74]. Therefore, membrane modification by amine dendritic polymers can help to optimize adhesion by increasing hydrophilicity and thus improve anti-fouling property. Hydrophilicity is not the only effective factor in reducing the adsorption of foulants, other factors such as roughness, surface charge, and orientation of functional groups are also important [75].

Moreover, dendrimer's generation plays an important role in adhesion phenomenon. With increasing the number of generation, the hydrophilicity increases due to the greater number of terminal amine groups, but it is also necessary to pay attention to the fact that the increase in the number of generation makes the surface rougher. After a certain generation number, functional groups make a dense shell [76] and some functional groups become inaccessible. Jiang *et al.* [77] reported the effects of charge and hydrophilicity after surface modification on adhesion. The results show that hydrophilicity provides an energy barrier that prevents fouling and increases charge density causing increased electrostatic repulsion between the foulant and the membrane surface. It seems that the surface charge of the membrane is significantly affected by the protonation of its amine groups. Examining the zeta potential at different pH confirms this hypothesis. It is worth mentioning that with the increase of generation, the zeta potential of the surface increases. Therefore, the use of dendritic material with amine-terminated groups increases the anti-fouling property of the membrane [70, 73].

The results of the investigations indicate that the positive surface charge of polyamidoamine leads to adhesion of bacteria, which often have a negative charge, and this prevents growth and sometimes even causes toxicity and kills bacteria. Therefore, PAMAM contributes to the significantly decreased live and dead bacterial attachment to membranes modified by PAMAM [79].

In the case of adsorbents, the efficiency depends on their adsorption power. Therefore, it is essential to establish proper interaction and optimal adhesion. For instance, halloysite is an adsorbent that can adsorb many chemical compounds, such as heavy metal ions, dyes, contaminants, and drugs. Also, hydroxyl groups on the surface have made it possible to improve the adsorbent capacity by modifying the surface. The use of dendrimers in two ways, convergent and divergent, can be effective for this purpose [80]. Due to the protonation of amines and high cationic charge density, especially in low pH and acidic conditions, it is possible to adsorb anionic dyes through electrostatic interaction [81]. On the other hand,

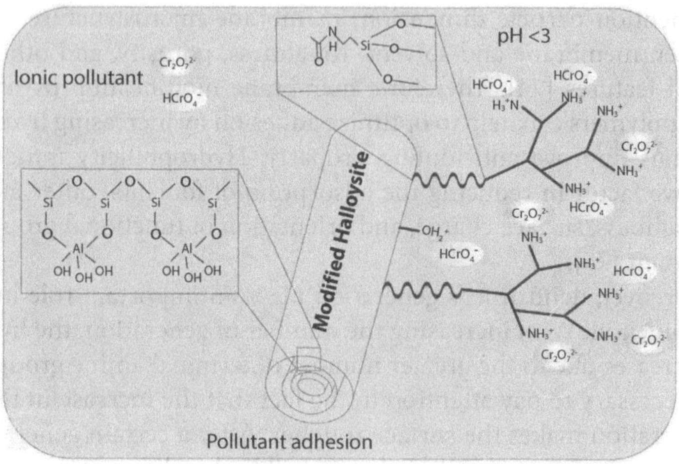

Figure 11.4 Interactions between Cr ions and halloysite functionalized by amine-dendritic groups.

Cr (VI) is also adsorbed by the electrostatic interaction between $HCrO_4^-$ [27] and cadmium [81]. Research shows that besides the increase in adsorption by electrostatic interaction in the adsorption of acidic dyes, the increase in dendrimer generation also has a positive effect [82]. The empty spaces between the branches of the dendritic structures also make it possible to load effluent between branches and eliminate pollutants through the guest-host chemistry [81]. As a result, strong interaction between pollutants and the functionalized minerals by amine-dendritic groups enhances adhesion of pollutants to the adsorbent. For instance, strong interactions between Cr ions and halloysite functionalized by amine-dendritic groups are illustrated in Figure 11.4.

Strong interaction of functionalized HNT in thin film nanocomposite membranes increases interactions between the HNTs and the matrix and hence prevents them from leaching out and entering the stream while the membrane is in use [83, 84].

11.4.5 Coatings Adhesion

Coatings are used for decorative or protective purposes, such as anti-corrosion coatings. The coating/substrate interface is crucial for determining the adhesion strength of the coating [85]. Before applying coatings, various steps are necessary to prepare the surface for better adhesion. After cleaning, a layer of primer is used to protect the surface from further contamination and increase the surface wettability by coating.

For instance, to increase the adhesion between the Au electrode and a polypyrrole (pPy) conductive polymer coating, PEI dendrimer was applied to the Au electrode surface as a primer. Although this PEI insulating layer delays the initial growth of the pPy, it improves adhesion up to 250%, allowing an increase in thickness without delamination of the coating after drying, ultimately leading to an enhancement in the electrochemical storage capacity of the electrode. It is believed that PEI interacts with Au surface through polar/ionic groups and van der Waals interactions [21].

Ramezanzadeh *et al.* [86] developed an epoxy resin-based composition improved with PAMAM-G2 modified graphene oxide (GO) as a corrosion-resistant coating. Incorporation of PAMAM not only improved GO dispersion and enhanced anticorrosive property but also strengthened the interface.

Mei *et al.* [85] investigated PAMAM's durability on an aluminum alloy. Since, an oxide layer forms on aluminum alloys and protects against further corrosion, alumina wafers were used instead of aluminum alloys. Dendrimers were adsorbed on alumina via acid-base chemical interactions and maintained the primer layer even after five rounds of sonication washing. The authors stated that GO-PAMAM could be a suitable primer for metals and carbon-fiber-reinforced polymeric composites.

Epoxy coatings have high chemical and corrosion resistance, good mechanical properties, and excellent thermal properties. There is a drawback due to high cross-linking and toughening of these coatings that leads to brittleness and poor resistance to crack. Adding hyperbranched (HB) amine-terminated dendrimers results in more efficient toughening and internal stress reduction due to the phase separation ability of dendrimers. The high cross-link density of fast-cure epoxy resins can be controlled and decreased with dendritic structures [87].

As PAMAM is soluble in common organic solvents and is compatible with many polymers such as epoxy resins, it provides a plasticizing effect to adhesives. As a result, the viscosity of the solution reduces. Therefore, an optimum concentration or generation number of PAMAM needs to be determined to obtain the best effect on adhesion. Although, the simultaneous increase of shear and peel strengths is desirable for an adhesive, an increase in shear strength, which means higher cross-link density, leads to a decrease in peel strength, making it less tough. Dendritic structures can increase both shear and peel strengths due to their 3D energy-absorbing structure. In a study, the effect of PAMAM was investigated on thermal and mechanical properties of polyurethane adhesive bonded Al, Mg, plastic joints, and on fiber after heat and humidity treatments. Polyurethane systems are composed of an isocyanate/polyol mixture, and replacing 1-2%

of the polyol with HB-PAMAM increased shear strength by 40-78% after room temperature curing. After post-curing at 80°C for five hours, both shear and peel strengths increased by 64% and 100%, respectively due to enhanced diffusion and micromechanical interactions [88].

11.5 Summary

Amine-terminated dendritic structures, among other dendritic structures, have been used widely to alter the adhesion property of polymeric materials. Here, polypropylenimine (PPI), polyamidoamine (PAMAM), and polyethylenimine (PEI) were chosen for their effects on adhesion in different platforms. Cell adhesion, bacteria adhesion, composites, water treatment, adhesives, and organic-inorganic interfacial phase were the main topics described in this chapter.

In the case of cell adhesion, these compositions provide numerous amine functional groups that can alter the inertness of polymers to a more cell-compatible surface as well as their adhesion, proliferation, and migration. Improving mechanical properties via cross-linking, enhancing water uptake, and biodegradation are other desirable features for scaffolds. With increasing generation of dendrimers, they show higher cytotoxicity. This property can be exploited in antibacterial applications to reduce bacterial adhesion and eliminate adhered bacteria.

Incorporation of amine-terminated dendritic structures in polymeric membranes and adsorbents endows non-fouling property that can reduce cost due to increased membrane lifetime by increasing hydrophilicity. The protonation of amine groups can attract bacteria with negatively charged membranes and inactivate them via disrupting bacteria membrane integrity.

In composites, homogeneous dispersion and interfacial interactions between filler and matrix are of great importance in the mechanical properties of composite structures. Using amine-terminated dendritic systems for filler surface modification not only helps to obtain a more homogeneous dispersion of filler particles due to the presence of numerous amine functional groups but also can increase interfacial interactions between filler and matrix via the creation of physical entanglements by their long chains.

Amine-terminated dendritic structures can either be a surface primer or an additive for adhesion. They produce a homogeneous layer on organic materials and enhance adhesion through their functional groups as primers. As additives, they serve as a 3D cross-linker with energy-absorbing property to improve peel strength, shear strength, and viscosity.

As a conclusion, dendritic materials due to their abundant functional groups and lot of cavity between branches enhance adhesion between two surfaces. Among the various types of dendritic materials, amine-terminated ones are beneficial for adhesion enhancement and offer new possibilities for further applications.

References

1. J. Schultz and M. Nardin, Theories and mechanisms of adhesion, in: *Handbook of Adhesive Technology*, A. Pizzi and K.L. Mittal (Eds.) pp. 19-34, Marcel Dekker, New York (1994).
2. D.E. Packham, The mechanical theory of adhesion, in: *Handbook of Adhesive Technology*, second edition, A.Pizzi and K.L. Mittal (Eds.) pp. 69-93, CRC Press, Boca Raton, FL (2003).
3. M.M. Chehmini, A. Aziune and E. Cabet-Deliry, Acid-base interactions: relevance to adhesion and adhesive bonding, in: *Handbook of Adhesive Technology*, second edtition, A. Pizzi and K.L. Mittal (Eds.) pp. 95-144, CRC Press, Boca raton, FL (2003).
4. K.L. Mittal (Ed.) *Acid-Base Interactions: Relevance to Adhesion Science and Technology*, Vol. 2, CRC Press, Boca Raton, FL (2000).
5. M. Thomas and K.L. Mittal (Eds.) *Atmospheric Pressure Plasma Treatment of Polymers, Relevance to Adhesion*, Wiley-Scrivener, Beverly, MA (2013).
6. K.L. Mittal and W. -S. Lei (Eds.) *Laser Technology: Application in Adhesion and Related Areas*, Wiley- Scrivener, Beverly, MA (2018).
7. J. N. Israelachvili, *Intermolecular and Surface Forces*. third edition, pp 151-166, Academic Press, San Diego, California (2011).
8. S. J. Marshall, S. C. Bayne, R. Baier, A. P. Tomsia, and G. W. Marshall, A review of adhesion science. *Dental Mater.* 26, e11–e16 (2010).
9. D. A. Tomalia, H. Baker, J. Dewald, M. Hall, G. Kallos, S. Martin, J. Roek, J. Ryder, and P. Smith, A new class of polymers: starburst-dendritic macromolecules. *Polymer Journal* 17, 117-132 (1985).
10. S. J. Teertstra and M. Gauthier, Dendrigraft polymers: Macromolecular engineering on a mesoscopic scale. *Prog. Polym. Sci.* 29, 277–327 (2004).
11. E. N. Augustus, E. T. Allen, A. Nimibofa, and W. Donbebe, A review of synthesis , characterization and applications of functionalized dendrimers. *Amer. J. Polym. Sci* 7, 8-14 (2017).
12. M. Delyanee, S. Akbari, and A. Solouk, Amine-terminated dendritic polymers as promising nanoplatform for diagnostic and therapeutic agents' modification: A review. *Eur. J. Medical Chem.* 221, 113572 (2021).

13. A. Fakhar, M. Sadeghi, and M. Dinari, Stepwise surface modification of mesoporous silica and its use in poly(urethane-urea) composite films. *Polymer International.* 71, 107-116 (2021).
14. D. A. Tomalia, Birth of a new macromolecular architecture: Dendrimers as quantized building blocks for nanoscale synthetic polymer chemistry. *Prog. Polym. Sci.* 30, 294–324 (2005).
15. S. Akbari, The application of dendritic materials in textile engineering. *Science Bulletin. ESCORENA* 7, 11–28 (2013).
16. P. J. Flory, Molecular size distribution in three dimensional polymers. VI. Branched polymers containing A—R—Bf-1 type units. *J. Am. Chem. Soc.* 74, 2718–2723 (1952).
17. J. M. J. Fre, Discovery of dendrimers and dendritic polymers. *Drug Discovery* 40, 2719–2728 (2002).
18. A. M. Holmes, J. R. Heylings, K. W. Wan, and G. P. Moss, Antimicrobial efficacy and mechanism of action of poly(amidoamine) (PAMAM) dendrimers against opportunistic pathogens. *Int. J. Antimicrobial. Agents* 53, 500–507 (2019).
19. J. Song, H. Kong, and J. Jang, Bacterial adhesion inhibition of the quaternary ammonium functionalized silica nanoparticles. *Colloids Surfaces B* 82, 651–656 (2011).
20. L. Jia, S Tao, J Yang, K Liang, Z Yu, Y Gao, M Fan, M Zhang, L He, and J Li, Adhesion of Streptococcus mutans on remineralized enamel surface induced by poly(amido amine) dendrimers. *Colloids Surfaces B* 197, 111409 (2021).
21. K. G. Kim and S. Y. Kim, Increase in interfacial adhesion and electrochemical charge storage capacity of polypyrrole on Au electrodes using polyethyleneimine. *Sci. Reports.* 9, 1–9 (2019).
22. A. Oner, E. Tufek, I. Yezer, A. Birol, M. Demir, S. Er, and D. O. Demirkol, High generation dendrimer decorated poly-Ɛ-caprolactone/polyacrylic acid electrospun nanofibers for the design of a bioelectrochemical sensing surface. *React. Funct. Polym.*161, 104853 (2020).
23. R. L. Zhang, B. Gao, J. Zhang, H. Z. Cui, and D. W. Li, Propagation of PAMAM dendrimers on the carbon fiber surface by *in situ* polymerization: a novel methodology for fiber/matrix composites. *Appl. Surf. Sci.* 359, 812–818 (2015).
24. D. Cakara, J. Kleimann, and M. Borkovec, Microscopic protonation equilibria of poly(amidoamine) dendrimers from macroscopic titrations. *Macromolecules* 36, 4201–4207 (2003).
25. J. Bugno, H.-J. Hsu, R. M. Pearson, H. Noh, and S. Hong, Size and surface charge of engineered poly (amidoamine) dendrimers modulate tumor accumulation and penetration: A model study using multicellular tumor spheroids. *Molecular Pharmaceutics.* 13, 2155–2163 (2016).
26. R. Shabanloo, S. Akbari, and M. Mirsalehi, Hybrid electrospun scaffolds based on polylactic acid/ PAMAM dendrimer/gemini surfactant for

enhancement of synergistic antibacterial ability for biomedical application. *Biomedical Materials.* 17, 045009 (2022).

27. M. H. Kanani-Jazi, S. Akbari, and M. Haghighat Kish, Efficient removal of Cr (VI) from aqueous solution by halloysite/poly(amidoamine) dendritic nano-hybrid materials: kinetic, isotherm and thermodynamic studies. *Advanced. Powder Technol.* 31, 4018–4030 (2020).
28. M. Delyanee, A. Solouk, S. Akbari, and M. Daliri Joupari, Engineered hemostatic bionanocomposite of poly(lactic acid) electrospun mat and amino-modified halloysite for potential application in wound healing. *Polym. Adv. Technol.* 32, 3934–3947 (2021).
29. S. Akbari, A. Bahi, A. Farahani, S. Milani, and F. Ko, Fabrication and characterization of lignin/dendrimer electrospun blended fiber mats. *Molecules* 26, 518 (2021).
30. S. Akbari, N. Eslahi, and M. Haghighat Kish, Evaluation of hydrophilic properties of acrylonitrile/acrylic acid copolymer films dendrigrafted with citric acid. *Polyolefins J.* 4, 213–220 (2017).
31. L. H. Sperling, and H. Leslie, *Introduction to Physical Polymer Science.* fourth edition, pp. 4-6, John Wiley & Sons (2005).
32. M. Abbasian, B. Massoumi, R. Mohammad-Rezaei, H. Samadian, and M. Jaymand, Scaffolding polymeric biomaterials: Are naturally occurring biological macromolecules more appropriate for tissue engineering? *Int. J. Biol. Macromol.* 134, 673–694 (2019).
33. S. Cai, C. Wu, W. Yang, W. Liang, H. Yu, and L. Liu, Recent advance in surface modification for regulating cell adhesion and behaviors. *Nanotechnol. Rev.* 9, 971–989 (2020).
34. A. Carré and K. L. Mittal (Eds.), *Surface and Interfacial Aspects of Cell Adhesion*, CRC Press, Boca Raton, FL (2011).
35. A. Ganjalinia, S. Akbari, and A. Solouk, Tuning poly (L-lactic acid) scaffolds with poly(amidoamine) and poly(propylene imine) dendrimers: surface chemistry, biodegradation and biocompatibility. *J. Macromol. Sci. Part A Pure Appl. Chem.* 58, 433–447 (2021).
36. S. S. Patil, K. C. Nune, and R. D. K. Misra, Alginate/poly(amidoamine) injectable hybrid hydrogel for cell delivery. *J. Biomater. Appl.* 33, 295–314 (2018).
37. A. Solouk, B. G. Cousins, H. Mirzadeh, M. Solati-Hashtjin, S. Najarian, and A. M. Seifalian, Surface modification of POSS-nanocomposite biomaterials using reactive oxygen plasma treatment for cardiovascular surgical implant applications. *Biotechnol. Appl. Biochem.* 58, 147–161 (2011).
38. A. Ganjalinia, S. Akbari, and A. Solouk, PLLA scaffolds surface-engineered via poly (propylene imine) dendrimers for improvement on its biocompatibility/controlled pH biodegradability. *Appl. Surf. Sci.* 394, 446–456 (2017).
39. L. J. Fox, R. M. Richardson, and W. H. Briscoe, PAMAM dendrimer - cell membrane interactions. *Adv. Colloid Interface Sci.* 257, 1–18 (2018).

40. X. Duan and H. Sheardown, Dendrimer crosslinked collagen as a corneal tissue engineering scaffold: Mechanical properties and corneal epithelial cell interactions. *Biomaterials*. 27, 4608–4617 (2006).
41. P. Razmshoar, S. H. Bahrami, and S. Akbari, Functional hydrophilic highly biodegradable PCL nanofibers through direct aminolysis of PAMAM dendrimer. *Int. J. Polym. Mater. Polym. Biomater.* 69, 1069–1080 (2020).
42. S. Maji, T. Agarwal, and T. K. Maiti, PAMAM (generation 4) incorporated gelatin 3D matrix as an improved dermal substitute for skin tissue engineering. *Colloids Surfaces B*. 155, 128–134 (2017).
43. A. Shaikh, P. Kesharwani, and V. Gajbhiye, Dendrimer as a momentous tool in tissue engineering and regenerative medicine. *J. Controlled. Release*. 346, 328–354 (2022).
44. T. Agarwal, A. Rustagi, J. Das, and T. K. Maiti, PAMAM dendrimer grafted cellulose paper scaffolds as a novel *in vitro* 3D liver model for drug screening applications. *Colloids Surfaces B*. 172, 346–354 (2018).
45. M. Delyanee, A. Solouk, S. Akbari, and E. S. Jafari, Modification of electrospun poly (L-lactic acid)/ polyethylenimine nanofibrous scaffolds for biomedical application. *Intl. J. Polymeric Materials and Polymeric Biomaterials* 67, 247-257 (2017).
46. M. Delyanee, A. Solouk, S. Akbari, and M. Daliri, Hemostatic electrospun nanocomposite containing poly(lactic acid)/halloysite nanotube functionalized by poly(amidoamine) dendrimer for wound healing application: *In vitro* and *In vivo* assays. *Macromol. Biosci*. 22, 2100313 (2022).
47. P. Memarian, A. Solouk, Z. Bagher, S. Akbari, and M. H. Nazarpak, Ionic conductive nanocomposite based on poly(l-lactic acid)/poly(amidoamine) dendrimer electrospun nanofibrous for biomedical application. *Biomedical Materials*. 17, 015007 (2021).
48. R. A. Bapat, S. Dharmadhikari, T. V. Chanubal, M. C. I. M. Amin, P. Bapat, B. Goraine, H. Choudhury, C. Vincent, and P. Kesharwani, The potential of dendrimer in delivery of therapeutics for dentistry. *Heliyon* 5, e02544 (2019).
49. X. Xue, X. Chen, X. Mao, Z. Hou, Y. Zhou, H. Bai, J. Meng, F. Da, G. Sang, Y, Wang, and X. Luo, Amino-terminated generation 2 poly(amidoamine) dendrimer as a potential broad-spectrum, nonresistance-inducing antibacterial agent. *AAPS J*. 15, 132–142 (2013).
50. S. Saedi and J. Rhim, Synthesis of Fe 3 O 4 @ SiO 2 @ PAMAM dendrimer @ AgNP hybrid nanoparticles for the preparation of carrageenan-based functional nanocomposite film. *Food Packaging Shelf Life* 24, 100473 (2019).
51. D. Mitra, E. T. Kang, and K. G. Neoh, Applications and challenges of smart antibacterial coatings, in: *Applications and Challenges of Smart Antibacterial Coatings*, A. S. Hamdy Makhlouf, and N. Y. Abu-Thabit (Eds.), pp 537-556, Elsevier (2020).
52. N. B. Pitts, R. H. Selwitz, and A. Ismail, Dental caries. *The Lancet* 3, 51-59 (2017).

53. Y. Ge, B. Ren, X. Zhou, H. H. Xu, S. Wang, M. Li, M. D. Weir, M. Feng, and L.Cheng Novel dental adhesive with biofilm-regulating and remineralization capabilities, *Materials (Basel).* 10, 1–11 (2017).
54. D. G. Moussa, A. Fok, and C. Aparicio, Hydrophobic and antimicrobial dentin: A peptide-based 2-tier protective system for dental resin composite restorations. *Acta Biomater.* 88, 251–265 (2019).
55. M. Eichler, V. Katzur, L. Scheideler, M. Haupt, J. Geis-Gerstorfer, G. Schmalz, S. Ruhl, R. Müller, and F. Rupp, The impact of dendrimer-grafted modifications to model silicon surfaces on protein adsorption and bacterial adhesion. *Biomaterials* 32, 9168–9179 (2011).
56. M. Gholami, R. Mohammadian, M. Arzanlou, F. Akbaridourbash, E. Kouhsari, G. Majidi, S. M. Mohseni and S. Nazari, *In vitro* antibacterial activity of poly (amidoamine)-G7 dendrimer. *BMC Infectious Diseasees.* 17, 1–11 (2017).
57. T. Dai, C. Wang, Y. Wang, W. Xu, J. Hu, and Y. Cheng, A nanocomposite hydrogel with potent and broad-spectrum antibacterial activity. *ACS Appl. Mater. Interfaces* 10, 15163-15173 (2018).
58. X. Yang, Q. Tu, X. Shen, M. Pan, C. Jiang, P. Zhu, Y. Li, P. Li, and C. Hu, Surface modification of Poly(p-phenylene terephthalamide) fibers by polydopamine-polyethyleneimine/graphene oxide multilayer films to enhance interfacial adhesion with rubber matrix. *Polym. Testing.* 78, 105985 (2019).
59. W. Luo, B. Zhang, H. Zou, M. Liang, and Y. Chen, Enhanced interfacial adhesion between polypropylene and carbon fiber by graphene oxide / polyethyleneimine coating. *J. Ind. Eng. Chem.* 51, 129-139 (2017).
60. L. Boogh, G. Jannerfeldt, and B. Pettersson, Dendritic-based additives for polymer matrix composites. Proc. ICCM-12 Conference, Paris, France, 5-9 (1999).
61. Y. Liu, Y. Fang, X. Liu, X. Wang, and B. Yang, Mussel-inspired modification of carbon fiber via polyethyleneimine / polydopamine co-deposition for the improved interfacial adhesion. *Composites. Sci. Technol.* 151, 164–173 (2017).
62. A. K. M. Moshiul Alam, R.M. Yunus, Mohd Bijirami, M. F. Mina, K. H. Maria, and T. Mieno, Modification of structure and properties of well-dispersed dendrimer coated multi-walled carbon nanotube reinforced polyester nanocomposites. *Polym. Testing.* 68, 116–125 (2018).
63. L. Ma, L. Meng, G. Wu, Y. Wang, M. Zhao, C. Zhang, and Y. Huang, Effects of bonding types of carbon fibers with branched polyethyleneimine on the interfacial microstructure and mechanical properties of carbon fiber/epoxy resin composites. *Composites. Sci. Technol.* 117, 289–297 (2015).
64. L. Ma, L. Meng, G. Wu, Y. Wang, M. Zhao, C. Zhang, and Y. Huang, Improving the interfacial properties of carbon fiber-reinforced epoxy composites by grafting of branched polyethyleneimine on carbon fiber surface in supercritical methanol. *Composites. Sci. Technol.* 114, 64–71 (2015).

65. J. Zhang, D. Zhang, A. Zhang, Z. Jia, and D. Jia, Dendritic polyamidoamine-grafted halloysite nanotubes for fabricating toughened epoxy composites. *Iranian Polymer Journal* 22, 501–510 (2013).
66. L. Ma, N. Li, G. Wu, G. Song, X. Li, P. Han, G. Wang, and Y. Huang, Interfacial enhancement of carbon fiber composites by growing TiO2 nanowires onto amine-based functionalized carbon fiber surface in supercritical water. *Appl. Surf. Sci 433*, 560-567 (2017).
67. K. L. Mittal, The role of the interface in adhesion phenomena. *Polym. Eng. Sci.* 17, 467–473 (1977).
68. B. Gao, R. Zhang, and C. Wang, Enhanced mechanical properties of carbon fiber composites by grafting different structural poly(amido amine) onto fiber surface. *Polym. Testing* 56, 192-199 (2016).
69. L. Ma, L. Meng, D. Fan, J. He, J. Yu, M. Qi, Z. Chen, and Y. Huang, Interfacial enhancement of carbon fiber composites by generation 1 – 3 dendritic hexamethylenetetramine functionalization. *Appl. Surf. Sci.* 296, 61–68 (2014).
70. M. S. Diallo, Water treatment by dendrimer-enhanced filtration: Principles and applications, in: *Nanotechnology Applications for Clean Water*, Second Edition, A. Street, R. Sustich, J. Duncan, and N. Savage (Eds.), pp. 227-239, William Andrew Publishing (2014).
71. C. Liu, Advances in membrane technologies for drinking water purification, in: *Comprehensive Water Quality and Purification,* Vol 2, S. Ahuja (Ed.), pp 75–97, Elsevier (2014).
72. R. Singh and N. P. Hankins, Introduction to membrane processes for water treatment. in: *Emerging Membrane Technology for Sustainable Water Treatment*, R. Singh and N. P. Hankins (Eds.), pp 15–52, Elsevier (2016).
73. J. J. Leam, M. R. Bilad, Y. Wibisono, M. D. H. Wirzal, and I. Ahmed, membrane technology for microalgae harvesting, in: *Microalgae Cultivation for Biofuels Production,* A. Yousuf (Ed.), pp 97–110, Academic Press (2020).
74. L. Liu, X.-B. Luo, L. Ding, and S.-L. Luo, Application of nanotechnology in the removal of heavy metal from water, in: *Nanomaterials for the Removal of Pollutants and Resource Reutilization*, X. Luo, and F. Deng (Eds.), pp 83–147, Elsevier (2019).
75. V. Vatanpour and A. Sanadgol, Surface modification of reverse osmosis membranes by grafting of polyamidoamine dendrimer containing graphene oxide nanosheets for desalination improvement. *Desalination* 491, 114442 (2020).
76. P. G. de Gennes and H. Hervet, Statistics of « starburst » polymers. *J. Phys. Lettres.* 44, 351–360 (1983).
77. L. Jiang, J. Yun, Y. Wang, H. Yang, Z. Xu, and Z. Xu, High-flux, anti-fouling dendrimer grafted PAN membrane: fabrication, performance and mechanisms. *J. Membrane. Sci.* 596, 117743 (2020).

78. Y. Mansourpanah and Z. Jafari, Efficacy of different generations and concentrations of PAMAM-NH2 on the performance and structure of TFC membranes. *React. Funct. Polym.* 93, 178–189 (2015).
79. S. Zhang, G. Qiu, Y. P. Ting, and T. Chung, Silver – PEGylated dendrimer nanocomposite coating for anti-fouling thin film composite membranes for water treatment. *Colloids Surfaces A* 436, 207–214 (2013).
80. S. Akbari, Comparison between divergent and convergent synthesis routes of amine-terminated dendrimer functionalized onto the halloysite nanotubes. *Aspects Mining Mineral Sci.3* , 1–4 (2019).
81. M. H. Kanani-Jazi and S. Akbari, Amino-dendritic and carboxyl functionalized halloysite nanotubes for highly efficient removal of cationic and anionic dyes: Kinetic, isotherm, and thermodynamic studies. *J. Environ. Chem. Eng.* 9, 105214 (2021).
82. F. Shahamati Fard, S. Akbari, E. Pajootan, and M. Arami, Enhanced acidic dye adsorption onto the dendrimer-based modified halloysite nanotubes. *Desalination Water Treatment* 57, 26222–26239 (2016).
83. F. Asempour, S. Akbari, M. H. Kanani-Jazi, A. Atashgar, T. Matsuura, and B. Kruczek, Chlorine-resistant TFN RO membranes containing modified poly(amidoamine) dendrimer-functionalized halloysite nanotubes. *J. Membrane. Sci.* 623, 119039 (2021).
84. F. Asempour, S. Akbari, D. Bai, D. Emadzadeh, T. Matsuura, and B. Kruczek, Improvement of stability and performance of functionalized halloysite nano tubes-based thin film nanocomposite membranes. *J. Membrane. Sci.* 563, 470–480 (2018).
85. L. Mei, X. He, Y. Li, Q.Peng, R. Wang, and J. Xu, Enhancement of composite-metal interfacial adhesion strength by dendrimer. *Surf. Interface Anal.* 43, 726–733 (2011).
86. M. Ramezanzadeh, B. Ramezanzadeh, M. G. Sari, and M. R. Saeb, Corrosion resistance of epoxy coating on mild steel through polyamidoamine dendrimer-covalently functionalized graphene oxide nanosheets. *J. Ind. Eng. Chem.* 82, 290–302 (2020).
87. H. A. Essawy, H. A. Mohamed, and N. H. Elsayed, Upgrading the adhesion properties of a fast-curing epoxy using hydrophilic/hydrophobic hyperbranched poly(amidoamine)s. *J. Appl. Polym. Sci.* 127, 4505–4514 (2013).
88. H. Dodiuk-Kenig, A. Buchman, and S. Kenig, Novel adhesion promoters based on hyper-branched polymers. *Composite Interfaces* 11, 453–469 (2004).

12

Arginine–Glycine–Aspartic Acid (RGD) Modification of Polymer Surfaces to Enhance Cell Adhesion

Yawen Li

Biomedical Engineering Department, Lawrence Technological University, Southfield, MI, USA

Abstract
Arginine–glycine–aspartic acid (RGD), a cell recognition motif present in many extracellular proteins, binds with integrin to regulate cell adhesion and many other cellular processes. Stable attachment of synthetic RGD peptides to polymers is essential to its use in promoting cell adhesion. This chapter reviews various techniques to immobilize RGD on polymers, and the application of RGD modified polymers in tissue engineering.

Keywords: RGD, cell adhesion, immobilization, tissue engineering

12.1 Introduction

Modifying polymer surfaces with cell recognition motifs is a highly effective approach to enhance cell adhesion. While extracellular matrix (ECM) proteins such as collagen, fibronectin and laminin are commonly used as cell adhesive molecules, they may increase the risk of infection, illicit immune response, and are prone to proteolytic degradation [1, 2]. Their orientation and conformation are also affected by the host surface texture, charge, topography and wettability, leading to variation in the cell adhesion behavior [3, 4]. One approach to solve this problem is to use small cell recognizing peptides, which are easier to synthesize, highly stable against

Email: yli@ltu.edu

K.L. Mittal and A.N. Netravali (eds.) Polymer Surface Modification to Enhance Adhesion: Techniques and Applications, (487–516) © 2024 Scrivener Publishing LLC

enzymatic degradation, and are generally highly selective in terms of recognizing a specific type of integrin receptor [5–7]. In this chapter, RGD, the most commonly used peptide to modify polymer surfaces in mediating cell adhesion, is introduced, followed by description of various RGD immobilization techniques, and the application in tissue engineering.

12.2 RGD Peptides

RGD was identified in 1984 as a minimum essential cell adhesion peptide sequence in fibronectin [8]. Subsequent studies have shown RGD to be present in over 100 other extracellular matrix (ECM) proteins such as collagen, laminin, fibrinogen, vitronectin, osteopontin, tenascin, bone sialoprotein, neurotoxin and disintegrins [9–11]. RGD mediates cell adhesion through binding with integrin, a superfamily of cell adhesion receptors. The binding leads to bidirectional inside-out and outside-in signal transduction, causing changes in both the cell behavior and ECM structure [12]. Integrins are heterodimeric (α–β) transmembrane proteins expressed at the cell surface. Of the 24 types of integrins, at least seven recognize the RGD sequence (Figure 12.1) [13].

A variety of peptides with RGD sequence have been synthesized with different flanking amino acids and conformations. These synthetic RGD peptides mediate cell adhesion in two ways: when they are immobilized on the material surface, they promote cell adhesion; when they are in solution, they act as decoys and prevent cell adhesion [15]. The integrin binding

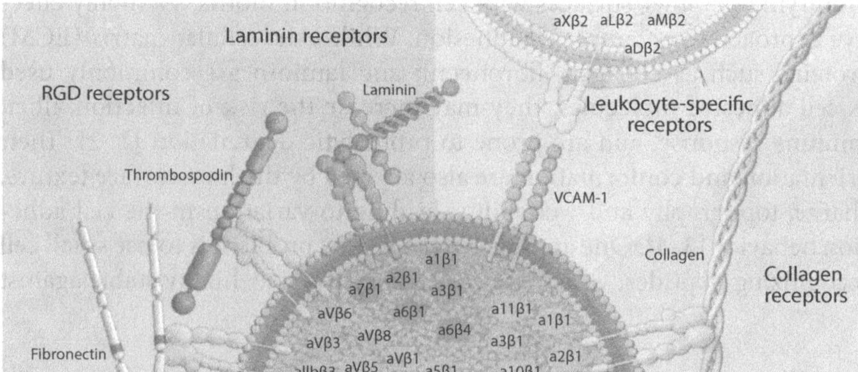

Figure 12.1 Binding of Integrins to a wide range of ligands. At least seven types of integrins bind to the RGD sequence. Figure adapted from [14].

Table 12.1 Examples of synthetic RGD peptides used for promoting cell adhesion to polymers. (Adapted from [21] with permission.)

Amino acid sequence	Polymer	Cell line/tissue
RGD	PLLA copolymer	3T3 FB (mouse)
	PEA	CHO-K1, NRK, A431 (epidermoid carc.), HeLa.S3 (carc. EC)
	PEG copolymer	Brain (rat)
	PS	Macrophages
	PCL, PLLA	Hepatocytes AML12
	PE	Rat, subcutaneous and intraperitoneal
	PEG based network	Hepatocytes (rat)
	PET	Skin FB (human)
	PCPU	Hepatocytes
RGDS	p(AA-PEG-AA)	Aortic SMC (rat), dermal FB (human)
	PEA	CHO-K1, NRK, A431 (epidermoid carc.), HeLa.S3 (carc. EC)
	PVA	Aortic EC (bovine)
	PEG copolymer	Dermal FB (human)
	PET	CaCo2 (colonic carcinoma)
	PMLG	MRC-5 FB (human)
	PS/PAA	STO FB (mouse)
	UHMWPE	HL-60 macrophages
	PVA	L-929 FB (mouse), corneal EC (rabbit)
	PEG based hydrogel	SD-SMC (rat)
	PVA hydrogel	Dermal FB (human)

(*Continued*)

Table 12.1 Examples of synthetic RGD peptides used for promoting cell adhesion to polymers. (Adapted from [21] with permission.) (*Continued*)

Amino acid sequence	Polymer	Cell line/tissue
(RGDS)n	pNIPAM	U 937 (human promonocytic leukemia)
GRGD	PAXY/dextran	L 929 FB (mouse)
	PU-PRG	HUVEC
	PEG-PPF	—
	PHPMA, PHEMA	NIH/3T3 FB, porcine aortic EC, transformed FB, rat (spinal cord, cortex, optic track)
	PET, PHEMA	NIH/3T3 FB, EC (porcine)
RGDV, RGDT	PEA	CHO-K1, NRK, A431 (epidermoid carc.), HeLa.S3 (carc. EC)
GRGDG	PCEVE-PU	HUVEC
	PEG based network	IC-21 macrophages (mouse)
	PS, PEG based network	Macrophages (rat, human), FBGC
GRGDS	Poly(norbornen)	HFF
	p(AA-PEG-AA)	HFF
	PEG based network	HFF, osteoprogenitor cells (human)
	PLA, PLGA	HUVEC
	PUR/dextran	Aortic EC (bovine)
	PMLG	MRC-5 FB (human)
	PLA/PLL	BAE-1 aortic EC
	PU/PVA	—
	Elastin-like polymer	Aortic EC (bovine), LNF (bovine)
	PCU/PHBA	HUVEC, HEP-2 (epidermoid carc.)
	FEP	Hippocampal neurons (mouse)
	p(VMA-co-VBRGD)	Hepatocytes

(*Continued*)

Table 12.1 Examples of synthetic RGD peptides used for promoting cell adhesion to polymers. (Adapted from [21] with permission.) (*Continued*)

Amino acid sequence	Polymer	Cell line/tissue
GRGDY	p(LA-co-lysine)	—
	p(LA-co-lysine)	Aortic EC (bovine)
	PET, PTFE	HUVEC
	Linear branched. PEG, PAA, PAMAM	—
	Linear PEG	Fetal brain cells (rat), PC12 pheochromacytoma cells (rat), Neuro-2a cells
(PepTite 2000™)	PTFE	HUVEC
	PLGA	Rat
	PTFE, PET	Vascular device (dog, sheep)
Fibrinogen chimera	Protein-PEG hydrogel	HFF

A single letter code for amino acids is used. Note that in peptide chemistry D-amino acids are depicted often as small, underlined or italic letters, e.g. D-Ala: a, <u>A</u>, *A*. To highlight the RGD sequence in the respective sequences, bold letters are used.

Polymers: AA: acrylic acid; AAm: acrylamide; AMA: 2-aminoethyl methacrylate; FEP: poly(tetrafluoroethylene-co-hexafluoropropylene; MACEE: methacrylic acid-(7-coumaroxy) ethyl ester; NIPAAm: N-isopropyl AAm; PAA: poly AA; PAAm: poly AAm; PAMAM: poly(amidoamine); PAXY: poly(amino)-p-xylene; PCEVE: poly(chloroethylvinylether); PCL: poly(caprolactone); PCLTMC: poly(caprolactone-co-trimethylene carbonate); PCPU: poly(carbonate urea)urethane; PCU: poly(carbonate urethane); PE: polyethylene; PEA: PE-AA copolymer; PEG: poly(ethylene glycol); PEGMA: PEG methacrylate; PET: poly(ethylene terephthalate); PHBA: poly(4-hydroxybutyl acrylate); PHEMA: poly(hydroxyethyl methacrylate); PHPMA: poly(N-2-hydroxypropyl methacrylate); PLGA: poly(lactic acid-co-glycolic acid); PLLA: poly(l-lactic acid); PMLG: poly(gamma-methyll-glutamate); PMMA: poly(methyl methacrylate); PPF: poly(propylene fumarate); PPO: poly(propylene oxide); PPy: polypyrrole; PS: polystyrene; PTFE: polytetrafluoroethylene, PU: polyurethane; PVA: poly(vinyl alcohol); UHMWPE: ultrahigh molecular weight PE; VBRGD: 6-(p-vinylbenzamido)-hexanoicacid-GRGDS; VMA: N-p-vinylbenzyl-d-maltonamide.

Cells: EC: endothelial cell; FB: fibroblast; AMEC: adipose microvascular EC; CHO: Chinese hamster ovary cell; FBGC: foreign body giant cells; HFF: human foreskin fibroblasts; HSVEC: human saphenous vein EC; HUVEC: human umbilical vein EC; LNF: ligamentum nuchae FB; NRK: normal rat kidney cell; PAEC: pulmonary aortic EC; SMC: smooth muscle cell.

affinity and activity to RGD peptides is affected by many factors, such as the RGD structure and conformation, RGD density and distribution on the polymer surface, cell type and culturing conditions. With the large number of synthetic RGD peptides, polymers and cell types (see Table 12.1), it will be difficult to obtain a comprehensive ranking of integrin-RGD binding activity, although some qualitative conclusions have been reported in literature usually based on comparison of a subset of RGD peptides and one cell type. For example, a more linear conformation of RGD promotes binding to the αIIbb3 integrin, whereas a more kinked configuration favors the avb3 integrin binding [16, 17]. Several studies showed that the RGD density on surface could affect the cell adhesion [18–20].

12.3 RGD Immobilization Techniques

Attaching RGD peptides to a polymer surface is an effective approach to utilize the integrin-RGD binding for modulating cell adhesion. Stable linking of RGD to the surface is essential since the cell contractile force could displace weakly attached ligands, leading to weak cell adhesion [22, 23]. Detached ligands may also be internalized by cells through endocytosis [24].

It is important to note that the RGD immobilization can be either a long-term or transient attachment of the peptide on or within the polymer. For inert solid polymers, the long-term retention of the RGD peptide is expected to mediate cell adhesion. For biodegradable polymers, the immobilized RGD peptide may be released as the matrix degrades.

Many techniques have been developed to immobilize RGD peptides through either surface modification or bulk incorporation. The surface modification usually leads to enhanced cytocompatibility of the polymer without altering its bulk properties, and has been widely used for a variety of biomaterials and tissue engineering applications. The bulk modification allows incorporation of more cell recognition sites in the material, which can be a useful strategy to add functionalization to injectable biomaterials. Choosing an appropriate immobilization strategy depends on the chemistry and geometry of the substrate, the availability of reactive groups in RGD, and the intent for bulk or surface modifications.

Immobilization of biomolecules can be achieved by physical adsorption, physical entrapment, covalent binding, or micro/nano-patterning. Physical adsorption is rarely used for the RGD immobilization due to the relatively weak binding between RGD and the polymer surface, and the difficulty to control the amount and density of adsorbed peptides. Studies of RGD immobilization using physical adsorption as control usually showed very low amount of bound RGD and less effective cell adhesion compared to covalent attachment.

12.3.1 Physical Entrapment

It may be desirable to "trap" RGD peptides in a polymer to achieve transient localization of the cell recognition motif on and within the material. This could also help to achieve continuous provision of signals to modulate cell interaction as the hydrogel swells and/or polymer degrades.

One method to trap RGD in polymers is to use copolymerization. Usually a polymer is co-polymerized with another polymer that can provide functional side chains or pendant groups for stable attachment of RGD. For example, acrylic acid is co-polymerized with polyethylene [25], polystyrene [26], and N-isopropylacrylamide [27] to introduce COOH group as a side chain to the polymer. Biodegradable polymers such as PLA, PGA, PLGA, and PCL can be co-polymerized with lysine [28, 29], poly(ethylene glycol) (PEG) [30–32], and acrylic acid [33, 34] to achieve effective RGD immobilization.

PEG-based interpenetrating network (IPN) is another approach to immobilize RGD on hydrogels. PEG has low toxicity and can be easily chemically modified. IPNs are synthesized either by blending a polymer in a crosslinked mechanical network, or by sequential polymerization [35-37]. One limitation of IPNs is the potential toxicity of the crosslinkers used and the complicated fabrication process [38, 39].

12.3.2 Covalent Attachment

Covalent immobilization of RGD is a preferred method in many applications with improved stability and persistence of RGD delivered to cells. The amine group on RGD is the most commonly used site to form a covalent bond with the substrate. The following chemistries can be used for immobilizing RGD (Table 12.2).

12.3.2.1 Water-Soluble Carbodiimide (WSC) Chemistry

1-(3-Dimethylaminopropyl)-3-ethylcarbodiimide hydrochloride (EDC), a WSC, is commonly used to achieve covalent attachment of RGD. EDC is a zero-length crosslinking agent that couples carboxyl group to amine group,

Table 12.2 Commonly used immobilization chemistries for RGD peptides.

Chemistry	Functional groups needed on substrate	Comments
Water-soluble carbodiimide coupling reaction	Amine or carboxyl	Simple and effective; Variable RGD binding efficiency
Click chemistry	Not needed	Robust and mild reaction with high yield, bioorthogonal; Cu is incompatible for *in vivo* applications
Photo-immobilization	Double bonds, C-H or N-H	Simple and effective; Potential risk of UV affecting RGD activity

resulting in a stable amide bond, as shown in Figure 12.2 [40]. The o-iso-acylurea intermediate is highly unstable in aqueous solution. Therefore, sulfo-N-hydroxysuccinimide (sulfo-NHS) is usually added to create a more stable amine-reactive sulfo-NHS ester intermediate and improve the reaction efficiency [41]. The WSC immobilization is a simple and effective approach performed under mild conditions. The main limitation of this approach is the lack of differentiation between the terminal reactive groups and reactive groups within the RGD structure, which could impede the RGD recognition activity, and lead to large variability in RGD binding efficiency and reproducibility [21]. Another potential issue is that the EDC could also crosslink the polymer substrate and alter its mechanical and chemical properties [42].

12.3.2.2 Click Chemistry

A set of general criteria proposed by Kolb, Finn and Sharpless in 2001 for a perfect reaction to "be modular, wide in scope, give very high yields, generate only inoffensive byproducts that can be removed by nonchromatographic methods, and be stereospecific" led to the definition of "click chemistry" [43]. Cu(I)-catalyzed alkyne-azide cycloaddition (CuAAC) is the most widely used click chemistry. The products of CuAAC can function as rigid linking units that mimic an amide bond with decreased

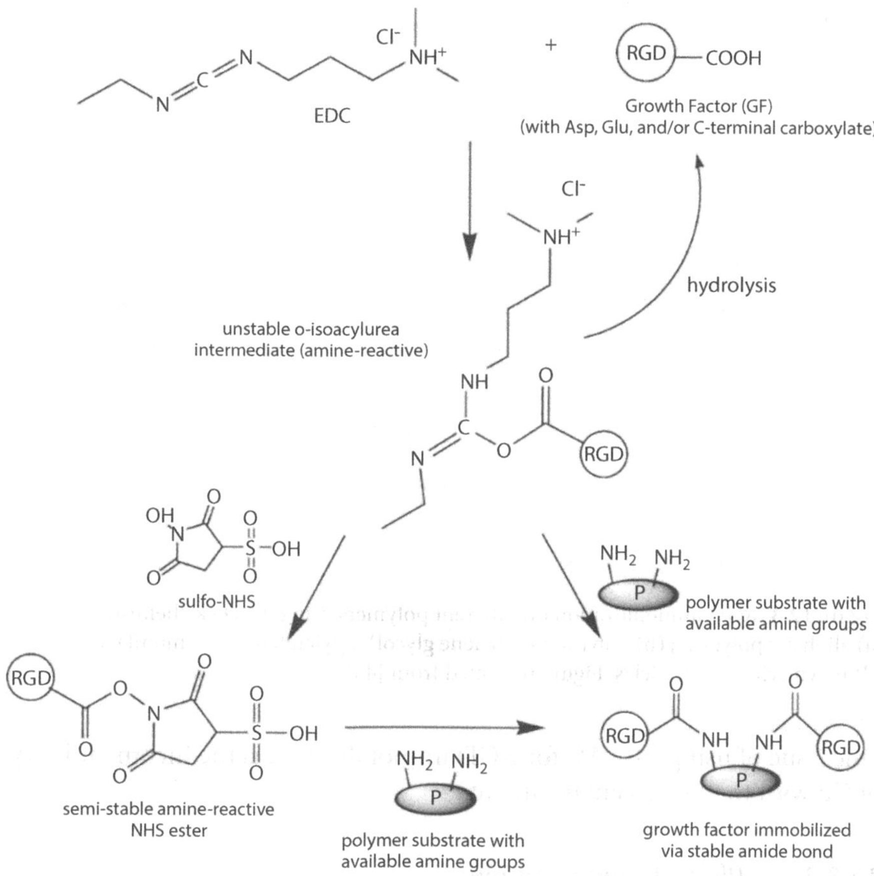

Figure 12.2 General mechanism for covalent immobilization of RGD to a substrate via WSC chemistry. This reaction scheme depicts an example where an amide bond is formed between a carboxylate on the RGD and an amine on the substrate, but the converse may also be performed (i.e., amine on the RGD reacted with carboxylate on substrate). Sulfo-NHS is usually added to form a more stable amine-reactive intermediate and improve the reaction efficiency. Figure adapted from [42] with permission.

susceptibility to enzymatic hydrolytic cleavage [44]. This makes the click chemistry an attractive strategy to link RGD to a variety of polymer surfaces, see Figure 12.3.

The click chemistry creates covalent rapid linkage of two dissimilar components under mild conditions. It is also biorthogonal, that is, the two components do not interact to affect the functionality in the biological system [46]. All these advantages make the click chemistry an attractive alternative to the traditional chemistry in many biomedical applications.

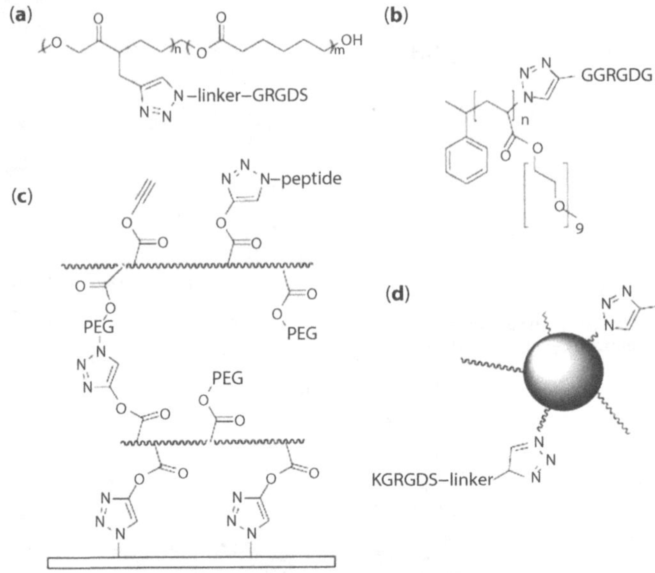

Figure 12.3 RGD immobilization on different polymers using the click chemistry. (a) aliphatic polyester; (b) poly(oligo(ethylene glycol) acrylate); (c) PEG multilayer; (d) polymeric nanoparticles. Figure reprinted from [45].

One issue of using CuAAC for RGD immobilization is the incompatibility of Cu with the *in vivo* environment.

12.3.2.3 Photo-Immobilization

Photo-immobilization is another commonly used method to covalently attach RGD onto the polymer surface. Usually the RGD is first functionalized with a photo-reactive group (such as the phenyl azide group), followed by exposure to a long-wave ultra-violet light to induce crosslinking. Phenyl azides are a family of heterobifunctional crosslinking molecules that have one terminus to bind with RGD and another terminus containing a photoreactive moiety to react with the polymer substrate [47]. Figure 12.4 illustrates the reaction of the N-hydroxysuccinimide end of the heterobifunctional crosslinking molecule Sulfo-(sulfosuccinimidyl 6-(4'-azido-2'-nitrophenylamino)hexanoate) (SANPAH) with a primary amine in RGD, followed by covalent immobilization of the modified RGD onto a substrate upon application of UV light to initiate reaction via the phenyl azide.

Photo-immobilization is also a relatively simple and effective process, with the added advantage of being able to create patterned RGD on surface,

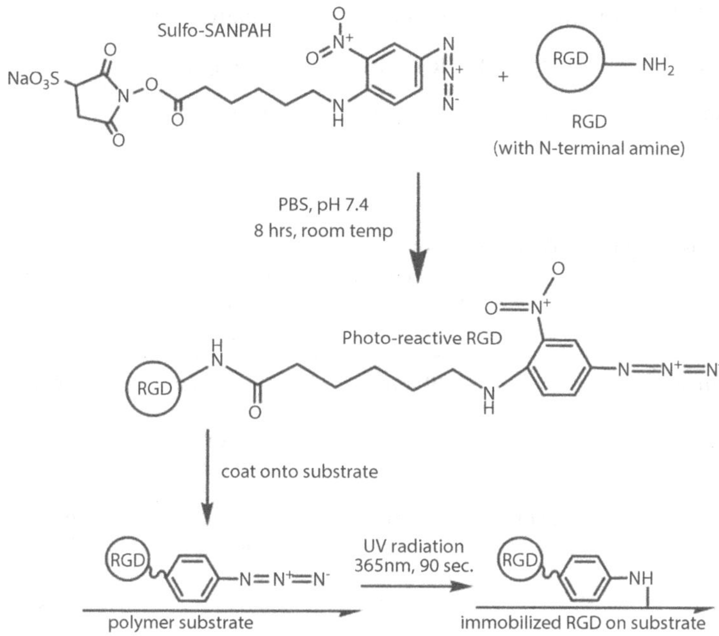

Figure 12.4 General reaction scheme for modifying an RGD peptide with a photo-reactive phenyl azide group and subsequent immobilization on a biomaterial substrate. Figure adapted from [42] with permission.

thus enabling greater control of the cell functions. One issue with the photo-immobilization is the potential risk to affect the RGD activity and cell viability [42]. Using long wavelength UV and shorter exposure time can help to minimize the risk.

12.3.3 Micro/Nano-Patterning

Patterning of RGD peptides at the micro and nanoscale provides a potentially more *in vivo* like microenvironment to study and manipulate the cell behavior. The photo-immobilization process described above can be utilized with the addition of a photomask to pattern RGD both in 2D and 3D systems. Here we focus on the use of micro and nanofabrication processes to pattern RGD on and within polymers. Table 12.3 summarizes some commonly used techniques for patterning RGD with micro/nano-scale resolution.

Table 12.3 Micro/nano fabrication techniques for patterning RGD peptides.

Technique	2D or 3D	Resolution	Advantages	Disadvantages
Photolithography	2D	100 nm	Well established process	Not compatible with rough, biodegradable substrates
Focused laser light	2D or 3D	Not reported	Precise modulation in time and space	Need multiphoton laser for 3D structures
Microcontact printing	2D	40 nm	Easy and rapid, applicable for curved surfaces	Less control of RGD density
Microfluidic patterning	2D or 3D	100 nm	RGD gradient possible	Potential cell damage due to fluid flow stress
Dip pen nanolithography	2D	15 nm	Precise geometry and pattern	Low writing speed
Bioprinting	2D or 3D	200 μm	Multifunctional cell/RGD possible with simultaneous printing of different inks	Limited resolution, effect of printing pressure and crosslinker on cells and RGD unknown

12.3.3.1 *Photolithography*

When photolithography is used to pattern RGD, a simple approach is to create defined patterns with micrometer or submicrometer resolution on polymer surface using photolithography, followed by immobilization of RGD peptides in selected patterns. Similarly, RGD can be grafted uniformly on the polymer surface, and then selectively removed using excimer laser ablation [48].

Another approach is to use photoremovable protecting groups, or caged compounds to achieve one-step patterning of RGD. For example,

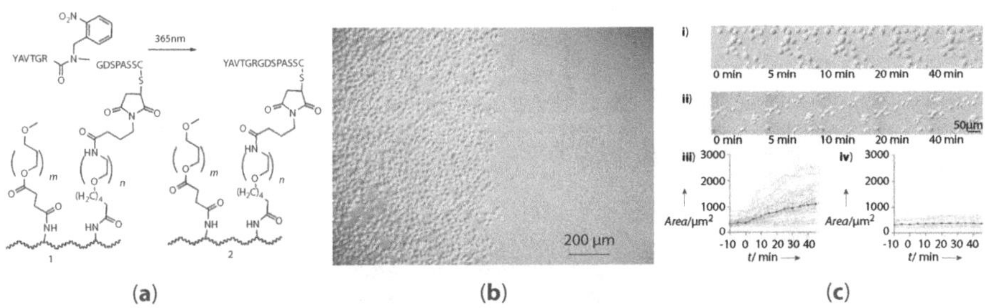

Figure 12.5 (a) Photochemical reaction of a caged RGD peptide attached to a culture dish. The caged RGD peptide was linked to poly-L-lysine (depicted as a wavy line). (b) Spatial control of cell adhesion. The left half of the dish was irradiated during cell cultivation. The micrograph was recorded after washing and fixing. (c) Temporal control of cell adhesion. Micrographs (i, ii) and temporal changes in cell areas (iii, iv) are shown. The dish was irradiated with UV light just after recording the micrograph at 0 min (i, iii). The results obtained in the absence of irradiation are shown in (ii) and (iv). Dotted lines indicate the areas of individual cells, and solid lines indicate the average area of the cells (iii, iv). Figures reprinted from [49] with permission.

3-(4,5-dimenthoxy-2-nitrophenyl)-2-butyl chromophore can be attached to the carboxylic side group of the aspartic acid to form the caged RGD, which can then be photolithographically patterned to study the effect of RGD density and temporal presentation on cell adhesion, proliferation and differentiation (Figure 12.5) [49].

12.3.3.2 Focused Laser Light

While the photolithography technique uses a lamp-based light source and masks to create 2D patterns, focused laser light uses either single or multiphoton to achieve either 2D or 3D patterning of biomolecules. This is particularly useful in creating hydrogels with spatially controlled distribution of biomolecules. In one study, single photon laser was used to immobilize RGD peptides within an agarose gel, which showed mediation of endothelia cell adhesion after a few hours (Figure 12.6) [50]. Two-photon laser scanning (TPLS) photolithographic technique was used to dictate the precise location of RGDS in collagenase-sensitive poly(ethylene glycol-co-peptide) diacrylate hydrogels, see Figure 12.7. Guided cell migration was demonstrated in the 3D hydrogel network [51].

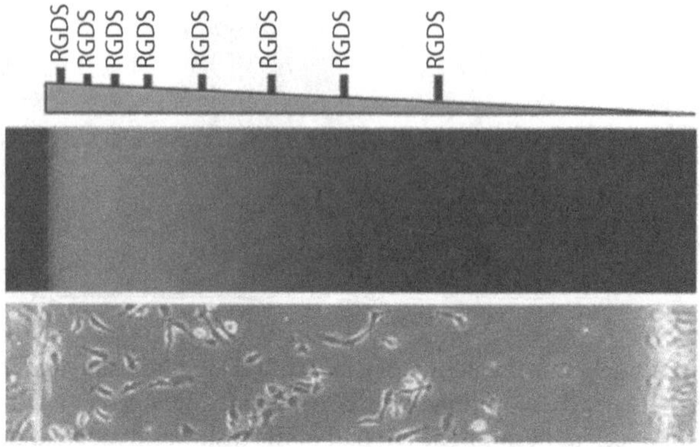

Figure 12.6 Agarose gel with RGD gradient on surface (Top). The shape of the gradient can be visualized using fluorescent molecules (Bottom). The effect of the gradient can be evaluated by examining the adhesion of endothelial cells (Bottom). Figure reprinted from [50], Copyright (2006) National Academy of Sciences, U.S.A.

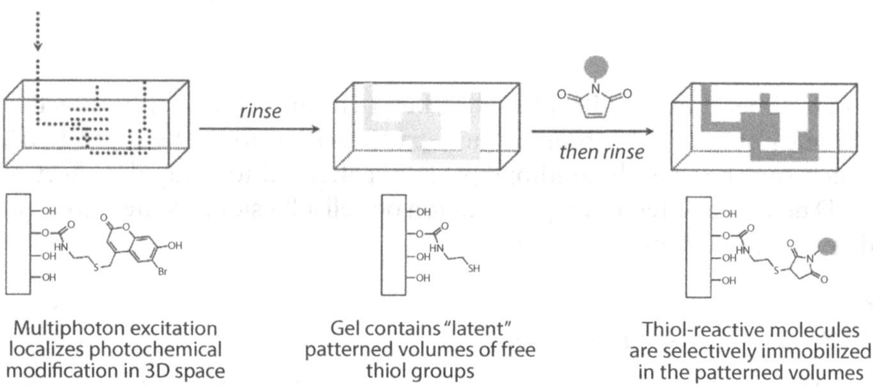

Figure 12.7 Schematic of multiphoton chemical patterning in the agarose hydrogel. Figure reprinted from [51]. Copyright (2008) American Chemical Society.

12.3.3.3 Soft Lithography

Soft lithography uses a patterned elastomer block as a stamp or mold to generate micropatterns. The patterned elastomer block is created by pouring and curing mixture of elastomer prepolymer and crosslinking agent against a master mold, usually a silicon substrate with SU-8 patterns created using photolithography. Poly(dimethylsiloxane) (PDMS) is the most commonly used elastomer. Of the many different soft lithography techniques, microcontact printing (μCP) and microfluidic patterning are the most commonly used in patterning biomolecules such as RGD [52].

In μCP, the PDMS block serves as a "stamp" and is inked with the target solution. The solution is brought into contact with a polymer substrate to create defined patterns. Although this is a simple process and many polymer inks have been successfully patterned on various polymer substrates, the ink selection and substrate preparation can affect the quality of the printed patterns [53]. Pure RGD ink is not used due to its weak binding with most polymer surfaces. RGD functionalized polymer, such as poly-l-lysine-g-poly(ethylene glycol) provided reproducible printed patterns with lateral resolution down to 1 μm. Coating the stamp with a layer of detergent helps to achieve optimized loading and ink transfer properties [54]. μCP can also be used to induce reactions to pattern RGD, as illustrated in Figure 12.8.

Microfluidic patterning utilizes controlled laminar flow through microchannels to create defined patterns of biomolecules. In one study, RGD was patterned on the PEG-PMMA surface and showed enhanced corneal stromal cell attachment and migration [56]. By adding a backflow step, an RGD peptide gradient was generated on polyelectrolyte multilayers of HA and poly(l-Lysine), see Figure 12.9. Pre-osteroblasts responded to the gradient in stiffness and RGD concentration, as shown in Figure 12.10 [57].

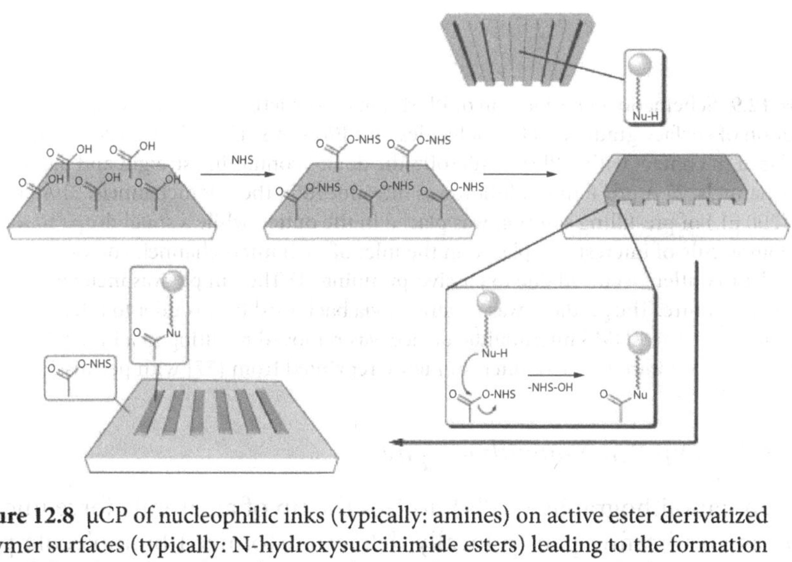

Figure 12.8 μCP of nucleophilic inks (typically: amines) on active ester derivatized polymer surfaces (typically: N-hydroxysuccinimide esters) leading to the formation of amide bonds exclusively in the area of contact between stamp and substrate. Figure reprinted from [55] with permission.

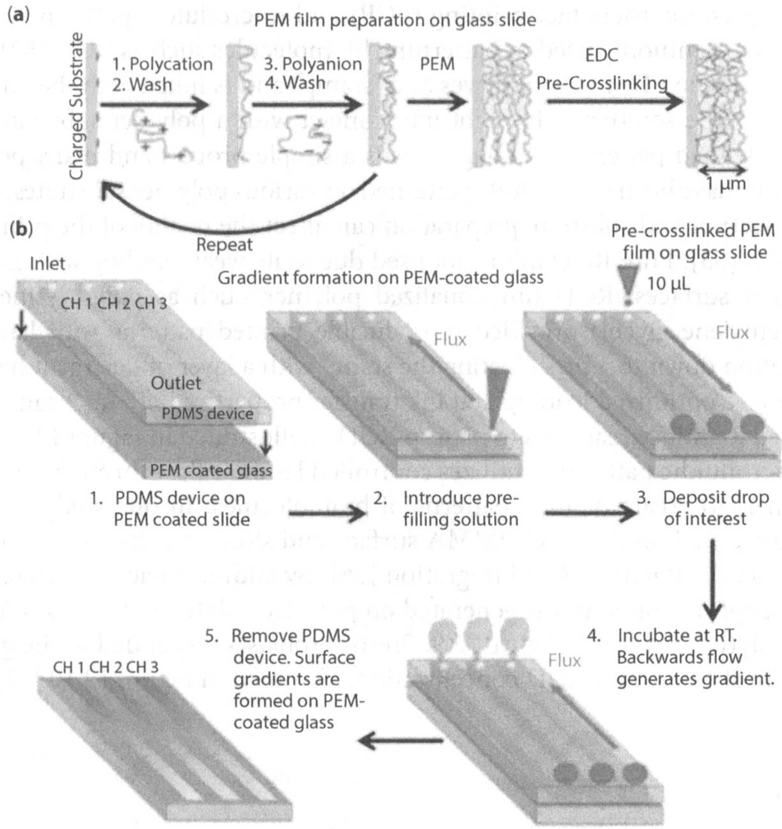

Figure 12.9 Schematic representation of PEM film construction on a glass slide (a) and (b) formation of surface gradient of biomolecules on PEM films. 1) A PEM-coated glass slide was placed in contact with a PDMS microfluidic device containing straight and parallel microchannels. 2) A pre-filling solution was introduced in the microchannels. 3) A large drop (200 μL) of pre-filling solution was placed in the outlet, while a small drop (10 μL) of a biomolecule of interest was placed in the inlet of each microchannel. Forward flow, from inlet to outlet, occurred due to passive-pumping. 4) The sample was incubated at room temperature. The gradient was generated via backward flow (outlet to inlet) due to evaporation. 5) The PDMS microfluidic device was removed resulting in a PEM film with a gradient of the biomolecule of interest. Figure reprinted from [57] with permission.

12.3.3.4 *Dip Pen Nanolithography*

Dip pen nanolithography (DPN) utilizes the tip of an atomic force microscope as an ultrafine pen to locally deliver customized biomolecule patterns onto the surface. The ink tip forms a small water meniscus, serving as a channel for the molecule transport. The resolution of the pattern is

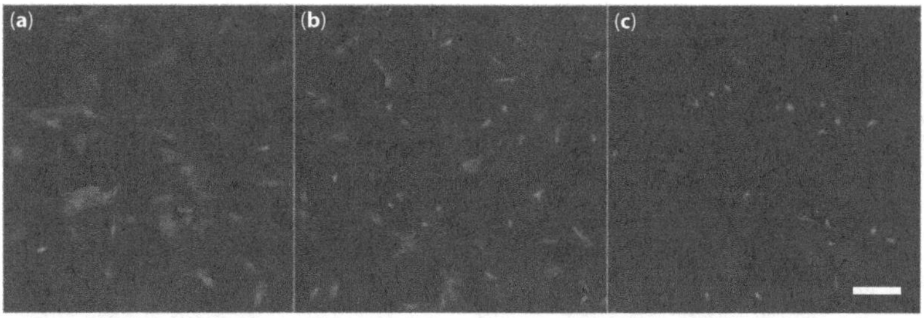

Figure 12.10 Seeded pre-osteoblasts at a distance of 0 mm (a), 5 mm (b), and 17 mm (c) away from the inlet. Cells spread better in regions of higher stiffness and higher RGD density (a), and their spreading decreases along the length of the gradient. Scale bar: 200 μm. Figure reprinted from [57] with permission.

related to the interaction between the coated tip and surface, and the writing conditions such as dwell time, temperature and humidity [58].

Advantages of DPN include its high resolution, direct writing capability, and room temperature processing. One main limitation of DPN is its slow speed of writing. DPN is normally performed on gold and oxide surfaces. Self-assembled monolayer (SAM) molecules have also been patterned using DPN, which can subsequently be functionalized with bioactive molecules such as RGD [59]. Generating a large array of non-SAM molecules remains a hurdle to practical use of DPN in cell biology applications [32]. Nevertheless, the high resolution of DPN generated patterns allows the study of single cell adhesion without the influence of cell-cell interactions. One study showed that the spatial presentation and affinity of RGD made by DPN had a strong effect on the polarization of adhered fibroblasts, as shown in Figure 12.11 [60].

12.3.3.5 Bioprinting

Bioprinting is an extension of the 3D printing process that utilizes a layer-by-layer method to create a biomimetic construct containing biomaterials, growth factors, and/or cells [61]. Hydrogels are commonly used as the bioink for printing. Bioactive molecules such as RGD can be easily incorporated into the bioink to tailor the physical and biochemical properties of the printed structure [62].

Alginate is commonly used in bioprinting to encapsulate cells because of its adjustable viscosity, cytocompatibility, and quick and safe gelation [63]. However, mesenchymal stem cells encapsulated in alginate exhibited round morphology due to the lack of cell adhesion sites in alginate. Modifying alginate with RGD showed pronounced effect in enhancing the

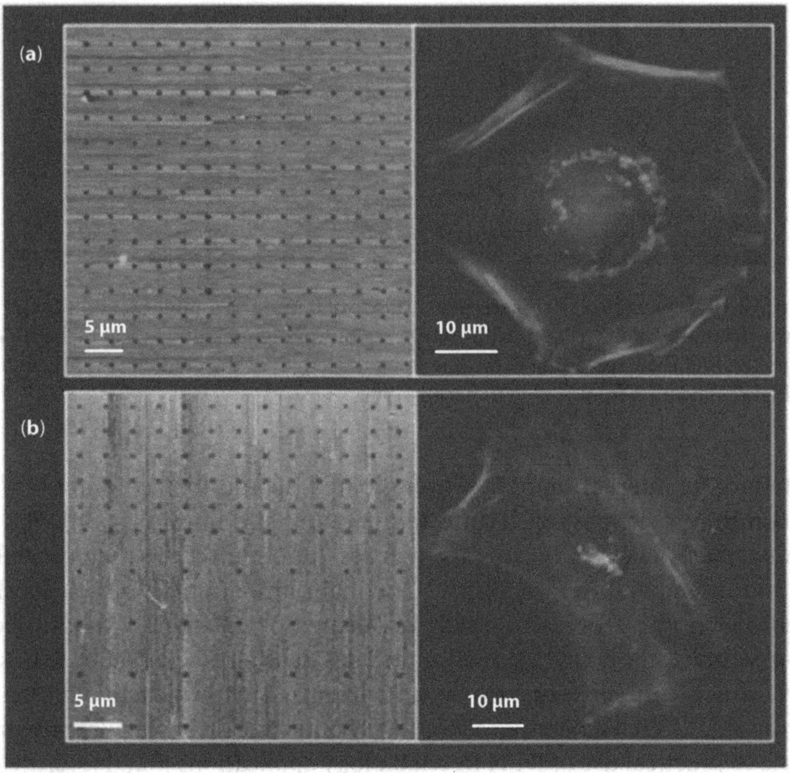

Figure 12.11 (a) (Left) Lateral force microscopy image of an expanded region of the control symmetric nanoarray (20 X 20, 500 nm spots spaced 3 μm apart). (Right) Representative fluorescent micrograph of 3T3 Swiss Albino mouse fibroblast adhered to a symmetric nanoarray of immobilized linear RGD peptide. The diffuse distribution of the Golgi surrounding the nucleus indicates that the cell is not polarized on the symmetric nanoarray. (b) (Left) Lateral force microscopy image of an expanded area of an asymmetric nanoarray composed of two arrays of 500 nm spots 20 X 10 (spaced 3 μm apart) and 10 X 5 (spaced 6 μm apart). The vector between the concentrated Golgi with respect to the nucleus and centrosome indicates that the cell is polarized toward the higher density region of the array. The cells were stained for nuclei (blue), Golgi apparatus (green), and actin (red). Figure reprinted from [60]. Copyright (2008) American Chemical Society.

proliferation and spreading of the encapsulated human adipose derived stem cells in the construct without affecting its mechanical properties [64]. In another study, RGD modified alginate gel was photo-crosslinked while printing (Figure 12.12). The amount of photoinitiator affects the stiffness of the gel with higher stiffness promoting the angiogenic differentiation of the encapsulated adipose progenitor cells while inhibiting its adipocyte differentiation, see Figure 12.13 [65].

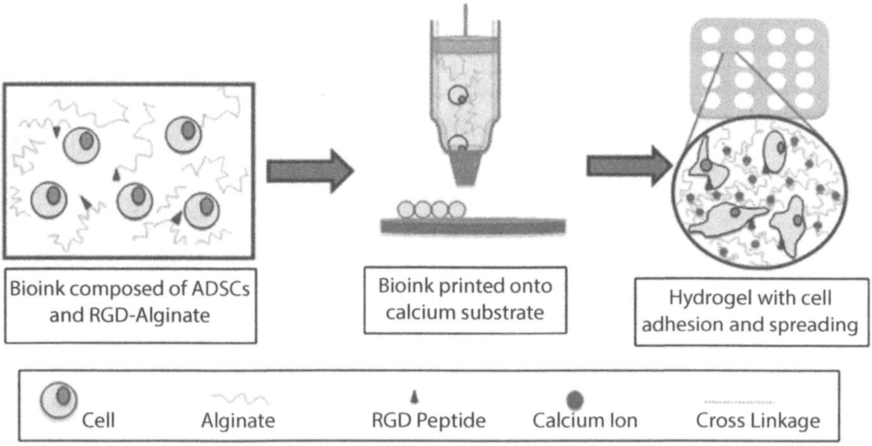

Figure 12.12 Schematic representation of biodegradable oxidized alginate bioink for bioprinting. A bioink consisting of RGD-modified oxidized alginate human adipose stem cells (hADSCs) was printed in a defined lattice structure on a gelatin substrate to crosslink the hydrogel. Figure reprinted from [65] with permission.

12.4 Characterization

It is important to analyze the RGD modified polymers to confirm and quantify the presence of RGD on the polymer, to evaluate its effectiveness in cell adhesion, and to assess its applicability for *in vivo* and clinical applications.

RGD immobilization may alter the wettability of some polymers. Water contact angle is a simple method to analyze if the hydrophilicity of the polymer has changed [66]. Amide bond is often used as an indicator of the RGD presence on the polymer surface. Fourier transform infrared spectroscopy (FTIR) and X-ray photoelectron spectroscopy (XPS) can both be used to detect the amide bond [67, 68]. XPS also provides surface elemental analysis. Usually an increase in the nitrogen content can be attributed to the addition of RGD [69].

The RGD peptide quantification can be conducted using the amino acid analysis (AAA). Radioactivity is also employed to quantify the peptide if radioiodination is used [15]. The Sakaguchi assay is used to detect the arginine group [30]. The bicinchonic acid (BCA) assay is another general protein assay that can be used for RGD quantification [67].

In vitro evaluation is important to assess the effect of immobilized RGD on cell adhesion and other behavior, and to predict the *in vivo* performance and potential medical uses. Most commonly conducted tests include the cell adhesion, cell spreading, cytoskeletal reorganization and formation

(a)

Viability after printing (day 0)						
Ox. \ Conc.	2	5	8	10	15	20
0	-	>95%	>95%	~70%	-	-
1	-	-	>95%	>90%	-	-
3	-	-	>95%	>95%	<20%	-
5	-	-	-	>95%	>95%	<20%
10	-	-	-	-	>95%	>90%

(b) Post -Printing (day 0)

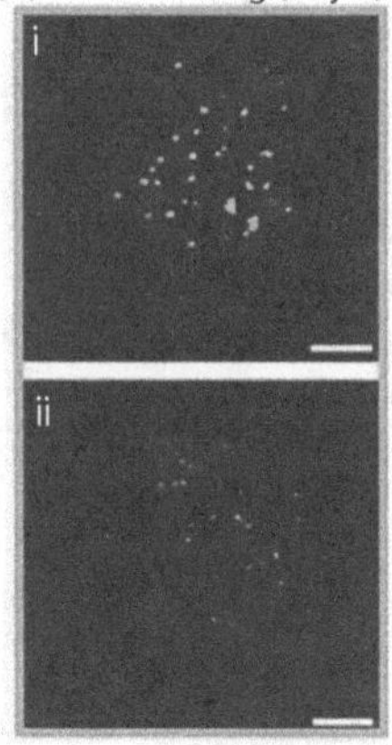

(c) After 8 days culture

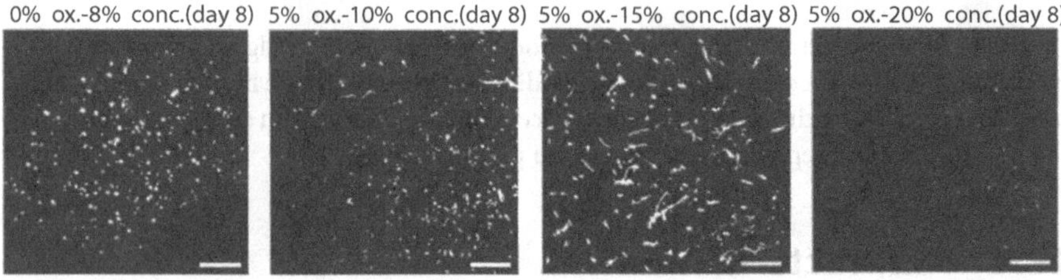

Figure 12.13 Cell viability assay of density and viscosity criterion-filtered samples. (a) Samples of high viability (>90%) after printing (green). (b) The fluorescent pictures of live–dead assay after printing: (i) high cell viability sample (e.g. 5% ox.–15% conc.); (ii) low cell viability sample (e.g. 5% ox.–20% conc.) (scale bar = 100 μm). (c) Cell viability assay at day 8. Except for the 5% ox.–20% conc. sample (0% viability), the remaining three samples showed high viabilities (>95%) after 8 days in culture (scale bar = 100 μm). Figure reprinted from [65] with permission.

of focal adhesion [70]. Numerous studies have shown the beneficial effect of RGD immobilization on promoting cell adhesion. Many studies also showed that the RGD type, density and arrangement could influence the cell adhesion and migration [21, 71]. Discussion of the *in vivo*/clinical studies will be included in the "Applications" section below.

12.5 Applications

RGD peptides have attracted considerable attention during the past decade as a molecular imaging probe for diagnostics, as a drug and gene delivery system, as a targeting unit for cancer theranostics, and as a cell adhesion promoter for tissue engineering [21, 72–74]. The first three applications are not covered in this chapter as these involve very little RGD immobilization on polymers. This section will focus on tissue engineering application.

The general framework of tissue engineering consists of a biodegradable scaffold, cells and soluble and insoluble regulators towards regeneration of a new tissue [75]. Cell adhesion to the scaffold is essential to almost all tissue engineering applications. RGD has been reported to be effective to promote the attachment of numerous cell types to the biodegradable scaffolds. An example is given below to describe how RGD modification is used for bone tissue engineering.

Bone contains osteoblasts (forming new bone), osteoclasts (resorbing bone), and osteocytes (responding to mechanical strain). Osteopontin and sialoprotein, two key proteins involved in the osteoblast and osteoclast activities, both contain the RGD binding motif [76]. RGD peptides isolated from the bone sialoprotein have been found to enhance the attachment, spreading and focal adhesion formation of osteoblasts.

To regenerate new bone using the tissue engineering approach, RGD has been used to modify the surface of various types of scaffolds. *In vitro* studies showed that the cyclic RGD peptide had a 10-fold higher activity in osteoclast retraction than the linear counterpart [77]. In one animal study, PLGA scaffolds coated with RGD showed higher osteocompatibility and more osseous ingrowth into the scaffolds compared to uncoated controls when implanted in the cortical wounds in the rat tibiae, although no significant difference was observed after 3 weeks [78]. In another study, RGD-coated porous PMMA implants were shown to regenerate bone tissue faster with higher strength in a rabbit patella model, see Figure 12.14 [79].

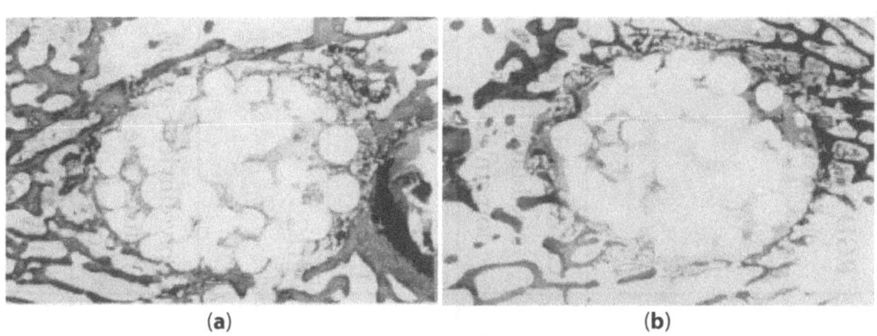

Figure 12.14 Cross-section of implanted PMMA implants (staining according to Goldner-Masson, magnification 16). (a) RGD-peptide-coated PMMA implant. (b) Uncoated PMMA implant. Color index: white: PMMA beads; green ± blue: already existing bone and newly formed bone (visible in the top image around the PMMA beads), light brown: newly formed osteoid, a bone precursor, dark brown: fibrous tissue.

Table 12.4 *In vivo* evaluation of RGD modified polymers for tissue engineering.

Polymer	RGD immobilization chemistry	Tissue engineering application	Experimental model	Results
PCL	EDC	Vascular graft	Rabbit carotid artery [83]	Significantly improved hemocompatibility; enhanced cell infiltration and homogeneous distribution within the PCL-RGD grafts after 4 weeks
Chitosan-alginate-hyaluronan	Crosslinked through a cellulose binding domain	Cartilage	Rabbit cartilage [84]	Complete repair of cartilage in 6 months
PHBV/PCL	EDC	Arterial graft	Rat abdominal aorta replacement model [85]	Enhanced formation of an endothelial cell monolayer one-month post implantation
Alginate	EDC	Neural tissue	Chick chorioallantoic membranes [86]	Adhesion and differentiation of bone marrow stromal cells into neuron and glia cells after 7 days
Hyaluronan	Poly(ethylene glycol) diacrylate as crosslinker	Soft tissue	Mouse subcutaneous dorsal injection [87]	Modestly accelerated growth of fibrous tissue after 4 weeks
PEG	Photo-polymerization	Soft tissue	Mouse subcutaneous pocket [88]	Less inflammatory response, normal foreign body reaction after 4 weeks

Compared to the numerous *in vitro* studies of RGD modified polymers for the tissue engineering application, the number of *in vivo* studies in literature is much smaller. For example, RGD modified alginate has been frequently proposed as a promising material for corneal regeneration, with many *in vitro* studies showing encouraging results, i.e., excellent biocompatibility and enhanced cell attachment [80, 81]. However, no successful *in vivo* studies have been reported yet using RGD modified alginate.

The efficacy of RGD is likely influenced by many different processes *in vivo*. Numerous *in vitro* studies performed in serum-free medium almost always showed improved cell attachment to the substrate, whereas blood and body fluids contain many different types of proteins, which quickly adsorb onto the biomaterial surface upon implantation. Some of these proteins are native integrin-binding proteins, which stimulate more robust integrin signaling than the synthetic RGD. Moreover, the adsorption of other proteins may also impact the integrin signaling. For example, albumin, an abundant protein *in vivo*, has been shown to inhibit cell adhesion [82]. All this complexity could explain the variation in the effectiveness of RGD *in vivo*. Table 12.4 lists some examples of *in vivo* studies using RGD modified polymers for non-bone tissue engineering applications.

12.6 Summary

The prevalence of RGD expressed in many ECM molecules, and the ease of creating different synthetic RGD peptides with flanking amino acids, make the RGD peptides a versatile motif for mediating cell adhesion. Most polymers do not exhibit functional groups to bind with RGD. Many immobilization chemistries have been developed to stably attach RGD on and within different polymers. Advanced micro- and nano-fabrication processes have enabled the creation of polymers with precise spatial control of adhesive molecules such as RGD. RGD modified polymers have tremendous potential in tissue engineering applications where cell adhesion to the scaffold is essential. The effectiveness of RGD in promoting cell adhesion has been shown in many *in vitro* studies. Encouraging results have also been reported in some, but not all *in vivo* studies, likely due to remodeling of the biomaterial surface upon implantation. With continued development in protein engineering and biomaterial processing techniques, it is possible to achieve more favorable *in vivo* and even clinical outcomes by combining RGD and other functional domains to better mimic and control the cell-ECM interactions.

References

1. J. Kaehler, P. Zilla, R. Fasol, M. Deutsch and M. Kadletz, Precoating substrate and surface configuration determine adherence and spreading of seeded endothelial cells on polytetrafluoroethylene grafts, *J. Vascular Surgery,* 9, 535-541 (1989).
2. R. J. Wade and J. A. Burdick, Engineering ECM signals into biomaterials, *Materials Today,* 15, 454-459 (2012).
3. G. B. Fields, J. L. Lauer, Y. Dori, P. Forns, Y. Yu and M. Tirrell, Proteinlike molecular architecture: Biomaterial applications for inducing cellular receptor binding and signal transduction, *Biopolymers,* 47, 143-151 (1998).
4. G. Altankov, F. Grinnell and T. Groth, Studies on the biocompatibility of materials: Fibroblast reorganization of substratum-bound fibronectin on surfaces varying in wettability, *J. Biomed. Mater. Res.,* 30, 385-391 (1996).
5. Y. Ito, M. Kajihara and Y. Imanishi, Materials for enhancing cell adhesion by immobilization of cell-adhesive peptide, *J. Biomed. Mater. Res.,* 25, 1325-1337 (1991).
6. J. A. Neff, K. D. Caldwell and P. A. Tresco, A novel method for surface modification to promote cell attachment to hydrophobic substrates, *J. Biomed. Mater. Res.,* 40, 511-519 (1998).
7. M. Aumailley, M. Gurrath, G. Müller, J. Calvete, R. Timpl and H. Kessler, Arg-Gly-Asp constrained within cyclic pentapeptides: Strong and selective inhibitors of cell adhesion to vitronectin and laminin fragment P1, *FEBS Lett.,* 291, 50-54 (1991).
8. M. D. Pierschbacher and E. Ruoslahti, Cell attachment activity of fibronectin can be duplicated by small synthetic fragments of the molecule, *Nature,* 309, 30-33 (1984).
9. K. M. Yamada and D. W. Kennedy, Dualistic nature of adhesive protein function: fibronectin and its biologically active peptide fragments can autoinhibit fibronectin function, *J. Cell Biology,* 99, 29-36 (1984).
10. T. K. Gartner, The tetrapeptide analog of the cell attachment site of fibronectin inhibits platelet-aggregation and fibrinogen binding to activated platelets, *J. Biological Chem.,* 260, 11891-11894 (1985).
11. E. F. Plow, M. D. Pierschbacher, E. Ruoslahti, G. A. Marguerie and M. H. Ginsberg, The effect of Arg-Gly-Asp-containing peptides on fibrinogen and von Willebrand factor binding to platelets, *Proc. Natl. Acad. Sci. (USA),* 82, 8057-8061 (1985).
12. R. O. Hynes, Integrins: bidirectional, allosteric signaling machines, *Cell,* 110, 673-687 (2002).
13. E. Ruoslahti and M. D. Pierschbacher, New perspectives in cell adhesion: RGD and integrins, *Science,* 238, 491-497 (1987).
14. What ligands bind to integrin? https://www.mechanobio.info/what-is-mechanosignaling/what-is-the-extracellular-matrix-and-the-basal-lamina/what-is-integrin/what-ligands-bind-to-integrin/.

15. E. Ruoslahti, RGD and other recognition sequences for integrins, *Annu. Rev. Cell Dev. Biol.*, 12, 697-715 (1996).
16. C. S. Chen, M. Mrksich, S. Huang, G. M. Whitesides and D. E. Ingber, Geometric control of cell life and death, *Science*, 276, 1425-1428 (1997).
17. R. Haubner, R. Gratias, B. Diefenbach, S. L. Goodman, A. Jonczyk and H. Kessler, Structural and functional aspects of RGD-containing cyclic pentapeptides as highly potent and selective integrin αVβ3 antagonists, *J. Am. Chem. Soc.*, 118, 7461-7472 (1996).
18. J. A. Neff, P. A. Tresco and K. D. Caldwell, Surface modification for controlled studies of cell–ligand interactions, *Biomaterials*, 20, 2377-2393 (1999).
19. B. K. Mann and J. L. West, Cell adhesion peptides alter smooth muscle cell adhesion, proliferation, migration, and matrix protein synthesis on modified surfaces and in polymer scaffolds, *J. Biomed. Mater. Res.*, 60, 86-93 (2002).
20. S. P. Massia and J. A. Hubbell, An RGD spacing of 440 nm is sufficient for integrin alpha V beta 3-mediated fibroblast spreading and 140 nm for focal contact and stress fiber formation. *J. Cell Biol.*, 114, 1089-1100 (1991).
21. U. Hersel, C. Dahmen and H. Kessler, RGD modified polymers: biomaterials for stimulated cell adhesion and beyond, *Biomaterials*, 24, 4385-4415 (2003).
22. E. Zamir and B. Geiger, Molecular complexity and dynamics of cell-matrix adhesions, *J. Cell. Sci.*, 114, 3583-3590 (2001).
23. K. Gaebel and I. A. Feuerstein, Platelets process adsorbed protein: a morphological study, *Biomaterials*, 12, 597-602 (1991).
24. L. M. Memmo and P. McKeown-Longo, The alphavbeta5 integrin functions as an endocytic receptor for vitronectin, *J. Cell. Sci.*, 111, 425-433 (1998).
25. S. H. Patel, J. Tsang, G. Harbers, K. E. Healy and S. Li, Endothelial cell function on a poly(acrylamide-co-polyethylene acid) interpenetrating polymer network: Cardiovascular applications, in: *Proc. 26th Annual International Conference of the IEEE Engineering in Medicine and Biology Society*, pp 5040-5043 (2004).
26. M. Hamidi, M. Shahbazi and K. Rostamizadeh, Copolymers: Efficient carriers for intelligent nanoparticulate drug targeting and gene therapy, *Macromol. Biosci.*, 12, 144-164 (2012).
27. R. A. Stile and K. E. Healy, Thermo-responsive peptide-modified hydrogels for tissue regeneration, *Biomacromolecules*, 2, 185-194 (2001).
28. J. S. Hrkach, J. Ou, N. Lotan and R. Langer, Synthesis of poly(L-lactic acid-co-L-lysine) graft copolymers, *Macromolecules*, 28, 4736-4739 (1995).
29. S. Gu, Z. Wang, C. Zhang and J. Ren, Synthesis and evaluation of a biodegradable material with cell recognition motives, *Carbohydr. Polym.*, 74, 572-578 (2008).
30. Z. Zhang, Y. Lai, L. Yu and J. Ding, Effects of immobilizing sites of RGD peptides in amphiphilic block copolymers on efficacy of cell adhesion, *Biomaterials*, 31, 7873-7882 (2010).
31. O. Diou, E. Fattal, V. Delplace, N. Mackiewicz, J. Nicolas, S. Mériaux, J. Valette, C. Robic and N. Tsapis, RGD decoration of PEGylated polyester

nanocapsules of perfluorooctyl bromide for tumor imaging: Influence of pre or post-functionalization on capsule morphology, *European J. Pharm. Biopharm.*, 87, 170-177 (2014).
32. H. J. Kim, S. J. You, D. H. Yang, H. J. Chun and M. S. Kim, Preparation of novel RGD-conjugated thermosensitive mPEG-PCL composite hydrogels and *in vitro* investigation of their impacts on adhesion-dependent cellular behavior, *J. Indus. Eng. Chem.*, 84, 226-235 (2020).
33. R. M. Rasal, A. V. Janorkar and D. E. Hirt, Poly(lactic acid) modifications, *Prog. Polym. Sci.*, 35, 338-356 (2010).
34. J. K. Oh, Polylactide (PLA)-based amphiphilic block copolymers: synthesis, self-assembly, and biomedical applications, *Soft Matter*, 7, 5096-5108 (2011).
35. P. D. Drumheller and J. A. Hubbell, Polymer networks with grafted cell adhesion peptides for highly biospecific cell adhesive substrates, *Anal. Biochem.*, 222, 380-388 (1994).
36. B. Li, A. Çolak, J. Blass, M. Han, J. Zhang, Y. Zheng, Q. Jiang, R. Bennewitz and A. D. Campo, Molecular stiffness cues of an interpenetrating network hydrogel for cell adhesion, *Materials Today Bio*, 15, 1-7 (2022).
37. A. P. Dhand, J. H. Galarraga and J. A. Burdick, Enhancing biopolymer hydrogel functionality through interpenetrating networks, *Trends Biotechnol.*, 39, 519-538 (2021).
38. R. Parhi, Cross-linked hydrogel for pharmaceutical applications: A review, *Adv. Pharm. Bulletin*, 7, 515-530 (2017).
39. C. O. Crosby, B. Stern, N. Kalkunte, S. Pedahzur, S. Ramesh and J. Zoldan, Interpenetrating polymer network hydrogels as bioactive scaffolds for tissue engineering, *Reviews Chem. Eng.*, 38, 347-361 (2022).
40. Z. Grabarek and J. Gergely, Zero-length crosslinking procedure with the use of active esters, *Anal. Biochem.*, 185, 131-135 (1990).
41. G. T. Hermanson, *Bioconjugate Techniques*. 2nd edition, Academic Press, Amsterdam (2008).
42. K. S. Masters, Covalent growth factor immobilization strategies for tissue repair and regeneration, *Macromol. Biosci.*, 11, 1149-1163 (2011).
43. H. C. Kolb, M. G. Finn and K. B. Sharpless, Click chemistry: diverse chemical function from a few good reactions, *Angew. Chem. Int. Ed.*, 40, 2004-2021 (2001).
44. V. D. Bock, H. Hiemstra and J. H. van Maarseveen, CuI-catalyzed alkyne–azide "Click" cycloadditions from a mechanistic and synthetic perspective, *Eur. J. Org. Chem.*, 51-68 (2006).
45. M. Colombo and A. Bianchi, Click chemistry for the synthesis of RGD-containing integrin ligands, *Molecules*, 15, 178-197 (2010).
46. J. Baskin and C. Bertozzi, Bioorthogonal click chemistry: covalent labeling in living systems, *QSAR Combinatorial. Sci.*, 26, 1211-1219 (2007).
47. M. F. Debets, van der Doelen, Christianus W. J., Rutjes, Floris P. J. T. and F. L. van Delft, Azide: a unique dipole for metal-free bioorthogonal ligations, *ChemBioChem*, 11, 1168-1184 (2010).

48. C. Chollet, S. Lazare, F. Guillemot and M. C. Durrieu, Impact of RGD micro-patterns on cell adhesion, *Colloids Surfaces B,* 75, 107-114 (2010).
49. Y. Ohmuro-Matsuyama and Y. Tatsu, Photocontrolled Cell adhesion on a surface functionalized with a caged arginine-glycine-aspartate peptide, *Angew. Chemie. Intl. Edition,* 47, 7527-7529 (2008).
50. A. Khademhosseini, R. Langer, J. Borenstein and J. P. Vacanti, Microscale technologies for tissue engineering and biology, *Proc. Natl. Acad. Sci. (USA),* 103, 2480-2487 (2006).
51. J. H. Wosnick and M. S. Shoichet, Three-dimensional chemical patterning of transparent hydrogels, *Chem. Mater.,* 20, 55-60 (2008).
52. Y. Xia and G. M. Whitesides, Soft lithography, *Angew. Chemie. Intl. Edition,* 37, 550-575 (1998).
53. A. Bernard, J. P. Renault, B. Michel, H. R. Bosshard and E. Delamarche, Microcontact printing of proteins, *Adv. Mater.,* 12, 1067-1070 (2000).
54. G. Csucs, R. Michel, J. W. Lussi, M. Textor and G. Danuser, Microcontact printing of novel co-polymers in combination with proteins for cell-biological applications, *Biomaterials,* 24, 1713-1720 (2003).
55. T. Kaufmann and B. J. Ravoo, Stamps, inks and substrates: polymers in microcontact printing, *Polymer Chemistry,* 1, 371-387 (2010).
56. S. Patel, R. G. Thakar, J. Wong, S. D. McLeod and S. Li, Control of cell adhesion on poly(methyl methacrylate), *Biomaterials,* 27, 2890-2897 (2006).
57. J. Almodóvar, T. Crouzier, Š Selimovi, T. Boudou, A. Khademhosseini and C. Picart, Gradients of physical and biochemical cues on polyelectrolyte multilayer films generated via microfluidics, *Lab on a Chip,* 13, 1562-1570 (2013).
58. R. D. Piner, J. Zhu, F. Xu, S. Hong and C. A. Mirkin, "Dip-Pen" nanolithography, *Science,* 283, 661-663 (1999).
59. C. Wu, D. N. Reinhoudt, C. Otto, V. Subramaniam and A. H. Velders, Strategies for patterning biomolecules with dip-pen nanolithography, *Small,* 7, 989-1002 (2011).
60. D. K. Hoover, E. W. L. Chan and M. N. Yousaf, Asymmetric peptide nanoarray surfaces for studies of single cell polarization, *J. Am. Chem. Soc.,* 130, 3280-3281 (2008).
61. A. B. Dababneh and I. T. Ozbolat, Bioprinting technology: a current state-of-the-art review, *J. Manuf. Sci. Eng,* 136, 1-11 (2014).
62. Y. Luo, X. Wei and P. Huang, 3D bioprinting of hydrogel-based biomimetic microenvironments, *J. Biomed. Mater. Res.,* 107, 1695-1705 (2019).
63. E. Axpe and M. L. Oyen, Applications of alginate-based bioinks in 3D bioprinting, *Intl. J. Molecular Sci.,* 17, 1976-1986 (2016).
64. E. M. Chandler, C. M. Berglund, J. S. Lee, W. J. Polacheck, J. P. Gleghorn, B. J. Kirby and C. Fischbach, Stiffness of photocrosslinked RGD-alginate gels regulates adipose progenitor cell behavior, *Biotechnol. Bioeng.,* 108, 1683-1692 (2011).
65. J. Jia, D. J. Richards, S. Pollard, Y. Tan, J. Rodriguez, R. P. Visconti, T. C. Trusk, M. J. Yost, H. Yao, R. R. Markwald and Y. Mei, Engineering alginate as bioink for bioprinting, *Acta Biomaterialia,* 10, 4323-4331 (2014).

66. H. Lin, C. García-Echeverría, S. Asakura, W. Sun, D. F. Mosher and S. L. Cooper, Endothelial cell adhesion on polyurethanes containing covalently attached RGD-peptides, *Biomaterials,* 13, 905-914 (1992).
67. Z. Guler, J. C. Silva and A. Sezai Sarac, RGD functionalized poly(ε-caprolactone)/poly(m-anthranilic acid) electrospun nanofibers as high-performing scaffolds for bone tissue engineering RGD functionalized PCL/P3ANA nanofibers, *Intl. J. Polymer Mater. Polymer Biomater.,* 66, 139-148 (2017).
68. A. Arabi, E. Boggs, M. R. Patel, J. Zwiesler-Vollick, T. Maerz, K. Baker, S. Tuck, J. Corey and Y. Li, Surface modification of electrospun polycaprolactone fibers and effect on cell proliferation, *Surface Innovations,* 2, 47-59 (2014).
69. L. Sabbatini and P. G. Zambonin, XPS and SIMS surface chemical analysis of some important classes of polymeric biomaterials, *J. Electron Spectroscopy and Related Phenomena,* 81, 285-301 (1996).
70. Z. Zhang, Y. Lai, L. Yu and J. Ding, Effects of immobilizing sites of RGD peptides in amphiphilic block copolymers on efficacy of cell adhesion, *Biomaterials,* 31, 7873-7882 (2010).
71. N. Huettner, T. R. Dargaville and A. Forget, Discovering cell-adhesion peptides in tissue engineering: beyond RGD, *Trends Biotechnol.,* 36, 372-383 (2018).
72. A. Meyer, J. Auernheimer, A. Modlinger and H. Kessler, Targeting RGD recognizing integrins: drug edvelopment, biomaterial research, tumor imaging and targeting, *Current Pharmaceutical Design,* 12, 2723-2747 (2006).
73. Y. Cheng and Y. Ji, RGD-modified polymer and liposome nanovehicles: Recent research progress for drug delivery in cancer therapeutics, *European J. Pharm. Sci.,* 128, 8-17 (2019).
74. J. Park, K. Singha, S. Son, J. Kim, R. Namgung, C. Yun and W. J. Kim, A review of RGD-functionalized nonviral gene delivery vectors for cancer therapy, *Cancer Gene Therapy,* 19, 741-748 (2012).
75. R. Langer and J. P. Vacanti, Tissue engineering, *Science,* 260, 920-926 (1993).
76. P. Schaffner and M. M. Dard, Structure and function of RGD peptides involved in bone biology, *Cellular Molecular Life Sciences (CMLS),* 60, 119-132 (2003).
77. F. P. Ross, J. Chappel, J. I. Alvarez, D. Sander, W. T. Butler, M. C. Farach-Carson, K. A. Mintz, P. G. Robey, S. L. Teitelbaum and D. A. Cheresh, Interactions between the bone matrix proteins osteopontin and bone sialoprotein and the osteoclast integrin alpha v beta 3 potentiate bone resorption, *J. Biological Chem.,* 268, 9901-9907 (1993).
78. K. Eid, E. Chen, L. Griffith and J. Glowacki, Effect of RGD coating on osteocompatibility of PLGA-polymer disks in a rat tibial wound, *J. Biomed. Mater. Res.,* 57, 224-231 (2001).
79. M. Kantlehner, P. Schaffner, D. Finsinger, J. ö Meyer, A. Jonczyk, B. Diefenbach, B. Nies, G. Hölzemann, S. L. Goodman and H. Kessler, Surface coating with cyclic RGD peptides stimulates osteoblast adhesion and proliferation as well as bone formation, *ChemBioChem,* 1, 107-114 (2000).

80. A. Kostenko, S. Swioklo and C. J. Connon, Alginate in corneal tissue engineering, *Biomed. Mater.,* 17, 1-17 (2022).
81. M. I. Neves, L. Moroni and C. C. Barrias, Modulating alginate hydrogels for improved biological performance as cellular 3D microenvironments, *Frontiers Bioeng. Biotechnol.,* 8, 1-16 (2020).
82. D. B. Horváthy, M. Simon, C. M. Schwarz, M. Masteling, G. Vácz, I. Hornyák and Z. Lacza, Serum albumin as a local therapeutic agent in cell therapy and tissue engineering, *Biofactors,* 43, 315-330 (2017).
83. W. Zheng, Z. Wang, L. Song, Q. Zhao, J. Zhang, D. Li, S. Wang, J. Han, X. Zheng, Z. Yang and D. Kong, Endothelialization and patency of RGD-functionalized vascular grafts in a rabbit carotid artery model, *Biomaterials,* 33, 2880-2891 (2012).
84. S. Hsu, S. W. Whu, S. Hsieh, C. Tsai, D. C. Chen and T. Tan, Evaluation of chitosan-alginate-hyaluronate complexes modified by an RGD-containing protein as tissue-engineering scaffolds for cartilage regeneration, *Artificial Organs,* 28, 693-703 (2004).
85. L. V. Antonova *et al,* Conjugation with RGD peptides and incorporation of vascular endothelial growth factor are equally efficient for biofunctionalization of tissue-engineered vascular grafts, *Intl. J. Molecular Sci.,* 17, 1920-1930 (2016).
86. M. K. Lee, M. H. Rich, J. Lee and H. Kong, A bio-inspired, microchanneled hydrogel with controlled spacing of cell adhesion ligands regulates 3D spatial organization of cells and tissue, *Biomaterials,* 58, 26-34 (2015).
87. X. Z. Shu, K. Ghosh, Y. Liu, F. S. Palumbo, Y. Luo, R. A. Clark and G. D. Prestwich, Attachment and spreading of fibroblasts on an RGD peptide–modified injectable hyaluronan hydrogel, *J. Biomed. Mater. Res.,* 68A, 365-375 (2004).
88. A. D. Lynn, T. R. Kyriakides and S. J. Bryant, Characterization of the *in vitro* macrophage response and *in vivo* host response to poly(ethylene glycol)-based hydrogels, *J. Biomed. Mater. Res.,* 93A, 941-953 (2010).

13

Adhesion Promotors for Polymer Surfaces

Thomas P. Schuman

Department of Chemistry, Missouri University of Science and Technology, Rolla, Missouri, USA

Abstract

There are many good books written about the topic of adhesion and adhesion to polymers/plastics. This chapter is an attempt to review and provide context into variables that affect adhesion obtained to polymer surfaces, including those of both thermoplastics and thermosets. Polymer surfaces pose special challenges regarding adhesion. Polymers, for instance, have a variety of structures and vary extensively in their surface energy, which has implications for both wetting by a layer adhering to the polymer and theoretical adhesion energy available to the layer. The need for adhesion promotion to a polymer surface is predicated on the need for surface protection or for improving aesthetic appearance and on the surface energy and the type of coating desired to be applied onto said polymer surface. Techniques to enhance surface energy and adhesion are reviewed along with relative performance of applications.

Keywords: Adhesion, polymer surface, adhesion promoter, surface energy, wetting, coating, interface

13.1 To Coat or Not to Coat Polymer Surfaces

One might question, why would anyone desire to coat a polymer surface at all since most coatings technologies themselves are comprised of polymeric binders and have their own polymer composition? In general, processed or molded plastics lack a high-quality surface finish that is expected of many consumer items. Molded items are often constructed of thermoplastics, i.e., polymers that are altered in shape using temperature and pressure,

Email: tschuman@mst.edu

K.L. Mittal and A.N. Netravali (eds.) Polymer Surface Modification to Enhance Adhesion: Techniques and Applications, (517–558) © 2024 Scrivener Publishing LLC

typically by a molding operation. Thermoplastic is often abbreviated as simply "plastic." These polymers generally have uniform compositions synthesized from building blocks called monomers, such as polyethylene or polypropylene (PE or PP, respectively), but can also be comprised of blends of monomers, such as (poly)acrylonitrile-butadiene-styrene (ABS) plastic that is a blend of these three monomers. Monomer content controls the composition and, therefore, primary structure and properties of the plastic material. Plastics are characterized by their ability to be reshaped by heating and re-molding. Thermoset polymers, on the other hand, are formed and "set," into fixed bonding structures by, often, heat-induced or "baked" reactions that build molecular weight through covalent bonds and crosslinking, into a final shape and cannot be remolded or reshaped without degrading the polymer. Most modern coatings are of thermoset type, e.g., epoxy-amine or polyurethane, that is characteristic of a class of coatings called enamels.

In general, the finish of a coated molded item is very much improved over that of the same molded item without a coating, i.e., as molded. The as-molded part will have less distinct color, gloss, and depth of appearance in contrast to what can be achieved by coatings. In addition, the finish of particular consumer items is usually expected to be of exceptional aesthetics, identically matched color and appearance, e.g., including gloss, for entire product surfaces including painted metal and plastic items [1].

The purpose of coatings, in general, is to protect the substrate. Therefore, another reason plastic items are coated is to protect them from their environment, similar to the reasoning for the painting of metal objects to protect them from corrosion and wear, as well as to endow durability. For instance, polycarbonate signal light covers are often coated with transparent organosilane coatings for added damage resistance [1]. Coatings are also made of polymer binder but pigments and additives are added that enhance protection from water and chemicals as well as enhance resistance to damage from ultraviolet (UV) light.

Once a decision that requires a plastic to be coated is made, which plastics can be coated directly and on what basis is the decision made? Higher surface energy plastics that have greater surface polarity can be coated and adhered to with durable adhesion [2]. Low energy plastics, on the other hand, have low interaction energies and are difficult to wet because of their low critical surface tensions. Hence, low energy plastics of less than about 40 mJ/m^2 require raising of their surface energy, improved wetting, and/or changes in topography/profile to obtain durable adhesion [3]. One of the largest plastics coating application industries in volume and types of plastics and, therefore, a driving force toward development of coatings technology for plastics is the automotive industry [1], where high performance coatings of excellent appearance and high-quality match of plastic and metal coatings in color and appearance are required. Other concerns besides appearance for this industry include mechanical abrasion, impact, and chemical, e.g., gasoline or human sunscreen product, resistance. However, in order for a coating to protect a substrate, much less be deemed a coating, it must possess adhesion to the substrate.

13.2 Theory of Adhesion: Adhesion Forces

Adhesion is a force of interfacial attraction between two substrate materials. A forced separation of the substrates results in an interfacial or interphasial separation [4]. Interfacial failure is the result where dewetting between two different phases occurs and thus interfacial separation between two different phases occurs. Interphasial failure, also known as cohesive failure, is a separation within a substance due to weak internal mechanical strength of that single phase or composition. Cohesive failure within a material produces two new surfaces comprised of the same material. Theoretical adhesion is the sum of mechanical and chemical bonding interaction strengths between the two substrates, whereas measured adhesion is a force per area required to separate the two parts, creating two new surfaces by fracture through some phase within the solid. To determine the mode of failure, interfacial or cohesive, compositional analysis of the created surfaces must be done and so identify the structure that failed within a phase (interphasial or cohesive failure) or between two different phases (interfacial failure). Adhesion failures always occur within the weakest link structure, known as a weak boundary layer (WBL) [5]. During a peel test for adhesion, the measurement results in force needed to move the line of contact of the two phases, which gives units of a force per width of interface created.

Total work or energy of adhesion W_A is equal to the force of adhesion per area F_A times the interfacial contact area (Equation (13.1)). One aspect of increasing adhesion, therefore, is to increase the contact surface area to be greater than the simple smooth surface dimensional area by roughening the surface through abrasion or etching. Thus for a given attraction force per area F_A, a higher contact area A produces stronger adhesion. A complete wetting of the substrate (see below) is necessary to obtain the theoretical adhesion strength.

$$W_A = F_A * A \tag{13.1}$$

Theoretical adhesion, also called basic, reversible work of adhesion [4] or intrinsic [6] or fundamental [7] adhesion is the work of adhesion that is obtained from theoretically available energy due to cross-interfacial bonding, including covalent bonding. Practical adhesion, on the other hand, is a measured value that can be smaller than theoretical adhesion or up to orders of magnitude higher than theoretical adhesion [6, 7]. Practical adhesion represents the adhesion strength observed in practice. The larger measured adhesion is due to loss mechanisms exhibited by a substance that can present a viscous response, such as an a viscoelastic polymer is separated from a substrate. Mechanical loss is not readily modeled by methods used to calculate theoretical adhesion. Practical adhesion can also be smaller than theoretical adhesion, as will be seen shortly, due to poor wetting/contact between phases or to the presence of WBLs. Theoretical adhesion arises from intrinsic interactions across an interface between two substrates that hold the two phases together as a combination of mechanical and chemical forces as a combined strength. W_A is thus these forces per unit area times the surface area between the phases, which directly affects the adhesion between the phases (as seen in Equation (13.1)).

13.3 Plastics

The word 'plastic' is a brief pseudonym for the term "thermoplastic polymer," which is in contrast to a "thermoset polymer." 'Thermoplastic' literally means "substance whose shape changes under the influence of temperature." The term "polymer" is a substance that derives its properties from entanglement of molecular chains through associative and/or covalent bonding. A 'thermoset polymer,' in contrast, is a substance whose form or shape is "set, cured or otherwise becomes fixed under influence of temperature" and the setting or curing process generally occurs through a bond forming process. Bond forming i.e., polymerization and crosslinking reactions, can vary in mechanism and scope but in general increases the number of bonds in the material per volume to produce a three-dimensional, insoluble network polymer. A network polymer is also known as a "gel."

In an industrial part manufacturing application, therefore, thermoplastic materials are polymeric meaning that they have an average molecular weight, which is controlled by the synthesis method, large enough to provide chain entanglement. Thermoplastics have their full molecular weight before any secondary shaping process occurs, such as by injection molding and are not increased in their molecular weight by processing. On the other hand, thermoset polymer materials begin the secondary shaping process at molecular weights less than those needed to be polymeric and can have mechanical properties ranging from rigid to fluid. Thermoset polymer will increase in molecular weight and/or bond density after or during the shape processing step, such as spraying or injection molding, to become polymeric and acquire chain entanglement from which it obtains mechanical properties.

The labels thermoplastic or thermoset have implications about the composition of the polymer material. Thermoset polymers will have polar bonding groups, e.g., hydroxyl, carbonyl, ester, amide, in their molecular structure because polar bond groups are the products of thermosetting reactions used to build molecular weight and form the polymer. Thermoplastic polymers may or may not have polar bond groups, e.g., may possess only non- or less-polar alkane, alkene, or aromatic groups, and may therefore be substantially non-polar chemically. Organic bond structures, also called functional groups, of a polymer have differing polarity and reactivity that are exhibited by a substrate to a coating on contact that interact, or do not interact, with the coating. These interactions, or lack thereof, are the cause of whether the coating can wet, adsorb or react, and adhere. Whether, and how, the coating can adhere to the polymer in terms of surface wetting, adsorption bonding, residual functional groups and reactivity, is dependent on the polarity and reactivity of bond structures present at the surface of the polymer substrate. Table 13.1 provides examples of some typical polymer substrate compositions, which is not inclusive of all polymer substrate types.

Polymerized thermoset materials have, in addition to high molecular weight that slows chain diffusion, crosslink bonds that further limit the mobility of the polymer chains to diffuse and entangle. A rigid molecular structure limits the thermoset polymer from diffusion and chain entanglement even between polymer materials of similar or identical composition. During a thermoset process, while molecular weight and crosslink density is low and still forming, it is possible for chains to diffuse and entangle with and covalently bond with other materials and presents the best opportunity to adhere to the thermoset polymer. For this reason, most thermoset materials have pot-life and re-coat window recommendations as a time window within which the thermoset polymer substrate will possess chain flow mobility and/or chemical reactivity to allow surface wetting, covalent bonding and/or entanglement with a subsequently applied coating. The term re-coat window is the permissible time window when a top-coating may be applied to the surface of another thermoset coating system, such as a primer coating, so that the primer, as substrate, has sufficient ability

Table 13.1 Examples of common thermoplastic and thermoset polymer substrates that are coated in various applications.

Thermoplastics	Thermosets
Polyethylene (PE, HDPE, LDPE, LLDPE)	Polyurethane, resin transfer molding (RTM)
Polypropylene (PP)	Polyurea, RTM
Poly(ethylene terephthalate) (PET)	Phenolic, phenol-formaldehyde (PF)
Polyamides (nylons)	Fiberglass, sheet molding compound (SMC)
Polyesters	Epoxy, epoxy composites
Poly(ether ether ketone) (PEEK)	Melamine, melamine-formaldehyde (MF)
Polycarbonate (PC)	Vinyl ester
Vinyl, poly(vinyl chloride) (PVC)	
Poly(acrylonitrile-co-butadiene-co-styrene) (ABS)	
Poly(phenylene sulfide) (PPS)	

to interdiffuse and/or react with the new top-coating and adhere but does not dissolve or damage the underlying primer. Applications earlier or later than allowed by a re-coat window can result in dissolution of the substrate or creation of a WBL interfacial defect near the interface of or between the coating layers. Misapplication outside of the re-coat window can result in interfacial failure.

Sheet molding compound (SMC), reaction injection molded (RIM) or resin transfer molded (RTM) polyurethanes possess polar bond, e.g., polyester or polyurethane, structures that as thermoset substrates have a capability to associatively bond to coatings. Thermoset parts will possess residual isocyanate groups on the part surface early after injection molding that offer a capability to form covalent bonds with a subsequently applied coating. A thermosetting primer coating, such as an epoxy or polyurethane, applied within a short time frame after the injection molding process would be expected to covalently bond to the RIM part and hence form durable covalent bond adhesion, perhaps equivalent to the cohesive mechanical strength of the molded polymer part. In practice, however, coating at short time after molding is rarely done because of a limited re-coat window time of the part that is required to retain reactivity of those surface groups. A lack of proximity of the coating operation to the RIM process requires shipping to a re-coat facility and time for that loading, shipping, and unloading processes. Even if these facilities are co-located, directly coating the part can also be affected by molding residues, such as lubricants, and/or surface defects on the molded part surface that surface coatings will not remove. Surface contaminants and/or mold defects adversely affect product quality as they result in potentially poor coating adhesion qualities or poor visual aesthetic, respectively.

13.4 Polymer Adhesion Mechanisms

The mechanisms of adhesion bonding are summarized into five categories: mechanical, electrical, adsorption, diffusion, and covalent, of bonding types. Also known as "lock-and-key", mechanical adhesion is literally the mechanical interlocking of two different materials. Mechanical adhesion and a general background of adhesion theories is being presented since surface profile, as affected by the surface treatment, can influence the adhesion obtained. Other adhesion mechanisms are presented so one can understand how adhesion promotors benefit or are detrimental to adhesion strength.

13.4.1 Mechanical

Bikerman was a strong advocate of a mechanical, rheological response of an adhesive joint [7]. Mechanical adhesion is akin to fiber reinforcement of polymer materials in that the mechanical entrainment of the fiber within the polymer, plus the adsorption interaction between the two materials, fiber and polymer, result in material strength. In typical examples, a substrate is mechanically abraded with wire brush or grit blasting to create a surface profile as a fractal with characteristic amplitude, height or surface area per square substrate. When another phase is placed in contact with the profiled substrate, wetting and penetration of the adhesive or coating into the profiled surface provides mechanical interlocking whose strength is dependent on elastic and viscous mechanical properties of the two interlocked phases. Abrasion is used to prevent interfacial failure for substrates that have exceeded their recoat window in order to create mechanical adhesion, instead of diffusion (section 13.4.4) or covalent (section 13.4.5) adhesion that would be obtained within the recoat window. This mechanism requires that the applied top phase has sufficiently low surface tension (see section 13.4.3) and viscosity during application to wet the substrate surface and quickly diffuse into the surface crevices, respectively. A surface tension less than that of the substrate provides spontaneous wetting [8].

Pigmentation of polymers to provide a rough surface is often used to enhance mechanical adhesion, e.g., of top-coatings to a primer coating surface. Good substrate wetting is critical to a lock-and-key approach to a rough or micro-rough surface. In this case, high interfacial contact area and penetration of the coating are obtained to give a durable mechanical bond. Abrasive profiling is, in most cases, the method of choice for adhesion promotion to metals but also applies to thermoset substrates where the substrate has good elastic mechanical integrity. When a re-coat window has been exceeded and therefore possesses limited ability to be entangled by diffusion (see section 13.4.4) or react with the coating, thermoset primer coatings are mechanically abraded to provide mechanical adhesion when curing has reduced the thermoset coating's capability to entangle with the top-coating. Creating a rough profile via abrasion can result in removal of WBLs and stronger adhesion. Abrasion can also create WBLs and decrease practical adhesion. If the surface is mechanically weak, soft and/or rubbery, the abrasion can be less effective at creating a uniform interfacial profile and/or can further weaken the near-surface region to create, rather than remove, a weak boundary layer. The pigmentation of a primer coating aids formation of strong cohesion near its interface with the substrate and aids formation of a strong, rough surface without WBL formation during abrasion when preparing for a top-coating.

The probability of an event occurring can be estimated as the number of actions causing the event divided by the total number of events possible. Visualization of probability of adhesion failure p of a theoretical bond between phases A and B is demonstrated by 3 parallel bond elements: A-A, A-B, and B-B (see Figure 13.1), any of which could fail as a single bond adhesion failure. Calculation of the probability of interfacial failure, breaking the interfacial bond A-B between phases A and B, p_{AB} = (number of events resulting in A-B bond failure)/(total possible events) = 1/3. There is one event resulting in interfacial failure, the breaking of A-B bond, where there are total possible bond breaking events to include cohesive failures of either A-A or B-B bonds or failure of the interfacial A-B bond, for a total of 3 events possible. For one adhesion bond being broken, the interfacial failure is one out of three possible bonds that could break. However, the probability of an adhesion failure within one of the pure phases, breaking either an A-A or a B-B bond, is $(p_{AA} + p_{BB})$/(total possible events) = (1+1)/3 or 2/3.

The probability of a series of events is the product of probabilities for all events in the series. The probability of interfacial failure (= p_I) progressively becomes smaller with each bond rupture as $p_I \sim (1/3)^n$, where the exponent n is total sequential number of interfacially, failed A-B bonds. Probability of interfacial failure will always be lower than that of cohesive failure [9]. An assumption in the calculation is that each of the three bonds has similar strength and therefore similar probability of being ruptured, each bond rupture has a normalized $p_i \sim 1$. If any one of these elements has much lower strength than the other two, failure probability p increases exponentially for that weakest element. This weakest element is called the weak boundary layer (WBL) and was originally proposed by Bikerman [10]. His WBL theory states that durable adhesion is produced where the interface between coating and substrate, as A-B (Figure 13.1), is mechanically as strong or stronger than either of the two pure phases, elements A-A or B-B, being adhered. Then, adhesion of A to B (as cohesion, since most probable failure occurs in the weakest interface) is governed by the element with weakest cohesion. That is, an adhesion failure will always occur in the WBL, whether located at the A-B interface or internally within the coating or substrate.

Mechanical adhesion can also be subject to a timescale of the disbonding event depending on glass transition temperatures of the substrate and top-coating phase, contributing to elastic and/or viscous deformations of the respective phase, upon mechanical de-entanglement, unlocking,

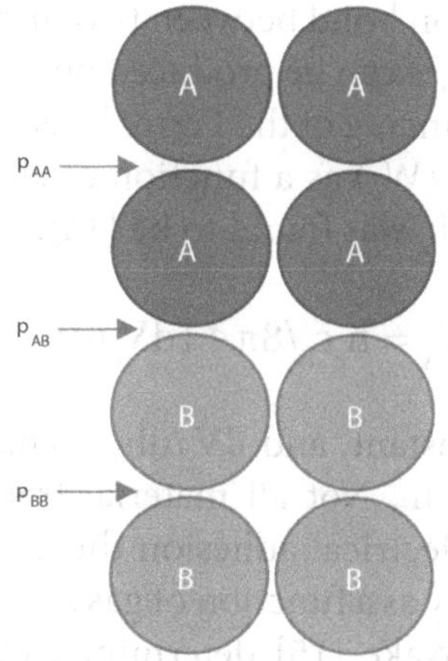

Figure 13.1 Visualization of probability of failure at theoretical interface between phases A and B comprised of 3 bonding elements A-A, A-B, and B-B. For equal bond strengths of A-A, A-B, and B-B bonds, probabilities p_{AA}, p_{AB}, and p_{BB} are approximately equal.

or disbonding. Viscous, mechanically lossy, energy-consuming phenomena create a larger gap between theoretical and practical adhesion values. Viscous energy losses are not reversible or conserved in energy and, hence, can result in more energy being expended to rupture the interface than exists as energy stored in elastic bonds, adsorption-associative and/or covalent. Bond breaking is a reverse, conserved energy, of bond formation whereas viscous flow, friction, or drag are non-conserved quantities and lost, not recovered.

Mechanical loss occurring during fracture is why measured practical adhesion can be orders of magnitude larger than theoretical adhesion [6, 7]. Increased strength and strain before failure corresponds to toughness, as an integrated function of applied stress and strain. Polymer chain mechanical behavior being more elastic or more lossy is a superposition function of temperature and strain rate. Diffusion and viscous flow are enabled above, at or near the glass transition temperature, but under fast induced strain rates mechanical behavior becomes more elastic and less lossy. At ambient temperatures near or above the glass transition temperature, adhesion is governed by viscoelasticity of the phase above its T_g as a function of its crosslink density or phase viscosity. There exists an optimal crosslink, i.e., bond, density to maximize viscosity and therefore adhesion as a function of strain rate [11, 12].

13.4.2 Electrical

Electrical charge adhesion applies to an interface where, in the process of destruction of the adhesion bond between two materials of different Fermi levels, electrical net charges can be produced on the resulting surfaces. The charges result from a balancing of the Fermi levels and can affect measured adhesion [13]. Adhesion (W_A) as a function of electrical potential change (dV) and separation gap h_c was found to be [14]:

$$W_A = h_c \varepsilon_d / 8\pi * (dV_c/dh_c)^2, \tag{13.2}$$

where ε_d = dielectric constant, and dV_c/dh_c = change in discharged voltage at the potential gap h_c. Not all material fractures result in surface-produced net charges. Electrical adhesion theory requires that measured work of adhesion will vary as a function of gas and gas pressure but no such variations were found. Wake [15] determined that Equation (13.2) had circular arguments and therefore demonstrated excellent agreement with theory [13, 14] where W_A was proportional to dV/dh_c was causal rather than predictive. Another issue with electrical discharge adhesion theory is that it does not account for other loss mechanisms of adhesion, e.g., viscous flow (see section 13.4.4: Diffusion).

13.4.3 Adsorption/Wetting

The strength of adsorption/wetting-associated chemical bonding across an interface is related to bond types, bond strengths, and number of bonds per area (bond density) present at and exhibited across the interface. Surface energy comes from the energy excess at the interface compared to that of the bulk [16]. The excess energy of the interface is the thermodynamic driving force for adsorption and wetting. Adhesion energy across the interface is maximized even when wetting is spontaneous, if the surface energies of the two contacting materials are nearly equal.

Wetting is spontaneous when the surface energy of the liquid or coating is less than that of the substrate, i.e., creates a new, lower energy surface upon wetting the higher energy substrate. Substrates that have functional group structures of higher polarity have greater interaction strengths and, hence, are capable of higher bond energies (Table 13.2). In contact with substrates or coatings that do not possess an ability to interact with the bond type, these higher surface energies provide potential bond energy but remain unrealized and are wasted bond energy in that, instead, the

Table 13.2 Bond energies (kJ-mole^{-1}) for different bond types. (Data from Kinloch [31]).

Bond type	Bond energy (kJ/mole)
Ionic	590-1050
Covalent	63-710
Metallic	113-347
Permanent dipole-dipole	
Hydrogen Bond w/ Fluorine	up to 42
Hydrogen Bond no Fluorine	10 - 26
Other dipole-dipole (no Hydrogen)	4 - 21
Dipole-Induced Dipole	< 2
Dispersion Force (London)	0.08 - 42

mismatch contributes to higher interfacial tensions and lower bond strengths [4]. Ionic, dipolar and London forces are physically attractive, associative interactions with no sharing of electrons. Metallic and covalent bondings occur through shared electron bonds between bonded atoms.

Adsorption adhesion is caused by associative bonding across an interface between a coating and the substrate. That is the mechanical work of adhesion (W_A) is equal to vector addition of interaction forces $\Sigma(\gamma_{SV}+\gamma_{LV})$ – Σ (γ_{SL}), where γ are surface tensions (typical units, dyne/cm = mJ/m^2) of the initial solid-vapor (SV), liquid-vapor (LV), and created solid-liquid (SL) interfaces giving adsorption adhesion equal to Equation (13.3):

$$W_A = \gamma_{SV}+\gamma_{LV} - \gamma_{SL} = \gamma_{LV}(1 + \cos\theta) + \pi_S, \tag{13.3}$$

where θ is liquid-solid contact angle and π_S is the change in surface tension due to an adsorbed solvent layer. The subscripts S and L can also refer to different solids in contact with each other after an interface is formed by, e.g., one of the phases being dissolved in solvent and applied onto the second phase and dried into a film or where the two polymers are co-extruded with mutual interface. The subscript L therefore represents surface tension of the solid coating material.

The forces in the work of adhesion, Equation (13.3), generally refer to a total surface tension, i.e., combining all interaction forces, dispersion with polar forces, into one term. These surface tensions are commonly referred to as Hildebrand surface tensions [17]. Another way to calculate work of adhesion is interfacial energy equal to the sum of the root-mean-square surface tensions on either side of the created interface:

$$W_A = 2 * (\gamma^d{}_A\gamma^d{}_S)^{1/2} + 2*(\gamma^p{}_A\gamma^p{}_S)^{1/2}, \tag{13.4}$$

where superscripts d and p refer to dispersion and polar forces, respectively, and subscripts A and S refer to each side/phase of the interface created upon interfacial failure, respectively, and where polar forces are independent of non-polar, dispersion forces. Equation (13.4) "unfortunately called the extended Fowkes equation"..."is quite incorrect" for predicting acid-base/polar interaction energies since the γ^p is not descriptive of the actual polarity of the material [17]. Fowkes, instead, preferred a cohesive energy density approach, analogous to that used for solubility parameter calculations, that inputs acid (A)-base (B) electrostatic (E) and covalent (C) polarity property contributions into a Drago equation format (of type provided in Equation (13.5)) in a calculation of acid-base enthalpic (–ΔH) interaction energy [18]. Fowkes has used the Drago equation format to correlate adhesion energy with spectroscopic and physical measurements of acid-base interaction energies [17].

$$-\Delta H = C^A * C^B + E^A * E^B \tag{13.5}$$

In short, surfaces having groups of higher association bond energies have greater potential bond strength. The Zisman plot of contact angles of liquids of different γ_{LV} measured on a single substrate surface results in a linear plot with an intercept at $\theta = 0$ [19]. The x-intercept value of γ_{LV} is called γ_C, the critical surface tension of the substrate, that is required for a liquid not to exceed for wetting of the substrate, which is also considered a measure of the ability of the solid substrate to bond with contacting materials. Practical adhesion was reported to correspond to a critical surface tension γ_C proposed by Zisman [19], as the limit of surface tension that permits full wetting of a substrate. A plot between practical adhesion of an epoxy coating of surface energy equal 50 mJ/m² and critical surface tension for various substrates produced a linear plot, demonstrating proportionality of practical adhesion as a function of critical surface tension (Figure 13.2) [20]. Levine showed good correlation between γ_C and measured adhesion between an epoxy film and various substrates.

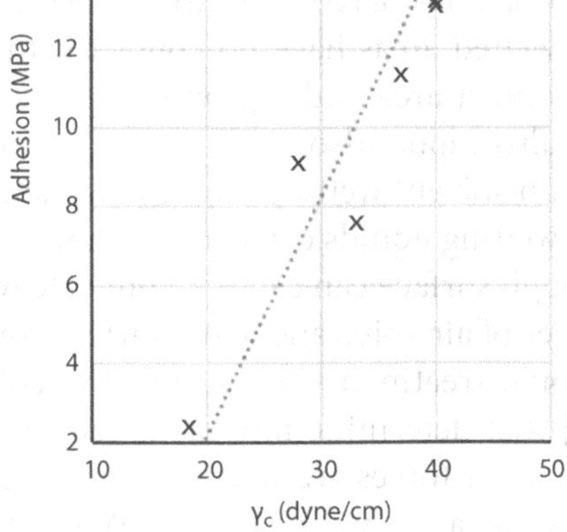

Figure 13.2 Practical adhesion of a 50 mN/m surface tension epoxy as a function of several substrates of critical surface tension γ_C (Replotted from Levine, *et al.* [20]).

Reasonable agreement between "ideal," i.e., theoretical, W_A value and practical adhesion, measured as intrinsic fracture energies, is generally observed [21] but mismatches in these values are also found [6]. Ideal and practical adhesion can match when there is a finite contact angle between 0 and 90 degrees but practical adhesion can become much less or much greater than theoretical adhesion for wetting contact angles at 0 or greater than 90 degrees. For systems where the substrate surface energy is greater than that of the liquid, contact angle will be zero for all cases. For systems of zero contact angle, i.e., fully spontaneous wetting, practical adhesion as a function of contact angle is no longer predictive.

Practical adhesion at zero contact angle will be best represented by Equation (13.3): $W_A = \gamma_{SV}+\gamma_{LV} - \gamma_{SL}$. Here interfacial tension γ_{SL} plays a significant role where optimum adhesion occurs for minimum interfacial tension, that is, where the surface energies of the two materials in contact are similar [4]. It should be noted that for polymer-polymer systems, a similarity in surface energy can also be represented by similar solubility parameter, $\delta_P = (E_{vap}/V_m)^{1/2}$, which leads to reduced enthalpy of mixing, ΔH_{mix}. A positive enthalpy of mixing presents an energetic barrier to mixing and thus a near-zero enthalpy of mixing, obtained by greater similarity of phases across the interface, results in a more spontaneous interfacial free energy. Interfacial mixing across the interface by interdiffusion also becomes energetically spontaneous and leads to greater practical adhesion (see diffusion theory, section 13.4.4, below) [22, 23].

Surface wetting is thus a measure of interaction forces of solid surface energy against liquid surface tension (Eq. 13.3), where θ is the contact angle of the liquid in contact with the substrate. Another consequence of this interaction is the effect on surface wetting itself. If the contact angle is equal to or greater than 90° the liquid is considered non-wetting and if its contact angle is less than 90° is considered as wetting. The ramifications of poor wetting is that non-wetted areas have no contact with the adhering coating, thus reducing contact area and a proportional reduction in practical adhesion. There are also vapor adsorption energy π_S and kinetic spreading contact time effects on solvent wetting angle. For fractal substrate profiles, even spontaneously wetting liquids can become non-wetting [24] such that a rough or micro-rough surface can cause incomplete wetting and produce a weak boundary layer of air voids and a low surface contact area.

There are additional treatments for adsorption adhesion and surface wetting energy [25] that determine interfacial tensions from sessile drop contact angle where polar forces are separated into independent acid and base interfacial forces as a γ^D, γ^+, γ^- scheme. While a good idea to consider acid-base effects on surface tensions and adhesion [17], the acid-base scheme has been shown to be non-self-consistent [26].

Practical adhesion values different than theoretical W_A are noted to be functions of both surface bond energy, as measured by W_A, and the structure dissipation/loss energy, i.e., energy lost through viscoelastic flow of the coating during separation from the substrate. Viscoelastic dissipation, e.g., as would occur for coatings at temperatures above their glass transition temperature (T_g), is less prominent for coatings at temperatures below their T_g since they are glassy and more elastic and usually fail elastically by crack propagation. Better agreement between practical and theoretical adhesion values should exist for elastic failures (elastic modulus-mediated fracture). Practical adhesion much larger than W_A occurs for cohesive failures that involve mechanically lossy (viscous deformation) mechanisms of fracture near or above glass transition temperature or at low rate of shear where plastic failure occurs (loss modulus-mediated fracture).

13.4.4 Diffusion

Viscous dissipation of energy relies on polymer chain diffusion. Chain diffusion theory describes chain behavior as series and parallel combinations of storage (elastic) and loss (viscous) elements. The relative size of the storage and viscous moduli is governed by the ratio of polymer chain viscoelastic relaxation rates and mechanical strain rates called a Deborah number. The general trend is that long relaxation times for chain motion or fast induced strain rates result in more elastic behavior, whereas short relaxation times or slow strain rates give time for chains to move and flow for a more lossy, viscous deformation but it is the relative size ratio of relaxation rate to shear rate that predicts behavior. Diffusion adhesion theory is reliant on an intertwining or de-intertwining of polymer chains of both the coating and substrate, as applicable, and their structurally limited relaxation times that control ratio of elastic to viscous deformation as a function of time, interfacial concentrations of chain ends, and chain reptation [27, 28]. Stiff polymer chain structures have long relaxation times (slower diffusion D_d); flexible ones can flow and deform more quickly with less energy storage and more loss, leading to larger practical adhesion than W_A. Vasenin [29] first developed quantitative models of viscous flow kinetics based on Fick's Laws to determine a relationship for peel energy (P) as a function of coating polymer molecular mass and contact time:

$$P = k_4 \left(\frac{2N\rho}{M} \right)^{2/3} D_d^{1/2} t_c^{1/4} \tag{13.6}$$

where P is a function of constant k_4 describing molecular characteristics, Avogadro's number **N**, density ρ and molecular weight M of the polymer, diffusion constant D_d and interface contact time t_c. Agreement between Equation (13.6) and experimental (practical) peel adhesion data was good, with peel energy P following proportionality to $t^{1/4}$ and 1/M. D_d was neither theoretically obtained nor measured but rather used as a fitting parameter. In Voyutski and Vasenin's reasonings, even very small chain diffusions, as small as 1 to 2 nm, were predicted to produce excellent adhesion values as a five to nine-fold increase in adhesion [30]. Indeed, larger practical adhesion values are thought to emanate from primarily viscoelastic (diffusive, mechanical) loss by the coating and/or polymer substrate where a good match of solubility parameters is a predictor of interpolymer miscibility.

Solvent also plays an important role in chain mobility. The presence of solvent plasticizes the coating polymer for increased polymer free volume and chain diffusion dynamics [31]. On the other hand, surface crystallinity and pigmentation reduce chain free volume, reduce diffusion rates, and will prevent chain entanglements and reduce peel adhesion. Interestingly, the same limiting rates of diffusion that provide entanglement and formation of diffusion bonding also apply to the interfacial separations, which produce ostensibly higher peel force at higher rates of interfacial separation r and longer interfacial contact time (see Equation (13.7)) [23]. Autohesion force between two polymer surfaces of identical composition is equal to:

$$F = 11\nu\left(\left(\frac{d(2+p)}{\bar{M}}\right)^{2/3} K^{1/2}\right) r\, t^{1/4} \tag{13.7}$$

where F = peel force, ν = vibrational frequency of a -CH- group, d = density of polymer, p = number of chain branches in the molecule, $\bar{M}$ = molecular weight, r = rate of interfacial separation, and t = contact time of the two surfaces prior to testing. In Equation (13.7), K is related to diffusion constant characteristic of the molecule structure (stiff or flexible) and r is related to strain rate. Plots of practical adhesion data show proportionality to inverse (molecular weight)$^{2/3}$, rate of separation r, and to interfacial contact time $t^{1/4}$ [23].

13.4.5 Covalent Bonding

The final mechanism to consider is covalent bonding across an interface, for example where the substrate has surface hydroxyl groups and an isocyanate functional coating is applied. Covalent bonding provides the strongest bond strengths compared to the above-mentioned mechanisms but is highly dependent on the interfacial bond density per contact area. For instance, covalent bonding between a coating isocyanate group and the surface hydroxyl leading to a urethane bond between the coating and substrate is in addition to mechanical, adsorption and diffusion mechanisms. Ionic bonding, while theoretically stronger than covalent (Table 13.2), is subject to solvation of the respective ions by ambient water that will lead to inconsistent bond strengths depending on concentration of water and amount of solvation. Solvation reduces the concentration of local charge and increases inter-ion separation distances on which ionic bond strength is solely reliant. Covalent bonding is stronger than adsorption and, depending on bond strength, bond density per area and surface contact area, can dominate adhesion strength. Covalent bonding in all cases will increase adhesion strength over an identical situation without covalent bonding.

13.4.6 Summary: Adhesion Mechanisms

Through mechanical, adsorption, diffusion, and/or covalent mechanisms, polymers attain adhesion to (polymer) coatings or other substrates. To reiterate, a review of adhesion theories is to support discussion about how polymers obtain adhesion of their own accord, or do not, and how adhesion promotion functions on polymer substrates that lack adhesion. In general, polymers that have critical surface energies greater than about 35 mJ/m^2 are capable to develop their own practical adhesion to adhesives or coatings [32] while those having γ_c below 35 mJ/m^2 can obtain excellent practical adhesion to higher energy substrates such as metals, for example, polyolefin powder coatings when applied by fluidized bed to metal wire-frame dishwasher racks. In part, the excellent practical adhesion of polyolefins of low theoretical adhesion is due to their low glass transition temperature and mechanical short relaxation time that give rise to a large viscous response to strain.

The issue of poor adhesion to polymers arises when the polymer substrate is of low surface energy, i.e., of low capability to be wetted by and bond to another relatively lower surface energy material, such as an organic coating. This is true especially for a coating that is of higher surface tension, such as waterborne coatings. Water has a high surface tension of ~72 mN/m, and despite the use of surfactants to lower interfacial tension, low bond strengths are usually obtained without utilizing one of the techniques described below to promote adhesion. Even for spontaneous wetting, a high interfacial tension borne out of a mismatch of material surface tensions on either side of an interface results in less adhesion than when material surface tensions across the interface are nearly equal, which produces low interfacial tension. Plasma, corona, flame, or UV/ozone are energetic treatments that react with and modify the surface composition to add polarity and raise surface energy that allows critical wetting and adhesion bond energy, as well as to remove WBLs from the near-surface [33].

A last point before moving on to chemical treatments of polymer surfaces is that polymer surfaces can reorient in response to their surrounding environment. The effects of reorientation are driven by lower system energy, through improved interaction across the interface, and a reduction of interfacial tension. Corona treatments, more so than plasma [34, 35] or flame [36] treatments, are known to undergo "hydrophobic recovery" [37]. Hydrophobic recovery of a plasma treated substrate is caused by one or more of these mechanisms: a thermodynamically-driven reorientation of the newly created, polar, high surface energy componants from the surface into the bulk; migration of low molecular weight additives internal to the plastic to its surface; or, post-treatment surface contamination that decreases the surface treatment's effectiveness to improve adhesion. Flame treatments, in particular, are more immune to reorientation-associated 'hydrophobic recovery' though surface recovery is complex and difficult to predict [37].

13.5 Pretreatments

Treatments are used for polymer surfaces that have low surface energies for wetting and adhesion. Polymer surfaces of higher surface energies allow good surface wetting and produce reasonable adhesion (~11-22 MPa). Even for the highest surface energy polymers, joint strengths are generally highest when pretreatments are used. Purposes for pretreatments include: 1) to remove, or prevent formation of weak boundary layers within the substrate; 2) to maximize intimate molecular contact between a coating layer and the substrate; 3) to engender chemical forces sufficient to obtain the desired adhesion and service life of the part; 4) to provide the desired surface profile; 5) to assist with hardening of the adhesive or coating; and 6) to protect the surface of the part prior to application of the adhesive or coating, which is particularly needed for substrates sensitive to the environment such as metals [2].

WBLs are caused by a variety of issues, which include structural integrity and surface contamination. In short, a WBL presents a lower strength region that limits adhesion strength since the strained article will readily fail in the WBL. Plasticizer, mold release, airborne contamination, oil/grease, lubricant or plasticizer diffusion, and water hydration are examples of surface contamination that can cause a WBL. Near-surface crystallinity or surface oxidation/slag are examples of surface structures that can have reduced cohesive strength near the molded surface. Metals are known to present similar issues, e.g., in oxidation slag of cast (black) steel. Water hydration can cause a coating or adhesive to react in a non-bonding fashion, resulting in poor surface contact or poor surface reaction and weak interfacial bond strength. In other instances, presence of interfacial moisture can help catalyze adhesive or coating curing, as in the case of cyanoacrylate adhesives, to enhance both curing speed and bond strength.

Maximizing interfacial molecular contact involves adjusting the surface profile to provide a large surface contact area. Larger area of contact for a constant adhesion per area results in overall larger total bond strength. Related to this is item 4), profiling, that increases surface roughness and surface contact area and may remove WBLs. Profiling can also provide surface structure such as a fiber-like penetration that involves a new failure mechanism, such as fiber pull-out, that provides additional dissipation of energy and hence greater practical adhesion [2].

13.5.1 Solvent Wipe Treatment

In general, a "solvent wipe" is used for extracting and cleaning a surface of its WBL contaminants or other small molecules, such as surfactant or mold lubricants. Solvent treatment of polypropylene with trichloroethylene, at short exposure times of up to 15 seconds, greatly improves adhesion of epoxy resin adhesive [38, 39]. The solvent treatment produces a "dull-white finish" and an improved adhesion was ascribed to removal of contamination on examination by SEM. Wetting studies showed worse wetting, while scanning electron microscopy (SEM) showed enhanced surface porosity, perhaps due to solvent preferentially attacking amorphous domains [38] but images show a change in surface profile (Figure 13.3). Interestingly, the dull white finish and increased porosity have another important connotation tied to the outcome of solvent treatment that will be discussed shortly. With no change caused in surface polarity, surface roughening could have a negative effect on wetting and surface contact area, which then could result in worse adhesion joint strength. This was not the case.

Studies of solvent effect on coatings adhesion to polypropylenes and ethylene-co-styrene interpolymers found that type of solvent greatly influenced coating adhesion, changing from no (0) adhesion to adhesion rating of 5 (ASTM D-3359B, no fail) [40]. In instances where a 'good' solvent,

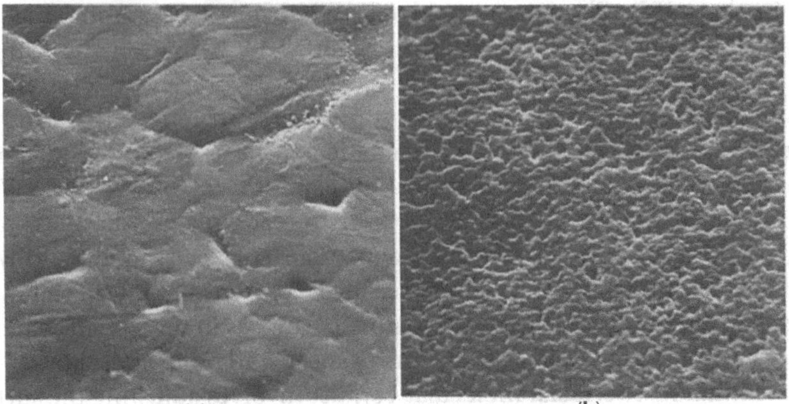

Figure 13.3 A change in (a) as received polypropylene surface topography (2500X) (b) after brief exposure to trichloroethylene solvent vapor (2500X) (Images taken from Garnish and Haskins [39]).

causing excellent volumetric swelling and capability to dissolve polymer substrate, was used, coating bond strengths were identical to that of poor solvents. Poor solvents cause no change in swelling or surface appearance and were non-solvents for the substrate polymer. Both good solvents and poor solvents alike produced low adhesion strength with interfacial failure between the substrate and coating. Then what coating solvent produces adhesion, if neither 'good' nor 'bad' solvent as determined based on volumetric swelling and solubility, did?

Strong, cohesive failures were observed when using solvents of low to moderate dissolution capability, i.e., solvents which produced some but minimal swelling of a molded substrate. Importantly, the solvents changed the surface appearance of the substrate. Adhesion engineers describe the change in surface appearance as solvent "bite". In these instances, 'bite' was observed as a change in surface topography and phase amplitude collected via tapping mode atomic force microscopy (AFM). Increased R_a and RMS (R_q) amplitudes and/or altered topography, also observed as a change in specific surface area (Surface area diff, Figure 13.4) were observed [39].

Garnish and Haskins [39] described a "white haze" appearance and in a more recent reference, Schuman and Thames [40] observed a similar effect as haze and/or a loss of as-molded surface gloss. Other studies have shown that morphology and entanglements at the interface are likely more important than equilibrium adhesion thermodynamics [41, 42]. These observations underscore the importance of mechanical interlocking and

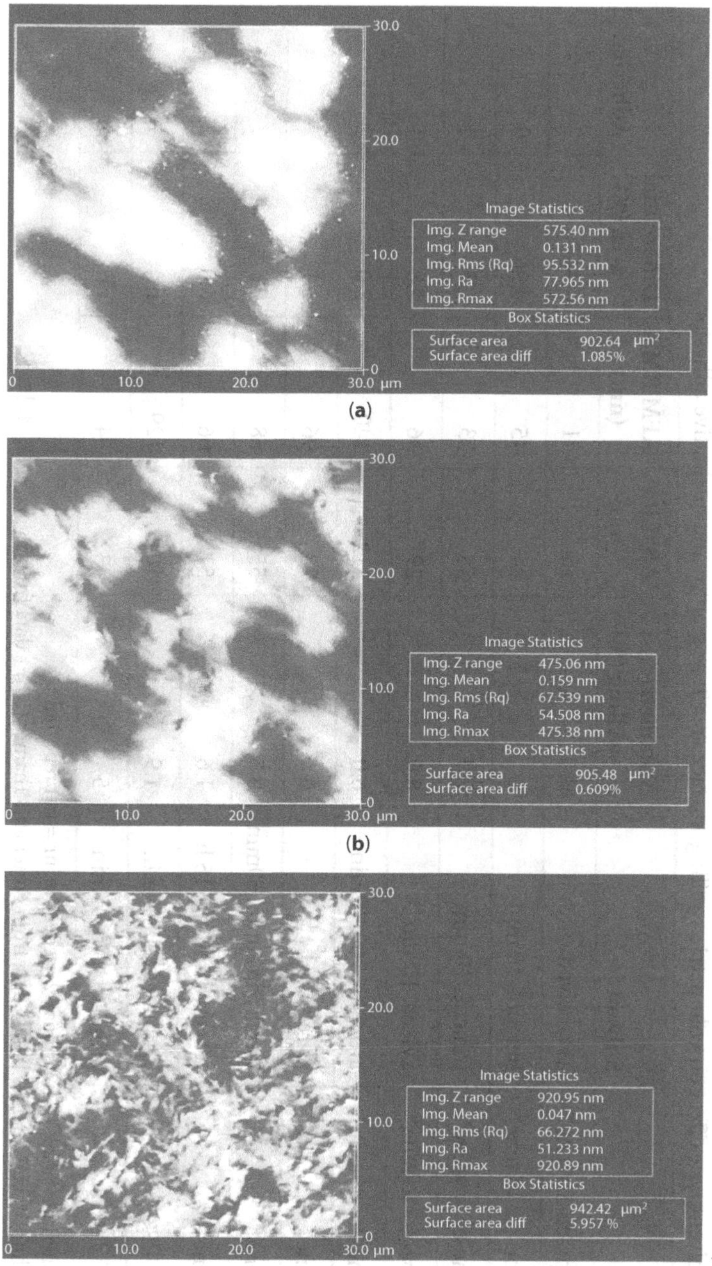

Figure 13.4 Scanning AFM tapping mode images of polypropylene substrate (a) as molded before exposure to solvent, (b) after 1 hr exposure, and (c) after 15 hour exposure to tetrahydrofuran solvent resulting in altered surface roughness.

Table 13.3 Solvent type effect on volumetric swelling, surface area/topography, and qualitative adhesion.

Substrate	Solvent	Volumetric swelling (%)	AFM Ra (nm)	AFM RMS (nm)	Differential area (%)	Adhesion
80%-Ethylene-co-styrene*	as received	0	16	21	0.44	--
80%-Ethylene-co-styrene*	butanol, 10 min	0	34	45	0.96	0
80% Ethylene-co-styrene*	toluene, 10min	7.9	58	73	1.3	1
80% Ethylene-co-styrene*	xylene, 10min	8.6	29	36	0.8	1
80% Ethylene-co-styrene*	aromatic 100, 10min	6.0	112	140	2.0	2
Polypropylene‡	as received	0	78	96	1.1	-
Polypropylene‡	tetrahydrofuran, 60min	0.3	55	68	0.6	2
Polypropylene‡	tetrahydrofuran, 15 h	1.6	51	66	6.0	2
Low rubber TPO†	aromatic 100, 45min	1.9	104	129	nr	1
High rubber TPO†	aromatic 100, 45min	5	432	54	nr	2

Legend: 0 = none/interfacial, 1 = marginal, 2 = effective, nr = not reported

* Data from Schuman and Thames [40]; ‡ data from Foster and Thames [58]; † data from Ryntz, *et al.* [41]

an increase of surface contact area that can be effected by solvent treatment. While solvent treatments can remove contaminants that can cause WBLs, solvent treatment can also initiate a WBL through differential swelling and enhanced local stress, creating sub-surface defects. It was important that swelling, but not optimal volumetric swelling, that induces changes in surface profile and roughness were observed to be best for promoting adhesion [40]. Ryntz [43] noted, however, that an aggressive solvent attack of amorphous regions can reduce the cohesive energy of the substrate, in effect creating a WBL, presumably by plasticization.

A comparison of swelling and solvent resulting in adhesion is noted in Table 13.3. The desired effects of solvent toward polymer adhesion would be to clean the substrate of WBL, increase the potential for diffusion mechanism, and increase surface area of contact and interface tortuosity that does not influence or, rather, enhances surface wetting. Wool [42] has shown that polymer interdiffusion rates are enhanced even by thermodynamically non-ideal solvents, which increase reptation rates and rate of interfacial entanglements. Sufficient drying depletes the interface from solvent so as not to plasticize the developed entanglements. One way to think about the solvent swelling effect on adhesion is that strong swelling appears to be indicative of a plasticization effect and therefore provides only weak adhesion strength and key to effective adhesion is a choice of solvent to be sufficiently aggressive to provide good wetting, surface contact time, and surface topographical changes, so as to result in adhesion, but not overly aggressive so as to induce overly large swelling, resulting in plasticization, reduced cohesive strength, and/or WBL formation [43].

13.5.2 Chemical Treatments

Despite the fact solvents do not change the composition of substrates, except through removal of contaminants, solvents have been shown to alter interfacial structure, nonetheless. Chemical treatments perform surface conversions akin to those induced by energetic treatments. Chemical treatments can include hydrolytic or oxidative etching by strong acid, base, or oxidizer such as chromic acid or peroxo free radical initiator. Chemical etching of fluoropolymers using sodium naphthalide, for instance, produces a defluorinated surface with enhanced oxygen and carbon contents with unsaturation, as -C=O, -COH, and -CO_2H upon exposure to air. Only anaerobic exposure post treatment results in an all-carbon surface [44]. A common theme of chemical treatments is a conversion of non-polar or low-polarity bonds composition substrates into substrates with polar, acid/base-rich surface bonding groups such as -OH, -COOH, and/or carbonyl.

The polar/polar protic groups increase surface energy, wetting, capability for covalent reaction and, hence, adhesion strength.

Ryntz [45] presented data that summarized surface compositions and practical adhesion after flame, plasma, benzophenone/UV light, and chromic acid treatments of polypropylene. She demonstrated the change in surface oxygen content after pretreatment and how a solvent wipe after the pretreatment affected the resulting surface composition. In effect, the post treatment wipe removed much of the added polar content but not all, as measured by X-ray photoelectron spectroscopy (XPS), and least polar content was removed from the benzophenone/UV light treatment. Also, no -C-O or -C=O were introduced by chromic acid etch so the oxide introduced was likely chromium-related species. These compositions showing increased surface polarity correspond to changes in practical adhesion (Figure 13.5) [46]. It should be noted that attenuated total reflectance infrared (ATR-IR) spectroscopy is commonly used for measuring surface compositions. However, surface-enhanced ATR-IR is still too insensitive toward measuring only the outer few nanometers of surface composition, responsible for adhesion. ATR-IR measures considerably deeper than a few outer nanometers of surfaces and therefore can provide misleading data on changes to surface composition with regard to predicting a change in adhesion. Surface treatment is best assessed by surface specific techniques such as XPS, time-of-flight secondary ion mass spectroscopy (TOF-SIMS), or laser microprobe mass spectroscopy (LMMS) techniques that are most sensitive to the outer few nanometers of surface composition.

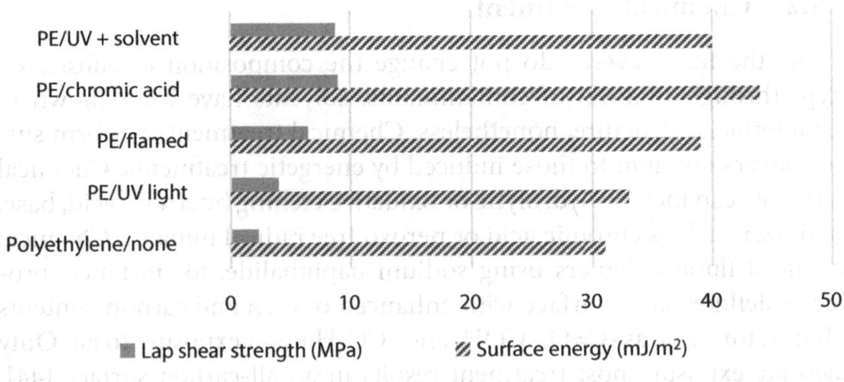

Figure 13.5 Surface energy (mJ/m^2) and practical adhesion (lap shear, strength MPa) as a function of polyethylene surface pretreatment: as-received (none), UV light, flaming, chromic acid etch, and UV plus solvent wipe (data from Ryntz [46]).

13.5.3 Chemical Etching

Use of chemical etching processes as a function of treatment time produces gradually increasing adhesion reaching a plateau in practical adhesion strength [47]. The treatment can remove WBLs if present, though where an interfacial failure occurs for an untreated substrate no WBL should be present [48]. Removal of oxidized groups on a treated surface produces poor adhesion akin to untreated substrates hence the oxidized groups are responsible for improved practical adhesion [49]. Fluorinated polymer etching, noted above, has polar protic groups introduced only after exposure to air and presumably moisture [44]. The strong base naphthalide causes fluoride etching that results in a carbon-enriched, reactive surface. Similarly, exposure of polyethylene or polypropylene to chromic or piranha acids results in etching with loss of mass but introduction of polar protic and oxygen-bearing chemical groups -OH, -COOH, -C=O, and sulfonyl on the outermost surface layers.

A perhaps more environmentally-friendly approach to chemical etching is by microorganisms. The key statement by the author is, "practically, there is no polymer that can resist attack by microorganisms (bacteria and/or microscopic fungi) [50]". Treatment time, composition of nutrient medium, and polymer structure all influence degree of modification that results in hydrolysis and cleavage of functional groups, chemical bond structures having a structure-specific reactivity. Functional group structures that resemble that of food to the organism, such as esters, amides, or ethers, are susceptible to etching by the organism. Polymers containing antimicrobial groups, such as nitro, halide, or sulfate, are resistant to microbial etching. The primary drawback to commercial utilization of microbial approaches is a treatment time of days to months as discussed in the article. The authors presented many different examples of microbial surface treatments, especially of fibers, and effect on practical adhesion and strength of composites but did not discuss organism batch-to-batch consistency that is a common issue when using living, agricultural commodities.

13.5.4 Adhesion Promoters

An adhesion promoter is a substance that, as a surface layer, increases the practical adhesion strength of subsequently applied adhesives or coatings. For this reason, adhesion promoters are also called "tie-coatings," securing the subsequent layer to the substrate. The adhesion promoting layers must avoid creation of a new WBL, either in the substrate [46, 51] or within the adhesion promoting layer [52, 53]. WBLs can occur as a function of composition, solvent, thermal treatment, substrate morphology,

Table 13.4 Comparison of chlorinated adhesion promoter practical adhesion strengths to polypropylene (PP) substrate and mode of failure as a function of surface preparation, CPO primer layer thickness, and top-coating thickness. (Data from Clemens, *et al.* [52]).

Substrate/ Preparation*	CPO primer† thickness (um)	Top-coat thickness (um)	Adhesion strength (g/cm)	Failure
PP/none	2-3	80-90	150-250	interfacial
PP/none	14	80	50-100	Cohesive in CPO
PP/flamed	0	80-90	>400	interfacial
PP/flamed	2	80-90	150-200	interfacial
PP/gently abraded	2	80	150	interfacial
PP/severely abraded	2	80	400	interfacial

*all substrates were cleaned with acetone
† Eastman CPO product 343-1 was used; top-coating composition was unspecified

or tie-coat thickness (Table 13.4) [52]. Interestingly, severely abraded substrates achieved similar practical adhesion as flamed substrates, which was thought to be due to mechanical adhesion. Flaming produces the best practical adhesion but with substrates of unusual shape and geometry (not regular or flat) produces varied performance compared to CPO treatments due to inconsistent distance from the flame source.

A different route to application of a tie-coat is through graft polymerization conducted directly onto the substrate [54]. CPOs comprise the primary industrial material for improving adhesion to polyolefins, such as polypropylene and thermoplastic polyolefin. The polyolefins have low capability to bond to a coating, being comprised of only non-polar covalent bonds. Not all polymers require adhesion promoters but all polymers can be improved in practical adhesion by the use of adhesion promoting techniques [32], which will include proper use of solvent in coating formulation [40].

13.5.4.1 *Chlorinated Polyolefin (CPO)*

Chlorinated polyolefins are tie-coatings applied to a low energy substrate, e.g., polyolefins, to provide adhesion for subsequent top coatings. CPOs are polypropylene that is maleic anhydride grafted, and subsequently chlorinated [55]. CPO products are supplied as either solvent or waterborne type coating formulations that provide application compatibility, tolerant or not, with volatile organic content (VOC). Chlorinated polyolefins (CPOs) are best applied as thin films since thicker (> 10μm) films tend to fail cohesively (Table 13.4). Solventborne CPO coatings do not appreciably change polyolefin surface energy, either dispersion or polar component, but waterborne CPO primers show an enhanced polar component at the primer surface (Table 13.5) [55]. CPO polymer is formulated at a specific chlorine content that affects not only solubility in the solvents of subsequent coating layers but also compatibility within its own formulation [55, 56]. Practical adhesion of CPOs tends to be negatively affected by polyurethane top-coatings, perhaps due to cosolvent incompatibility. Waterborne formulations are obtained by use of the polar acid monomer (maleic anhydride) grafted onto the polypropylene olefin backbone and blending with water-soluble resin, surfactant, and/or amine.

Unfortunately, CPO tie-coat layers do not autoadhere or interpenetrate into the substrate to confer practical adhesion by chain entanglement but appear to work by creation of induced dipoles within an otherwise non-polar substrate [52, 57]. Induced dipoles enhance surface interaction energy and thereby adhesion to low energy polyolefin substrates as well as to more polar, higher surface energy top-coats. Adhesion and compatibility properties are tradeoffs as a function of percent chlorination [55, 57] and percent polar grafting [57, 58]: Greater chlorine concentrations than about 25% in a CPO produce solubility and compatibility to organic solvents, whereas polar grafting of maleic anhydride improves polarity and

Table 13.5 Dispersion and polar surface energies determined for plastic substrates without and with CPO primer layers. Solventborne primers give no apparent change in total surface energy but waterborne CPO primers provide a large increase in polar component and total surface energy. (Data from Lawniczak, *et al.* [55]).

	No treatment			**CP343-1***			**CP310W†**		
Substrate	**D**	**P**	**S**	**D**	**P**	**S**	**D**	**P**	**S**
Polypropylene	30	12	42	28	13	41	30	43	73
TPO	30	3	32	29	4	33	29	44	73

Eastman CPO products: * - solventborne CPO; † - waterborne CPO
Legend: D= dispersion component; P= polar component; S= total surface energy (mJ/m^2)

Table 13.6 Changes in surface tension and wetting properties for polypropylene substrates caused by chlorinated adhesion promoters (data from Foster and Thames [58]). Surface contact angle data for the chlorinated maleinized guayule rubber (CMGR) primer surface and failure interface provides evidence of orientation of maleic groups at the air surface or interface with polypropylene as a function of maleic concentration.

Substrate	Adhesion promoter	Drying conditions	Air surface CMGR, water contact angle (degrees)	CMGR failure surface, water contact angle (degrees)
Polypropylene	CMGR 25%,	WB 80°C 20min	74	73
	CMGR 15%	WB 80°C 20min	73	87
	CMGR 10%	WB 80°C 20min	79	82
	CMGR 5%	WB 80°C 20min	75	76
	CPO*	SB 80°C 20min	95	95
	CPO†	WB 80°C 20min	66	-

CMGRs are lab-synthesized materials of listed wt.-% maleic anhydride grafting; * Eastman product 343-1; † Eastman product 347W; WB = waterborne primer; SB = solventborne primer

adhesion to top-coats but worsens adhesion to nonpolar substrates. Top-coat solubility of the CPO layer allows dissolution of the CPO into the top-coating layer and removes the CPO primer as a tie-coat. Within the CPO coating layer, there can be extensive orientation of non-polar groups toward the polypropylene to minimize interfacial tension whereas polar groups orient towards air (Table 13.6) [52, 58], which will maximize wetting and adhesion of top-coats.

13.5.4.2 CPO Cosolvent

Cosolvent of the CPO primer also plays a crucial role in practical adhesion strength [33, 40, 58] by providing solubility to CPO and by causing substrate effects delineated in Section 13.5.1. Anionic surfactant, used for waterborne formulation stability, limits adhesion performance [55]. Interfacial failures were evidenced by smooth failure surfaces where surface analysis of the adhesion failure found CPO only on the top-coat surface, regardless of the solvent or top-coat applied to the CPO [52]. CPO or polypropylene would have been found on both top-coat and substrate surfaces in the case of a cohesive failure within the CPO layer or substrate, respectively. That no diffusion of the CPO into the substrate had occurred was concluded from the interfacial failure and sharp change in composition from substrate to CPO at the adhesion interface. CPO may, however, entangle or dissolve into the top-coating layer depending on top-coat solvent content and percent CPO chlorine content in the tie-coat, where higher chlorine content renders the CPO more soluble and compatible in the top-coating layer [55] and also less compatible with a polypropylene substrate [56].

Waterborne CPOs can adhere to polypropylene and TPO as well or better than solventborne counterparts (see Table 13.7). Wetting of the CPO layer is likely further enhanced by its synthesis and formulation with polar graft monomers, e.g., maleic anhydride, and surfactants [52]. Contact angle of water is lowered from 87 to 68 degrees by application of CPO, which aids subsequent wetting of the top-coating compared to bare polypropylene or thermoplastic polyolefin substrate though cohesive, rather than adhesion, failures were more commonly observed for waterborne systems. Oddly, CPO tie-coats tend to be negatively influenced by isocyanate cured top-coats [52, 55].

13.5.4.3 Other Chlorinated Tie-Coats

Other coatings have been investigated as tie-coat matrices/adhesion promoters to polyolefin substrates. Chlorinated maleinized natural rubber (CMGR) [58] was dispersed in water using tetrahydrofuran as cosolvent (@ 1 wt.-% CMGR and 25 wt.-% THF) based on adhesion data of coatings applied to polypropylene [59]. Adhesion was measured as a response in Design of Experiments (DOE) as a function of baking temperature and percent maleinization, where it was observed that maximum practical (pull-off) adhesion was obtained at 10% maleinization and at the lowest baking temperature of 80 °C. Lowest baking temperature may correlate with maximizing cosolvent contact time with the substrate for maximized effect. A change in substrate topography is observed for the cosolvent tetrahydrofuran on polypropylene [40].

Table 13.7 Adhesion data and failure modes for chlorinated and non-chlorinated adhesion promoter treatments of polypropylene and thermoplastic polyolefin (TPO) substrates.

Substrate	Adhesion promoter	Drying conditions	Adhesion	Units
Polypropylene	CMGR 10% maleic*	WB 80°C, 4% THF	21	N/cm^2
	CMGR 10% maleic*	WB 80°C, 25% THF	57	N/cm^2
	CMGR 10% maleic*	SB 80°C, 75% THF	65	N/cm^2
	CPO*	WB 80°C	19	N/cm^2
	CPO*	SB 80°C	75	N/cm^2
	CPO$	SB 25% in xylene	0.15-0.25	kg/cm
	Maleinized polyolefin (non-chlorinated)‡	WB w/18% butyl carbitol, 120°C 30min	pass	JIS K5400
TPO†	WB maleic-co-acrylic modified PP (non-chlorinated)	L-WB 90°C L-WB 120°C H-WB 90°C H-WB 120°C	0.14 0.7 0.47 0.6	kg/cm kg/cm kg/cm kg/cm

Legend: WB= waterborne; SB= solventborne; THF= tetrahydrofuran; L= low [maleic] resin; H= high [maleic] resin; bake times = 20 min unless otherwise indicated; all failures were interfacial

(* Data from Foster and Thames [58]; § data from Clemens, *et al.* [52]; ‡ data from Ohmika, *et al.* [61]; † data from Fujino [62])

Use of 2-ethanolamine as a cosolvent/dispersant to reduce VOC was not as effective as tetrahydrofuran alone. The waterborne CMGR adhesion promoter was found to be as effective as a solventborne CPO in pull-off adhesion strength, though reported strengths are slightly less for the waterborne primer.

13.5.4.4 In-Mold Application of CPO

Chlorinated polyolefin coatings are typically applied as thin films by traditional coating processes, spray or dip, and baked to volatilize solvent and obtain good interfacial contact. CPOs were found to be capable of being applied to polypropylene substrates using the heat and pressure of an injection molding process [59]. Solventborne and waterborne formulations were applied to the injection mold by spray and then a series of injection molding cycles were conducted. To the molded surfaces, adhesion of a 2K-acrylic polyurethane coating impregnated into cloth strip was measured as a function of cycle number after CPO application. Failure surfaces were examined by X-ray photoelectron spectroscopy for composition and CPO transfer to the injection molded polypropylene was evidenced. A single application of CPO to the metal mold consistently gave 2 to 4 transfers of CPO to the molded polypropylene surface under repeated injection cycles depending on the CPO tie-coat composition. Interestingly, solventborne CPOs generally provided worse transfer and worse 2K polyurethane practical adhesion compared to the traditional spray or dip CPO application methods. In earlier studies, 2K top-coats were problematic for CPO adhesion [55]. Waterborne CPOs, on the other hand, gave equivalent adhesion to traditional solventborne or waterborne CPO application to polypropylene (0.43 to 0.45 kg/cm) and consistent mold transfer, at 3 injection transfers per treatment. Cohesive failure within the CPO tie-coating was observed for all samples, which is consistent with thicker CPO layers [52].

CPO compounds have also been applied as additives to paints analogous to an approach described in section 13.5.4.7 below. In this example, an acrylate-graft-modified CPO was utilized as a paint additive to promote adhesion to plastic substrates [60]. The additive composition was toluene free and allowed paints to achieve adhesion to the plastic substrates without substrate pretreatment. Quality of adhesion was reported to depend on ability of the CPO to orient at the plastic substrate-paint interface during application and drying process.

13.5.4.5 Non-Chlorinated Adhesion Promoters

Adhesion promoters can be homogeneous in composition or, more often, utilize co-polymer or complementary chemistries to create a tie-coat effect. Transesterification of a poly(vinyl acetate) backbone with different, liquid crystalline polyester side chains [54] provides adhesion for liquid crystal display devices. Similarly, modification of a hydrophobic polyolefin backbone with hydrophilic α,β-unsaturated carboxylic acids provides adhesion promotion for automobile plastic components (Table 13.7) [61]. The trans-esterified poly(vinyl acetate)s have a modified, liquid-crystalline, polyester side chain structure. The unsaturated carboxylic acid-modified polymer provides a waterborne application of the adhesion promoting film to polyolefin substrates. The adhesion promoting film when baked at 120 °C for 30 minutes had polyurethane top-coat stability, good gasoline resistance, and excellent adhesion to bumper-grade polypropylene substrates.

An acrylic-modified, chlorine-free polyolefin was synthesized by graft copolymerization of acrylic and maleic anhydride monomers onto polypropylene in the melt state to engender a self-water-dispersing capability, as an adhesion promoting tie-coat. The composition of the graft copolymer has hydrophilic and hydrophobic domains that provide interaction with water and polypropylene, respectively. Polypropylene was grafted with maleic anhydride, other select acrylic monomers, and free radical initiator in the melt state (see Table 13.7) [62, 63]. In particular, the authors sought to control crystallinity and melting temperature by varying the degree of maleic and acrylic monomer contents. Higher crystallinity enabled stronger interfacial strengths given that sufficiently high baking temperatures promoted good film formation. Better gasoline resistance was noted for films of higher crystallinity, as well. However, a tradeoff existed regarding bond strength and gasoline resistance versus bake temperature. Baking temperatures above the tie-coat crystalline melt temperature were required for wetting and enhanced gasoline resistance. Of course, lower baking temperature is to be preferred but these tie-coats provided lower bond strength and gasoline resistance. The authors projected that baking temperatures lower than 90 °C should be achievable with their system.

13.5.4.6 Environmentally-Benign Approaches

An unusual adhesion promoter approach relies on application of a natural protein to a substrate in conjunction with a separate "anchor" peptide [64]. The approach might be thought of as a variation of 'barnacle adhesion-to-ship bottom' mechanism, which is tenacious and protein-mediated, and as a result has found broad areas of application (see barnacle adhesion data in Table 13.8 and references [65, 66]). A film of adsorbed 'enhanced green fluorescent protein,' when combined/treated with 'anchor peptide' cecropin A, allows surfactant-resistant adhesion to polypropylene substrates. Four nanometer thick films were applied by dip-coating and were envisaged as viable alternatives to conventional surface modifications.

Table 13.8 Adhesion data for use of silane adhesion promoters on steel, aluminum and glass. Adhesion of barnacles on steel provided for comparison.

Substrate	Adhesion promoter	Conditions	Adhesion	Units	Failure
Steel*	0.4% diamine silane in polyurethane	dry (48h)	29.8	MPa	
	0.4% diamine silane in polyurethane	wet (1500h)	21.3		
	control	degreased, dry	20.8		
	control	degreased, wet	11.8		
	0.2% diamine silane in epoxy	dry	25.3		
	0.2% diamine silane in epoxy	wet (1500h)	23.8		
	control	degreased, dry	20.9		interfacial
	control	degreased, wet	9.2		interfacial
Steel†	barnacles, adult cement	seawater	0.2-12		
Aluminum†	0.4% diamine silane in polyurethane	dry	25.2	MPa	
	0.4% diamine silane in polyurethane	wet (1500h)	15.6		
	control	degreased	13.6		
	control	degreased, wet	8.5		interfacial
	0.2% diamine silane in epoxy	dry	28.7		
	0.2% diamine silane in epoxy	wet (1500h)	28.2		

(*Continued*)

Table 13.8 Adhesion data for use of silane adhesion promoters on steel, aluminum and glass. Adhesion of barnacles on steel provided for comparison. (*Continued*)

Substrate	Adhesion promoter	Conditions	Adhesion	Units	Failure
	control	degreased	13.7		
	control	degreased wet	8.5		interfacial
Glass‡	Aminopropylsilane	acetone water 24h	0.61	kJ/m^2	
	Aminobutylsilane	acetone water 24h	0.39	kJ/m^2	
	Control	heat cleaned	0.32	kJ/m^2	

* - data from Plueddemann [72]; † - data from Walker [65]; ‡ - data from Wang and Blum [76]

Another unusual treatment is the use of alkyl boranes as adhesion promoters to low energy polymer surfaces. Alkyl boranes promise improved adhesion without further pretreatments [67].

13.5.4.7 Melt-Blended Polymer Additives

The application of adhesion promoter as an additive within a molded substrate polymer relies on the ability, when mixed into substrate polymer, to locate polar/polar-protic groups at the molded part surface. Polar surface groups raise surface energy, allowing for good wetting, and can form a strong adsorption interface thus providing practical adhesion. One adhesion promoter additive was obtained by grafting maleic anhydride to polypropylene (PP) or thermoplastic olefin (EP) (Figure 13.6a), which was then mixed into PP or EP through melt-blending, and then extruded as blended polypropylene or 5% ethylene-co-propylene copolymer [68]. The authors varied the graft ratio from 2.7 to 4.4% maleic anhydride by weight of polypropylene and examined bond strengths of the modified polymer melt blended into polypropylene or 5% ethylene-co-propylene to aluminum substrates (Figure 13.6b) [68].

An optimum loading of 400 ppm total maleic anhydride, as diluted by unmodified polymer, was observed for polypropylene, while a 1000 to 2000 ppm maleic anhydride was most effective in ethylene-co-propylene polymer. Optimum adhesion was also a function of the molecular weight of the maleic anhydride modified additive polymer. Higher molecular weight

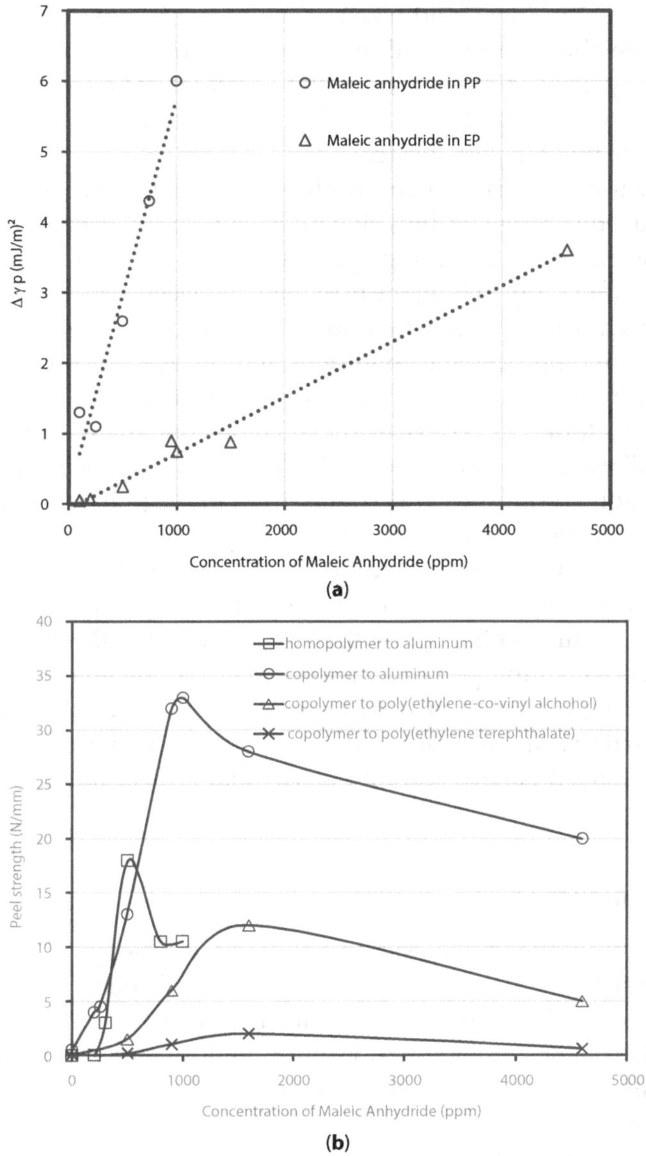

Figure 13.6 (a) Change in surface energy produced by incorporation of poly(olefin-graft-maleic anhydride) additive melt blended into polypropylene or thermoplastic olefin as a function of maleic anhydride concentration in blend; and (b) peel energy of aluminium foil adhered to melt-blended, adhesion promoted plastic substrate. Olefins are polypropylene (PP) and poly(ethylene-co-propylene) (EP) (data replotted from Schultz, *et al.* [68]).

additives gave the most optimized magnitude of adhesion but higher molecular weights also required lower percent polar modification due to high viscosity. Higher molecular weight increases cohesion strength but reduces ability of the additive to diffuse to and locate polar/polar-protic groups on the molded part surface.

A different melt extrusion approach improves adhesion between polyamides and polypropylene [69]. Poly(ethylene-co-ethylene oxide)s were melt extrusion blended with polypropylene. Surface composition was investigated after sandwiching the melt blend between glass and teflon substrates. Enrichment of copolymer additive on the polypropylene surface was dependent on processing conditions and surface tension of the mold. Practical adhesion between polyamide (nylon) and additive-containing polypropylene films after coextrusion and blow molding was improved.

Other adhesion promoter blended approaches include, e.g., the addition of phenolic resin to EPDM rubber to improve practical adhesion to polyester sheet molding compound (SMC) [70] and treatment of nylon 6,6 with epoxy-amine. The addition of phenolic additive to EPDM by melt extrusion enhanced the partitioning of ester groups to the surface of the SMC, in turn enhancing adhesion between EPDM and the SMC. Similarly, epoxy-amine polymer can be surface impregnated into nylon 6,6 fibers, using solvent to provide interpenetration within the nylon 6,6, that increased the lap-shear strength up to 3 times compared to untreated samples [71]. Surprisingly, solvent polarity had little or no effect on bond strengths reported.

13.5.4.8 Silanes

Silanes are a well-known adhesion promotion technology [72, 73]. Silanes are hydrolyzed by water and the produced siloxy groups condense with surface and/or silyl hydroxyl groups to form silyl ether bonds, thereby creating covalent tethering of adhesion-promoting silane coating directly to the substrate. This is also true for metal oxide reinforcing agents in polymer composites, including powders or fibers, to improve composite cohesion, environmental stability and strength of the resulting composites held together by covalent rather than merely adsorption bonds (see Table 13.8). Silanes can help achieve critical bonding in, e.g., medical devices [74, 75].

Strength provided by silanes is dependent on the silane linker structure and how much interfacial mobility the silane group allows between the substrate modified with silane and the matrix, such as glass fiber, or coating adhered to that substrate. For example, Wang and Blum [76] found correlation between surface mobility of the silane to measured bond strength of the glass-reinforced composite. Silanes with shorter hydrocarbon length provide lower diffusion rate that is observed as slower diffusion rate. The shorter chain and slower diffusion were ascribed to the silane layer being more mechanically stiff between the amino group and the substrate surface. The shorter silane layer and stiff interface between glass fiber and polymer produced a higher strength composite.

13.6 Summary

The strongest bond types are ionic and covalent bonds followed by strong dipolar, i.e., polar associative, bonding. Ionic bonds as well as associative bonds are prone to the influence of solvation by, e.g., water, which can severely reduce the bond strength. Covalent bonds are permanent and not subjected to environmental effects such as solvation. Polymers that possess sufficient polarity at their surface generally do not require adhesion promotion, such as reaction injection molded polymers, polyesters, acrylics, or urethanes, due to their providing more or less durable, strong bonds [2, 77] but are often coated anyway to provide superior aesthetic appearance. Strong bonds result from combinations of mechanical interlocking, interdiffusion, surface energy, and covalent and/or associative bonding [78].

Adhesion promotion is an attempt to effect these bonding mechanisms in a positive way and to remove and/or prevent the introduction of impurities and/or weak boundary layers. In particular, adhesion promotion techniques seek to increase surface polarity, and hence surface free energy, increase wetting and interfacial contact, and increase the interfacial chemical bonding components of the substrate surface, which have been approximated twice as important to practical adhesion as mechanical interlocking adhesion [15]. However, mechanical interlocking adhesion still controls adhesion outcome, that is why removal of weak boundary layers is so critical, and remains a strong consideration for any adhesion-promoting technology. In particular, polyolefinic substrates are in need of adhesion promotion due to their low surface energies, which is why their surface treatments are profuse and prominent in the chemical literature.

A solvent wipe is intended not only to clean but also can have a significant effect on surface topography and thus on mechanical interlocking bonding. Chemical treatments and flame and plasma treatments described earlier in this book, on the other hand, seek to convert the polymer chemical composition toward increased surface energy, and provide chemical component bonding, and to etch, prevent and/or remove weak boundary layers as needed. The chemical treatments have the capability to, and should, either strengthen or remove a WBL.

Adhesion promoters, on the other hand, coat the substrate to "tie" a top-coat to the substrate surface and thereby convert the outermost layer, typically to that of higher surface energy or polarity for increased wetting and interfacial associative bonding capability. For a solventborne coating, it is critical that adhesion promoting primers do not cause or exacerbate underlying weak boundary layers through solvent swelling effects [33]. Chlorinated or non-chlorinated adhesion promoter coatings are, in general, both simpler than flame or plasma treatments and are mostly amenable to complex surface geometries and fast production rates. CPOs have a long history of successful industrial use and user experience. As a surface coating, adhesion promoters are also least able to resolve the underlying weak boundary layer or other material problems and are most prone to cause, through swelling, weak boundary layers or other problems, rather than cure them. Silanes provide a means of introducing covalent bonding across interfaces with hydroxyl reactive groups.

References

1. S.J. Babinec and M.C. Cornell, Overview of the automotive plastics market, in: *Coatings of Polymers and Plastics*, R.A. Ryntz and P.V. Yaneff (Eds.), chapter 1, Marcel Dekker, New York (2003).
2. A.J. Kinloch, *Adhesion and Adhesives, Science and Technology*, pp. 101-103, Chapman and Hall, New York (1987).
3. M.V. Raaphorst, C.P. Senechal, L.A. Williams, M.J. Masini and P.A. Carbutt, Method for enhancing bonding surface of difficult to bond plastics, U.S. Patent Application 20220259459, assigned to Illinois Tool Works, Inc. (2022).
4. K.L. Mittal, The role of the interface in adhesion phenomena, *Polym. Eng. Sci.*, 17, 467-473 (1977).
5. J.J. Bikerman, *The Science of Adhesive Joints*, second edition, Academic Press, New York (1968).
6. A.J. Kinloch, *Adhesion and Adhesives, Science and Technology*, chapter 1, Chapman and Hall, New York (1987).
7. D.E. Packham, Adhesion – fundamental and practical, in: *Handbook of Adhesion*, second edition, D.E. Packham (Ed.), John Wiley, West Sussex, England (2005).
8. F.N. Jones, M.E. Nichols and S.P. Pappas, *Organic Coatings: Science and Technology*, fourth edition, chapter 6, John Wiley, Hoboken, NJ (2017).
9. J.J. Bikerman, *The Science of Adhesive Joints*, second edition, p. 140, Academic Press, New York (1968).

10. J.J. Bikerman, *The Science of Adhesive Joints*, second edition, p. 164, Academic Press, New York (1968).
11. A.W. Adamson and A.P. Gast, *Physical Chemistry of Surfaces*, sixth edition, p. 455, Wiley and Sons, New York (1997).
12. P. de Gennes, Adhesion promoters, *J.Phys.Chem.*, 98, 9405 (1994).
13. B.V. Deryaguin and V.P. Smilga, Adhesion: fundamentals and practice, in: *Adhesion, Fundamentals and Practice*, p. 152, McLaren, London, UK (1969).
14. A.J. Kinloch, *Adhesion and Adhesives, Science and Technology*, p. 74, Chapman and Hall, New York (1987).
15. W.C. Wake, *Adhesion and Formulation of Adhesives*, pp. 70-89, Applied Science, London, UK (1982).
16. D.K. Chattoraj and K. S. Birdi, *Adsorption and the Gibbs Surface Excess*, Springer, New York (2013).
17. F.M. Fowkes, Quantitative characterization of the acid-base properties of solvents, polymers and inorganic surfaces, in: *Acid-Base Interactions: Relevance to Adhesion Science and Technology*. K.L. Mittal and H.R. Anderson, Jr. (Eds.), pp. 93-115, CRC Press, Boca Raton, FL (1991).
18. R.S. Drago, G.C. Vogel and T.E. Needham, Four-parameter equation for predicting enthalpies of adduct formation, *J. Am. Chem. Soc.*, 93, 6014-6026 (1971).
19. W.A. Zisman, Relation of the equilibrium contact angle to liquid and solid constitution, in: *Contact Angle, Wettability, and Adhesion, Advances in Chemistry*, Vol. 43, pp. 1-51, American Chemical Society, Washington D.C. (1964).
20. M. Levine, G. Ilkka and P. Weiss, Relation of the critical surface tension of polymers to adhesion, *J. Polym. Sci., Polym. Lett. Ed.*, 2, 915-919 (1964).
21. A.J. Kinloch, *Adhesion and Adhesives, Science and Technology*, p. 86, Chapman and Hall, New York (1987).
22. J. Comyn, Compatibility, in: *Handbook of Adhesion*, second edition, D.E. Packham (Ed.), pp. 73-76, Wiley (2005).
23. K.W. Allen, Diffusion theory of adhesion, in: *Handbook of Adhesion*, 2nd edition, D.E. Packham (Ed.) , pp. 103-105, Wiley, New York (2005).
24. A.J. Crosby, M, Hageman and A. Duncan, Controlling polymer adhesion with "pancakes", *Langmuir*, 21, 11738-11743 (2005).
25. C.J. Van Oss, M. Chaudhury and R.J. Good, Estimation of the polar parameters of the surface tension of liquids by contact angle measurements on gels, *J. Colloid Interface Sci.*, 128, 313-319 (1989).
26. D.Y. Kwok, D. Li and A.W. Neumann, Evaluation of the Lifshitz-van der Waals/acid-base approach to determine interfacial tensions, *Langmuir*, 10, 1323-1328 (1994).
27. P.G. de Gennes, Simple views on adhesion and fracture, *Can. J. Phys.*, 68, 1049-1054 (1990).
28. S.S. Voyutskii, *Autohesion and Adhesion of High Polymers*, p. 15, Interscience, New York (1963).

29. R.M. Vasenin, Problems in the diffusion theory of adhesion of polymers, in: *Adhesion Fundamentals and Practice*, p. 29, Gordon and Breach, London (1969).
30. A.J. Kinloch, *Adhesion and Adhesives, Science and Technology*, p. 68, Chapman and Hall, New York (1987).
31. A.J. Kinloch, The science of adhesion—Part 1 Surface and interfacial aspects, *J. Mater. Sci.*, 15, 2153-2166 (1980).
32. A.J. Kinloch, *Adhesion and Adhesives, Science and Technology*, p. 103, Chapman and Hall, New York (1987).
33. R.A. Ryntz, *Adhesion to Plastics, Molding and Paintability*, p. 155, Global Press, Oxford, UK (1998).
34. J.B. Lynch, P.D. Spence, D.E. Baker and A. Postlethwaite, Atmospheric pressure plasma treatment of polyethylene via pulse dielectric barrier discharge: comparison using various gas compositions versus corona discharge in air, *J. Appl. Polym. Sci.*, 71, 319–331 (1999).
35. T. Schuman, B. Adolfsson, M. Wikström and M. Rigdahl, Surface treatment and printing properties of dispersion-coated paperboard, *Prog. Org. Coat.*, 54, 188–197 (2005).
36. D.M. Brewis and D. Briggs, Adhesion to polyethylene and polypropylene, *Polymer*, 22, 7–16 (1981).
37. K.S. Johansson, Surface modification of plastics, in: *Applied Plastics Engineering Handbook - Processing, Materials, and Applications*, second edition, chapter 20, M. Kutz (Ed.), Elsevier, Oxford, England (2017).
38. A.J. Kinloch, *Adhesion and Adhesives, Science and Technology*, p. 113, Chapman and Hall, New York (1987).
39. E.W. Garnish and G.G. Haskins, The effect of surface conditions when bonding with epoxy adhesives, in: *Aspects of Adhesion*, Vol. 5, p. 259, D.J. Alner (Ed.), University of London Press, London, UK (1969).
40. T. Schuman and S.F. Thames, Coating solvent effects producing adhesion to molded plastic parts, *J. Adhesion. Sci. Technol.*, 19, 1207-1235 (2005).
41. R. Ryntz, Q. Xie and A.C. Ramamurthy, The effects of coating solvents on the morphology of thermoplastic polyolefins, in: Proc. *Waterborne, Higher Solids and Powder Coatings Symposium*, pp. 812-849 (1994).
42. R.P. Wool, Polymer-polymer adhesion: Incompatible surfaces, in: *Handbook of Adhesion*, p. 341, second edition, D.E. Packham (Ed.) Wiley, New York (2005).
43. R.A. Ryntz, *Adhesion to Plastics, Molding and Paintability*, p. 165, Global Press, New York (1998).
44. A.J. Kinloch, *Adhesion and Adhesives, Science and Technology*, p. 107, Chapman and Hall, New York (1987).
45. R.A. Ryntz, *Adhesion to Plastics, Molding and Paintability*, p. 168, Global Press, Oxford (1998).
46. R.A. Ryntz, The effects of thermoplastic poly(olefin) (TPO) morphology on subsequent paintability and thermal shock performance, *Prog. Org. Coat.*, 27, 241-254 (1996).

47. A.J. Kinloch, *Adhesion and Adhesives, Science and Technology*, chapter 4, Chapman and Hall, New York (1987).
48. D.J. Briggs, Application of secondary ion mass spectrometry to adhesion studies, *J. Adhesion*, 21, 343-352 (1987).
49. A. Chew, R.H. Dahm, D.M. Brewis, D. Briggs and D.G. Rance, Adhesion to polyethylene studied by means of a reversible bromination reaction, *J. Colloid Interface Sci.*, 110, 88-95 (1986).
50. E.V. Pisanova, Microbial treatment of polymer surfaces to improve adhesion, in: *Adhesion Promotion Techniques: Technological Applications*, K.L. Mittal and E. Pizzi (Eds.) , pp. 323-346, Marcel Dekker, New York (2002).
51. R. Ryntz, Thermal and impact induced stress failure in painted TPO: the role of surface morphology, *J. Coatings Technol.*, 67(840), 35-46 (1995).
52. R.J. Clemens, G.N. Batts, J.E. Lawniczak, K.P. Middleton and C. Sass, How do chlorinated poly(olefins) promote adhesion of coatings to poly(propylene)?, *Prog. Organic Coatings*, 24, 43-54 (1994).
53. R.A. Ryntz, Coating adhesion to low surface free energy substrates, *Prog. Org. Coat.*, 25, 73-83 (1994).
54. H. Arasi, K. Adlem, S. Mulcahy and I. Gardiner, UV curable adhesion promoters based on functionalized polyvinyl alcohols, U.S. Patent Application 20220363992, assigned to Merck Patent GMbH (2022).
55. J.E. Lawniczak, P.J. Green and J.S. Witzeman, Water-reducible adhesion promoters for coatings on polypropylene-based substrates, *Surface Coatings Australia*, 30, 18-19, 21-22 (1993).
56. T.S. Ellis, Interactions of chlorinated polyolefins with ethylene-propylene copolymers and their relevance to painted TPOs, *Polym. Eng. Sci.*, 41, 2065-2072 (2001).
57. R.A. Ryntz, *Adhesion to Plastics, Molding and Paintability*, pp. 175-177, Global Press, Oxford (1998).
58. M.D. Foster and S.F. Thames, Chlorinated guayule rubber as an adhesion promoter for polypropylene, *J. Coatings Technol.*, 71(889), 91-99 (1999).
59. T. Schuman and J.O. Stoffer, An in-mold application of adhesion promoters to polyolefin substrates, in: *Polymer Surface Modification, Relevance to Adhesion*, Vol. 4, K.L. Mittal (Ed.), pp. 263-284, CRC Press, Boca Raton, FL (2007).
60. E. deWolf, T. Fu, H. Andringa, G. Haldankar, K. Weaver, W. Liu and M. Dai, A new CPO acrylic adhesion promoter resins for plastic coatings, *Eur. Coat. J.*, 10, 18-24 (2019).
61. H. Ohmika, M. Yamanaka and M. Ohno, Aqueous primer composition for polyolefin resins, U.S. Patent 5,102,944, assigned to Nissan Motor Co., Ltd. (1992).
62. K. Fujino, Waterborne chlorine-free adhesion promoter for TPO; effect of crystallinity on adhesion and gasoline resistance performance properties, *Proc. International Coatings for Plastics Symposium* (2005).

63. K. Usui, K. Fujino, T. Fujitaka, K. Masumoto and H. Yoshioka, Modified polyolefin resin composition and uses thereof, U.S. Patent 6,495,629, assigned to Nippon Paper Industries Co. (2000).
64. K. Ruebsam, B. Stomps, A. Boeker, F. Jakob and U. Schwaneburg, Anchor peptides: a green and versatile method for polypropylene functionalization, *Polymer*, 116, 124-132 (2017).
65. G. Walker, The adhesion of barnacles, *J. Adhesion*, 12, 51-58 (1981).
66. L. Khandeparker and A.C. Anil, Underwater adhesion: the barnacle way, *Intl. J. Adhesion Adhesives*, 27, 165-172 (2007).
67. J. Cognard, Some progress in adhesion science and technology, *Comptes Rendus Chemie*, 9, 13-24 (2006).
68. J. Schultz, L. Lavielle, A. Carre and P. Comien, Surface properties and adhesion mechanisms of graft polypropylenes, *J. Mater. Sci.*, 24, 4363-4369 (1989).
69. I. Ozen, C. Rustal, K. Dirnberger, H.-G. Fritz and C.D. Eisenbach, Modification of surface properties of polypropylene films by blending with poly(ethylene-b-ethylene oxide) and its application, *Polym. Bull.*, 68, 575-595 (2012).
70. A. Allue, K. Gondra, I. DeMarco and G. Diez, Improvement of the chemical adhesion of EPDM rubber to sheet moulding compound (SMC) by the addition of phenolic resin as adhesin promoter, *J. Adhesion*, 97, 1167-1188 (2021).
71. J. Zhang, M. deSouza, C. Creighton and R.J. Varley, New approaches to bonding thermoplastic and thermoset polymer composites, *Composites Part A*, 133, 105870 (2020).
72. E.P. Plueddemann, Silane adhesion promoters, in: *Coatings Technology Handbook*, third edition, A. A. Tracton (Ed.), chapter 70, CRC Press, Boca Raton, FL (2005).
73. E.P. Plueddemann, *Silane Coupling Agents*, Plenum Press, New York (1982).
74. S.U. Nielsen, M.H. Olsen, M.S. Kongsfelt, S.U. Pedersen and K. Daasbjerg, Hydrosilane-modified poly(2-hydroxyethyl methacrylate) brush as a nanoadhesive for efficient silicone bonding, *ACS Omega*, 4, 12130-12135 (2019).
75. K. Kilway, L. Bonewald and T. Schuman, Biomaterial compositions, U.S. Patent 9,186,302, assigned to The Curators of the University of Missouri (2015).
76. T.W.H. Wang and F.D. Blum, Interfacial mobility and its effect on interlaminar fracture toughness in glass-fibre-reinforced epoxy laminates, *J. Mater. Sci.*, 31, 5231 5238 (1996).
77. D. A. Berta, Formulating plastics for paint adhesion, in: *Coatings of Polymers and Plastics*, R.A Ryntz and P.V. Yaneff (Eds.), p. 85, Marcel Dekker, New York (2003).
78. A.J. Kinloch, *Adhesion and Adhesives, Science and Technology*, p. 60, Chapman and Hall, New York (1987).

Index

Also of Interest

Check out these published and forthcoming in the Adhesion and Adhesives series

Progress in Adhesion and Adhesives, Volume 8
Edited by K.L. Mittal
Forthcoming 2024. ISBN 978-1-394-23820-0

Polymer Surface Modification to Enhance Adhesion: Techniques and Applications
Edited by K.L. Mittal and Anil Netravali
Published 2024. ISBN 978-1-394-23100-3

Progress in Adhesion and Adhesives, Volume 7
Edited by K.L. Mittal
Published 2023. ISBN 978-1-394-19810-8

Adhesives in Biomedical Applications
Edited by K.L. Mittal and S. Neogi
Published 2023. ISBN 978-1-394-20920-0

Biobased Adhesives
Sources, Characteristics and Applications
Edited by Manfred Dunky and K.L. Mittal
Published 2023. ISBN 978-1-394-17463-8

Structural Adhesives
Properties, Characterization and Applications
Edited by K.L. Mittal and S.K. Panigrahi
Published 2023. ISBN 978-1-394-17472-0

Progress in Adhesion and Adhesives, Volume 6
Edited by K.L. Mittal
Published 2022. ISBN 978-1-119-84665-9

Rubber to Rubber Adhesion
By Dinesh K. Kotnees and Anil K. Bhowmick
Published 2021. ISBN 978-1-119-76889-0

Surface Science and Adhesion in Cosmetics
Edited by K.L. Mittal and H.S. Bui
Published 2021. ISBN 978-1-119-65482-7

Ice Adhesion
Mechanism, Measurement and Mitigation
Edited by K.L. Mittal and Chang-Hwan Choi
Published 2020. ISBN 978-1-119-64037-0

Structural Adhesive Joints
Design, Analysis and Testing
Edited by K.L. Mittal and S. K. Panigrahi
Published 2020. ISBN 978-1-119-73643-1

Progress in Adhesion and Adhesives, Volume 5
Edited by K.L. Mittal
Published 2020. ISBN 978-1-119-74806-9

Bioadhesives in Drug Delivery
Edited by K. L. Mittal, I, S. Bakshi and J.K. Narang
Published 2020. ISBN 978-1-119-64019-6

Advances in Contact Angle, Wettability and Adhesion
Volume 4
Edited by K.L. Mittal
Published 2019. ISBN 978-1-119-59254-9

Adhesives for Wood and Lignocellulosic Materials
By R.N. Kumar and A. Pizzi
Published 2019. ISBN 978-1-119-60543-0

Progress in Adhesion and Adhesives, Volume 4
Edited by K.L. Mittal
Published 2019. ISBN 978-1-119-62525-4

Advances in Contact Angle, Wettability and Adhesion Volume 3
Edited by K.L. Mittal
Published 2018. ISBN 978-1-119-45994-1

Laser Technology: Applications in Adhesion and Related Areas
K.L. Mittal and Wei-Sheng Lei
Published 2018. ISBN 978-1-119-18493-5

Progress in Adhesion and Adhesives, Volume 3
Edited by K.L. Mittal
Published 2018. ISBN 978-1-119-52629-2

Textile Finishing: Recent Developments and Future Trends
Edited by K.L. Mittal and Thomas Bahners
Published 2017. ISBN 978-1-119-42676-9

Progress in Adhesion and Adhesives, Volume 2
Edited by K.L. Mittal
Published 2017. ISBN 978-1-119-40638-9

Adhesion in Pharmaceutical, Biomedical and Dental Fields
Edited by K.L. Mittal and F.M. Etzler
Published 2017. ISBN: 978-1-119-32350-1

Interface/Interphase in Polymer Nanocomposites
Edited by Anil Netravali and K.L. Mittal
Published 2016. ISBN 978-1-119-18491-1

Progress in Adhesion and Adhesives
Edited by K.L. Mittal
Published 2015. ISBN 978-1-119-16219-3

Advances in Contact Angle, Wettability and Adhesion Volume 2
Edited by K.L. Mittal
Published 2015. ISBN 978-1-119-11698-1

Particle Adhesion and Removal
Edited by K.L. Mittal and Ravi Jaiswal
Published 2014. ISBN 978-1118-83153-3

Laser Surface Modification and Adhesion
Edited by K.L. Mittal and Thomas Bahners
Published 2014. ISBN 978-1-118-83163-2

Adhesion in Microelectronics
Edited by K.L. Mittal and Tanweer Ahsan
Published 2014. ISBN 978-1-118-83133-5

Advances in Contact Angle, Wettability and Adhesion
Volume 1
Edited by K.L. Mittal
Published 2013. ISBN 978-1-118-47292-7

Advances in Modeling and Design of Adhesively Bonded Systems
Edited by S. Kumar and K.L. Mittal
Published 2013. ISBN 978-1-118-68637-9

Atmospheric Pressure Plasma Treatment of Polymers
Edited by Michael Thomas and K.L. Mittal
Published 2013. ISBN 978-1-118-59621-0